BASIC MANUFACTURING PROCESSES

H. C. Kazanas
University of Illinois—Champaign

Glenn E. Baker
Texas A&M University

Thomas Gregor
Temple University

Gregg Division / McGraw-Hill Book Company
New York / Atlanta / Dallas / St. Louis / San Francisco / Auckland / Bogotá
Guatemala / Hamburg / Johannesburg / Lisbon / London / Madrid / Mexico / Montreal
New Delhi / Panama / Paris / San Juan / São Paulo / Singapore / Sydney / Tokyo / Toronto

Sponsoring Editor: Myrna Breskin
Editing Supervisor: Katharine Glynn
Design Supervisor: Eileen Kramer
Art Supervisor: Howard Brotman
Production Supervisor: Priscilla Taguer

Cover Designer: Jack Weaver

Library of Congress Cataloging in Publication Data
Kazanas, H   C
    Basic manufacturing processes.

    Bibliography: p.
    Includes index.
    1. Manufacturing processes.   I. Baker, Glenn E.,
joint author.   II. Gregor, Thomas G., joint author.
III. Title.
TS183.K39   670.42   80-16280
ISBN 0-07-033465-X

234567890 SMSM 88765432

PREFACE

The conversion of raw materials into useful products is achieved through manufacturing processes. Manufacturing processes constitute the core of the modern production industry. *Basic Manufacturing Processes* is written for individuals who are studying manufacturing processes at technical-vocational schools, community and junior colleges, technical institutes, as well as colleges and universities.

The purpose of this text is to introduce the study of basic manufacturing processes through a broad, conceptual framework rather than a narrow, specific one. This attempt reflects changes in modern industry in terms of specific processes and materials, that are becoming increasingly more rapid. Therefore, an introduction to the basic manufacturing processes in a conceptual framework will enable the reader to grasp the processes as they may apply to the conversion of not only one kind of material but rather many different materials. For example, the process of extrusion is presented conceptually, not only as it applies to metals, but in its broader sense, as it applies to metals, plastics, ceramics, and other materials. In addition, the reader is exposed to the manufacturing processes as a decision-maker: he or she chooses the most appropriate process for the product needed, based on the materials available.

This text helps the reader to understand the variety and complexity of this field. The organization of the text is unique. It presents the basic manufacturing processes in logical groups, based on similarities of purpose and application of the process involved. Related manufacturing processes have been grouped under common parts. Each part in turn is divided into chapters which discuss closely related processes. This organization allows the reader to conceptualize not only individual processes but also the interrelationships that may exist among related processes.

Basic Manufacturing Processes has been written at an introductory level, yet its coverage of the subject matter is technical. It will be of value to technicians, technologists, engineers, and teachers of technical subjects, as well as practitioners in modern industry. The review questions at the end of each chapter provide ample opportunity for challenging the student's comprehension of the subject matter.

The authors would like to express their sincere appreciation to many individuals and organizations who amply contributed their time, illustrations, and related subject matter to make this book possible. In particular, the authors would like to extend their acknowledgment to Professor Lyman D. Hannah for his contribution to the discussions of welding.

H. C. Kazanas
G. E. Baker
T. G. Gregor

CONTENTS

CHAPTER 1

ORGANIZING AND PLANNING FOR MANUFACTURING

DEVELOPMENT OF MODERN MANUFACTURING

Manufacturing is the process of coordinating manpower, tools, and machines to convert raw materials into useful products. The first steps toward manufacturing can be seen in primitive peoples' early efforts to convert such raw materials as stone and wood into tools like the digging stick and the lance. Human beings are earth's only inhabitants capable of thinking and of tool making, which are the two unique characteristics that have enabled them to dominate other animals.

The early tools were simple *hand tools,* but with the gradual accumulation of knowledge through the use of tools and other implements, human beings gradually developed the technology that enabled them to make the transition from *hand tools* to *machine tools* (see Fig. 1-1). It was the development of machine tools capable of producing themselves (and other machines) that resulted in the beginning of modern manufacturing.

During early manufacturing, products were produced largely on an individual basis, and their quality was dependent mostly on the worker's skill. This type of manufacturing had serious limitations in terms of production output, diversity of product design, product cost, and quality. With the development of modern manufacturing machines and methods, however, part of the worker's skill is built into the production machines. This makes possible the employment of a greater number of people with relatively little skill, while increasing production output, reducing production cost, improving diversity of product design, and providing more reliable product quality. By building part of the production skill into the machines, the human error in production is minimized, inasmuch as machines are not as susceptible to fatigue, carelessness, or inattention.

The develoment of modern manufacturing is also related to materials. Product diversification and high-production outputs invariably call for diversity in the materials used. On the other hand, the development of new and improved materials requires new and improved manufacturing processes. When special alloys were developed to meet the needs of the aerospace industry, it was discovered that traditional manufacturing processes could not be used to process these alloys efficiently. Thus, new and radically different processes such as *electrodischarge machining* and *electrochemical machining* had to be developed. There is a direct relationship between the types of materials used and the manufacturing processes required to produce a product for a specified function.

ORGANIZING FOR MANUFACTURING

Modern manufacturing is an industrial activity which requires such resources as human power, materials, machines, and capital. However, for efficient, economical, and competitive production, all the available resources must be carefully organized, coordinated, and controlled. This should result in an integrated organization capable of competitive production in terms of quality and profit. It is important, therefore, to understand that manufacturing activities, though specialized in nature, are an integral part of a larger system called the *organization* or the *company*.

Manufacturing organizations or companies may be divided into large companies which have two or more production plants, medium-size companies which usually have one production plant, and small companies or job shops.

The nature of organization and coordination re-

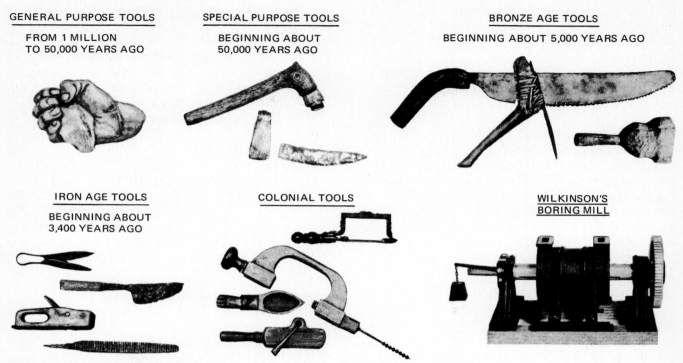

GENERAL PURPOSE TOOLS
FROM 1 MILLION
TO 50,000 YEARS AGO

SPECIAL PURPOSE TOOLS
BEGINNING ABOUT
50,000 YEARS AGO

BRONZE AGE TOOLS
BEGINNING ABOUT 5,000 YEARS AGO

IRON AGE TOOLS
BEGINNING ABOUT
3,400 YEARS AGO

COLONIAL TOOLS

WILKINSON'S
BORING MILL

Fig. 1-1 The evolution of hand tools resulted in the development of machine tools which led to modern manufacturing. (*Leighton A. Wilkie.*)

quired by manufacturing companies depends on the size of the company and also on the type of product or products manufactured. Thus, a large manufacturing company with five different production plants may be organized differently from a small job shop. But even though these companies may differ organizationally, there are certain elements and functions which are found in almost any manufacturing organization. Among these are *ownership, general administration, sales and marketing, purchasing, finance, personnel, product development, manufacturing,* and *quality control.* These and other functions are usually differently or-

ganized and coordinated in each company, and even in each plant of the same multiplant company.

Figure 1-2 shows a typical organization of a *small* or *job shop company.* In this type of company the owners, who usually are the managers of the company, perform a multitude of functions. They usually know all the essentials required to run the business, and the decision-making process is simple. The communication process is on a direct, one-to-one basis, and no written company policy is required. The specialized functions of design, cost estimating, sales, accounting, and purchasing are assigned to individual employees

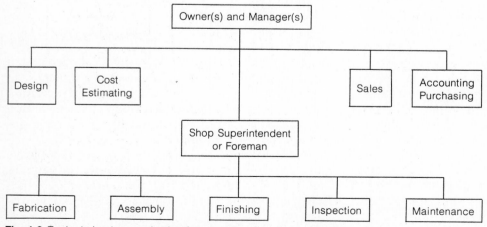

Fig. 1-2 Typical simple organization for a small or job shop company.

who are directly supervised by the owners. The production functions are usually assigned to a shop superintendent or shop supervisor directly responsible to the owners. The overhead cost (indirect cost in addition to material and labor costs) of operating the company is relatively low and the profit (or loss) is enjoyed entirely by the owners.

Unlike small companies, which can operate with a relatively simple organizational structure, large companies require a very complex organizational structure. Figure 1-3 shows a typical organization of a relatively large manufacturing plant. In this type of organization, functions are usually grouped together and assigned to various departments or divisions, and not necessarily to individuals. Individuals in large organizations are highly specialized in their functions. Since there is no direct contact between the top man-

agement and the operational levels of the organization, *written policy* becomes the guide in achieving the company's objectives. The manufacturing functions are dispersed throughout the organization and are coordinated by the objectives outlined in company policy. Coordination of resources is complex and often creates problems that hinder the functioning of the company. Overhead cost is high for this kind of an organization.

MANUFACTURING A PRODUCT: BASIC STAGES

The main purpose of manufacturing is to produce useful products from raw materials. The conversion of materials into products made available to the con-

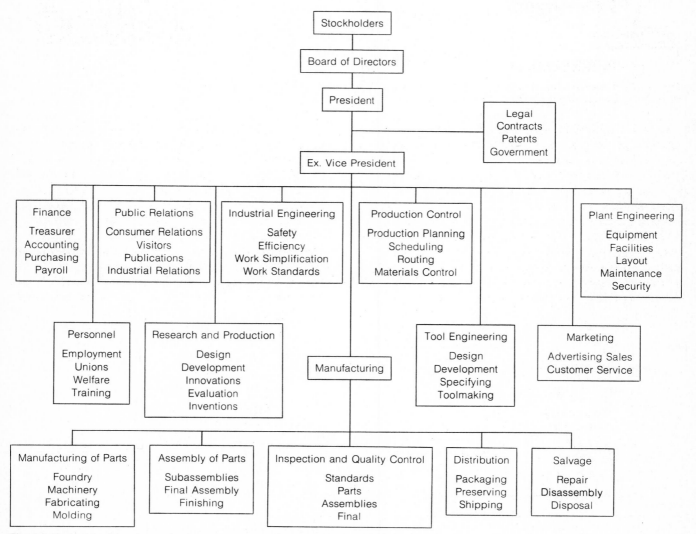

Fig. 1-3 Typical functional organization chart of a large manufacturing company.

sumer involves four basic manufacturing stages or steps (Fig. 1-4).

1. *Product research and development* or *product design:* This is where the concept of the product is formulated. It is also the decision-making stage where all basic decisions to "go" or "not go" with the idea are made.

2. *Production planning and tooling:* Most of the technical planning to produce a product which will be competitive in the market is done at this stage. Cost estimating and process design are elements of the planning stage of manufacturing.

3. *Manufacturing* or *production:* Once the product has been designed and its production has been planned, the next step is to produce it.

4. *Marketing:* After the product has been manufactured, it must be marketed to make it available to the consumer. The marketing stage will not be discussed in this text, which is concerned only with the production of a product.

It should be remembered that these four stages are organized in different ways in industry, depending on many factors. Nevertheless, they represent the four logical areas of manufacturing a product, from its conception through its delivery to the consumer.

PRODUCT RESEARCH AND DEVELOPMENT: PRODUCT DESIGN

It has been said that the heart of manufacturing is the product itself. Manufacturing exists to make and sell products to realize profits. For this reason, most product designs are subject to consumer demands, and usually reflect the changing likes and dislikes of society. Figure 1-5 illustrates two ways of designing an automobile. Consumer demand over the years has helped to bring about this radical change.

Among the things the consumer considers when buying a product are its appearance, convenience, durability, selling price, and maintenance cost. These are factors that the product designer must keep in mind in designing the product. Overlooking the consumer in designing a product is like driving an automobile without watching the road! Product design, therefore, combines the creativity of the design with consideration of consumer demands. Therefore, the five basic guideposts to product design are *function, appearance, quality, cost,* and *quantity.*

Function and appearance are crucial concerns of product design. Quality and cost are covered later in this chapter. Product design is a creative activity requiring curiosity, basic knowledge, interest, and motivation. Product design starts with a *new idea.* New ideas may originate in many different ways. Basic research is a source of new ideas for new products. New inventions may lead to new ideas for products. Ideas may also originate on the production floor with workers who are involved with production. Many companies use the practice of the "idea box" where employees can drop suggestions for product or process improvement or for the development of new products. The sales people, who are in constant contact with the customer, are a valuable source of new ideas. Materials purchasing and management people may offer important ideas. Many ideas originate because of a problem demanding a solution, which in turn leads to the design of a new product.

After the idea has been conceptualized and the

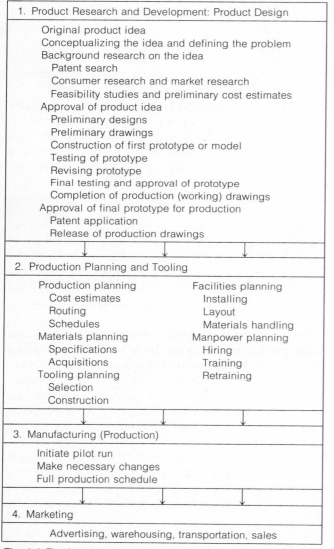

Fig. 1-4 The four basic stages of manufacturing a product.

Fig. 1-5 The design of many industrial products is changing as the attitudes of the consumer are changing. (*Ford Motor Co.*)

problem defined, the product designer initiates the *background research* on the idea. A patent search may be necessary to determine whether or not the product resulting from the idea will be free from patent infringement. A market research study may be necessary to determine whether or not there is consumer demand for such a product. The product designer may make a feasibility study to ascertain whether or not the product can be produced economically in order to be competitive in the market.

After completing the background research on the idea, the *preliminary designs* are produced. The product designer may already have developed a set of sketches prior to this point. These sketches must then be transformed into designs. Mention should be made that sketches, blueprints, and drawings constitute the principal means of communication in manufacturing. It is a good practice to prepare several preliminary designs and choose the best one. In developing the product design such factors as *function, appearance, quality, convenience,* and *cost* must be carefully built into the product. A product designed without careful consideration of these factors may be doomed to failure.

Functional design means that the product will perform to accomplish the purpose for which it was intended. A good functional design means that the designer has built into the product those characteristics which make the product responsive to its purpose.

Appearance is a basic characteristic that the product designer must build into the product. In the past, many industrial products were produced with little or no attention given to their appearance. The predomi-

nant considerations were function and cost. However, the demands of the consumer have changed; the consumer has become *appearance conscious.* While considerations of function and cost should never be sacrificed in the effort to produce an attractive product, the designer must keep in mind that products with limited eye appeal have limited sales value. For this reason, the product designer must be well acquainted with the principles that govern product appearance. Those principles are unity, interest, balance, and surface treatment.

Unity in product design refers to the form of the product. Appealing product forms, or unity, can be achieved by giving attention to the proportional relationships of the various parts of the product. Simplicity of form is a characteristic of unity. Repetition of similar lines or components also contributes to product unity.

Interest in product design refers to attracting and holding attention. This can be achieved by providing the product with a point of emphasis or by contrasting components of the product in matters of color, size, and location. Rhythm is another characteristic which contributes to product interest; components or elements on the product are arranged to attract and hold the eye.

Balance refers to the sense of stability the observer gets when looking at the product. Symmetry is one way to achieve balance. Symmetrical designs such as circles, cylinders, or spheres have balance built into them.

Surface treatment processes are used for a variety of functional and aesthetic purposes. The discussion here refers to their effects on product appearance. Color and texture are two means by which surface treatment can be used to improve the appearance of the product. Research indicates that most people are highly color-oriented; how the product looks in terms of color dynamics may affect its sales appeal more than any other single factor.

PRODUCIBILITY PRINCIPLES

Since product design is basically a creative activity, the product designer should be free from as many constraints as possible that may impede creativity. Yet it should be kept in mind that the products designed must not only be functionally sound and pleasing in appearance but also relatively inexpensive and easy to manufacture. Otherwise excellent designs may never be realized due to manufacturing difficulties. This is why a thorough knowledge of producibility principles is vital to the product designer.

Over the years producibility principles have evolved which have proved helpful in product design and development from the conceptual stage to the finished product. When properly applied, these principles may contribute distinct advantages in both product design and process design. It is the proper integration of these two manufacturing aspects (product and process design) which usually results in products of high quality at a reasonable cost.

In designing a product for manufacturing, the designer should keep in mind the principle of *elimination* (Fig. 1-6): Eliminate unnecessary parts, functions, and part characteristics. Avoid excessive scrap and the use of expensive or hard-to-find material and unwarranted product size. Scrapless designs are more economical than designs which waste a great deal of material and may require skillful stock layout. Designs with assemblies which are economical and easy to manufacture are recommended. Assemblies requiring no fasteners or screws are preferred.

Always try to *simplify* the design (Fig. 1-6). Keep in mind that the simpler the product design, the less expensive and easier it will be to manufacture. Therefore, simplify the design by modifyng such elements as costly tolerances and finishes, complicated assemblies, and difficult-to-produce physical characteristics of the product. The substitution of a small number of simple shapes for a single complex shape may reduce labor, reduce manufacturing complexity, and provide better use of materials (Fig. 1-7).

Skillful product designers know that *combining* functions into fewer parts and subassemblies invariably results in reduced cost and easier manufacturing. This may be especially true for products designed with few *master parts* or subassemblies (Fig. 1-8) that can fit in various products or create combinations of products.

Standardization is the heart of modern manufacturing (Fig. 1-9). Therefore, the product designer should try to design products that can use standard materials, standard parts, standard processes and methods, and standard equipment. Where possible, he or she should design a product that can be standardized by using *common parts,* across all models and

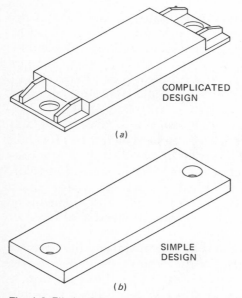

Fig. 1-6 Eliminating part characteristics may simplify production and lower cost. Parts *a* and *b* can perform basically the same function, yet Part *a* includes characteristics which could increase production cost and processing. (*General Electric.*)

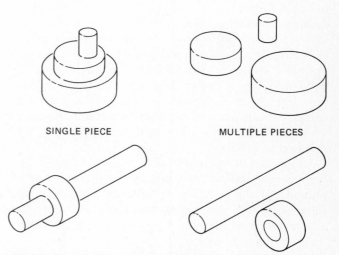

Fig. 1-7 The principle of two-for-one design may provide a small number of simple shapes rather than a single complex shape which may be difficult to produce. (*General Electric.*)

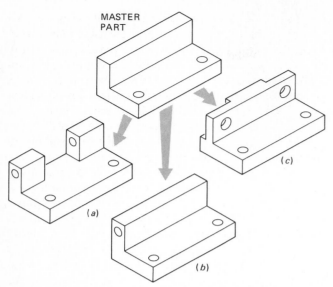

MASTER PART

(a)

(b)

(c)

Fig. 1-8 Parts *a, b,* and *c* can be produced from the *master part* with relatively little additional processing. (*General Electric.*)

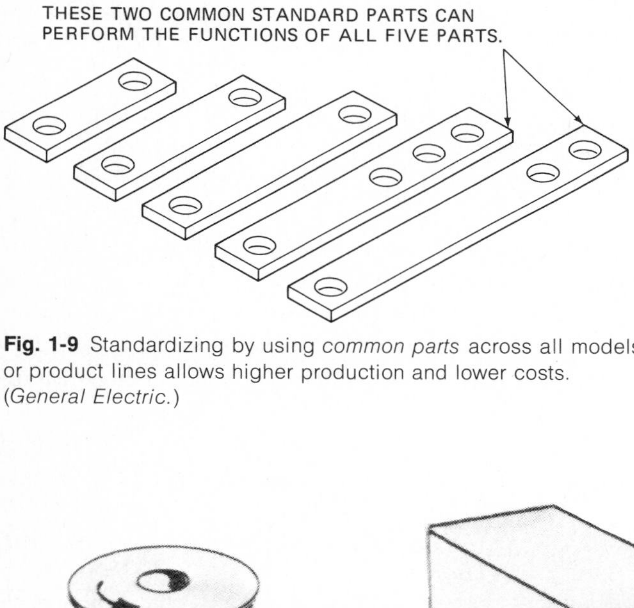

THESE TWO COMMON STANDARD PARTS CAN PERFORM THE FUNCTIONS OF ALL FIVE PARTS.

Fig. 1-9 Standardizing by using *common parts* across all models or product lines allows higher production and lower costs. (*General Electric.*)

possibly across product lines. This type of standardization leads to higher production volume, lowered product cost, and increased product quality.

Products are made from raw materials. Although there is a wide variety of materials produced to satisfy almost any need, relatively few materials have been standardized or are available in all desired forms. It follows, then, that the product designer should design products which utilize materials that *are* available. In addition, since materials have to be processed, they should have characteristics that will allow processing by existing machines and methods.

The product design should facilitate manufacturing with inexpensive processes and tooling. It has been said that any product designed can be produced, but relatively few products are designed that can be produced economically. If a product is designed with characteristics which will facilitate its production (by using relatively low-cost materials and processes and continuous production methods), it is only logical that the product will be inexpensive to manufacture and will be competitive in the market. Figure 1-10 illustrates a well-designed and economic operation.

The product design must *minimize production and materials-handling operations.* The product designer should keep in mind that the fewer the production operations required, the fewer the handling steps on the production floor. Materials handling is considered overhead (indirect cost). It does not add any value to the product. Therefore, the product de-

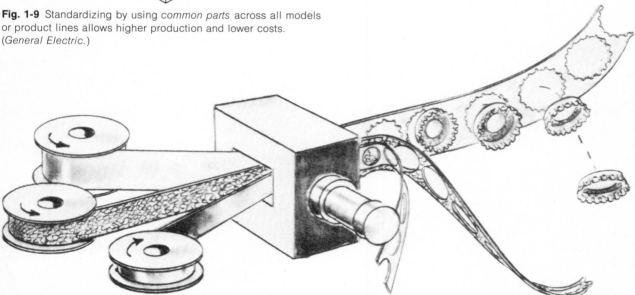

Fig. 1-10 Selecting the right continuous manufacturing process which may combine operations without manual handling or in-process delays can increase production rates and reduce cost. (*General Electric.*)

signer should look for a combination of production and materials-handling operations which provides the lowest possible cost for the processing of the product.

In the foregoing discussion, an attempt was made to emphasize the importance of producibility principles in designing new products (or modifying existing ones). The product designer is responsible for designing a product that has a pleasing appearance, is functional, and can be manufactured economically. To achieve these objectives, the product designer must work closely with the members of the production team. By cooperating closely with the production team, the product designer minimizes the possibility of designing a product that cannot be manufactured economically.

PRODUCTION PLANNING AND TOOLING

The purpose of production planning and analysis is to identify the most appropriate method(s) of manufacturing a product, economically and within required specifications. Production planning or process analysis is critical in modern manufacturing; it affects both the cost and quality of the product. Years ago, when manufacturing was relatively simple, process analysis was not as important as it is now. In those days the craftsmen (journeymen) were responsible for producing the entire product. They were thoroughly familiar with all the methods and equipment used in the shop and were able to use them effectively.

As manufacturing became more complex due to technological developments and requirements, however, the craftsmen in the shop were forced to produce only certain parts of the product. Therefore, the demand for part interchangeability increased.

Process analysis then became a necessity in the manufacturing industry.

The practices of process planning and analysis vary widely in modern industry, depending on such factors as type of product, the equipment available, and the volume of production. The procedure suggested by the Society of Manufacturing Engineers (SME) is the most widely used and is described later in this section.

The individual responsible for process analysis is the *process analyst* or *process engineer,* also known in some industries as the *manufacturing, production,* or *methods engineer.* To be effective on his or her job, the process analyst must be familiar with material characteristics. Knowledge of the nature, types, and properties of standard and new materials will assist the process analyst in selecting the most appropriate process,

equipment, and methods for manufacturing a particular product. The process analyst must also be familiar with engineering drawings and product design. Drawings provide the part configuration and the dimensional tolerances and specifications that need to be met by the manufacturing process selected.

In addition, the process analyst must be familiar with the operating characteristics and costs of the production and tooling equipment, either available in the plant or to be purchased.

Process analysis starts with a careful examination of the drawing or design of the part. The process analyst must be able to analyze the engineering drawing and visualize the three-dimensional part configuration. The part configuration must then be analyzed to determine its basic geometric components. All parts manufactured consist of one or more basic geometric elements that can be identified by *part analysis* (Fig. 1-11). Identifying these basic geometric elements assists the process analyst in selecting the most appropriate process to manufacture the part.

With the part analysis completed, the next step is to select the process to be used. According to SME, consideration should be given to the following factors for selecting a particular process:

1. Nature of part, including materials, tolerances, desired finishes, and operations required.
2. History of fabrication, including machining or assembling of similar parts or components.
3. Limitations of facilities, including the plant and equipment available.
4. Possibility of product design changes to facilitate manufacturing or cost reduction.
5. In-plant and out-of-plant materials-handling systems.
6. Inherent processes to produce specified shapes, surfaces, finishes, or mechanical properties.
7. Available qualified workers to be used for the production.

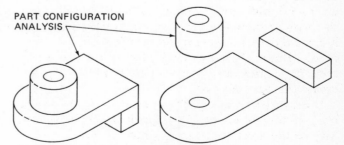

PART CONFIGURATION ANALYSIS

Fig. 1-11 Part configuration must be carefully analyzed to determine the basic geometric components that may aid in process planning and selection.

Sometimes the following additional factors affect the selection of a particular process:

1. Proposed or anticipated production requirements, including volume requirements, production rates, and short- or long-term production runs.
2. Total end-product costs.
3. Time available for tooling up.
4. Materials receipt, storage, handling, and transportation.

Careful consideration of these factors will result in the selection of the most appropriate process for the manufacture of a particular part.

As an illustration of how these factors can affect the selection of a particular process, assume that the part shown in Fig. 1-12 must be manufactured. The part serves as a spacer in a machine and therefore must be made of metal. The part analysis indicates that this part can be broken down into two rectangular shapes of $2 \times 1\frac{1}{2} \times \frac{1}{4}$ in. [50.8 × 38.1 × 6.4 mm] and $1\frac{1}{2} \times \frac{1}{4} \times \frac{1}{4}$ in, [38.1 × 6.4 × 64 mm], respectively. The part can be produced by *machining* from standard mill stock of 2 in [50.8 mm] wide and $\frac{1}{2}$ in [13 mm] thick cut to $1\frac{1}{2}$ in [38.1 mm] length. It can be produced by *welding* two pieces of standard mill stock (2 in [50.8 mm] wide and $\frac{1}{4}$ in [6.4 mm] thick cut to $1\frac{1}{2}$ in [38.1 mm] length, and $\frac{1}{4}$ -in-square [6.4 mm²] stock cut to $1\frac{1}{2}$ in [38.1 mm] in length). It can be produced by a *casting* process, either by single or multipart molds. It can be produced by *extrusion* and cut to $1\frac{1}{2}$ in [38.1 mm] length. It can be produced by *forging*. All these processes are capable of producing the required part, and the process analyst must select the most appropriate one. In making the selection, the factors previously outlined must be considered.

If 10 pieces are required in the job order and aluminum is specified as the material, machining seems to be the most feasible process. However, before the decision is reached, the analyst must determine whether a milling machine or a shaper is available in the plant. If these machines are not available, the analyst could consider welding or casting as the next best choice. But in these cases, a jig for welding or a pattern for casting must be prepared, which may increase the per-unit cost.

If the order calls for 100 pieces, neither machining nor welding is economically justified. Sand casting may be the most appropriate process.

If 1000 pieces are required, die casting or extrusion would be appropriate. However, in both cases, a die is needed, and the process analyst must consider the

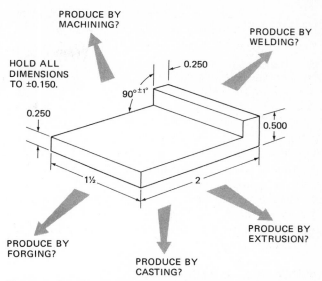

Fig. 1-12 Selecting the most economical process requires knowledge of part configuration, functions, materials, and manufacturing processes.

cost of the die and whether or not the facility provides for casting.

This simplified example illustrates that the selection of an appropriate manufacturing process depends on many factors and requires the considerable knowledge, skill, and competence of the process analyst.

Once the process has been selected, the next step is to list the operations. Operations are first listed without any sequence in a "laundry list," and then are sequenced in the *process* or *operation analysis sheets*. The process sheets are used as instruction sheets for the operator to process the part. Process sheets vary widely in industry, depending on such factors as type of product, type of industry, type of equipment, and type of manufacturing. However, most process sheets include such matters as description and numerical ordering of operations, manufacturing equipment used, jigs, fixtures, tools and gauges, speeds, feeds and depths of cut, material specifications, drawing specifications, and revisions. Figure 1-13 shows a typical process sheet used by a manufacturing company.

An example will be used to summarize the various steps in process planning and analysis. Assume that an order has been placed to manufacture the gear shaft shown in Fig. 1-14. The order specifies that five pieces are required and should be delivered within 30 days from receipt of the job order. Following the procedure outlined above, the process analyst can prepare the process sheet by examining the drawing as follows:

A. Part configuration analysis:
   1. Three concentric and adjacent cylinders

Precision Manufacturing Corporation

Oper. No.	Description of Operation	Machine Name or No.	Cutting Tool	Cutting Speed (Ft/Min)	rpm	Feed (irp)	Depth of Cut Inches	Locating Points on Drawing	Fixtures, Tools, Gauges, etc.	Remarks
10	Face end	Engine lathe 7071	Facing tool	80	300	Hand	—	A	Use one inch Jacobs lathe chuck	If Jacobs chuck is not available, use 3-jawed universal chuck
20	Center drill end	"	Combination center drill		650	"	—	B		Cutting fluid or oil may be used
50	Cut off to 3 9/16 length	"	Parting tool	80	300	"	—	A	Support work with line center to avoid chatter	Use coolant
90										
100	Remove sharp edges	"	File or emery cloth	—	650	Hand	—			

No.	Revisions Schedule	Date	Part Identification Details	
1	Added Oper. No. 100	18-1-79	Drawing No.: 76-10 GM	Stock No.: 48252
2			Part Name: Gear Shaft	Specification No.: 01-305GM
3			Materials: SAE1030 CRS	Process Eng: A.J.J.
4			APR: A. J. Jones Date: 16-1-79	Sheet No.: 1 of 2 sheets

Fig. 1-13 A typical sample of an operation (process) analysis sheet used in manufacturing.

with diameters of 0.898 to 0.902 in [22.80 to 22.91 mm], 0.598 to 0.602 in [15.18 to 15.29 mm], and 0.298 to 0.302 in [7.56 to 7.67 mm] and lengths of $1\frac{1}{2}$, $1\frac{1}{2}$, and $\frac{1}{2}$ in [38.1, 38.1 and 12.7 mm], respectively.

2. Four parallel, plain surfaces forming the ends of the cylinders.
3. One chamfer of 45° × $\frac{1}{8}$ and one 45° × $\frac{1}{16}$ at the ends of the shaft.
4. The closest dimensional tolerance is ±0.002 in [0.051 mm] and the angular tolerance is ±1°.

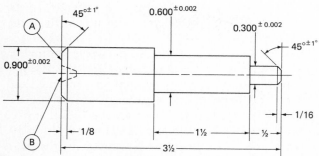

MATERIAL: 1 IN DIAMETER SAE 1030 COLD DRAWN STEEL
ALL TOLERANCE: ±0.150, UNLESS OTHERWISE SPECIFIED.
NUMBER OF PARTS REQUIRED: 5
Fig. 1-14 Gear shaft for water pump.

B. From job order and drawing, the following additional information is derived:
   1. Materials: SAE 1030 cold drawn steel.
   2. Quantity: five pieces.
   3. Delivery time: 30 days from receipt of job order.
C. Process selection. From the information derived, the process analyst could reach the following decisions:
   1. The most feasible process is machining on an engine lathe.
   2. The part could be machined from l-in [2.54-cm] diameter standard stock.
   3. One engine-lathe operator is needed to complete the order.
   4. The job must be scheduled within 20 days from receipt of order to meet delivery date.
D. Identify the operations. To produce the part the following operations are required: facing, center drilling, cutting off, turning, chamfering, and finishing (removing sharp edges). In addition, the analyst must estimate the time for each of these operations, the setup time, the cutting tools needed, jigs, fixtures and gauges required, the cutting speed, feed and depth of cut.
E. Completing the process sheet. The operations listed are sequenced in numerical order and

are listed in the process sheet along with other necessary information, as shown in Fig. 1-13.

MANUFACTURING COSTS

In modern industry, stockholders invest money by buying stock from manufacturing companies in order to make *profits*. Without profits there will be no investment, and without investment there may not be any manufacturing company. In a free enterprise system, profit is the basic reason for the existence of most companies. Therefore, knowledge of the basic elements of *cost* is essential. Since cost is such an important factor in manufacturing, companies have developed highly specialized methods and techniques to deal with it. A detailed discussion of these methods is beyond the scope of this book; however, an understanding of the basic elements of cost should be of great assistance to the manufacturing specialist, since cost directly affects every decision relative to the materials and processes selected for manufacturing a particular product.

The ultimate objective of cost estimating in manufacturing is to arrive at a *sales price* for the product that will be competitive and also to provide a reasonable profit on the investment of the company. To achieve this objective, the cost accountant or analyst must consider such factors as: (1) *primary costs,* consisting of direct-labor and material costs, (2) *factory costs,* consisting of primary costs plus factory expenses such as supplies, light, heat, rent, and power, (3) *manufacturing costs,* consisting of the factory costs plus general expenses such as purchasing, production methods, production engineering, office costs, and depreciation, (4) *total costs,* consisting of the manufacturing costs plus sales expenses, and (5) *product selling price,* consisting of the total cost plus the profit. Primary costs are often referred to as *direct labor* and *materials cost.* The other costs are referred to as *overhead* or *burden.*

In manufacturing, certain costs, such as labor and material costs, change in proportion to output or units produced and are considered *variable costs* because they vary with the output. Other costs, such as tooling, setup, depreciation, office costs, and utilities, do not vary directly with the output and are considered *fixed costs.* Variable costs increase with the total volume of output. Fixed costs decrease (per unit) because they are "spread over" a larger number of units produced.

An example may illustrate this point. Assume that the gear shaft shown in Fig. 1-14 is to be manufactured. To produce 100 shafts at one time, the company will pay the same rent, the same office expenses,

the same administrative and engineering personnel as it would in producing 100,000 shafts, because those fixed costs can be avoided only if the plant is shut down. In addition, the company stands to gain a much larger profit by producing 100,000 shafts than for producing only 100 shafts. Therefore, the overhead (fixed costs) could be spread over 100,000 rather than 100 units. On the other hand, the cost of labor and materials needed to produce 100,000 units would be considerably higher than that needed to produce only 100 units.

Cost accountants or analysts have developed many different methods (models) for estimating manufacturing costs. One of the methods that has been used effectively for many years is the *break-even analysis* (Fig. 1-15). This method assumes that (1) a linear relationship exists among the three basic elements of output, costs, and revenue so that a change in any one of these elements results in a proportional change in the other two, (2) fixed and variable costs can be identified and estimated with reasonable accuracy, and (3) the sales price of the product is not affected by the volume sold. These relationships of output, costs, and revenue can be shown graphically, as in the chart illustrated in Fig. 1-15. They can also be expressed mathematically as follows:

1. *Total revenue* = unit sales price × number of units produced

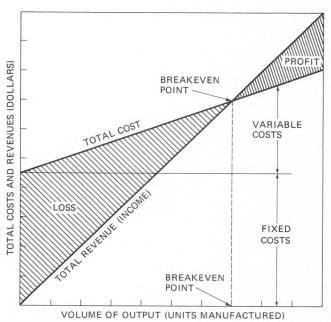

Fig. 1-15 Break-even chart used with break-even analysis model or method.

2. *Total cost*   = total fixed cost + (variable cost per unit × number of units produced)

3. *Profit*   = total revenue − total cost

4. *Break-even point* = $\dfrac{\text{total fixed cost}}{\text{unit sales price − variable cost per unit}}$

To illustrate the concepts discussed in this section, assume that a manufacturing company would like to determine whether or not it will break even or make a profit by manufacturing the gear shaft shown in Fig. 1-14. The job order is for 200,000 units. The unit sales price has been estimated at $0.40 in order to be competitive in the market. The cost accountant or analyst estimated that for that job order, the variable costs in labor and materials will be $20,000 and the fixed costs $30,000. Variable costs per unit:

$$\frac{20,000}{200,000} = \frac{2(10^4)}{2(10^5)} = 1(10^{-1}) = \$0.10 \text{ per unit}$$

Using a break-even analysis, the total revenue, total cost, profit, and break-even point can be calculated as follows:

1. *Total revenue* = $0.40 × 200,000 = $80,000

2. *Total cost*   = $30,000 + (200,000 × 0.10) = $50,000

3. *Profit*   = $80,000 − $50,000 = $30,000

4. *Break-even point* = $\dfrac{\$30,000}{\$0.40 - \$0.10} = 100,000 \text{ units}$

This information is shown graphically in Fig. 1-16.

The gear shaft will be manufactured with a profit

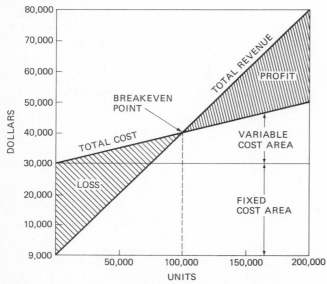

Fig. 1-16 Break-even chart for manufacturing the gear shaft shown in Fig. 1-14.

of $30,00 or a unit profit of $0.15. The company can break even at 100,000 units. At this point of production, the expenditure equals the revenue. Above this point the company makes a profit; below this point it loses money. This also indicates that the unit sale price could be considerably reduced to make the product more competitive in the market by increasing the volume of production, by using cheaper materials, or by employing a more productive manufacturing method which could increase the volume of production and decrease the direct labor costs.

All these possibilities must be carefully considered by the manufacturing specialists before a decision is made in terms of the material and processes selected for a particular product. This is the main reason why manufacturing specialists need to understand the basic elements of cost in manufacturing.

CLASSIFICATION OF MANUFACTURING PROCESSES

Manufacturing processes can be classified in different ways on the basis of such factors as types of materials processed, types of equipment used, and type of manufacturing. A system of classification based on the basic functions of manufacturing processes in the production of parts will be presented here. Although these processes are described in detail in later chapters of this book, a survey should help you to visualize the entire spectrum of manufacturing processes and perceive some of the relationships among different processes.

Manufacturing processes can be classified in four broad categories, based on the way parts are produced: shaping, assembling, finishing, and miscellaneous. *Shaping* of parts is accomplished by forming and material-removal processes. *Assembling* is accomplished by joining, either permanently or semipermanently, two or more parts. *Finishing* is accomplished by either cleaning or coating the surface of the part. Such processes as heat treating and quality control are classified as *miscellaneous*. Figure 1-17 gives a breakdown of these manufacturing processes.

Another classification system based on established categories of manufacturing processes is shown in Fig. 1-18. Both of these systems have been incorporated into this book and are explained in greater detail in later chapters. In this section, however, the main purpose is to give the reader an overview of the manufacturing processes "landscape."

SHAPING PROCESSES

Parts can be formed by casting or molding, forming, and material removal (cutting or machining).

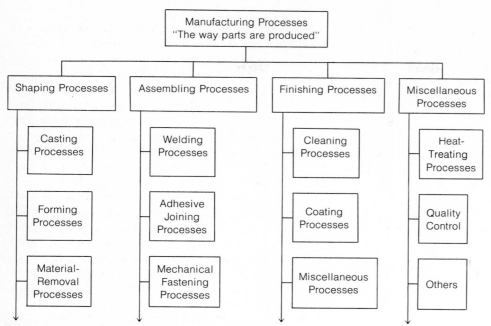

Fig. 1-17 Classification of manufacturing processes.

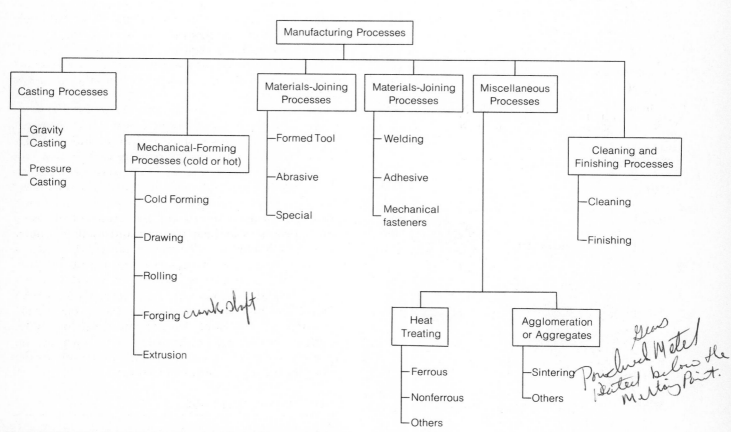

Fig. 1-18 Classification of basic categories of manufacturing processes.

Casting

Casting is relatively simple and inexpensive compared to some other processes. In casting or molding, the material in the form of liquid, (in the case of plastics, the material usually is in the form of powder or granules) is introduced into a preshaped cavity called the *mold*. The mold is shaped with the exact configuration of the part to be molded or cast. After the material fills the mold and sets or solidifies, it assumes the shape of the mold, which is the shape of the part. The mold is broken or opened and the part is removed. Casting processes are used to cast or mold such materials as metals, plastics, and ceramics. Casting processes can be classified by the type of mold used (either permanent or nonpermanent) or by the way in which material enters the mold (*pressure* or *gravity casting*).

Figure 1-19 shows the various casting processes based on gravity and pressure casting and the materials usually cast with these processes.

The word "casting" is always used for metals, but the process differs in no substantive respect from molding (the word generally used for plastics). For example, *injection molding* is the term for a pressure process for molded thermoplastic parts. The machine used is an *injection molding machine,* which injects the melted plastic into a metal *mold*. Basically the same process, but operated at higher temperatures, produces *die castings* in a *die casting machine,* which injects molten zinc or aluminum, for example, into a steel *die*.

Parts produced by casting processes vary in size, accuracy, surface smoothness, complexity of configuration, finish required, production rate, and production cost and quality. The size of parts cast may vary from a few ounces for those produced by die casting, to several tons for those produced by sand casting. Dimensional tolerances may vary from 0.005 to 0.250 in [0.127 to 6.35 mm], with the most accurate parts produced by die, shell molding, injection, and investment casting. Less accurate parts are produced by sand and continuous casting. However, continuous casting is mostly used to produce mill stock: slabs, billets and rounds as opposed to finished parts.

Die, shell, injection, transfer, vacuum, and investment casting produce parts with relatively smooth surfaces. Sand, continuous, centrifugal, and mold casting produce parts with the roughest surfaces. Relatively simple configurations are usually produced with form, sand, and continuous casting, while more complex configurations are produced by investment and die casting. Die casting is considered a high-production-rate process, while sand casting is considered a "one at a time" or relatively slow production process.

Casting is a relatively inexpensive process. However, molds for compression molding and injection molding, and dies for die casting, are very costly.

Mechanical-Forming Processes

Forming of parts by application of mechanical force is considered the most important shaping process in terms of value of production and method of production. Forming parts can be accomplished with the material cold (*cold forming*) or with the material hot (*hot forming*). The forces used to form parts may be of the bending, compression, shearing, or tension type. Forming processes can be classified on the basis of how the force is applied (Fig. 1-20).

Bend forming is accomplished by forcing the material to bend along an axis. Among the bending processes are bending, folding, corrugating, and spinning. *Shear forming* is actually a material-separation process in which a cutting edge or two edges are forced through a fixed part. Shearing includes such processes as punching or piercing, stamping, blanking, and trimming. *Compression forming* is accomplished by forcing the material, hot or cold, to form

Fig. 1-19 Classification of casting processes based on gravity and pressure casting.

Category	Processes	Materials Cast
Gravity casting	Investment	Metals
	Sand (pit, floor, and bench)	Metals
	Shell	Plastics, ceramics, metals
	Slush and slip	Metals, ceramics
	Dip	Plastics
	Form	Ceramics (concrete)
Pressure casting	Centrifugal	Metals
	Rotational	
	Die	Metals, ceramics
	Injection	Plastics, rubber
	Blow	Plastics, ceramics
	Continuous	Metals
	Compression	Plastics
	Transfer	Plastics
	Vacuum	Plastics, metals

Fig. 1-20 Classification of mechanical forming processes based on the type of force used

Bending	Shearing	Compression	Tension
Bending	Piercing	Forging	Drawing
Folding	Shearing	Extrusion	Stretch forming
Corrugating	Stamping	Upsetting	Flaring
Embossing	Blanking	Rolling	
Spinning	Trimming	Coining	
	Turning	Swaging	
	Drilling	High energy rate	
	Milling	Hydroforming	
		Magnetic forming	

into a desired shape with the help of a die, a roll, or a plunger. Compression forming includes such processes as forging, extrusion, rolling, and coining. *Tension forming* is accomplished by stretching the material to assume the desired configuration. It includes such processes as drawing, stretch forming, and flaring. Figure 1-21 shows the most representative of the mechanical-forming processes.

Material-Removal (Machining) Processes

These processes are used to shape parts of such materials as metals, plastics, ceramics, and wood. Machining is a relatively time-consuming and material-wasting process. It is, however, very accurate and can

Fig. 1-22 Classification of material-removal processes, traditional, and nontraditional

Traditional	Nontraditional (special)
Boring	Abrasive-jet machining
Broaching	Ultrasonic machining
Drilling	Liquid-jet machining
Facing	Electrochemical machining
Grinding	Chemical machining
Grooving	Chemical milling
Honing	Electrodischarge machining
Lapping	Electron beam machining
Milling	Laser beam machining
Sawing	Ion beam machining
Shaping	Plasma arc machining
Planning	
Turning	

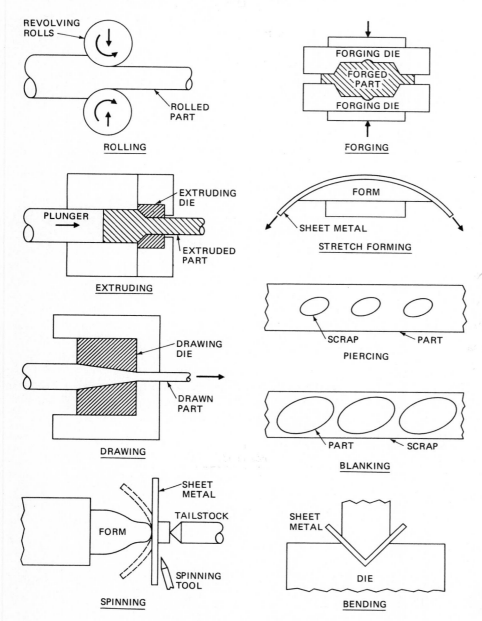

Fig. 1-21 Schematics of the most common mechanical forming processes.

produce surface smoothness difficult to achieve with other shaping processes. Traditional machining is accomplished by using a cutting tool which removes material from the workpiece in the form of chips, thus shaping the piece to the desired configuration. Material-removal processes are classified as *traditional* or *chip forming,* and *nontraditioanl* or *chipless,* processes (see Fig. 1-22 on page 15).

In all traditional material-removal processes the three basic elements are the *workpiece,* the *cutting tool,* and the *machine tool.* The basic functions of the machine tool are (1) to provide the relative motions between the cutting tool and the workpiece in the form of *feeds* and *speeds* and (2) to maintain the relative positions of the cutting tool and workpiece so that the resulting material removal produces the required

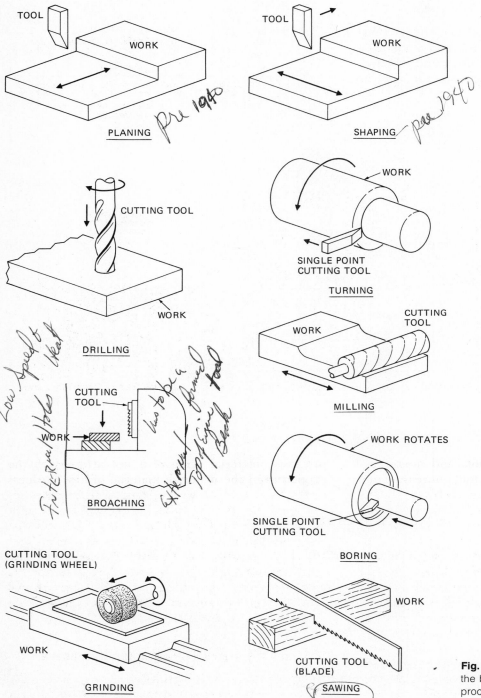

Fig. 1-23 Schematic representations of the basic traditional material-removal processes.

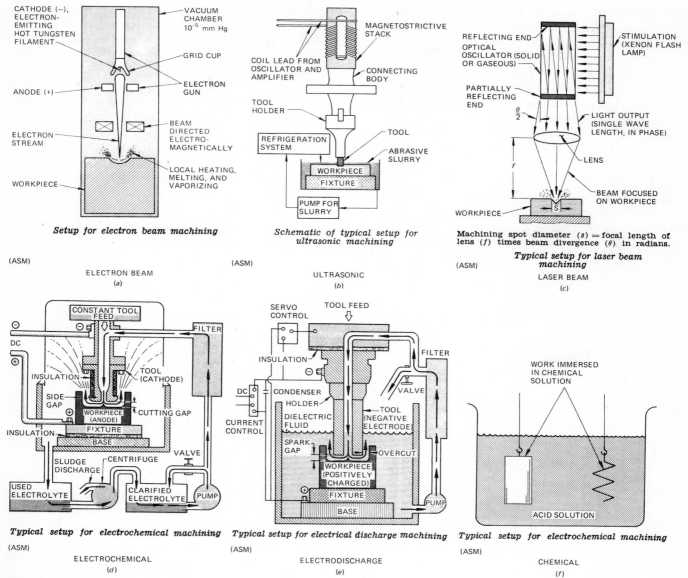

Fig. 1-24 Schematic representations of the basic nontraditional material-removal processes. (*SME.*)

shape. By changing various positions and motions between the workpiece and cutting tool, more than one operation can be performed by the machine tool. The cutting tools used are either *single point* (edge) or *multiple point*. These basic processes are illustrated in Fig. 1-23 on page 16.

With the progress of technology, stronger and harder materials have been developed. The effective and efficient processing of such materials was not possible with the traditional material-removal processes. Therefore, several new and specialized processes have been developed. Unlike traditional processes, where material removal requires a cutting tool, nontraditional processes are based on the phenomena of ultrasonic, chemical, electrochemical, electrodischarge, electron, laser, and ion beams. In these

processes, material removal is not affected by the properties of the material; material of any hardness can be machined. However, some of these processes are still in the developmental stage and do not lend themselves to high-production rates. With most of these processes parts are shaped one at a time. Nontraditional processes are relatively complex, and it requires considerable skill and knowledge to operate them effectively and efficiently. The nontraditional material-removal processes most commonly used are shown schematically in Fig. 1-24.

Material Joining and Assembling Processes

The basic function of assembling processes is to join two or more parts together into a complete assembly or subassembly. Joining of parts can be accomplished by welding, brazing, or soldering or by using mechanical fastening or adhesives.

Mechanical fastening has been used for centuries and is still one of the principal means for joining parts. Mechanical fastening can be accomplished by means of screws, bolts, rivets, pins, keys, and press-fit joints. Riveted and press-fit joints are considered semipermanent, while the other mechanical fasteners are nonpermanent. Fastening parts with mechanical fasteners is relatively costly and requires skill in the preparation of parts for joining. However, there are many situations in modern industry where mechanical fastening is still the only practical method, particularly where maintenance requires disassembling.

With *welding, brazing,* and *soldering, and some adhesives,* parts are joined permanently. Welding is used extensively and is the most important joining process in manufacturing. Welding is accomplished by using heat or pressure or both. The heat will have some effect on the properties of the joined parts.

To satisfy the wide variety of needs in manufacturing, many welding processes have been developed and used. Figure 1-25 shows the various welding processes used. Arc, gas, resistance, and brazing are the most widely used welding processes. Schematic representations of arc and gas welding are shown in Fig. 1-26 *a* and *b*. In both processes, the materials joined are melted where they interface and form a continuous joint.

The use of *adhesives* to join parts together is considered permanent joining. Adhesives can be applied to join such materials as metals, plastics, wood, rubber, and ceramics. Both natural and synthetic adhesives are used, but synthetic adhesives are gradually eliminating the natural ones. Epoxy-resin-based adhesives are among the strongest, and have been very widely used in recent years. Adhesive joints are relatively strong and have a clean appearance.

Finishing Processes

The main functions of finishing processes are to clean, protect, and decorate the surface. Cleaning of

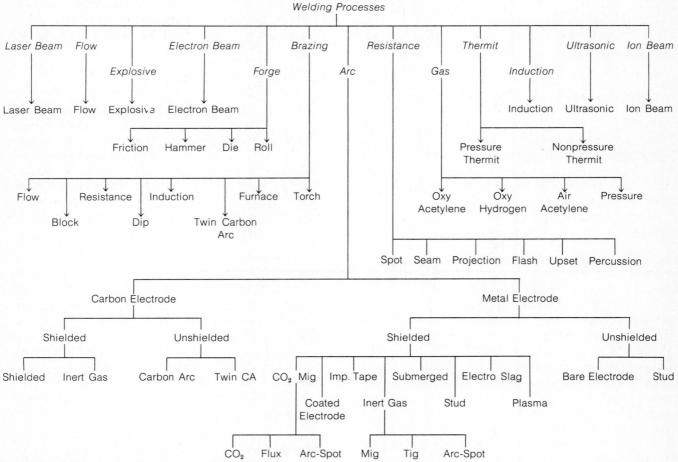

Fig. 1-25 Classification of common welding processes.

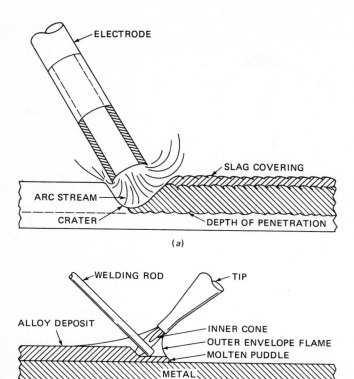

Fig. 1-26 Schematic showing typical arc (*a*) and gas (*b*) welding.

the surface is usually the first step. Cleaning removes dirt, oils, grease, scales, and rust, thus preparing the surface for further treatment. Cleaning can be accomplished by mechanical means, such as abrasive blasting, or by chemical means, such as alkaline cleaning. However, some cleaning processes may serve both cleaning and finishing purposes. Other purposes of finishing processes are to protect the surface from deterioration and to decorate it to increase its

aesthetic appeal. These finishing purposes can be accomplished by covering the surface with a suitable coating. Surfaces can be coated with organic coatings (paints), metallic coatings, phosphate coatings, porcelain enamels, and ceramic coatings. Figure 1-27 lists the most commonly used cleaning and finishing processes.

Miscellaneous Processes

Heat treatment and quality control are processes which can be classified as miscellaneous. *Heat treatment* is accomplished by heating and cooling the material to change certain characteristics such as softness, hardness, ductility, and strength. It is also used to relieve stresses that develop in the material from other processes. Heat-treatment processes are applied to such materials as metals, plastics, glass and ceramics. Heat treatment of metals is an especially important process and is extensively used in manufacturing metallic parts. Heat treating of tools is equally important.

Quality control is an integral part of manufacturing and is applied throughout the production and assembling of parts. Its basic function is to inspect, control, and improve the quality of the product and the processes.

MATERIALS HANDLING

A great deal of time is often wasted and many production problems are caused by inefficient materials-handling systems. About 80 percent of production time is spent in transferring the materials from one place to another on the production floor, and less than 20 percent is spent on the actual processing. In most cases it takes more handling than processing operations to produce a part. Thus, inefficient materials

Fig. 1-27 Classification of cleaning and finishing processes.

Cleaning Processes		Finishing Processes
Mechanical	Chemical	
Abrasive blasting	Alkaline cleaning	Organic finishes (paints, enamels, lacquers, etc.)
Mass finishing or tumbling	Solvent cleaning	Powder coatings
Belt sanding	Solvent-vapor degreasing	Metallic coatings
Wire brushing	Acid cleaning (pickling)	Electroplating
Polishing and buffing	Molten-salt descaling	Electroforming
Ultrasonic cleaning	Electropolishing	Oxide coating
		Vacuum metalizing
		Metal spraying
		Oxide coatings anodizing
		Phosphate coatings
		Porcelain enamels and ceramic coatings

handling may create production problems, such as insufficient or excessive supply at the processing station or the transporting of wrong materials. Most materials-handling operations do not add value to the manufactured part; on the contrary, they may substantially reduce the value, thus increasing manufacturing cost.

In relatively small production shops, the responsibility for materials handling is given to the shop supervisor. However, in medium and large production plants, materials handling is the responsibility of the *materials-handling department.* This department is responsible for handling the materials during receiving, storing, processing, assembling, warehousing, and shipping. Therefore, it is important that the materials-handling department should be involved in the planning and organizing for manufacturing.

During the actual production, materials handling is usually confined to moves between and within each process of manufacture. Thus, in selecting the materials-handling system, consideration should be given to the volume of production, the types and sizes of materials being processed, and the nature of processing operations. For handling materials between processing operations in high-volume production, such systems as gravity or power conveyors and elevators may be most appropriate. On the other hand, for relatively low-production volume, such equipment as hand lifts and fork trucks may be used. When parts are large and heavy, monorail and crane equipment is more appropriate.

Handling of materials or parts during processing is by far the most important and difficult aspect of materials-handling. Where conventional machines are used, the machine operator usually performs the materials-handling function. However, using skilled personnel who receive relatively high wages to handle materials that can be handled by a materials-handling system is very costly. In the long run, perhaps a machine with a built-in automatic materials-handling system will be less expensive. If so, an automatic materials-handling system should be designed and used.

When such a system is designed and used, the product size, weight, shape, production rates, permanence, and dimensional tolerances will determine the system's complexity and cost. For relatively compact parts, which lend themselves to *hopper* or *magazine* type feeding, various such systems can be designed. For parts which can be handled by *chuting,* this system has proved very reliable.

REVIEW QUESTIONS

1-1. Define manufacturing processes.

1-2. Why are machine tools considered the foundations of modern manufacturing? Explain.

1-3. How is modern manufacturing related to materials?

1-4. Identify and explain some of the production elements or functions found in most manufacturing organizations.

1-5. Why is organizing for manufacturing so important? Explain.

1-6. Compare the organizational structure of a multiplant manufacturing organization with a job shop company.

1-7. To what extent is product design affected by manufacturing processes?

1-8. Why is it important that a market survey be conducted before a final decision to manufacture a product is reached?

1-9. To what extent is product design conditioned by consumer attitudes?

1-10. What is the most important characteristic that a buyer may look for in a manufactured product?

1-11. What is the relationship of producibility principles and product design?

1-12. Explain some of the producibility priciples.

1-13. Why is it important that a part analysis be performed before the product can be produced?

1-14. What costs vary with the output, and how can those costs be reduced in manufacturing?

1-15. Explain the break-even cost model used in manufacturing.

1-16. Is the cost of materials handling considered direct or overhead cost in manufacturing?

1-17. Differentiate between shaping and finishing processes as used in manufacturing.

1-18. Why is forming considered the most important process in terms of production value?

1-19. Compare the traditional with nontraditional machining processes.

CHAPTER 2

MANUFACTURING SPECIFICATIONS AND QUALITY CONTROL

A prime concern of modern manufacturing is to produce products of high quality. In producing high-quality products, all components must be manufactured in accordance with the design and specifications developed by designers and engineers. Predetermined and acceptable standards, statistical quality control, and a variety of measuring, gauging, and testing devices all aid manufacturers in achieving high quality. This chapter will discuss the various aids and aspects of *quality control*.

Specifications used in manufacturing include size, location, operation, surface texture, materials, finishes, electrical requirements, roundness, and so on. These are specified by the designers. If and when a product is used as a component in other products or by various manufacturers, the specifications become a standard for that particular item. For example, a bolt can be manufactured having any number of threads, diameters, and pitches. If a bolt's specifications meet their particular needs, other manufacturers may want to produce it. In order for the bolt to be interchangeable with a nut, the thread sizes and designs must be compatible. Hence, this particular thread specification becomes standardized throughout the industry.

QUALITY CONTROLS

Quality-control operations are necessary to maintain desired precision and accuracy as well as to maintain the material quality required by the customer. The production of quality products is an important factor in customer relations in that customers satisfied with good products will continue to award contracts to the producer. Bad products tend to discourage customers because the materials do not fit their needs. Consider the reputation of the Rolls Royce automobile. Although it is an automobile with four wheels much the same as other automobiles, its reputation for qual-

ity, accuracy, and durability is maintained through the strictest adherence to the production of only the finest quality product.

Inspection methods are important to the control of quality. Inspection helps detect faulty parts and helps to locate and remedy problems in the manufacturing processes. Repeating flaws of a particular type can determine whether a faulty process is occurring in the casting material, in the hydraulic fluidity controls, in the mold material, or in the mold construction. Thus, problems can be eliminated by constant inspection and quality-control processes.

Inspection stations may be automatic or may be operated by individuals. They may be placed at the end of each operation or they may be placed at the end of a series of operations. The quality-control inspection may also be done by the operators as they make the molds and by various other operators and artisans as they perform the manufacturing process. Quality control inspections are also made by individuals who have no other responsibility than the inspection and control of quality.

Defective parts that may be repaired are generally drawn from the production line and recycled through a repair shop. The repair shop or station repairs the defect by welding, filling, or various other operations. If the material or object cannot be repaired, it must be scrapped. Scrap plays an important part in metal manufacturing because the molten metals that are used to cast the objects are largely composed of scrap metal from gates, risers, sprues, defective castings, and other scrap metals.

It is important to remove defective pieces from production lines as soon as possible. If defective parts are allowed to be cleaned, machined, finished, and then scrapped, the cost of the cleaning, machining, and finishing has been wasted and represents a factor which can increase total production costs.

SPECIFICATIONS AND STANDARDS

Specifications and standards were used as far back as ancient Egypt. Construction of the pyramids and other architectural wonders would have been impossible without some form of standardization. Egyptian attempts at standardization of measurement were based on the human body, since instrumentation was lacking.

Early measurement practices included using the length of the arm from the elbow to the tip of the longest finger, the width of an open hand, the closed hand, the thumb, and the length of the human foot. However crude, these did provide a means of standardizing measurement.

Advances proceeded in standardization of measurement, leading to the development of the *metric system*. The metric system uses the *meter* as the standard for length. A meter was initially defined as one ten-millionth of the distance from the North Pole to the equator on an imaginary line passing through Paris, France. A prototype meter standard, a platinum and iridium bar, was produced in 1875. Fine lines were etched on the bar to denote the length of the meter when measured at 0°C [32°F]. Recently, through advancing technology, the meter was redefined using the wavelength of monochromatic light from krypton gas.

Congress officially recognized the metric system in 1866 but did not make its use mandatory in the United States. Action has been taken in the United States to convert from the English, or conventional American, system to the metric system (Table 2-1). More details about the metric system are given in the Appendix.

Manufacturing prior to the sixteenth century consisted of individually constructing and assembling parts. Products were hand-fitted and finished. This method not only was time consuming, but it also required excessive human energy. Even replacement parts needed for repair work had to be obtained from their original source. Eli Whitney recognized the drawbacks to this method and proceeded to develop specialized jigs in order to assure standard sized and shaped parts. Whitney demonstrated the interchangeability of parts for the first time in 1794 in Washington, D.C. The experiment proved so successful that Whitney received a contract to manufacture several thousand muskets. Through his success, Whitney introduced the use of interchangeability and manufacturing standardization.

SPECIFICATIONS AND STANDARDIZATION

Awareness of the principle of interchangeability provided the impetus needed for standardization of manufactured parts. In 1918, a number of engineering associations met to form an organization that would be responsible for the development and coordination of national standards. Their efforts led to the formation of the American Standards Association (ASA), later changed to the United States of America Standards Institute. Today the institute is called the *American National Standards Institute* (ANSI). The number of ANSI affiliates has grown to include virtually all engineering and technical societies. Their objective is to develop, approve, and coordinate standardization of practices and manufactured parts among industries.

International trade and U.S. industrial holdings and sales in foreign countries have caused a need for international standards. Recently, over 40 countries formed the International Organization for Standardization (ISO), responsible for the development and coordination of such standards.

American National Standards Institute

The American National Standards Institute (ANSI) standards include preferred practices, testing methods, safety, design, and sizes for use in American industries. Table 2-2 lists the major groups of standards that have been developed. Each group is further subdivided into major components. For example, mechanical (B) standards are divided into over 150 categories, which themselves have additional subdivisions. Table 2-3 lists the major categories for section B of the ANSI standards. A coding system classifies each standard and identifies the date when the standard was approved or revised. For example, the elements of standard B94.19-1968 refer to:

B	Mechanical
94	Cutting tools, holders, drives, and bushings
.19	Milling cutters and end mills
1968	Date

Modern mass-production techniques require that parts be manufactured according to predetermined

Table 2-1 **Metric Conversion Table: Metric to American Conventional Units**

0.0394	×	Millimeters	=	Inches
3.2809	×	Meters	=	Feet
0.6213	×	Kilometers	=	Miles
0.1550	×	Sq. centimeters	=	Sq. inches
10.7641	×	Sq. meters	=	Sq. feet
247.1098	×	Sq. kilometers	=	Acres

size specifications. Designers and engineers designate the size and allowable variations that are to be maintained. Producing a part to an exact size is economically unfeasible, since, as the exact size is approached, the manufacturing cost increases rapidly. Most parts do not have to be of an exact size, but within a certain range of sizes. Therefore, when a part is designed, allowable sizes are determined, depending on the proposed use of the part.

An understanding of the terminology involved in specifying various sizes and fits is important for the student of manufacturing technology. Terminology relative to size and allowable limits includes:

Basic Size. The basic size is the size from which the limits of size are derived through the application of allowances and tolerances.

Tolerance. The tolerance is the total amount of size variation allowable.

Limits. The limits are the absolute minimum and maximum sizes allowable.

Clearance. The clearance is the difference in sizes between mating parts when the internal dimension of the female part is larger than the external dimension of the male part.

Interference. The interference is the difference in sizes between mating parts when the internal dimension of the female part is smaller than the external dimension of the male part.

Bilateral Tolerance. Bilateral tolerance is a designation used when variation in size is permitted in both directions from the basic size.

Table 2-2 Major Categories of ANSI Standards

A	Construction
B	Mechanical
C	Electrical and electronic
D	Highway traffic safety
F	Food and beverage
G	Ferrous materials and metallurgy
H	Nonferrous materials and metallurgy
J	Rubber
K	Chemical
L	Textiles
M	Mining
MC	Measurement and automatic control
MD	Medical devices
MH	Materials handling
N	Nuclear
O	Wood
P	Pulp and paper
PH	Photography and motion pictures
S	Acoustics, vibrations, mechanical shock, and sound recording
SE	Security equipment
W	Welding
X	Information systems
Y	Drawing, symbols, and abbreviations
Z	Miscellaneous

Table 2-3 ANSI Mechanical (B) Standards

Number	Category
B 1	Screw threads
B 3	Ball and roller bearings
B 5	Machine tools and components
B 6	Gears
B 16	Pipe flanges and fittings
B 18	Bolts and nuts
B 29	Transmission chains
B 31	Pressure pipe
B 36	Wrought iron and wrought steel pipe
B 72	Plastic pipe
B 74	Abrasives
B 93	Fluid power systems and components
B 94	Cutting tools, holders, drivers, and bushings
B 125	Iron and steel pipe
B 141	Aerospace

Unilateral Tolerance. Unilateral tolerance is a designation used when variation in size is permitted in only one direction from the basic size.

Running or Sliding Fits. Designated as RC1 through RC9, fits allowing for accurate location of parts with minimum play through fits with a free running or sliding fit where accuracy is not important.

Location Fits. Location fits are designated as *Location Clearance, LC, Location Transition, LT,* and *Location Interference, LN.* Location Clearance, LC, fits range from tight to free moving and are intended mainly for stationary parts. Location Transition, LT, fits have minimum amounts of clearance for parts requiring accuracy in location. Finally, Location Interference, LN, fits are very tight, with no clearance, and are used on parts requiring precision alignment.

Force Fits. Designated as FN fits, ranging from FN1 through FN5. Force fits, as the name implies, need force or pressure for assembly. FN1 requires little pressure, while FN5 requires high pressure for assembly.

Working drawings, in addition to showing the shape of the part being manufactured, also give the dimensions and allowable variations. Standard practices for lines, section symbols, surface textures, and joining methods are used in preparing the drawings. Special instructions and specifications such as parallelism, flatness, concentricity, angularity, and so forth, should be included. It would be impossible to list all the types of specifications that can be included on a drawing; however, Table 2-4 lists examples of common specifications.

Surface Texture Specifications

The American National Standards Institute, in conjunction with the American Society of Mechanical Engineers (ASME) and the Society of Automotive Engi-

Table 2-4 Examples of Common Specifications for Manufactured Parts

Finish all over
Sandblast
Lap
Grind
Break all sharp edges
Broach
Heat treat
Anneal
Pickle
Galvanize
0.012 mil. coating
Fillets and Rounds $^1/_{16}$®
Surface parallel
45°25′6″ ± 0.5″
Straight within 0.005 in total
Square within 0.003 in total
Concentric with 0.002 in
SAE 3051 steel
Brinnel 510-530
Chamfer 0.05 × 45°
5 HOLES Equispaced
Class RC-1 fit
Peen all over
$^1/_4$ UNC Hex H.D. Cap SC. 3 REQ'D
Spring 2 REQ'D 4 Full coils #12 piano wire 0.38 long 0.250D
$^{39}/_{64}$ DRILL 0.6245/0.6255 REAM
Woodruff #240 KEY
Keyway $^1/_4$ in wide × $^1/_8$ in deep

neers (SAE), developed the first specifications for surface texture in 1940. Subsequent revisions and refinement led to revised standards, which were approved in 1962. The current accepted surface texture standard is ANSI B46.1-1962.* This standard was reviewed in 1971, and may be designated ANSI B46.1-1962(R1971). It is concerned with definitions, rating, and standardized symbols.

Definitions. Definitions as specified in ANSI B46.1-1962 deal only with the height, width, and direction of surface irregularities. Here is a list of terms and definitions prepared by the ANSI in 1962 (Fig. 2-1).

Surface Texture. Surface texture refers to repetitive or random deviations from the nominal surface which form the pattern of the surface. Surface texture includes roughness, waviness, lay, and flaws.

Surface. The surface of an object is the boundary which separates that object from another object, substance, or space.

Profile. The profile is the contour of a surface in a plane perpendicular to the surface, unless some other angle is specified.

Nominal Profile. The nominal profile is the profile disregarding surface texture.

* American National Standards Institute, *Surface Texture,* ANSI B46.1-1962. The American Society of Mechanical Engineers, United Engineering Center, 345 East 47th Street, New York, N.Y. 10017.

Measured Profile. The measured profile is a representation of the profile obtained by instrumental or other means.

Center Line. The center line is the line about which roughness is measured and is a line parallel to the general direction of the profile within the limits of the roughness-width cutoff, such that the sums of the areas contained between it and those parts of the profile which lie on either side of it are equal.

Microinch. A microinch is one millionth of an inch (0.000001 in) and is abbreviated as μ inch.

Roughness. Roughness consists of the finer irregularities in the surface texture usually including those irregularities which result from the inherent action of the production process. These are considered to include traverse feed marks and other irregularities within the limits of the roughness-width cutoff.

Common machines and manufacturing processes produce various surface finishes. Table 2-5 shows the range of surface roughness that is achieved through various production methods. It should be noted that a range of finishes can be achieved by any one production process.

Roughness Height. For the purpose of the standard, roughness height is rated as the arithmetical average deviation expressed in microinches measured normal to the center line. The preferred series of roughness-height values is given in Table 2-6.

Roughness Width. Roughness width is the distance parallel to the nominal surface between successive peaks or ridges which constitute the predominant

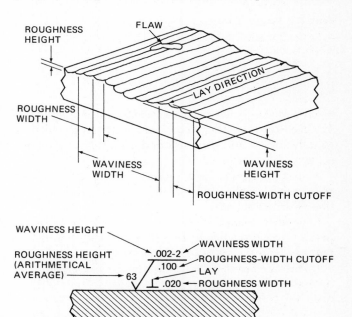

Fig. 2-1 Surface characteristics and relationship to identification symbol. (*Source: ANSI B46. 1-962.*)

Table 2-5 **Surface Roughness Produced by Common Production Methods**

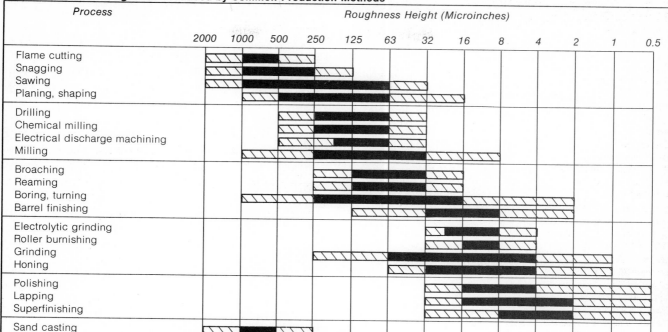

Key

■ Average application
▨ Less frequent application

The Ranges shown above are typical of the processes listed.
Higher or lower values may be obtained under special conditions.

SOURCE: ANSI B46.1-1962.

pattern of the roughness. Roughness width is rated in inches.

Nominal Surface. Nominal surface is the intended surface contour, the shape and extent of which is usually shown and dimensioned on a drawing or descriptive specification.

Measured Surface. The measured surface is a representation of the surface obtained by instrumentation or other means.

Roughness-Width Cutoff. The greatest spacing of repetitive irregularities to be included in the measurement of average roughness height. Roughness-width cutoff is rated in inches. It must always be greater than the roughness width in order to obtain the total roughness-height rating (Table 2-5).

Waviness. Waviness is the usually widely spaced component of surface texture and is generally of wider spacing than the roughness-width cutoff. Wavi-

Table 2-6 **Roughness-Height, Waviness-Height, and Roughness-Width Cutoff Values**

Preferred Series Roughness-Height Values (microinches)			
5	20	80	320
6	25	100	400
1	8	32	125	500
2	10	40	160	600
3	13	50	200	800
4	16	63	250	1000

Preferred Series Waviness-Height Values (inches)					
0.00002	0.00008	0.0003	0.001	0.005	0.015
0.00003	0.0001	0.0005	0.002	0.008	0.020
0.00005	0.0002	0.0008	0.003	0.010	0.030

Standard Roughness-Width Cutoff Values (inches)					
0.003	0.010	0.030	0.100	0.300	1.000

SOURCE: ANSI B46.1-1962.

SURFACE TEXTURE

	LAY SYMBOLS	
LAY SYMBOL	DESIGNATION	EXAMPLE
‖	LAY PARALLEL TO THE LINE REPRESENTING THE SURFACE TO WHICH THE SYMBOL IS APPLIED.	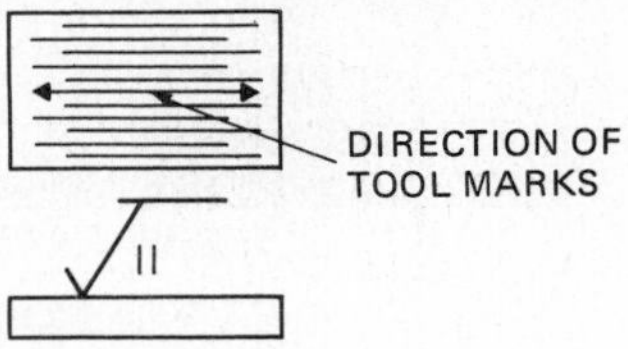
⊥	LAY PERPENDICULAR TO THE LINE REPRESENTING THE SURFACE TO WHICH THE SYMBOL IS APPLIED.	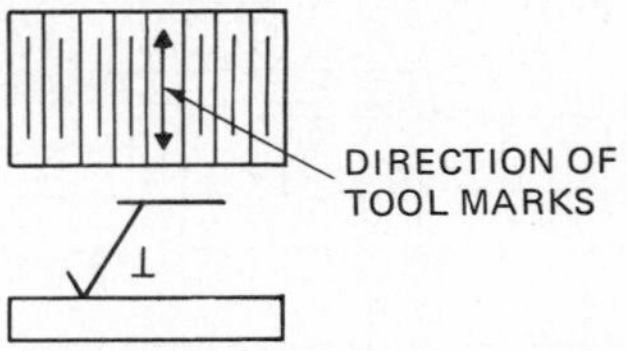
X	LAY ANGULAR IN BOTH DIRECTIONS TO LINE REPRESENTING THE SURFACE TO WHICH SYMBOL IS APPLIED.	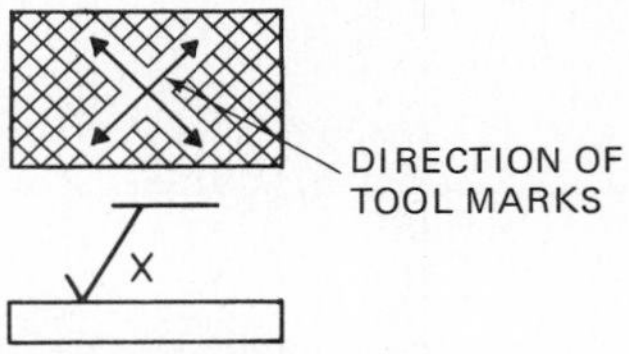
M	LAY MULTIDIRECTIONAL.	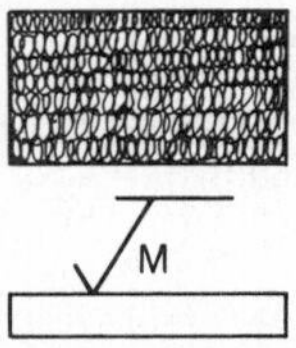
C	LAY APPROXIMATELY CIRCULAR RELATIVE TO THE CENTER OF THE SURFACE TO WHICH THE SYMBOL IS APPLIED.	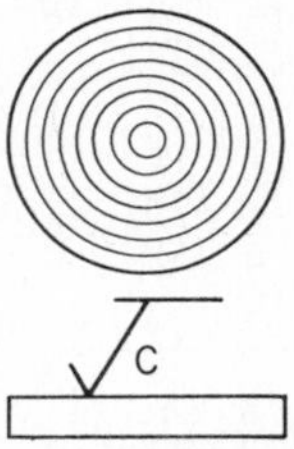
R	LAY APPROXIMATELY RADIAL RELATIVE TO THE CENTER OF THE SURFACE TO WHICH THE SYMBOL IS APPLIED.	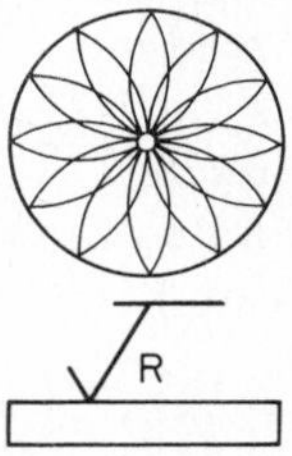

Fig. 2-2 Surface-texture lay symbols. (*Source: ANSI B46.1-1962.*)

ness may result from such factors as machine or work deflections, vibration, chatter, heat treatment, or warping strains. Roughness may be considered as superposed on a "wavy" surface.

Waviness Width. Waviness width is rated in inches as the spacing of successive wave peaks or successive wave valleys. When specified, the values shall be the maximum permissible.

Lay. Lay is the direction of the predominant surface pattern, ordinarily determined by the production method used. Figure 2-2 shows surface-texture lay symbols.

Flaws. Flaws are irregularities which occur at one place or at relatively infrequent or widely varying intervals in a surface. Flaws include such defects as cracks, blow holes, checks, ridges, and scratches. Unless otherwise specified, the effect of flaws are not be included in the roughness-height measurements.

Contact Area. Contact area is the area of the surface required to effect contact with its mating surface. Unless otherwise specified, contact area is distributed over the surface with approximate uniformity.

Screw-Thread Standards

The first attempts at standardizing screw threads were made by Sir Joseph Whitworth of England. The *Whitworth thread,* having a 55° angle, was widely accepted in Great Britain but not in the United States. It was not until 1864, when William Sellers proposed a standard to a committee appointed by the Franklin Institute, that the United States had an acceptable standard. The *Sellers thread* had a 60° angle and a different number of threads than the Whitworth thread. Because of their differences, the Sellers thread and the Whitworth thread were not interchangeable.

The United States Navy adopted the Sellers thread and changed its name to the *United States Standard Thread USS.* The USS thread series was used in the majority of American industries. The Society of Automotive Engineers (SAE) proposed a thread series having finer threads in 1911. A number of other thread series were in use during World War I, causing a considerable problem in interchangeability of parts.

Lack of standardization lead Congress to establish the *National Screw Thread Commission* in 1918. The commission's work lead to the development of the *American National Screw Thread Standard* in 1924. The new standard provided for both a coarse and fine thread series.

Again, in 1933, the commission modified these standards by changing the designations to the *National Coarse (NC)* and *National Fine (NF)* series. Additional modifications were added at that time to include the 8, 12, 16, and *National Extra Fine (NEF)* series.

World War II and a lack of international standards led the United States, Canada, and Great Britain to formulate a unified standard. In 1949, agreement was reached between these countries making an interchangeable standard called the *American Standard for Unified Screw Threads.*

Screw threads, regardless of the standard, have the same basic parts. Differences exist in standard size, fits, and angles. Standard sizes can be found in copies of the standards and in machinist handbooks.

Following is a list of commonly accepted terms and definitions necessary in understanding screw-thread specifications. These are shown in Fig. 2-3.

Major Diameter. The major diameter is the largest diameter of an internal or external thread.

Minor Diameter. The minor diameter is the smallest

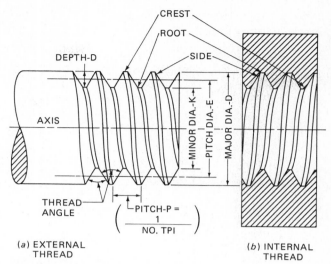

Fig. 2-3 Screw-thread nomenclature.

diameter of an internal or external thread (formerly known as the root or inside diameter).

Pitch. Pitch is the distance from a point on a screw thread to an identical point on the adjacent thread.

Lead. Lead is the distance that a screw thread moves along its axis in a mating part during one complete revolution on its axis. The lead is equal to the pitch on a single thread, twice the pitch on a double thread, and three times the pitch on a triple thread.

Angle of Thread. Angle of thread is the included angle measured on an axial plane between the sides of the thread.

Number of Threads. Number of threads is the number of threads per each inch length.

Lead Angle. Lead angle is the angle of a thread made by the helix at the pitch diameter when measured in a plane perpendicular to its axis.

Crest. The crest is the top surface of the thread at the major diameter of an external thread and minor diameter of an internal thread.

Flank. The flank is either surface of the thread that connects the crest and root of the thread.

Root. Root is the bottom surface of the thread at the minor diameter of an external thread or bottom surface of the thread at the major diameter of an internal thread.

Height or Depth of Thread. The height or depth of thread is the distance from the major diameter to the minor diameter of the thread.

Depth of Engagement. The depth of engagement is the distance of thread contact when internal and external threads mate.

Fit. Fit is the relationship that exists between tightness or looseness of mating threads.

Length of Engagement. The length of engagement is the distance of contact measured parallel to the axis.

Unified and American National Screw Thread Standards

Although the Unified and American National Screw Thread series are standardized, a slight difference still exists between them. The diameter and the root of the unified version are rounded, not pointed as in the American National version. Even though there is a difference, the threads are interchangeable because allowable tolerances are provided. Figure 2-4 shows the profile of the Unified internal and external threads and thread formulas.

Screw Thread Series. The Unified Screw Thread Standard, as adopted by the American National Standards Institute (ANSI B1.1-1974), identifies the following thread series:*

UNC—*Unified National Coarse,* a coarse thread for general use where fine threads are not required.

UNF—*Unified National Fine,* a fine thread primarily used in the automotive and aircraft industries.

UNEF—*Unified National Extra Fine,* a very fine thread having a required maximum number of threads per length usually for thin-walled or precision parts.

In addition, the Unified Screw Thread Standard includes the constant-pitch thread series. The constant pitch series contains the same number of threads per inch regardless of the diameter. The series includes: UN4, UN6, UN8, UN12, UN16, UN20, UN28, and UN32.

Classes of Unified Threads. The class of Unified thread refers to the amount of tolerance and allowance specified for each thread. Internal and external threads do not have to be of the same class to be interchangeable. Any combination of thread classes can be mated as long as the threads are of the same size. Classes 1A, 2A, and 3A are used to designate *external threads,* while 1B, 2B, and 3B identify *internal threads.*

Thread classes for Unified threads are:

1A and 1B: loose fit having some play, used for nonprecision applications.

2A and 2B: free fit, commonly for commercial use.

3A and 3B: close fit, used for precision work.

Thread Identification and Specification. Unified and American National threads are identified by giving the diameter, number of threads per inch, thread series, and class. For example, $^1/_4$-28UNF-2A designates a $^1/_4$ in diameter thread, 28 threads per inch,

* ANSI-B1 Report ISO Metric Threads contains dimensional information on ISO metric screw threads.

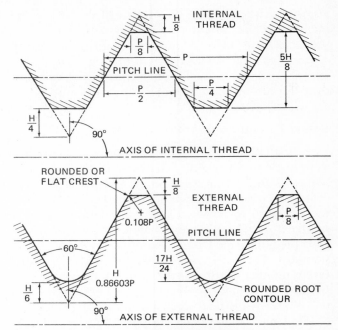

Fig. 2-4 Profile of unified internal and external thread.

Unified National Fine series with a class 2A (external) thread. This designation also forms the abbreviation for a thread having:

Major diameter = 0.2500

Pitch diameter = 0.2268

Minor diameter = 0.2062

Lead angle = 2°52′

The diameters and lead angles for the various thread series are obtained from tables available from the American National Standards Institute, or in various machinery handbooks, such as *Machinery's Handbook.*

Other Thread Series. In addition to the Unified Thread Series, other series are available for special uses. These include the Acme, Buthers, Square, 29° Worm, American Standard Pipe, and International Metric Series. Figure 2-5 shows the profile of some of their threads.

Gear Specifications

The American Gear Manufacturers Association (AGMA) and the American Society of Mechanical Engineers (ASME), in conjunction with the American National Standards Institute, developed standards for gears. A few of the standards for gears are:

B6.1-1968(R1974) Tooth Proportions for Coarse-Pitch Involute Spur Gears

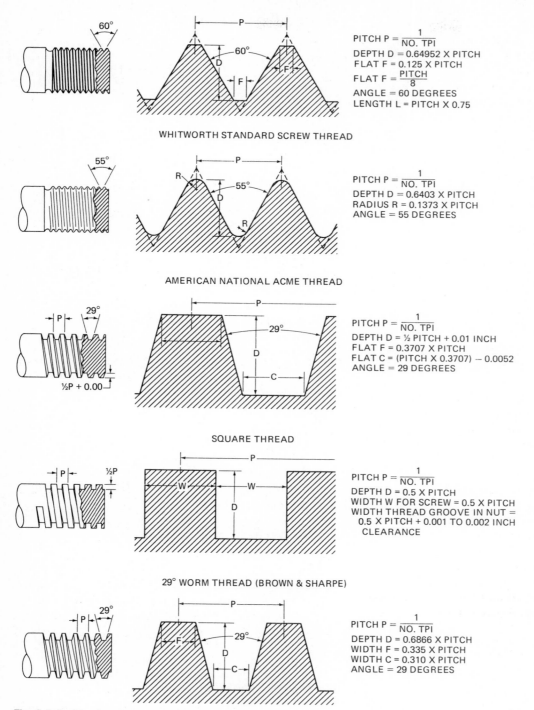

AMERICAN NATIONAL SCREW THREAD FORM

PITCH $P = \dfrac{1}{\text{NO. TPI}}$
DEPTH D = 0.64952 X PITCH
FLAT F = 0.125 X PITCH
FLAT $F = \dfrac{\text{PITCH}}{8}$
ANGLE = 60 DEGREES
LENGTH L = PITCH X 0.75

WHITWORTH STANDARD SCREW THREAD

PITCH $P = \dfrac{1}{\text{NO. TPI}}$
DEPTH D = 0.6403 X PITCH
RADIUS R = 0.1373 X PITCH
ANGLE = 55 DEGREES

AMERICAN NATIONAL ACME THREAD

PITCH $P = \dfrac{1}{\text{NO. TPI}}$
DEPTH D = ½ PITCH + 0.01 INCH
FLAT F = 0.3707 X PITCH
FLAT C = (PITCH X 0.3707) — 0.0052
ANGLE = 29 DEGREES

SQUARE THREAD

PITCH $P = \dfrac{1}{\text{NO. TPI}}$
DEPTH D = 0.5 X PITCH
WIDTH W FOR SCREW = 0.5 X PITCH
WIDTH THREAD GROOVE IN NUT =
  0.5 X PITCH + 0.001 TO 0.002 INCH
  CLEARANCE

29° WORM THREAD (BROWN & SHARPE)

PITCH $P = \dfrac{1}{\text{NO. TPI}}$
DEPTH D = 0.6866 X PITCH
WIDTH F = 0.335 X PITCH
WIDTH C = 0.310 X PITCH
ANGLE = 29 DEGREES

Fig. 2-5 Profile of various screw threads.

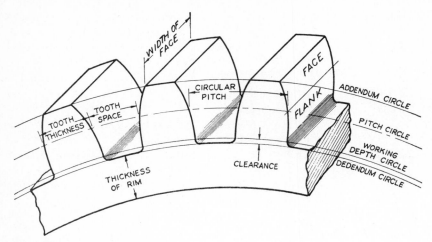

Fig. 2-6 Basic elements of a gear.

Only terms commonly used in gear standards will be discussed in this section. Students desiring additional information are referred to the above standards and Fig. 2-6.*

The following terminology is commonly used to describe gear elements.

Pitch Circle. Pitch circle is the curve of intersection of a pitch surface of revolution and a plane of rotation.

Addendum Circle. The addendum circle coincides with the tops of the teeth in a cross section.

Root Circle. The root circle is tangent to the bottoms of the tooth spaces in a cross section.

Addendum. Addendum is the height by which a tooth projects beyond the pitch circle or pitch line.

Dedendum. Dedendum is the depth of a tooth space below the pitch circle or pitch line.

Clearance. Clearance is the amount by which the dedendum in a given gear exceeds the addendum of its mating gear.

Pitch Diameter. Pitch diameter is the diameter of the pitch circle.

Number of Teeth. The number of teeth include the teeth in the whole circumference of the pitch circle.

Additional Standards

The standards discussed above dealt with size and shape designation. Additional standards are available relative to various materials and their properties. For example, standards such as specifications for masonry cement, shrinkage and thermal expansion for motors, compressive strength of cement, effect of air and heat on asphaltic materials, ceramic glazed struc-

* Further information can be obtained from the Internal Organization for Standardization (ISO) on gear terminology. See R1122-1969, *Glossary of Gears.*

tural tiles, load test for refractories, and so on, are included in construction material standards of the American National Standards Institute. In addition, numerous associations and societies, such as American Ceramic Society, Society of Plastics Industries, American Glass Manufacturers, American Iron and Steel Institute, American Society for Testing and Materials, and others, formulate standards for their specific concerns.

QUALITY CONTROL

Quality control is concerned with producing manufactured parts of high quality within the designated standards and specifications. Numerous methods are employed in order to maintain high quality, including inspection and statistical controls.

Inspection methods aim at determining whether or not the product meets the required specifications and standards.

Quality control probability theory is based on the premise that manufactured parts will be close to the specified size, within the confines of a normal distribution. For example, assume that 50 parts are taken from a machining process and measured according to their outside diameter. Each part is then classified and grouped by this measurement (Fig. 2-7). The 50 parts tend to group toward the center (specified size), forming a bell-shaped curve. Classification and grouping of parts in this manner are not physically laid out but are plotted on graph paper. This distribution of parts is referred to as a *normal frequency distribution.* A normal frequency distribution is divided into six equal sections. Three sections are above the average and three sections below the average (Fig. 2-8). Each line is called a *standard deviation of sigma* (σ). The three sigmas above the average are referred to as

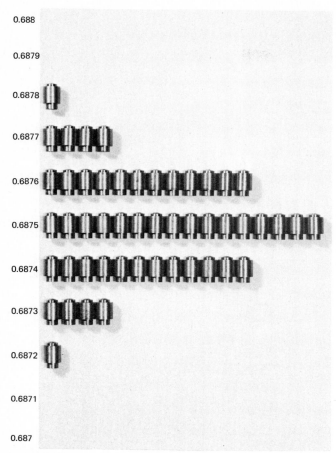

Fig. 2-7 Distribution of random sample. (*Source: Federal Products Corp.*)

area between the $\pm 3\sigma$ lines is generally accepted as being an economic or practical limit for quality control work. Therefore, limits on manufactured parts should be designated between the $\pm 3\sigma$ lines. If limits are narrow or hard to achieve in manufacturing, the cost of production and the number of rejects increases.

Actual practice uses worksheets and control charts for plotting, rather than the curves. Preparation of the control chart requires careful planning and recording of inspection data. Inspection is conducted on a periodic or sampling basis, with each observation accurately recorded.

Samples

Statistical quality control uses statistical methods (methods based on statistical inference) and sampling techniques rather than 100 percent inspection. During inspection, a small sample of the parts is selected at random for evaluation. Samples should consist of from 5 to 20 units selected at random at predetermined intervals. Intervals may be based on time (i.e., every half hour) or they may be based on the number produced, as from every 100 produced. Whatever interval is employed, consistency is important. Each sample should be selected from a different batch.

Worksheet

To aid in recording and analyzing data, worksheets are used for recording observations made during the inspection. Figure 2-9 shows a worksheet completed for 15 samples of five parts each. Specifications and a description of the part are recorded prior to the inspection process. Sample number 1 contains five dimensions, as observed in the parts being inspected.

$+1\sigma$, $+2\sigma$, and $+3\sigma$, and below the average as -1σ, -2σ, and -3σ. According to rules of the distribution, 68 percent of the parts fall between the $\pm 1\sigma$ lines, 95.5 percent of the parts between the $\pm 2\sigma$ lines, and 99.75 percent between the $\pm 3\sigma$ line. The

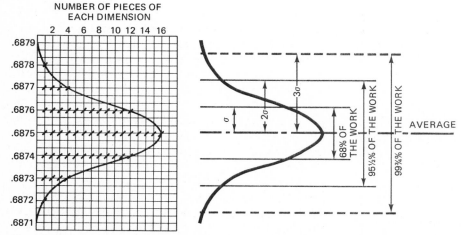

Fig. 2-8 Normal sample distribution. (*Source: Federal Products Corp.*)

DIMENSIONAL QUALITY CONTROL
WORK SHEET
AVERAGES AND RANGES

PART NUMBER PL 3421 A	DESCRIPTION Plunger - finished O.D.					
LOT NUMBER 17	ORDER NO. 22-185	MACHINE NO. G 4		DEPT. E-GM		
OPERATOR Jones	SHIFT 2	DATE 9-18	INSPECTOR Smith			

SAMPLE OR SUB-GROUP NUMBER			1	2	3	4	5	AVERAGES
DIMENSION AND TOLERANCES: .6875 ± .0005	SAMPLE READINGS OR MEASUREMENTS X		.6875	.6877	.6875	.6875	.6877	.6874
			6877	6879	6877	6877	6874	.6877
			6873	6877	6877	6873	6874	.6876
			6871	6875	6876	6875	6878	.6875
			6874	6878	6874	6873	6877	.6876
SUM OF SAMPLE READINGS			3.4370	3.4386	3.4379	3.4373	3.4380	
AVERAGE of each Sample X (Plot on Chart)			.6874	.6877	.6876	.6875	.6876	
LARGEST VALUE			6877	6879	6877	6877	6878	10.3139
SMALLEST VALUE			6871	6875	6874	6873	6874	TOTAL .6876
RANGE of each Sample R (Plot on Chart)			.0006	.0004	.0003	.0004	.0004	GRAND AVERAGE X

6	7	8	9	10	11	12	13	14	15	RANGES
.6871	.6877	.6876	.6875	.6878	.6880	.6883	.6874	.6876	.6875	.0006 / .0004
6874	6878	6877	6876	6880	6873	6879	6876	6877	6877	.0003 / .0004
6871	6875	6875	6876	6876	6879	6880	6877	6875	6876	.0004 / .0007
6872	6873	6875	6875	6877	6882	6879	6875	6875	6876	.0006 / .0003
6878	6872	6878	6877	6876	6874	6876	6876	6873	6876	.0002 / .0004
3.4366	3.4375	3.4381	3.4379	3.4387	3.4388	3.4397	3.4378	3.4376	3.4380	.0009 / .0007
.6873	.6875	.6876	.6876	.6877	.6878	.6879	.6876	.6875	.6876	.0003 / .0004
6878	6878	6878	6877	6880	6882	6883	6877	6877	6877	.0002 / .0068
6871	6872	6875	6875	6876	6873	6876	6874	6873	6875	TOTAL .0004
.0007	.0006	.0003	.0002	.0004	.0009	.0007	.0003	.0004	.0002	GRAND AVERAGE R

LIMITS FOR RANGES CHART = $D_4\bar{R}$ and $D_3\bar{R}$ 2.114 x .0004 = .00085 UPPER: .00085 LOWER: 0	LIMITS FOR AVERAGES CHART = $\bar{\bar{X}} \pm A\bar{R}$ .6876 ± .577 x .0004 UPPER .6878 LOWER .6874	

ADDITIONAL WORK SHEETS AVAILABLE FROM FEDERAL PRODUCTS CORP. PROVIDENCE R I

COPYRIGHT BY FEDERAL PRODUCTS CORP

A 368 10 11 45 PRINTED IN U S A

Fig. 2-9 Sample quality control worksheet. (*Source: Federal Products Corp.*)

The dimensions are added vertically (3.4370) and averaged. The range (largest size to smallest size) is calculated and recorded. This procedure is repeated for each of the samples observed. Averages and ranges for the samples are recorded on the right side of the worksheet and totaled. The totals are then averaged, yielding the *Grand Average* $\bar{\bar{X}}$ and *Grand Range* $\bar{R}$.

Control Limits

Control limits, also called sigmas, are calculated from the obtained data. Upper and lower limits ($\pm 3\sigma$) represent the standard or guide for comparison in future inspections. These limits are calculated for both the averages and ranges.

Control limits for averages are calculated by multiplying the Grand Range $\bar{R}$ by a constant (A_2) based on the sample size. The resulting number is added to the Grand Average $\bar{\bar{X}}$ for the upper control limit and subtracted from the Grand Average $\bar{\bar{X}}$ for the lower control limit.

Control limits for the range also use constants derived from the sample size. Upper range limits are calculated by multiplying the Grand Range $\bar{R}$ by the constant D_4 (Table 2-7). The lower range limits are

Table 2-7 **Values of Constants A_2, D_4, and D_3 for Control Limits**

	Constants		
Sample Size	A_2 for Averages	D_4 for Upper Range Limits	D_3 for Lower Range Limits
5	0.577	2.114	0
8	0.373	1.864	0.136
10	0.308	1.777	0.233
12	0.266	1.717	0.284
15	0.233	1.656	0.348

SOURCE: ANSI B46.1-1962.

calculated in the same manner as the upper range limits, but using the constant D_3. Formulas used in setting control limits are shown in Table 2-8.

Control Chart

Information from the worksheet and the resulting calculations are transferred to a control chart in an interpretable form. Figure 2-10 shows the control chart developed from the previously described worksheet (Fig. 2-9). Variations in size are recorded on the left side of the chart. A solid line is drawn through the graph indicating the specified size (0.6875). The Grand Average $\overline{\overline{X}}$ is indicated by a dash and dotted line. Finally, dotted lines are drawn representing values of the upper and lower control limits. Ranges are listed at the bottom of the chart, and a line is drawn representing the upper control limit. The control chart is now ready for the individual averages to be plotted. Each average is plotted on the graph in relationship to the time of the observations. In the example, samples were collected every 15 minutes and represented by a square on the graph paper. Small x's are used to identify the averages and small o's used to identify the ranges on the control chart.

Use of Control Charts

The averages plotted on the control chart represent trends in the quality of the inspected workpieces. It can be observed from the chart that at the end of the third and sixth hours, the parts were out of the control limits. It is possible that after three hours, the machine operators need a break, a tool needs replacement or the machines need adjustments. On the basis of interpretation of the chart, quality control experts can identify the reasons for parts being out of the control limits. When the reasons are identified, the problems can be foreseen and corrected.

AIDS FOR QUALITY CONTROL

In order to maintain products of high quality, inspections are performed using instruments, gauges, and

Table 2-8 **Statistical Quality Control—Formulas**

Average of samples	$\bar{X}n = \sum \dfrac{X}{N_1}$
Grand average	$\bar{\bar{X}} = \sum \dfrac{\bar{X}n}{N_2}$
Range	$R = X_1 - X_s$
Grand range	$\bar{R} = \sum \dfrac{R}{N_2}$
Upper average control limit	$UAL = \bar{\bar{X}} + A_2\bar{R}$
Lower average control limit	$LAL = \bar{\bar{X}} - A_2\bar{R}$
Upper range control limit	$URL = D_4\bar{R}$
Lower range control limit	$LRL = D_3\bar{R}$

X = individual observations
N_1 = number of observations in sample
N_2 = number of samples
$\bar{X}n$ = average of specific sample
X_1 = largest observation in sample
X_s = smallest observation in sample
A_2 = constant for average control limit
D_4 = constant for upper range limit
D_3 = constant for lower range limit

nondestructive testing techniques. These aids are used to measure distance, angularity, surface texture, and roundness; to check allowances; and to detect imperfections in the workpiece's structure.* Inspection procedures use these aids to compare an unknown quality with a predetermined standard.

Measuring Instruments

Instruments used to measure manufactured products are classified as direct or indirect measuring devices. *Direct devices* have some means or scale that enables the individual to directly read or determine the measurement. Scales, micrometers, verniers, and so forth, are examples of direct measuring instruments. *Indirect measuring instruments* do not have the provision to allow a direct measurement but require the individual to transfer the measurement to a direct reading instrument. Calipers, dividers, telescoping gauges, and so on, are indirect measuring devices.

Discrimination refers to the smallest measurement unit that can be read reliably with an instrument. Precision instruments have a high degree of discrimination; units as small as a millionth of an inch or a thousandth of a millimeter can be read reliably.

Linear Measuring Instruments

Steel Rules. The simplest and most common measuring instrument is a steel rule or scale. Rules are

---

* A temperature-controlled room should be used since various materials expand and contract, which would affect accuracy in quality control.

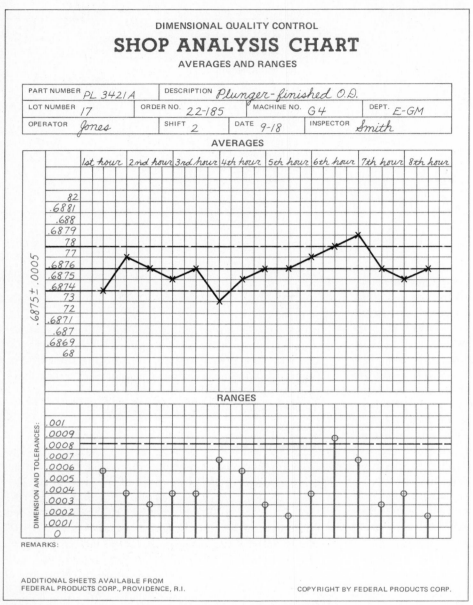

Fig. 2-10 Sample quality control chart. (*Source: Federal Products Corp.*)

available in lengths from ¹/₄ in to over 100 in, and have various measuring gradations and numbers of scales. Common gradations of the inch are 8ths, 16ths, 32nds, and 64ths. However, 10ths, 50ths, and 100ths can also be obtained. Rules are available with English, metric, or English-metric scales.

Rule-Depth Gauge. Rule-depth gauges are useful for measuring slots, recesses, and depth of holes. A rule-depth gauge is a steel rule fitted with a sliding head which provides a reference surface for measuring the depth.

Combination Square. Combination squares consist of a rule with three attachments: center head, square head, and protractor (Fig. 2-11). Each attachment is used independently of the others. The square-head attachment is used when measuring the depth of a hole, checking squareness, checking 45° angles or making linear measurements. When laying out or measuring angles, the protractor head attachment is used. Various angles can be obtained by rotating the protractor head. Finally, the center-head attachment enables the user to determine the center of a cylindrical object.

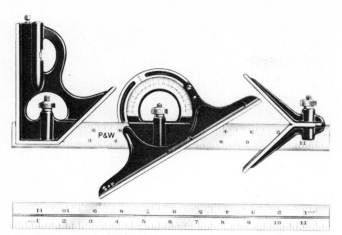

Fig. 2-11 Combination set consisting of a blade, square head, bevel, protractor, and center head. (*Source: Pratt and Whitney.*)

Micrometer Calipers. A useful and widely used measuring instrument is the micrometer caliper, also called a *Mike*. Standard micrometers are capable of measurements to a thousandth (0.001) of an inch. Measurements accurate to one ten-thousandth (0.0001) of an inch are attainable when using a vernier micrometer.

Micrometers consist of a frame, anvil, spindle, thimble, and barrel (see Fig. 2-12). Measurements are made when the anvil and spindle contact the work. Each turn of the thimble moves the spindle 0.025 in due to an accurately machined screw-thread mechanism. Gradations on the barrel and thimble yield the measurement. A number of different types of micrometers are available for specialized applications.

Inside micrometers are used to measure the width of slots and diameter of holes. Spindles come in various lengths and are interchangeable, providing the capacity for measurements from 1 to 12 in.

Depth micrometers have measuring rods of various length that replaces the conventional spindle. The rods extend through an accurately machined barrel into the area being measured. Readings are made as on the standard micrometer.

Screw-thread micrometers are similar to the standard micrometer except for the shape of the anvil and spindle. The spindle is pointed, while the anvil has a double-V shape. Standard sized screw-thread micrometers range from 0 to 2 in and from 8 to 30 *threads per inch* (TPI). Other forms of micrometers have variously shaped anvils and spindles to facilitate special applications. Blade shaped anvils and spindles facilitate measurement of keyways. Rounded anvils allow for measuring curved surfaces, such as tubing or pipe. Special shapes provide contact surfaces for measuring paper thickness and edge to hole distances.

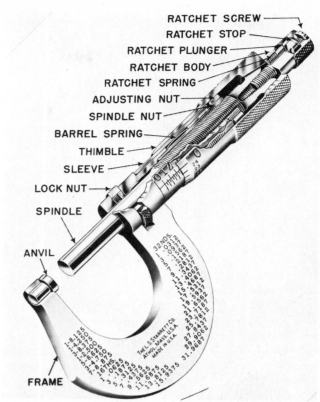

Fig. 2-12 Typical 0–1 in micrometer, with the principal parts identified. (*Source: L. S. Starrett Company.*)

Fig. 2-13 Bench micrometer for metrology laboratory use. (*Source: Pratt and Whitney.*)

Bench micrometers are used mainly in metrology laboratories, for direct measurements to 0.0001 in. The bench micrometer provides adjustments to ensure standard contact pressure. Figure 2-13 shows a bench micrometer having a readout meter which allows comparisons for production parts to 0.00002 in.

Vernier Calipers. Vernier calipers are capable of internal and external measurements to within a thousandth of an inch. The caliper consists of two scales: a main scale and sliding vernier scale. An English-metric combination vernier scale is also available.

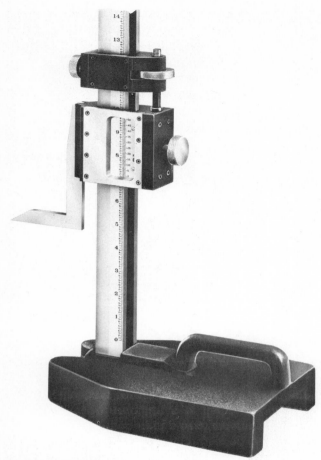

Fig. 2-14 Vernier height gauge. (*Source: Pratt and Whitney.*)

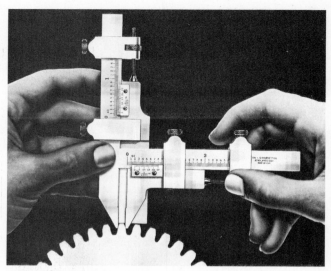

Fig. 2-15 Vernier gear-tooth caliper. (*Source: L. S. Starrett Company.*)

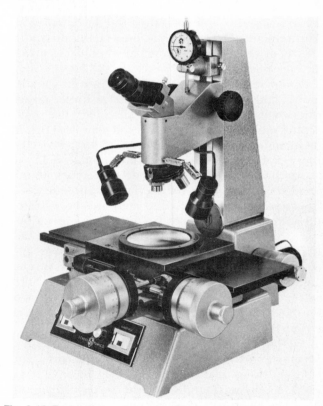

Fig. 2-16 Toolmaker's microscope. (*Source: Scherr-Tumico.*)

The vernier scale concept is also used on vernier height gauges, depth gauges, gear-tooth calipers, and vernier protractors. The vernier protractor reads in degrees and minutes. Figures 2-14 and 2-15 show the vernier height gauge and vernier gear-tooth caliper respectively. The vernier height gauge is available in 12 to 72 in sizes, or 300 to 600 mm (millimeter) on the metric scale. Vernier gear-tooth calipers measure the thickness of teeth, distance from the top of the tooth to the cord, and the chordal thickness of the gear tooth. The instrument consists of two vernier scales at right angles to each other, and measures from 1 to 20 diametral pitch in English and metric units.

Toolmaker's Microscope. Toolmaker's microscopes provide for accurate measurement of small parts using optical magnification methods (Fig. 2-16). Workpieces are placed on a viewing stage which moves longitudinally and crosswise. Micrometer heads attached to the stage provide a means of controlling the movement of the stage. A reference point for the measurement is located on the workpiece, and is aligned by means of cross hairs in the microscopic lens. Readings are taken from the micrometer heads at the reference point. The workpiece and stage are then moved to the point where the measurement is to be taken. A second reading from the micrometers is taken and the differences calculated. The differences represent the desired dimensions in two directions.

Angular measurements can also be made by using a rotating stage. Toolmaker's microscopes come in English and metric versions, with an accuracy of 0.0001 in [0.0025 mm]. The stage size accommodates workpieces up to approximately 10 in. In addition to the cross-hair ocular, other oculars are available with templets of screw-thread silhouettes.

Computer-Controlled Measuring. Computer-numerical control is used to position a probe, take a measurement, perform mathematical computations, and process the data automatically. Standard numerical-control tapes are punched with the prescribed inspection sequence and coordinate location for either point to point or continuous-path operation.

A probe is programmed to perform linear measurements, locate holes, or record contour. The computer calculates the difference between the specified and actual position or measurement to within ±0.0001 in [±0.0025 mm]. Computer programs are also available to compensate for misalignment of the workpiece on the table and locate and measure the radius or diameters of holes. The machine is also capable of translating linear coordinates to prepare punched tapes directly from prototype parts. Computer-controlled measuring machines come in vari-

Fig. 2-17 Computer-controlled measuring machine. (*Source: Bendix Corporation.*)

ous sizes, with accuracy up to 0.0003 in [0.0076 mm] (Fig. 2-17). Additional equipment includes a sine table, optical viewers, layout punches, and specially shaped probes.

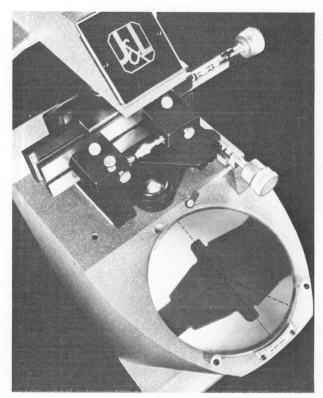

Fig. 2-18 Optical comparator and typical applications. (*Source: Jones and Lamson.*)

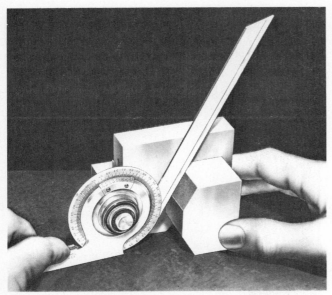

Fig. 2-19 Application of universal bevel protractor with vernier scale. (*Source: Starrett Company.*)

Optical Comparator. Optical comparators (see Fig. 2-18 on page 37) are useful for either direct or comparative measurements. Optical comparators consist of a mounting device, micrometer adjustable table, optical system, and projection screen. A workpiece is mounted on the moveable stage and through the optical system the contour of the workpiece is magnified and projected on the screen. The comparator has the capability of magnifying the contour from 5 to 100 times its size, allowing direct measurement from the screen. Comparison of the contour of the workpiece to a standard is accomplished by drawing an outline of the desired contour on a sheet of plastic that can be attached to the screen. The optical comparator can also be used like the toolmaker's microscope because the table has micrometer adjustments. Cross hairs are contained in the projection screen rather than in the lens. Angular measurements are made with the rotary table and angular graduation along the circumference of the projection screen.

Angular Measurement

As discussed, instruments such as the combination square, toolmaker's microscope, and comparator are used for both linear and angular measurements. Additional instruments used for angular measurements include the protractor, sine bar, and gauges.

Protractors. Protractors are instruments used to measure angular rather than linear measurements. Protractors of varying degrees of accuracy and price are available.

Bevel protractors consist of two adjustable blades, which conform to an angular surface. Once the blades are set, the instrument's setting is compared to a graduated protractor or gauge for assessing the angular measurement.

Universal bevel protractors, used for layout and measurement, have a vernier-type angular scale. Readings accurate to $1/2°$ are easily made using the universal bevel protractor (Fig. 2-19).

Sine Bar. Sine bars provide a precision method to measure indirectly or establish an angle. The sine bar consists of an accurately ground steel bar and two pins of common diameter attached. Centers of the pins are placed at an accurately predetermined length from 5 to 20 in [127 to 508 mm] at 5-in [508-mm] intervals for each measurement. Use of the sine bar is based on the trigonometric function of sine, or the relationship of the angles and sides of a right triangle. If the desired angle is known, calculations are made to determine the height of the angle side. Gauge blocks or adjustable machine tables are used to set the height calculated. The sine bar is then set on the gauge blocks and the workpiece mounted on the bar. The workpiece can then be machined or inspected using instruments or gauges. Figure 2-20 shows the sine bar and gauge blocks for the following example:

Given:	30° angle
	5-in sine bar
Formula:	$\sin B = b/a$
Solution:	$\sin 30° = b/5$
	$b = 5 \times \sin 30$ Sin of 30° from
	trigonometric table is 0.500
	$b = 5 \times 0.500$

This answer ($b = 2.500$ in) indicates the gauge-block height required to elevate the sine bar to a precise 30° angle. The angle can also be calculated if the

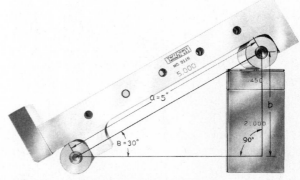

Fig. 2-20 Sine bar and gauge-block setup. (*Source: Taft and Pierce.*)

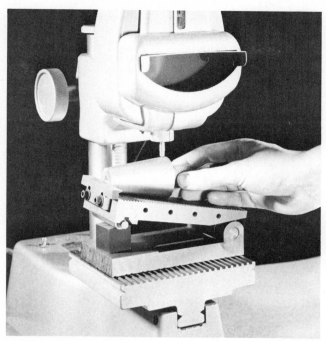

Fig. 2-21 Measuring a taper with a sine bar and visual gauge. (*Source: Automation and Measurement Division, Bendix Corporation.*)

height and size of the sine bar are known. Figure 2-21 illustrates the use of a sine bar and visual gauge.

Sine Plates. Sine plates consist of accurately finished cylindrical pins attached to a precision surface plate. Calculations for use and gauge blocks are employed as with the sine bar. Sine plates are capable of holding workpieces at a predetermined angle for machinery or inspection. Magnetic sine plates hold the workpiece without conventional clamps or holding devices.

Angle Gauge Blocks. Angle gauge blocks are built up and wrung (slid) together to produce angles of various sizes. Angle gauge blocks allow angles to be measured in one-second increments and are available in various categories of accuracy.

Angle Gauge. Another means of checking angles is with angle gauges. Angle gauges consist of a series of various leaf gauges, each having a different angle. Useful in inspection, the gauges range from 1 to 45°.

Roundness

Roundness is usually specified by a diameter and acceptable allowances. Measuring instruments and gauges, such as the micrometer, vernier caliper, telescoping gauge, and so on, are used to measure the diameter. The round shape is probably the easiest and most commonly produced shape in industry. However, accurate measurement of roundness requires precision instruments.

Profiling Instruments. Profiling instruments provide accurate measurements of roundness through a tracer device and recording mechanism. A tracer stylus passes over the surface being inspected. Variations in the surface cause the stylus to move and produce an electrical signal. The signal is amplified and recorded on a polar graph. Minor variations in roundness are easily detected from the graph. Two types of profilers allow for measurement of virtually any size and shape of workpiece. Figure 2-22 shows an instrument where the workpiece remains stationary, while the stylus moves over the surface being inspected. The instrument records roughness, waviness, or contour with a magnification of 1 million times. The spindle rotation is accurate to within 0.000003 in [0.000076 mm]. Figure 2-23 shows the second type of roundness profile. The workpiece is placed on a table that revolves while the stylus remains stationary.

Surface Texture

Visual inspection is a common and long standing technique for evaluating surface texture. The sense of touch is also helpful in assessing the surface texture of an object. While serving as a general guide, however, these methods are inconsistent and imprecise. A number of mechanical devices have been developed to aid in accurate surface texture evaluation.

Calibrated Blocks or Gauges. Calibrated blocks or gauges facilitate closer evaluation of the surface. The blocks have surface textures of varying degrees, thus providing a visual standard for evaluation. Figure 2-24 shows a set of roughness-comparison specimens.

Optical Flats. Optical flats are instruments that utilize the principle of *interferometry*. Measurements are made from variations in fringe patterns that result

Fig. 2-22 Profiling a stationary workpiece. (*Source: Automation and Measurement Division, Bendix Corporation.*)

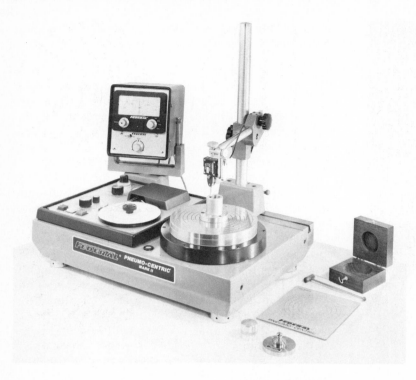

Fig. 2-23 Profiling a moving workpiece. (*Source: Federal Products Corp.*)

from light wave interference. Interferometric measurements require a light source and an observation instrument.

Monochromatic light is used because it contains a single wavelength. Helium light having a wavelength of 23.2 microinches permits calculations of surface texture to a millionth of an inch. The observation instrument used is an *optical flat*. Optical flats are fused

quartz disks, 1 to 12 in in diameter. Flats are from $1/2$- to 2-in thickness, with a flatness accuracy of 1, 2, or 4 millionths of an in.

The principle involved in using optical flats is illustrated in Fig. 2-25. When the optical flat is placed over the workpiece, a thin sloping space of air separates the surfaces. Monochromatic light rays enter the optical flat and are reflected from the surface of the workpiece. The light rays are reflected from the surface of the workpiece. When the light rays are reflected, interference bands (dark bands) are visible through the optical flat. Each band represents a distance of 11.6 microinches (μ in) between the flat and the workpiece. Figure 2-26 shows the bands as they appear through an optical flat.

Fig. 2-24 Set of standardized roughness-comparison specimens indicating average surface roughness. (*Source: Scherr-Tumco.*)

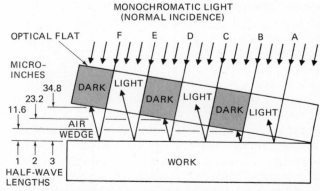

Fig. 2-25 Principle of light-wave interference. (*Source: Van Keuren Co.*)

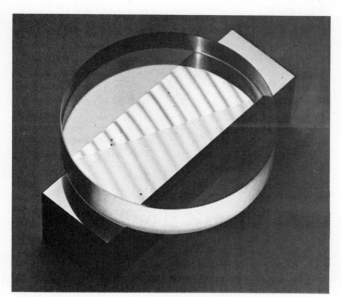

Fig. 2-26 Application of optical flat. Note the band pattern (*Source: Van Keuren Co.*)

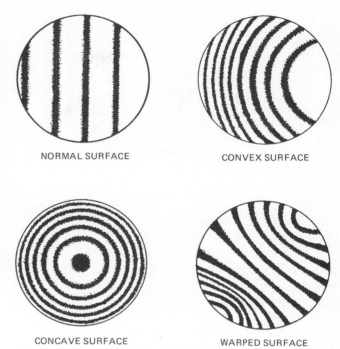

NORMAL SURFACE CONVEX SURFACE

CONCAVE SURFACE WARPED SURFACE

Fig. 2-27 Various band patterns as viewed through an optical flat. (*Source: Van Keuren Co.*)

If the surface of the workpiece and the flat are in perfect planes, the bands appear perfectly straight and equidistant. Curved bands represent various surface conditions in the workpiece. Figure 2-27 illustrates the difference in bands produced by various surfaces. The number and type of bands produced provide an accurate measure of surface irregularities.

Profilometer. The profilometer is a tracer instrument similar to the roundness profiler and is used to inspect surface texture. The stylus (comparable to the arm on a record player) is drawn across the workpiece, detecting variations on the surface height. Figure 2-28 shows a profilometer used to measure the average surface roughness and a microrecorder used to record the actual profile of the surface. After amplification of the signal, the profile is graphed for a permanent record of the inspection.

Autocollimator. The autocollimator is a precision instrument used to measure the inclination or distortion of a surface rather than the surface texture. Components of the autocollimator are a light source, *micrometer-microscope eyepiece* and optical system. Angular measurements are made when parallel beams of light are aimed at the workpiece's surface and reflected back through the instrument. If the beams are not parallel when reflected back through the instrument, the surface is distorted. Differences in the reflected light waves are measured, yielding the angular measurement. Autocollimators are accurate within 0.15 seconds (s) of an arc. Figure 2-29 shows an autocollimator setup.

Fig. 2-28 A profilometer used to measure directly the surface roughness. Note the permanent record of the actual profile as recorded on the microrecorder. (*Source: Micrometrical Division, Bendix Corporation.*)

Gauge Blocks

Gauge blocks provide a standard to calibrate measuring instruments that is accessible throughout the industrial setting. Gauge blocks are produced in English and metric units, in three grades of varying accuracy. They are made from an alloy steel, machined,

Fig. 2-29 Application of autocollimator. (*Source: L. S. Starrett Company.*)

Fig. 2-30 Typical set of gauge blocks. (*Source: Pratt and Whitney.*)

hardened, heat treated to relieve internal stress, ground, and lapped to the designated accuracy.

Grade AA, called *laboratory gauge blocks,* are accurate within ±0.000002 in for blocks one inch or less in length. Metric Grade AA are accurate within ±0.000005 mm. Laboratory gauge blocks are used to check and calibrate other grades of gauge blocks. This type of gauge block is usually used in research laboratories or temperature controlled environments.

Grade A, called *inspection gauge blocks,* are accurate to +0.000006 in and −0.000002 in. Metric-equivalent-type blocks are accurate to +0.00015 mm and −0.00005mm. Inspection-grade gauge blocks are primarily used for checking working gauge blocks.

Grade B, called *working gauge blocks,* are accurate to +0.000010 in and −0.000006 in, while the metric working gauge blocks are accurate to +0.00025 mm to −0.00015 mm. Working gauge blocks are used for checking and setting routine measurement and inspection devices.

In addition to the standard gauge blocks, angle gauge blocks are also produced. Grade AA are accurate to within ¼ sec, Grade A accurate to ½ sec, and Grade B accurate to 1 sec. Gauge blocks can be used separately or in combination to build to the desired sizes. Therefore, they are purchased in various sized sets. After the needed size blocks are selected, they are slid together (wrung), forming a strong bond. To develop a block combination equivalent to 3.8356 in, the following gauge blocks are wrung together:

0.1006 in block from the first series

0.135 in block from the second series

0.600 in block from the third series

3.000 in block from the fourth series

The same procedure is followed for both the metric gauge blocks and angle gauge blocks. Figure 2-30 shows a typical set of gauge blocks used for shop applications.

Gauges and Indicators

Inspection of manufactured products can be a time-consuming and costly aspect of production. Since a majority of products have allowances and tolerances, gauges can be used to inspect them. Gauges provide an easy method for determining if the workpiece is within the specifications or standards specified. The two major types of gauges are fixed and comparative. *Fixed gauges* are designed and constructed with a single setting, while *comparative gauges* have a range of settings.

Fixed Gauges. Fixed gauges include many different types, for inspection of common and standardized production operations.

Plug gauges, as the name implies, are shaped like a plug to check internal diameters. Straight, tapered, and threaded plug gauges are available. For cylinders with a constant cross section, a *straight plug gauge* of the appropriate size is inserted into the work. Usually

Fig. 2-31 Go–No-Go gauge. (*Source: Bendix Corporation.*)

Fig. 2-32 Various gauges including plain cylindrical plug gauges, master setting discs, plug and ring gauge, adjustable limit plug and snap gauges, dial indicator snap gauges. (*Source: Bendix Corporation.*)

the plug gauge has two different dimensions. The smaller size is used to check the minimum allowable dimension and is called a *Go* gauge. The larger size is used to check the maximum allowable size and is called a *No-Go* gauge (see Fig. 2-31 on page 42). Go and No-Go gauges can be purchased separately or with both limits on the same instrument. Tapered holes are checked with a *tapered plug gauge.* The angle and size of the taper can be inspected at the same time. *Threaded plug gauges* are used to check the limits for the pitch diameter of internal threads.

Ring gauges are cylindrical gauges that have Go and No-Go limits in their internal surface. Ring gauges are

used to check the outside surfaces of cylindrical workpieces. Ring gauges, like plug gauges, are also available with tapers and threads.

Snap gauges are used to check external dimensions. Most snap gauges consist of a U-shaped frame and at least two gauging surfaces. One surface is fixed while the other surface is adjustable over a limited range. Figure 2-32 shows a variety of gauges, including plug, ring, snap, and adjustable snap gauges.

Form gauges are used to check common manufacturing specifications and standards. Form gauges usually are available in sets of various sizes and standards. Common form gauges include:

Drill-size gauge	Filler gauge
Drill-point angle	Angle gauge
Wire-size gauge	Taper gauge
Sheet and plate gauge	Ball and diameter gauge
Center gauge	Screw-pitch gauge
Radius gauge	Fillet gauge

Figure 2-33 shows a screw-pitch gauge.

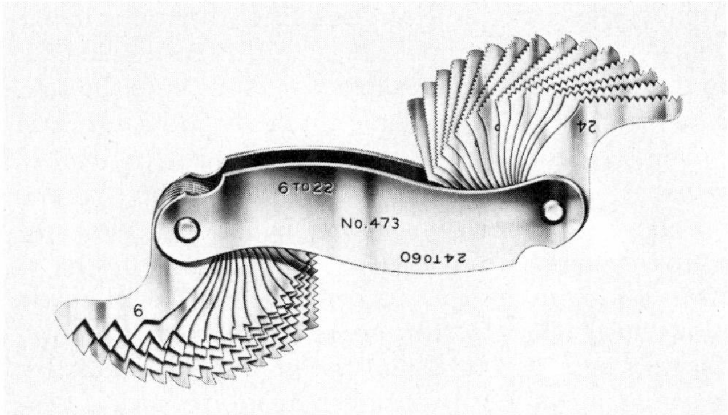

Comparative Gauges. Comparative gauges include telescoping gauges, small-hole gauges, and dial indicators. *Telescoping gauges* are used to measure holes and slots from $1/2$ to 6 in in size. The gauge consists of

Fig. 2-33 Typical screw-pitch gauge for V, American National and U.S. Standard 60° threads. (*Source: L. S. Starrett Company.*)

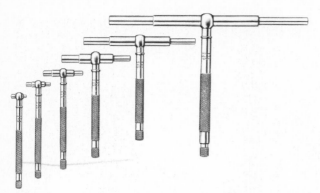

Fig. 2-34 Typical set of telescoping gauges. (*Source: L. S. Starrett Company.*)

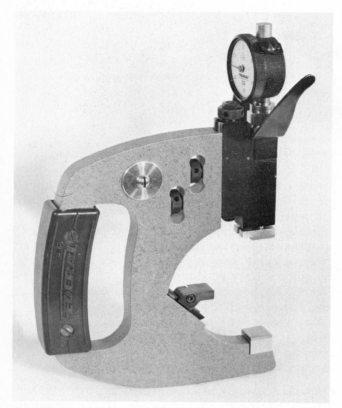

Fig. 2-35 Dial indicator on a snap gauge. (*Source: Federal Products Corp.*)

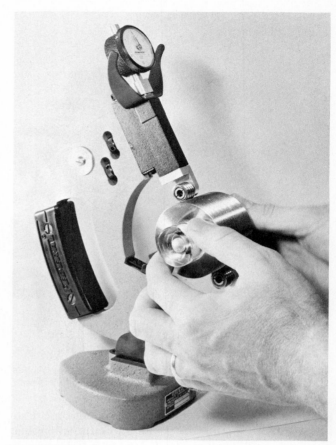

Fig. 2-36 Dial indicator on a thread-snap gauge. (*Source: Federal Products Corp.*)

a handle and spring-loaded telescoping plunger. The ends of the telescoping plunger make contact with the surface being measured and are secured. The gauge is then removed, and an instrument such as vernier caliper or micrometer is used to measure the distance between the ends of the plunger. Telescoping gauges come in a set, with each gauge capable of a range in sizes (Fig. 2-34).

Small hole gauges are used for measuring small holes, grooves, and recesses too small for telescoping

gauges. Small-hole gauges come in sets with sizes from 0.125 to 0.500 in. Each gauge has a range of sizes from 0.075 to 0.100 in per gauge. The gauge consists of a split-ball-shaped end, with a mechanism within the ball that causes it to expand. Final size is determined by measuring the expanded end.

Dial indicators show the amount of difference in size or alignment. The dial indicators consist of an indicating hand enclosed in a case, spring-loaded plunger, dial face, and an assortment of detachable contact points. The indicating hand is controlled by the plunger, which moves in accordance with the size being measured. Two common types of indicators are available: balance and continuous. *Balance-type indicators* use a scale with both positive and negative directions from zero. *Continuous-dial indicators* have only positive or negative directions. Some indicators have revolution counters that record the number of revolutions made by the continuous indicator, thus increasing the range of the instrument. Dial indicators are available with either English or metric scales. Figures 2-35 and 2-36 show the use of dial indicators on a snap gauge and a thread-snap gauge.

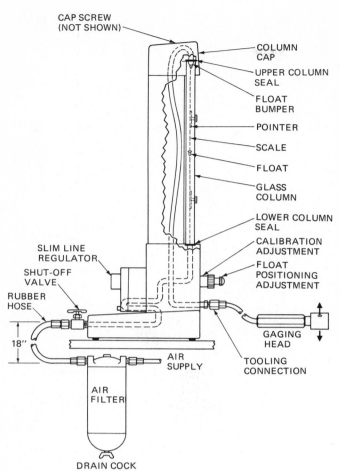

Fig. 2-37 Principle of air-gauge operation. (*Source: Automation and Measurement Division, Bendix Corp.*)

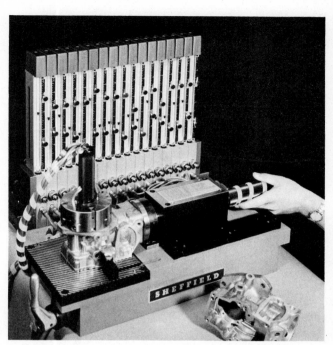

Fig. 2-38 Multiple air-gauge setup. (*Source: Automation and Measurement Division, Bendix Corp.*)

Air Gauges. Air gauging utilizes an air flow indicator to record dimensional changes based on variations of air flow through various styles of gauging heads. Air is supplied to the unit at 50 to 100 psi (pounds per square inch) [3.5 to 7.0 kg/cm² (kilograms per centimeter squared)] and passed through a filter. Working pressures around 10 psi [0.7 kg/cm²] are obtained by a regulator in the line. Regulated air flows through a cylinder containing a float suspended in the column of moving air (Fig. 2-37). Air flows from the cylinder to the gauging head, where it exhausts through a clearance between the gauge and the workpiece. The rate of air flowing through the cylinder is proportional to the air flow through the clearance. Differences in the clearance and exhausted air flow are indicated directly by the position of the float in the cylinder. Air gauges are calibrated by establishing the lower and upper float settings with master gauges of the appropriate tolerance. Gauging heads are similar to standard mechanical gauges with the addition of air passages. Common air-gauging

heads include the plug and ring gauges. Other variations of air gauges employ various systems of operation. Gauging heads may contain a spring-loaded plunger that acts as a valve to regulate the air flow. Variations in the workpiece cause the plunger to open and close the air valve. Changes in the air flow are indicated by the floats. Air gauging is used for both internal and external inspection. Diameters, tapers, concentricity, parallelism, squareness, clearance, contour, and flatness are only some of the many applications. A number of air gauges can be used in a multiple setup to increase efficiency and versatility (Fig. 2-38).

Electric Gauging. Electric gauges indicate whether dimensions are within tolerances. If the measurements are not within desired limits, a light identifies the limit which is violated. A probe consists of a plunger that is calibrated with master gauges. When the workpiece is tested, the plunger is activated and makes the desired measurement. If the limit is not acceptable, the plunger works like a switch to illuminate the warning light. Tolerances can be set between 0.00005 and 0.010 in [0.0012 and 0.254 mm] in electric gauges. Electric gauges are assembled in combinations for making an accurate test of many dimensions simultaneously. A meter light will automatically go on when any of the tolerances is not acceptable. Figure 2-39 shows a multidimensional gauging ma-

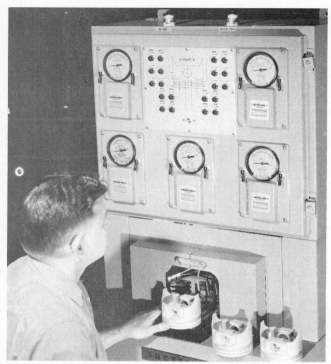

Fig. 2-39 Electronic multidimensional gauging machine used to inspect and classify pistons into acceptance or rejection categories. (*Source: Automation and Measurement Division, Bendix Corp.*)

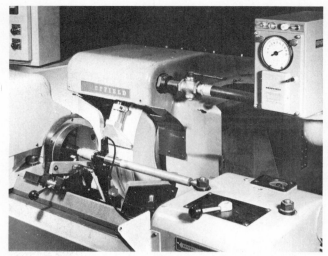

Fig. 2-40 In-process gauge controls grinding operation. (*Source: Bendix Corporation.*)

Fig. 2-41 Final gauging station segregates parts according to size. (*Source: Bendix Corp.*)

chine. Note the drawing on the machine which illustrates the dimensions being inspected on the piston.

Automatic Gauging. The gauging devices previously described can be used separately or on automatic systems. Automatic gauging provides an opportunity to inspect parts during or after production. Air and electric gauges lend themselves readily to automatic systems.

In-Process Gauging. In-process gauging utilizes pneumatic or electric gauges to measure critical dimensions while a part is being manufactured. Signals from a sensor are received in a control unit that commands the machine to perform a predetermined operation. Figure 2-40 illustrates the principle of in-process gauging.

Final-Inspection Gauging. Final-inspection gauging is used to check multiple dimensions simultaneously on finished products. Parts are fed to the final station by hand or by automatic transfer devices. Gauges are used to check and segregate the finished parts. Figure 2-41 illustrates a final gauging station that checks the outside diameter, taper, inside diameter, out of round, and length. Parts are separated into appropriate bins or transfer lines. Gauging instruments are also used to signal the machine operator automatically when the machine is malfunctioning. Some units even adjust or stop the machine in case of tool failure.

Other Quality Control Aids

A number of additional inspection and quality controls are available to insure that specifications and standards are maintained. These are commonly referred to as *nondestructive tests*. Nondestructive tests used for quality control include: visual, magnetic, ultrasonic, laser, and radiograph tests (see Chap. 4).

REVIEW QUESTIONS

2-1. List as many types of specifications used in manufacturing as you can.

2-2. Trace the history of industrial standardization.

2-3. Explain the various components of the screw thread.

2-4. Distinguish between basic size, allowance, and tolerance.

2-5. Distinguish between bilateral and unilateral tolerance.

2-6. Explain the difference between the UNC, UNF, and UNEF screw-thread series.

2-7. Distinguish between quality control and statistical quality control.

2-8. Explain probability theory relative to statistical quality control.

2-9. Explain how to use the sine bar in angular measurement.

2-10. What is the purpose of optical flats? Explain the principle of interferometry.

2-11. Explain the operation principle of air gauges.

2-12. Differentiate between in-process gauging and final-inspection gauging.

CHAPTER 3

NATURE, PROPERTIES, AND TYPES OF MATERIALS

Technology can be defined as the application of scientific knowledge to practical problems. When scientific principles are used to convert materials into useful products, the result can be called *manufacturing processes technology*. Since materials are to be converted into useful products by manufacturing processes, knowledge of the nature, properties, and types of materials is necessary to make their conversion more successful and efficient. Therefore, the purpose of this chapter is to provide the reader with an understanding of (1) the nature of materials, (2) properties of materials, and (3) the most commonly used types of materials.

NATURE OF MATERIALS

In past centuries, knowledge of the structure and composition of materials was relatively limited. However, with the development of sophisticated methods and techniques such as X-ray diffraction and electron microscopy (Fig. 3-1) for the study of materials, knowledge of materials has been increased tremendously during this century. Under the general topic of *science of materials*, men and women are attempting to study and understand the nature, composition, and properties of the natural materials and be able to develop a variety of manufactured materials. Through studies of the atom in physics and chemistry, humans have come close to understanding the nature of matter, which is the basis of all materials.

Atomic Theory

All matter consists of very small particles called *atoms*. Atoms represent the basic elements which are the most fundamental forms of materials. At the subatomic level, atoms are composed of three basic components called *protons, neutrons,* and *electrons*. These three subatomic particles do not have a specific chemical identity like the atoms, but their arrangement in the atom determines the chemical properties of the element. Therefore, a knowledge of the atomic structure of elements is important in understanding materials.

Although protons, neutrons and electrons are basically the same in all elements, when they combine in different arrangements they produce different atoms having distinct chemical properties. Each atom is composed of a nucleus, which consists of neutrons and protons, surrounded by electrons (Fig. 3-2). These different atoms provide an almost infinite variety of materials, each of which has different structure, composition, and properties.

The protons in the nucleus are positively charged and the neutrons are neutral, having no electrical charge. The number of protons in the nucleus determines the *atomic number* of the atom, and the sum of the neutrons and protons determines the *atomic weight*. Therefore, since the nucleus of atoms in each element contains a different number of protons and neutrons, the atomic number, weight, and chemical properties of each element are different. Scientists have arranged the 104 elements (92 occurring naturally and 12 artificially produced) into a *periodic table* by atomic number and by similar properties. The periodic table indicates the repetitive nature or periodicity of the common properties of the elements.

Electron Shell Theory. Electrons are negatively charged and orbit the nucleus at high speeds like the various planets orbit the sun in the solar system. While the orbits of planets can be easily determined in terms of distances from the sun, the position of electrons is difficult to determine. Therefore, it is easier to think of electron "orbits" as *energy shells* with the electrons in each shell possessing a different energy level from those in another shell. Electrons in shells closer to the nucleus of the atom have lower energy levels than electrons located further from the nucleus. However, electrons can be made to change energy shells in the atom.

An electron changing from a lower to a higher energy shell absorbs a certain amount of energy whereas an electron changing from a higher to a lower energy shell releases a discrete quantity of energy called a *quantum*. Each atom has a predetermined number of electrons in each of the energy shells. The shell closest to the nucleus can retain two electrons, while the shells further away from the nucleus usually retain more than two.

The electrons at the outermost shell are the highest energy electrons called *valence electrons*. These electrons are the least tightly held to the nucleus but are responsible for the chemical reactivity of the atom. The chemical reactivity of an atom indicates the ease with which the atom enters into electron reactions by losing or gaining valence electrons. For any atom to achieve a stable electronic structure, eight valence electrons are usually required for its orbit.

Valence electrons are the most important in the study of materials because they determine the properties and structure of materials. When an atom loses a valence electron, it becomes positively charged, or a *positive ion*. When it gains an electron in its outermost shell, it becomes negatively charged, or a *negative ion*.

Atomic Bonding

Many atoms bonded together by interatomic forces form the basic structure of materials. Interatomic bonding of materials can occur on two levels, depending on the interatomic forces present. Some materials are composed of individual atoms directly aggregated into the structure. In these materials the atoms are bonded together by *primary* or *strong bonds*. In other materials, the atoms are bonded into similar groupings called *molecules*. The molecules are then bonded together to form the material characteristic of *secondary* or *weak bonds*.

The three common types of strong bonds are *covalent, ionic,* and *metallic*. Weak bonds are formed by *Van der Waals* forces, relatively weak forces acting over several atomic diameters. Whereas the strong bonds are based primarily on the sharing and interchange of valence electrons, weak bonds are based on weak electrostatic attraction of asymmetric molecules. Van der Waals forces may be created when the motion of the valence electrons in one atom influences valence electron motion in nearby atoms.

Materials bonded by weak bonds are usually char-

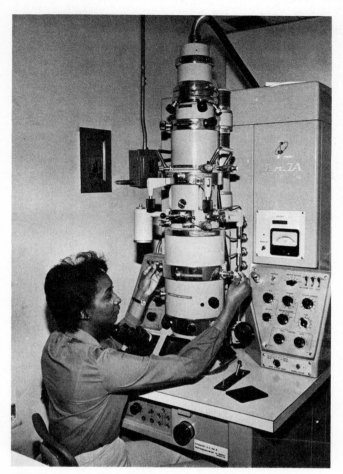

Fig. 3-1 Electron microscope used to study and investigate the structure of materials.

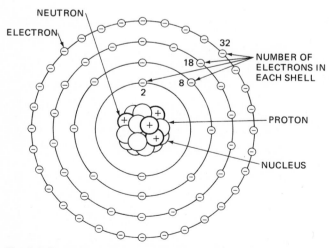

Fig. 3-2 Structure of the atom consisting of the nucleus, with protons and neutrons and the electrons orbiting around the nucleus at specified orbits called *shells*.

acteristic of molecular structure and are relatively soft, plastic, and can be deformed easily. Among the materials bonded primarily by weak bonds are some types of plastics, some ceramics, rubbers, and the inert gases. It should be emphasized, however, that although some materials may be bonded primarily by strong or weak bonds, most materials are bonded by more than one type of bond.

Covalent Bonding. Covalent bonding is an important strong bond based on the sharing of valence electrons. It occurs when one or more valence electrons are shared by adjacent atoms to achieve a stable electronic structure. Figure 3-3 shows the sharing of electrons by the atoms in the chlorine (top) and methane (bottom) molecules, forming a covalent bond. In the chlorine molecule, each of the chlorine atoms has only seven valence electrons, but it requires eight to achieve a stable electronic structure.

Many of the materials bonded covalently are strong, relatively hard, and are corrosion resistant. Among the materials bonded covalently are some types of plastics; wood; some types of ceramics; the diatomic molecules of oxygen, hydrogen, and nitrogen; and many other organic compounds.

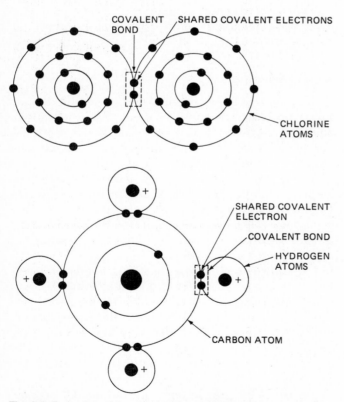

Fig. 3-3 Formation of the covalent bond of the chlorine molecule (*top*) and the methane molecule (*bottom*).

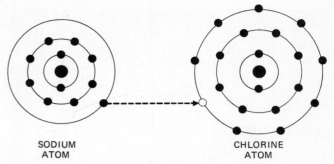

Fig. 3-4 The interchange or transfer of valence electrons between sodium and chlorine atoms creates a stable sodium chloride (table salt) molecule bonded ionically.

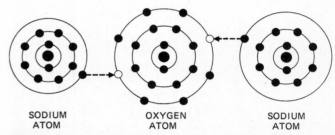

Fig. 3-5 Interchange or transfer of valence electrons between the two sodium atoms and the oxygen atom creates a stable electronic structure in the sodium oxide molecule.

Ionic Bonding. Ionic bonding is based on the interchange of valence electrons which is due to positive and negative charges of the atoms. For example, the chlorine atom needs only one valence electron to achieve a stable electronic structure (Fig. 3-4). If this atom is brought close to a sodium atom which has only one valence electron, the sodium will donate the valence electron to the chlorine, achieving a stable electronic structure. For this interchange to take place the atoms must be *ions*.

In Fig. 3-4, after the chlorine atom accepts the valence electron it becomes a negative ion. Conversely, the sodium which donates the valence electron becomes a positive ion. The electrical attraction created by the ionic condition of the atoms holds these atoms together as a molecule by a strong bond called an *ionic bond*. It is possible, however, to have more than two atoms bonded ionically to form molecules, as illustrated with the sodium oxide molecule in Fig. 3-5.

Many of the materials bonded ionically are relatively hard, good insulators, have high melting points, and are corrosion resistant and heat resistant. Among materials bonded ionically are table salt (sodium chloride), ceramics, and sand.

Metallic Bonding. Metallic bonding is based on the ability of metal atoms to donate valence electrons to a

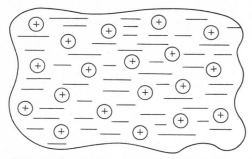

Fig. 3-6 The donation of valence electrons in the electron cloud results in metal ions positively charged and the electron cloud negatively charged forming the metallic bond holding the metal ions (atoms) together.

common pool of electrons called the *electron cloud* or *gas*. The resulting atomic structure consists of positively charged ions, with a complete outer shell, and a negatively charged electron cloud uniformly distributed throughout the structure (Fig. 3-6). The valence electrons donated to the electron cloud are the farthest from the nucleus and move freely within the metal structure. It is this movement and the interatomic distance of the valence electrons in the electron cloud that make these electrons subject to attraction by other atoms, providing metals with the characteristic properties of high thermal and electrical conductivity.

STRUCTURE OF INDUSTRIAL MATERIALS

The preceding discussion outlined some of the general concepts of atomic theory applied to materials at an elemental or pure state. However, materials are rarely used in the pure or elemental state; they are mostly used in combinations of elements to produce desirable compositions and properties. The most common materials used in modern industry are metallic, ceramic, and organic. The organic group includes such materials as plastics, rubbers, and woods. The following sections will discuss the structures of these materials and the relationship of structure to material properties.

The Crystal Structure of Metallic Materials

Metals are probably the most important group of materials in modern industry. A metal can be defined by its properties, but a better definition would be in terms of atomic structure. A *metal* is an element with the ability to donate its valence electrons when bonded with another atom. The arrangement of a large number of atoms in a repetitive, regular geometric pattern corresponding to a lattice results in a

crystal or grain structure which is the basic structure of metallic materials.

Although metallic materials crystallize in many different forms of crystal lattice structures, three of these are the most important ones: (1) *body-centered cubic* (BCC), (2) *face-centered cubic* (FCC), and (3) *hexagonal close-packed* (HCP). Some of the metals crystalizing in these three structures are iron, chromium, molybdenum, and tin in BCC structure; aluminum, copper, lead, and silver in FCC structure; and magnesium and titanium in HCP structure. Every crystal lattice has its own *unit cell*. A unit cell is the simplest model of a crystal lattice exhibiting all the characteristics of the lattice.

The BCC-lattice structure shown in Fig. 3-7 is a unit cell consisting of nine atoms arranged one on each of the corners of the cube and one in the center of the cube. The FCC lattice (Fig. 3-8) is a unit cell consisting of fourteen atoms arranged on each of the corners of the cube and one at the center of each cube face. The HCP lattice (Fig. 3-9) is a unit cell consisting of seventeen atoms arranged one on each of the corners of the hexagonal configuration, one at the center of each hexagonal face and three atoms spaced equally at the center of the lattice at alternating sides.

However, it is interesting to note that some metals, notably iron, can exist in two or more crystal-lattice structures. These metals are said to be *allotropic* and the change from one lattice structure to the other is

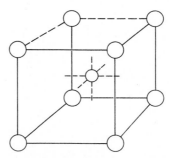

Fig. 3-7 A schematic representation of the body-centered cubic lattice BCC.

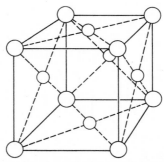

Fig. 3-8 A schematic of the face-centered cubic lattice FCC.

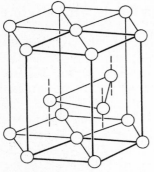

Fig. 3-9 A schematic of the hexogonal close-packed lattice HCP.

called *allotropic change* or *allotropic transformation*. Allotropic changes depend upon temperature and pressure.

The allotropic changes of pure iron are shown in Fig. 3-10. Pure iron solidifies at about 2800°F (1538°C) into a BCC crystal-lattice structure known as *delta iron*. At about 2552°F (1400°C) the atoms rearrange themselves into a FCC crystal-lattice structure known as *gamma iron*. As the temperature drops further to about 1670°F (910°C) pure iron changes again

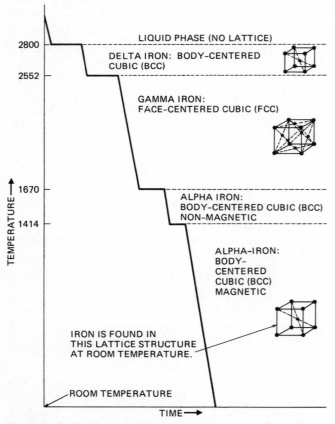

Fig. 3-10 Cooling part of the heating-cooling curve of pure iron showing the allotropic changes and lattice formations.

into a BCC-crystal-lattice structure known as *alpha-iron*. Iron at room temperature is found in this structure.

Grain Structure in Metals. The crystal structure discussed in the preceding section pertains to individual crystals. However, metallic materials used in modern industry consist of many crystals bonded together to form a polycrystalline structure sometimes referred to as *grain structure*. Grain structure or crystal orientation in metals develops through a process called *nucleation* (nucleus formation). As a melted metal cools, its structure begins to change by first forming extremely tiny particles called nuclei usually minute impurities. The nuclei gradually grow and spread three-dimensionally, forming the grain structure.

The growth of the nuclei resembles a tree pattern called *dendrite*. Each nucleus formed develops into a grain in the final structure of the solid metal. However, the nuclei formed are not aligned uniformly, resulting in the formation of *grain boundaries*. Grains vary in size, depending on such factors as cooling rate, pressure, and composition.

Grain size is an important variable because it influences the strength and formability of metals. At room temperature, a metal with small (fine) grain size is generally stronger and tougher than the same metal with courser grain. Grain size can be changed by controlled temperatures, chemical composition (alloying), and processing techniques.

Alloying Metallic Materials. An alloy is a metallic material consisting of two or more elements. In forming alloys, the size of the atom and arrangement of atoms in the crystal-lattice structure are important variables. It has been established that, for proper alloying, the atoms of the elements must be within 8 percent of each other in size. If the size difference of the atom is more than 15 percent, proper alloying is retarded.

When two or more elements with atoms of about the same size are alloyed, a *substitutional solid solution* results. In this case the similar size atoms *substitute for each other in the crystal-lattice structure* (Fig. 3-11). If, however, the size difference is too great, an *interstitial*

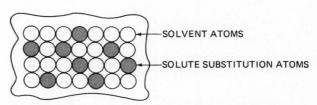

Fig. 3-11 Arrangement of atoms in a substitutional solid solution.

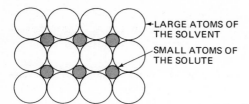

Fig. 3-12 Arrangement of atoms in an interstitial solid solution.

LARGE ATOMS OF THE SOLVENT

SMALL ATOMS OF THE SOLUTE

solid solution results. In this case the small-size atoms occupy the voids created by the larger atoms in the crystal-lattice structure (Fig. 3-12). The elements in solid solutions are soluble in each other. *Solid solutions* constitute one form of alloying metallic materials. The second form of alloying is accomplished through *intermetallic compounds*. These are chemical compounds formed when atoms from different elements interact chemically and form a strong bond (ionic or covalent). The third form of alloying is accomplished through *mechanical mixtures*. A mechanical mixture can be formed when two or more elements are alloyed but retain their individual identities in the alloy.

Equilibrium Diagrams

These diagrams have been developed to record and study changes occurring during the heating-cooling cycles of pure or alloyed metallic materials. The diagrams show the temperatures at which changes occur and the products of these changes. Equilibrium diagrams are important aids in studying the complex alloying and heat treating processes of metallic materials.

A relatively simple equilibrium diagram is shown in Fig. 3-13. Several important facts can be learned by studying this diagram. The two alloying elements of copper and nickel are completely soluble (one dissolves in the other) in each other. At room temperature the two elements form a solid solution having a single phase (a phase is a physically distinct state of a substance). The alloy system can exist in the liquid and solid states. The melting temperatures of both elements are shown, and the percent of each element at any particular temperature can be calculated in a given alloy (e.g., Monel).

One of the most important equilibrium diagrams is the *iron-carbon equilibrium diagram* (Fig. 3-14). The alloy resulting from carbon added to pure iron is called *plain carbon steel* or *steel*. The allotropic changes of iron make the diagram very complex. However, several important facts can be seen by studying the diagram.

Line *ABC* represents the temperature (critical point) at which transformation from a liquid to a mix-

ture of solid and liquid begins. Line *PSY* represents the A_2 points, the temperatures below which only alpha iron exists. Line *GSEC* represents the upper critical points, the temperatures at which the transformation to 100 percent *austenite* is complete.

As the temperature is increased above the upper critical point, the grains increase in size until melting of the steel begins at the temperatures represented by line *NJE*. This line is called the *solidus curve*. The line slightly above the solidus curve represents the temperatures at which the steel is completely liquid and is referred to as the *liquidus curve*. Between the solidus and liquidus curves, the steel is in a mushy condition. When molten steel is cooled, it starts to solidify at the temperatures represented by the liquidus curve and is completely solid at those represented by the solidus curve.

It can be seen in Fig. 3-14 that iron changes from *austenite* (FCC) to *ferrite* (BCC). Austenite (Fig. 3-15) is a solid solution of carbon in FCC gamma iron. It can hold about 1.7 percent carbon in solution at 2066°F (1130°C). This indicates that carbon is an interstitial alloy in gamma iron. However, above or below that temperature the solubility of carbon in FCC changes. On the other hand, *ferrite* or alpha iron has a BCC-lattice structure, and it can hold about 0.025 percent carbon in solid solution. Pearlite is a mixture of cementite (iron carbide) and alpha iron (ferrite) (Fig. 3-16). In the austenitic FCC structure, the lattices are

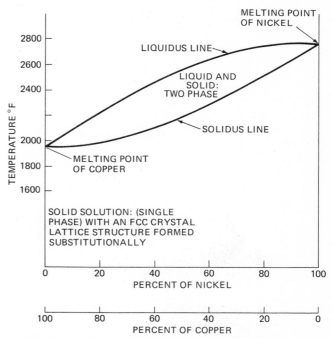

Fig. 3-13 A typical equilibrium diagram of a copper-nickel alloying system.

MELTING POINT OF NICKEL

LIQUIDUS LINE

LIQUID AND SOLID: TWO PHASE

SOLIDUS LINE

MELTING POINT OF COPPER

SOLID SOLUTION: (SINGLE PHASE) WITH AN FCC CRYSTAL LATTICE STRUCTURE FORMED SUBSTITUTIONALLY

TEMPERATURE °F

PERCENT OF NICKEL

PERCENT OF COPPER

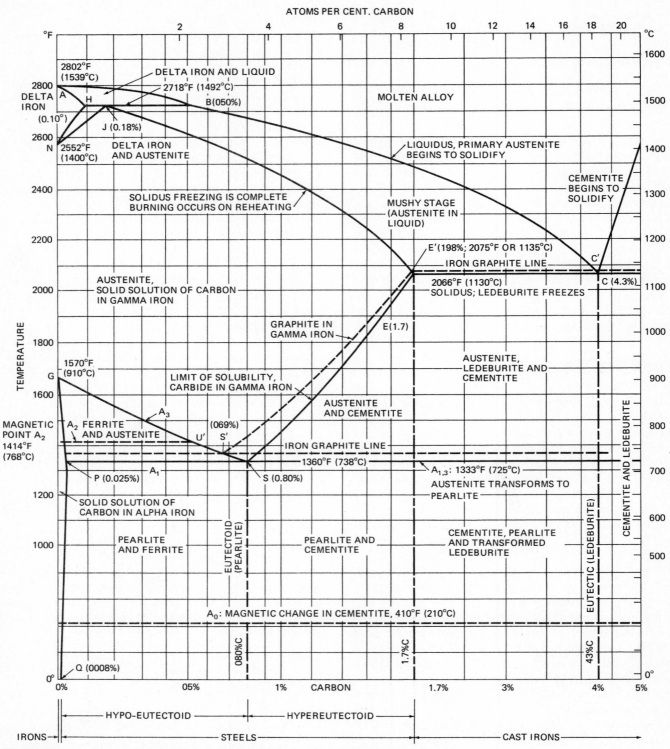

Fig. 3-14 A typical iron-carbon equilibrium diagram indicating the various structural changes in carbon steels based on carbon content and temperature.

large and can hold carbon up to 1.7 percent, while in the ferrite BCC structure the lattices are small and cannot hold much carbon.

The solubility of carbon at 1333°F (722.8°C) is at maximum 0.8 percent, but below that temperature it is extremely low. As the temperature drops below 1333°F (722.8°C), the excess carbon may form a *eutectoid* structure called *pearlite* (the eutectoid mixture of ferrite and cementite). *Cementite,* or iron carbide, is an intermetallic compound of iron and carbon (Fe_3C). The compound is very hard and brittle and forms below 1333°F (722.8°C) with the iron in the BCC structure.

Refer again to Fig. 3-14 (equilibrium diagram), where you will see that line *PS*, which represents the lower critical points in carbon steels, is horizontal. This shows that the lower critical points do not vary with carbon content. The A_2 points also remain unchanged as the carbon content increases up to about 0.7 or 0.8 percent carbon. This is shown by the horizontal line *O*.

The eutectoid point *S* is of interest in the iron-carbon equilibrium diagram. The point indicates the lowest temperatures at which the transformation of gamma to alpha iron can occur. The eutectoid point *S* occurs at about 0.8 percent carbon; thus carbon steels of this composition are referred to as *eutectoid steels.* Therefore, on the basis of point *S*, carbon steels can be classified as *hypoeutectoid* steels (with less than 0.8 percent carbon) and *hypereutectoid* steels (with more than 0.8 percent carbon).

In a hypoeutectoid steel, an austenitic structure exists whenever the steel is heated above the upper

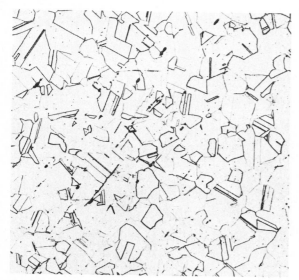

Fig. 3-15 Austenite in an AISI type 304 stainless steel, ×100, Vilella's reagent. (*U.S. Steel Corp.*)

Fig. 3-16 Pearlite formations in a 0.004 percent carbon steel, ×100, nital etch. (*U.S. Steel Corp.*)

critical point. On the other hand, the austenite formed in a hypereutectoid at the upper point contains only the eutectoid percentage of carbon in solution. The remaining carbon has formed cementite, which is mixed with the austenite. To produce a complete austenitic structure in hypereutectoid steel, the steel must be heated to the temperatures indicated by line *EC*. This line indicates the solubility of cementite in austenite. The temperatures at which all the cementite is dissolved vary from 1333 to 2066°F (722.8 to 1130°C), depending on the carbon content. Above these points, the excess cementite goes into an interstitial solid solution in gamma iron to form a complete austenitic structure.

STRUCTURE OF CERAMIC MATERIALS

Unlike metallic materials, which are bonded primarily by metallic bonds, ceramic materials are bonded by both covalent and ionic bonds, depending on the size of the atoms and interacting valence electrons, and they are not simply metallic but contain other compounds (oxides, etc.). As with metallic material, ceramics have the characteristic of a crystal structure. However, the ceramic crystal structure is more complex than the metallic. Because of this com-

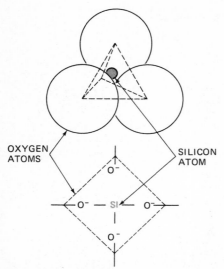

OXYGEN ATOMS

SILICON ATOM

O⁻

O⁻ — SI — O⁻

O⁻

Fig. 3-17 The interstitial position of the silicon atom in the silicate tetrahedron.

plex crystalline structure, ceramic materials have high melting points and low conductivity, and are chemically inert.

The basic ceramic materials are based on the silicate unit, which consists of four oxygen atoms and one silicon atom arranged in a *silicate tetrahedron* (Fig. 3-17). The silicon atom occupies an interstitial position in the void created by the four oxygen atoms. The structure is bonded together by both ionic and covalent bonds to form the basic unit, which can react in several ways to form different types of ceramics.

The basic silicate tetrahedrons join together in a *chain structure* to form the *fibrous ceramics,* such as asbestos. Or the chains can be bonded into two dimensions, forming a *sheetlike structure* characteristic of clay-based ceramic materials.

STRUCTURE OF ORGANIC MATERIALS

Unlike metallic and ceramic materials characteristic of crystalline structures, organic materials are com-

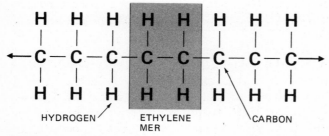

HYDROGEN

ETHYLENE MER

CARBON

Fig. 3-18 Structure of polymeric materials. The repeated pattern of ethylene monomers results in the formation of long chains polyethylene molecule or polymer.

posed of very large covalently bonded molecules generally based on carbon. The large molecules are composed of a repetitive pattern of structural units called *mers* (Fig. 3-18). When many *mers* are joined together in long chains, they form *polymers* or *polymeric materials* called *plastics.* Among the most commonly industrially used organic materials are *plastics, woods, rubbers, fibers,* and *leather.*

PRODUCTION OF INDUSTRIAL MATERIALS

The many types of products manufactured by modern industry require a wide variety of material production processes. The basic function of manufacturing processes is to process raw materials into semifinished or finished products, but most industrial materials have to be processed before they can be made into useful products. The purpose of this section is to describe the production of the most commonly used materials.

Production of Metallic Materials

Metallic materials constitute one of the most important group of industrial materials. They are distinguished from other materials in terms of their good conductivity of heat and electricity, formability, weldability, castability, and machinability. Metals are relatively strong, heavy, opaque, and lustrous in appearance. Metals are classified as *ferrous,* such as pig irons, cast irons, and steels and as *nonferrous* metals, such as aluminum, magnesium, copper, zinc, tin, lead, gold, and silver.

The ferrous metallic materials are extracted from *iron ores* found near the earth's crust. Iron ores are

Fig. 3-19 Partial view of an open-pit mine. Power shovel loading iron ore. (*U.S. Steel Corp.*)

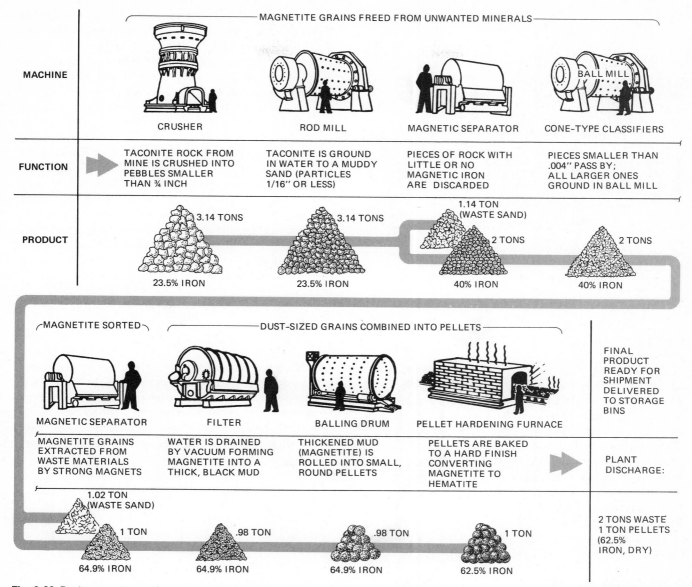

Fig. 3-20 Basic steps in producing taconite for the blast furnace. (*Reserve Mining Co.*)

mined by underground (shaft) or open-pit mining (Fig. 3-19). The principal iron ores are *hematite* (Fe_2O_3), *magnetite* (Fe_3O_4), *lemonite* (Fe_2O_3), and *siderite* ($FeCO_3$). After mining, the iron ore is improved by concentration methods aiming to increase the percentage of iron by eliminating impurities. In an effort to prolong the life of areas rich in iron deposits, the iron-ore industry has developed methods to utilize economically such lower-grade iron ores as *taconite* and *jaspar* from Minnesota and Michigan. The *agglomeration* process used to concentrate the taconite iron ore from 25 to 63 percent iron is illustrated in Fig. 3-20.

The concentrated iron ore is transported via railroad cars, barges, or ocean ships to the steelmaking plants where it is converted into *pig iron* by melting it in the *blast furnace* (see Figs. 3-21 and 3-22). The iron ore, mixed with coke and limestone in predetermined amounts, is *charged* (loaded) by *skip cars* into the blast furnace.

Hot air from the stoves is blown into the furnace near the bottom, causing the coke to burn at about 3000°F (1649°C), melting the charge in the furnace. The molten iron settles at the bottom of the furnace from where it is tapped and cast into *pigs* weighing 100 lb each, which are transported by special cars to the steel mill. Pig iron contains about 3 to 4 percent carbon, 0.06 to 0.10 percent sulphur, 0.10 to 0.50 percent phosphorus, 1 to 3 percent silicon, and cer-

Fig. 3-21 Typical installation of two blast furnaces and their heating stoves in between. The tall structures at right and left are the actual furnaces. Figure 3-22 shows a cross section of the furnace. (*Inland Steel Co.*)

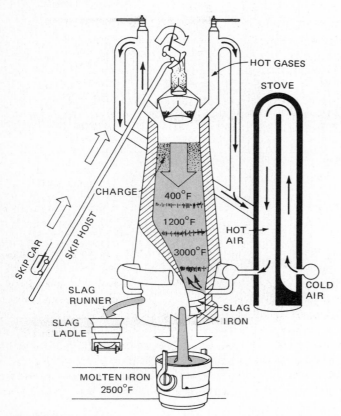

Fig. 3-22 A schematic showing the blast furnace. (*Inland Steel Co.*)

tain impurities. It is soft, brittle, and has low tensile strength. Pig iron is used to produce cast iron and steel.

Production of Cast Iron

Cast iron is an alloy of iron, carbon, silicon, and other elements in insignificant amounts. It is produced mostly in the *cupola furnace* (Fig. 3-23) and used to make castings. The three primary types of cast iron produced by the cupola furnace are *gray, white,* and *nodular* or *ductile* cast irons. *Malleable* and *alloy* cast irons are secondary types of cast irons and are produced by heat-treating and alloying the three primary types. The carbon content and its form in the cast iron determines the structure of the cast iron. Most of the carbon content in *gray cast iron* is in the form of free graphite flakes, which give gray cast iron its color and desirable properties. Gray cast iron is easy to machine, has high damping capacity, and has good fluidity for casting. However, it is brittle and of low tensile strength. It has high resistance to corrosion and heat, and relatively good wear resistance. Gray cast iron is extensively used for such applications as machine tool bases, frames of heavy machinery, and engine blocks in automotive industry. It is the most widely used cast iron.

White cast iron is produced in the cupola furnace, but its composition and solidification rate are rigidly controlled, resulting in a microstructure in which most of the carbon is in the combined form of cementite. Because of its structure, white cast iron is hard, brittle, and very difficult to machine. It is used primarily to make castings to be transformed by heat-treatment into malleable cast iron. White cast iron is also used for applications requiring good wear resistance such as crushing equipment and roll and grinding mills.

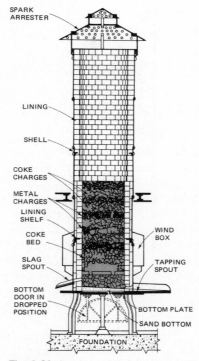

Fig. 3-23 A schematic of the cupola furnace. (*U.S. Steel Corp.*)

Nodular or *ductile cast iron* is also produced in the cupola furnace by melting pig iron and scrap mixed with coke and lime. Most carbon content in nodular cast iron is in the nodular form (small ball-shaped formations). To produce the nodular structure the molten cast iron from the furnace is innoculated with a small amount of such material as magnesium and/or cerium. The innoculants prevent the formation of cementite and changes the graphite flakes into nodules. This microstructure produces such desirable properties as high ductility, strength, good machinability, good fluidity for casting, good hardenability, and toughness. It cannot be as hard as white cast iron unless it is specially heat-treated on the surface. It is extensively used by the automotive industry for car and truck parts, by machinery builders, and by agricultural implements manufacturers.

Malleable and *alloy cast irons* are special types of cast irons produced by heat-treating and alloying white cast iron. These irons are of good machinability, strength, toughness, and wear resistance. They are extensively used for special applications in automotive, agricultural, tool, and hardware industries.

Steel Production

Steel is the most important of the ferrous materials and is produced basically from pig iron and steel scrap. The production of steel entails removing the impurities, adjusting the carbon content, and facilitating the addition of required alloy elements in steel. Steel is produced today by three major steel-making processes: *open hearth, basic oxygen,* and *electric.* In terms of tons of steel produced, the open hearth and basic oxygen are the two principal processes, while the electric has been used primarily to produce alloy and high-grade steels. Figure 3-24 is a simple flowchart showing the various stages in steel making.

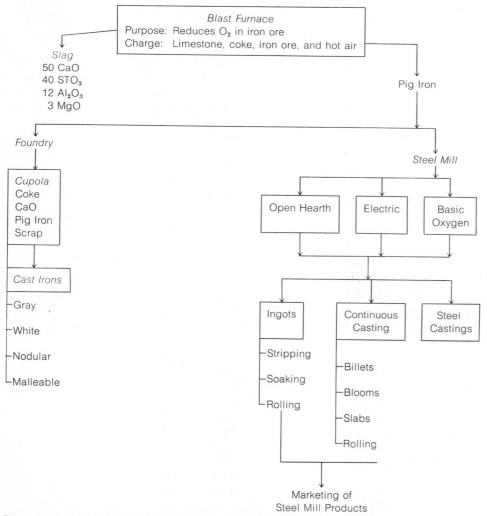

Fig. 3-24 Flowchart showing the various stages of steelmaking.

Open-Hearth Steel Making. The open-hearth process has been for over a century the principal steel-making process, producing over 90 percent of the carbon steel. The open-hearth furnace is a relatively shallow, rectangular basin about 90 ft [27 m] long and 30 ft [9 m] wide in which the metal is open to the sweep of the flames across the hearth (Fig. 3-25). The open-hearth steel-making process is divided into *basic* and *acid* processes, depending mostly on the type of furnace lining used. The acid process is used mostly for foundry steel to make steel castings, while the basic process is used for other types of steel. The furnace is lined inside with firebricks that can withstand the high operating temperatures and pressures.

A single furnace charge (load) is called a *heat.* Heats vary in size from 100 to 600 tons, depending on the size of the furnace. The materials used in open-hearth steel making are pig iron, scrap steel, limestone, iron ore, fuel (gas, oil or metallurgical coke or coal), oxygen, and air. Limestone is used as flux, which combines with the impurities to form the *slag.* Steel scrap, constituting about 50 percent of the heat, is charged cold while pig iron is charged in pigs or as molten iron. Oxygen serves to reduce the carbon content by oxidation and to increase the flame temperature to about 3000°F (1649°C). Oxygen is introduced into the molten metal in the hearth under high pressure by a retractable pipe called the *lance.*

The average cycle from charge to tap takes about 8 to 10 hours. After the melting and refining is completed, the heat is tapped into a ladle ready to be poured into ingots. The open-hearth process has been used mostly for making carbon steel; however, open hearths have also been used to produce some alloy steel.

Fig. 3-25 Front portion of an open-hearth furnace showing the charging doors, charging car at rear. (*Inland Steel Co.*)

Basic-Oxygen Steel Making. Basic-oxygen steel making is the most recent of the steel-making processes, currently producing about one-third of the world's steel with prospects to completely replace the open hearth in the future. Its main advantage is the short time cycle (about 55 minutes) to complete a heat as compared with 8 to 10 hours in the open hearth. The efficiency of basic-oxygen furnace (BOF) is due to high-purity (99.0 percent) oxygen used as the sole oxidizing agent in refining. The heat consists of molten pig iron, steel scrap, limestone, and oxygen.

The furnace is a pear-shaped steel vessel lined with firebricks in the inside and supported on trunnions, enabling 180° tilting of the furnace. The furnace is tilted and scrap steel is charged first, followed by the molten pig iron. The furnace is raised in the vertical position and, through a water-cooled retractable lance, the oxygen is introduced. Coming in contact with the molten pig iron, the pure oxygen reacts violently and combines with the carbon of the charge, forming carbon monoxide. The carbon burns and the gas escapes from the top of the furnace. During this violent oxidation, a great deal of heat is generated, raising the temperature of the charge to about 3000°F (1649°C). While oxidation is progressing, a certain amount of limestone is added to clean the melt and form the slag. The heat is then discharged into a ladle by tilting the furnace. Alloy elements may be added in the ladle. The BOF is mainly used for carbon steel though some alloy steel is also made. The BOF varies in size from 35 to 200 tons, which is also the hourly production of the furnace.

Electric Steel Making. Electric steel making is used to produce special grades of steel, such as tool and die steel, stainless, and heat-resistant steels. Three-phase current is usually used with the furnace, and the heat is generated by the arc resulting from the electric current (Fig. 3-26). The heat may be generated between the electrodes and the charge or between the electrodes only. Therefore, steel produced by the electric furnace is considered the cleanest of all.

The electric furnace may be of either the arc or the induction types, but most steel is produced by the arc furnace. The electric furnace is made in a circular teacup-shaped-steel shell lined with firebrick in the inside. The furnace is mounted on rockers to enable tilting to discharge the molten steel.

The size of electric arc furnaces vary from 2 to 200 tons. The time cycle per heat varies from 3 to 6 hours, depending on the size of the heat and type of steel produced. The charge consists of highly selected steel scrap, lime, and mill scale, all charged in the furnace from the top through the swivel door. Small amounts

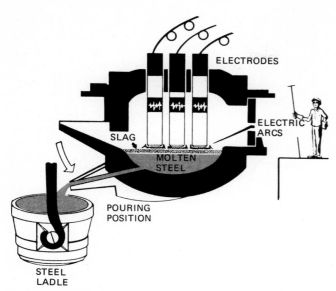

Fig. 3-26 Schematic diagram of a cross section of the electric-arc furnace. (*U.S. Steel Corp.*)

ELECTRODES

ELECTRIC ARCS

SLAG

MOLTEN STEEL

POURING POSITION

STEEL LADLE

of alloy elements are added in the ladle, while large amounts are added in the charge. Both acid and basic methods are used, with acid usually for foundry steel castings and basic for special steels. The electric arc furnace provides high temperatures, close composition control, and minimum contamination. However, because of the amount of electricity used, the operation of the furnace is more expensive than the other steel-making processes.

Steel Mill Products. After the steel is tapped from the furnace, it may be poured into *ingot forms* (Fig. 3-27), used directly in a continuous casting mill, or used directly in the foundry for making steel castings. The steel made into ingots and that used in the continuous mill is further processed by rolling to produce such mill products as slabs, plates, bar, sheet, strip, pipe and tubes, structural shapes, wire, and rails, to name only a few.

Rolling is the broad generic term used for all processes involved in transforming ingots into finished mill products. In rolling, the metallic material is passed through two rolls revolving at the same speed but in opposite directions. Since the opening between the rolls is smaller than the thickness of the ingot, rolling reduces its cross-sectional area, increases its length and thus shapes the piece into a desired mill product. Metallic materials can be rolled cold, producing *cold-rolled steel products* or can be rolled hot producing *hot-rolled steel products.*

The basic steps in transforming an ingot into semi-finished mill products are shown in Fig. 3-28. After

the ingot has been stripped from its mold, it is placed in a *soaking pit* to be heated to the rolling temperature. The ingot is then removed from the soaking pit and passes through the *blooming* or *slabbing* or *billet mills* rolled into *slabs, blooms,* or *billets,* depending on the need. Blooming and slabbing are quick and efficient operations which can reduce an ingot of about 25×27 in $[63.5 \times 68.6$ cm] in cross section into a bloom of 9×9 in $[22.9 \times 22.9$ cm] or a slab of 4×25 in $[10.2 \times 63.5$ cm] in cross section in less than 10 minutes.

Structural shapes such as angles, beams, and rails are usually rolled from blooms. The ingot is first rolled into a bloom in the blooming mill and then the bloom is finished into the desired structural shape in the finishing mill (Fig. 3-29). In rolling plates and sheets, the ingot is first rolled into a slab in the slabbing mill. Then the slab is reheated and rolled into plates or sheets in the finishing mill. Sheets are usually rolled in *continuous strip mills* (Fig. 3-30) where the sheet enters from one side in slab form and exits from the other in finished sheet form. Some blooms are rolled into billets in the billet mill. The billets are reheated and rolled into bars of various sizes. Bars are used in the drawing and seamless mills to produce wire and seamless tubes of different sizes.

Blooms, slabs, and billets can also be produced by the continuous casting process. Continuous casting is a relatively recent steel-shaping process and holds great promise for the future. In continuous casting, the molten steel from the furnace is poured directly in the continuous casting machines and cast into a

Fig. 3-27 Steel ingots stripped from their molds. (*Inland Steel Co.*)

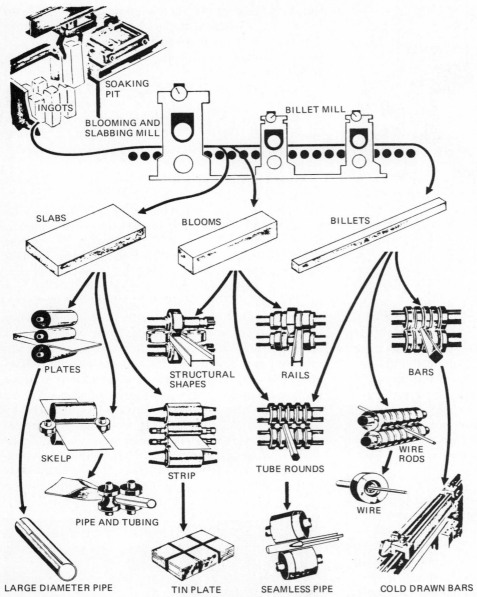

Fig. 3-28 Basic processes involved in the production of semifinished steel products from ingot to semifinished product. (*American Iron and Steel Institute.*)

continuous slab, bloom, or billet. Figure 3-31 on page 64 shows a typical continuous casting installation for casting blooms, slabs, and billets.

Types, Properties, and Uses of Steel

The wide variety of steels used by modern industry presents a serious classification problem. Thus steels have been classified in different ways depending on such factors as composition, production and pro-

cessing methods, strength, and application. In general, steels may be classified in *carbon*, *alloy*, and *special* steels. This classification is satisfactory for simple purposes. But a more complete classification system based on composition has been developed by the Society of Automotive Engineers (SAE) and adopted by the American Iron and Steel Institute (AISI) and the American Society for Testing and Materials (ASTM).

The SAE-AISI system is based on a four- or 5-digit numbering system. The first digit represents the percent carbon. Table 3-1 on pages 65 gives the SAE-AISI system in detail.

Carbon Steels. Carbon steels constitute about 85 percent of all steels used by modern industry. They are classified as *low carbon* (0.08 to 0.35 percent carbon), *medium carbon* (0.35 to 0.50 percent carbon), and *high carbon* (those steels with higher than 0.55 percent carbon). The carbon content plays the most important role in determining the properties of carbon steel. Carbon steels are made in the *hot-* and *cold-rolled* mill products. Hot-rolled carbon steels have the lowest cost per pound of all steels and are used economically in many applications.

Carbon-steel mill products are produced in a wide range of compositions and properties. Among the standard forms of carbon-steel mill products are wire, rod, bar, sheet, strip, plate, tube, pipe, and structural shapes. The tensile strength of common grades vary from 40,000 to 80,000 psi [2800 to 5600 kg/cm²]. Table 3-2 on page 65 gives typical applications of the three basic classes of carbon steel.

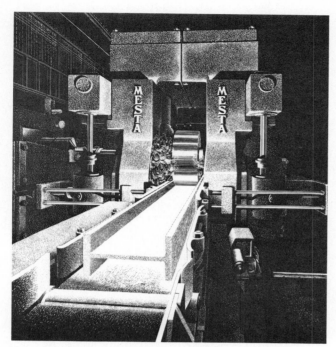

Fig. 3-29 A schematic showing the finishing stand of a 28-in I-structural beam. (*Inland Steel Co.*)

principal alloy element, the second digit gives the alloy content, and the last two or three digits indicate the carbon content. For example, *SAE-AISI 1049* indicates a plain carbon steel with 0.49 percent carbon. *SAE-AISI 5050* indicates a chromium steel with 0.50

Alloy Steels. Alloy steels are produced with specified alloy elements. The properties of alloy steels depend on the content and number of the alloying elements in the steel. Steels are alloyed to increase strength, hardness, resistance to corrosion, wear and temperature, and uniformity of properties. Alloy

Fig. 3-30 Partial-view showing of roughing stands of an 80-in mill where a slab is rolled into a strip 80-in wide. (*Inland Steel Co.*)

1. MOLTEN STEEL POURS FROM A LADLE INTO A RESERVOIR CALLED A TUNDISH.

2. THE METAL FLOWS OUT THE BOTTOM OF THE TUNDISH AT A CAREFULLY REGULATED RATE INTO THE MOLD, WHICH IS MOVING UP AND DOWN TO PREVENT THE HOT METAL FROM STICKING. THE INTERIOR OF THE MOLD IS HOLLOW—JUST THE SIZE, IN WIDTH AND THICKNESS, OF THE SLAB TO BE FORMED. LINING THE WALLS ARE PIPES THROUGH WHICH WATER FLOWS, CHILLING THE METAL. A THIN SHELL OF STEEL BEGINS TO SOLIDIFY AROUND THE MOLTEN METAL.

3. THE GRADUALLY SOLIDIFYING SLAB MOVES DOWN THROUGH THE SECONDARY COOLING ZONE. A SERIES OF ROLLERS SUPPORT THE SLAB AND GRADUALLY TURN IT INTO A HORIZONTAL POSITION. SPRAYS OF WATER UNDER HIGH PRESSURE COOL AND HARDEN THE METAL STILL FURTHER.

4. A FLAME-CUTTING TORCH SLICES DOWN THROUGH THE METAL. WHEN THE SLAB IS CUT OFF, IT IS CARRIED ON ROLLERS TO A COOLING BED. THE ENTIRE TRIP FROM THE LADLE HAS TAKEN LESS THAN ONE-HALF HOUR.

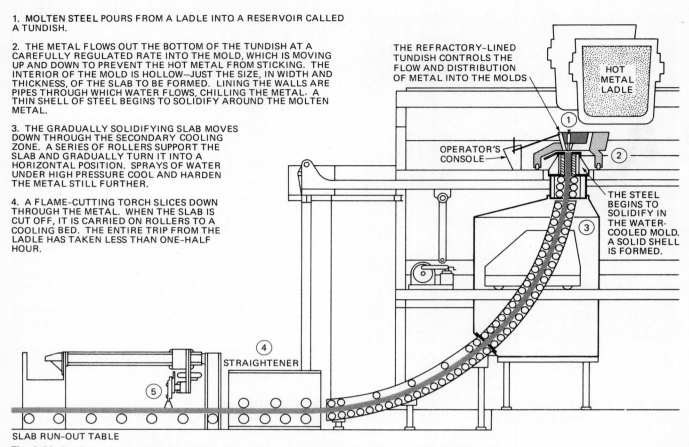

Fig. 3-31 A schematic showing the continuous casting process. (*American Iron and Steel Institute.*)

steels are classified as *low-medium-,* and *high-*alloy or *special* steels.

Low-alloy steels contain less than 4 percent of alloy elements and are produced in two grades: low-alloy high-strength and low-alloy heat-treated construction steels. The principal alloying elements in low-alloy steels are carbon, phosphorus, manganese, chromium, and nickel. These steels have a minimum of 50,000 psi [3500 kg/cm²] tensile strength and are easily cold-formed, welded, machined, and heat-treated. They are available in all standard forms of mill products.

Medium-alloy steels have been developed to improve hardenability, especially for deep heat-treating purposes. The principal alloying elements in these steels are nickel, molybdenum, vanadium, chromium, manganese, and silicon, with nickel as the most common element. The effects of the various alloying elements on the properties of steel are summarized in Table 3-3 on page 66. Medium-alloy steels are used for such applications as tools, gears, ball bearings, machinery parts, and high temperature steam lines.

Special steels (high alloy) are produced by steel makers in many varieties to satisfy the demands of modern industry. Special steels are produced to function satisfactorily in extreme conditions of temperature, wear, corrosion, and strength that cannot be satisfied by other types of steels. The two most important classes of special steels are *stainless steels, heat-resistant steels,* and *tool steels.*

Chromium is the principal alloying element in stainless steels. Any steel containing more than 11.5 percent chromium is considered stainless steel; those with less than 4 percent are considered low-alloy steels. The most outstanding characteristic of high-chromium steels is their high resistance to corrosion and to scaling at high temperatures. Stainless steels are produced in such standard forms of mill products as sheet, plate, strip, bar, structural shapes, wire, tube, casting, and forgings.

Stainless steels are classified in three groups: *martensitic,* or 400 and 500 series containing 11.5 to 18 percent chromium; *ferritic,* or 400 series containing 17 percent chromium; and *austenitic,* or 200 and 300

Table 3-1 SAE-AISI Alloy Steel Designation System

SAE-AISI Number	Alloy Elements and Approximate Percentages
	Carbon steels
10xx	Plain carbon steels (0.05–0.90 C)
11xx	Free-cutting carbon steels
	Manganese steels
13xx	(1.75 Mn)
	Nickel steels
23xx	(3.50 Ni)
25xx	(5.00 Ni)
	Nickel-chromium steels
31xx	(1.25 Ni and 0.65 Cr)
33xx	(3.50 Ni and 1.57 Cr)
303xx	(Corrosion and heat resisting)
	Molybdenum steels
40xx	(Carbon-molybdenum; 0.25 Mo)
41xx	(Chromium-molybdenum; 0.95 Cr)
	Nickel-chromium-molybdenum steels
43xx	(1.82 Ni; 0.50 Cr; and 0.25 Mo)
47xx	(1.05 Ni; 0.45 Cr; and 0.20 Mo)
86xx	(0.55 Ni; 0.50 Cr; and 0.20 Mo)
87xx	(0.55 Ni; 0.50 Cr; and 0.25 Mo)
93xx	(3.25 Ni; 1.20 Cr; and 0.12 Mo)
98xx	(1.00 Ni; 0.80 Cr; and 0.25 Mo)
	Nickel-molybdenum steels
46xx	(1.57 Ni and 0.20 Mo)
48xx	(3.50 Ni and 0.25 Mo)
	Chromium steels
50xx	(Low chromium: 0.27–0.50 Cr)
51xx	(Low chromium: 0.80–1.05 Cr)
51xxx	(Medium chromium: 1.02 Cr)
52xxx	(High chromium: 1.45 Cr)
514xx	(Corrosion and heat resisting)
	Chromium-Vanadium Steels
61xx	(0.95 Cr and 0.15 V)
	Silicon-Manganese Steels
92xx	(0.65–0.87 Mn and 0.85–2.00 Si)
xxBxx	Boron steels
xxLxx	Leaded steels

Table 3-2 Typical Applications of the Three Basic Classes of Carbon Steels

SAE-AISI Number	Typical Applications
	Low-Carbon Steels (0.08–0.35% carbon)
1006–1012	Used where soft and plastic steel is needed, as in soft sheet, strip, tubing, pipe, and welding.
1015–1022	Used where soft and tough steel is needed, as in rivets, screws, tubing, wire, rods, structural shapes, and strips.
1023–1032	Used for such parts as pipes, gears, shafts, bars, and structural shapes.
	Medium-Carbon Steels (0.35–0.50% carbon)
1035–1040	Used for large sections as in forged parts, shafts, axles, rods and gears.
1041–1050	Used for heat-treated machine parts, such as shafts, axles, gears, and spring wire.
1052–1055	Used for heavy-duty machine parts, such as gears and forgings.
	High-Carbon Steels (more than 0.55% carbon)
1060–1070	Used where good shock resistance is needed, as in forge dies, rails, and set screws.
1074–1080	Used for tools requiring toughness and hardness, such as shear blades, hammers, wrenches, chisels, and cable wire.
1084–1095	Used for cutting tools, such as dies, milling cutters, drills, taps, lathe tools, files, knives, and other woodworking tools.

and *high-speed* steels. However, SAE-AISI developed a more complete system for classifying tool steels in 13 standard groups based on composition and application. Tool steels are used in such applications as dies, cutting tools, bearings, rollers, jet blades, truck parts, and tools.

Production of Nonferrous Metallic Materials

Among the most widely used nonferrous metallic materials are aluminum, copper, zinc, lead, tin, magnesium, silver, and gold. Aluminum is considered the second most important metallic material after steel. It is produced by melting and refining the aluminum ore *bauxite*.

Bauxite is extracted, ground and mixed with water, soda ash, and lime to dissolve and separate the alumina, which is an aluminum oxide (Fig. 3-32). To separate the aluminum from oxygen, the dry alumina is charged in electrolyte furnaces called *cells*. The cells, consisting of the anode and cathode, are filled with the molten electrolyte called *cryolite*. Alumina is charged into the molten cryolite and due to the direct current of the cell, it separates into the metallic aluminum and oxygen. The melted aluminum settles to the bottom of the cell, then is tapped and poured into

series containing chromium-nickel or chromium-nickel-manganese. The martensitic steels are primarily heat-resistant steels and can be hardened by heat-treating, but their corrosion resistance is relatively low; ferritic steels cannot be hardened by heat treatment but have good corrosion resistance and are adaptable to high-temperature uses. Austenitic steels are by far the most corrosion resistant of all stainless steels. They have high tensile strength and good ductility but cannot be hardened by heat treatment.

Tool steels (high alloy, high carbon) are produced primarily for tools to cut and shape other metallic and nonmetallic materials. These steels are extremely expensive and have good wear and corrosion resistance, high strength, and hardenability. Based on application, tool steels are classified as *general purpose, die,*

Table 3-3 **Effects of Principal Alloying Elements on the Properties of Steel**

Effects on Steel	Boron (B)	Carbon (C)	Chromium (Cr)	Cobalt (Co)	Columbium (Cb)	Copper (Cu)	Lead (Pb)	Manganese (Mn)	Molybdenum (Mo)	Nickel (Ni)	Silicon (Si)	Sulfur (S)	Titanium (Ti)	Tungsten (W)	Vanadium (V)
Improve:															
Abrasion resistance	X	X	X					X							
Corrosion resistance			X			X				X					
Dioxidizing capability									X		X				
Ductility										X					
Elastic limit			X						X		X				
Electrical and magnetic properties											X				
Fatigue resistance										X					X
Grain structure					X			X					X		X
Hardenability	X	X	X					X	X	X					X
Hardness	X	X	X						X						
High-temperature service properties					X				X	X			X	X	
Impact strength		X													X
Machinability							X					X			
Magnetic properties											X				
Shock resistance			X						X						X
Strength (tensile)	X	X	X		X			X	X	X			X	X	
Toughness		X						X	X	X				X	X
Wear resistance	X	X	X	X				X						X	
Workability			X					X							

Table 3-4 **The Aluminum Association (AA) Four-Digit Aluminum Designation System**

This diagram shows the significance of the numbers in the Aluminum Association standard designation for wrought alloys.

X X X X

This digit identifies alloy type.

This digit identifies alloy modification. Modifications were formerly indicated by letters. With the change to the new system, the letter is replaced by the digit corresponding to its position in the alphabet. For example, A17S becomes 2117. Zero is the original alloy.

These two digits identify the aluminum purity or the specific aluminum alloy. For alloys in use prior to the adoption of the four-digit system, the digits are the same as the numbers in the old designation. For example, 24S becomes 2024.

Type of Aluminum Alloy

	Number Group					Number Group		
Aluminum—99.00% minimum and greater	1	X	X	X	Magnesium....................	5	X	X	X
Copper	2	X	X	X	Magnesium and silicon.........	6	X	X	X
Manganese	3	X	X	X	Zinc.........................	7	X	X	X
Silicon.......................	4	X	X	X	Other element	8	X	X	X
					Unused series	9	X	X	X

SOURCE: Aluminum Company of America.

REFINING

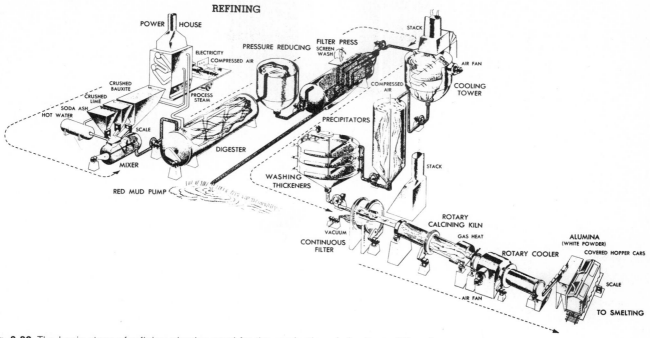

Fig. 3-32 The basic steps of refining alumina used for the production of aluminum. (*Alcoa.*)

ingots. The ingots are rolled into standard forms of mill products as described for the production of steel-mill products.

The problem of designating aluminum is even more complicated than that of steel. At least five classification systems exist in limited practice today. However, the Aluminum Association (AA) developed a four-digit classification system which is the most widely used and will be the universal system of the future. In this system the first digit denotes the alloy group or major alloy element, the second digit indicates the variation of alloy composition, and the last two digits identify the alloy or aluminum purity.

In addition to the four-digit number, the *temper* of aluminum is designated with a letter following the four-digit number. The letters O, F, H, T, and W are used. The letter O indicates annealed temper, T followed by one or more numbers indicates heat treatment, F indicates the as-fabricated condition, H followed by a number indicates cold-worked temper, and W indicates quenching of the alloy. Thus AA 1065-O designates an aluminum 99.65 percent pure with annealed temper; AA 1165-F designates an aluminum of 99.65 percent purity with some special control of the impurities in as-fabricated condition. Table 3-4 (see page 66) gives the AA classification system in detail.

Aluminum is one of the most versatile metallic materials in its properties and applications. Aluminum is light, relatively strong, workable, and soft. Its tensile strength is about 13,000 psi [910 kg/cm²] for commercially pure wrought aluminum but it can be substantially increased (up to 82,000 psi [5740 kg/cm²]) by heat treatment, alloying, and cold-working. It can be hardened by precipitation hardening. Table 3-5 summarizes some of the unique properties of aluminum. Aluminum has had so many different applications that it would be impossible to list them; Table 3-5 gives some of the most common uses.

Copper is the second most important nonferrous metallic material in tonnage used and variety of applications. Production of copper involves five steps: mining, milling, smelting, refining, and fabricating. Figure 3-33 on page 69 shows the step-by-step procedure for producing copper.

The most important properties of copper are high electrical and thermal conductivity, corrosion resistance, wear resistance, and ductility. Copper has a relatively low tensile strength (32,000 psi [2400 kg/cm²]), but it can be improved by alloying, heat treatment, and cold-working. Copper is used extensively in the electrical industry, chemical industry, and heating and air conditioning industries, to name only a few.

Magnesium is produced from *magnesium chloride* found in sea water and from magnesium ores. It is separated from magnesium ores by chemical reduction or it is separated from magnesium chloride by electrolysis.

Table 3-5 **Unique Properties of Aluminum**

Ease of Working	Aluminum can be fabricated economically by all the common processes.
Weight	Aluminum weighs far less than many other common industrial metals.
Tensile Strength	Some aluminum alloys have tensile strength higher than 80,000 psi.
Corrosion Resistance	Aluminum needs no protection in most ordinary environments.
Coatability	Oxide coatings of many colors and hard, wear-resisting surface finishes can be applied.
Electrical Properties	Pound for pound, aluminum has twice the conductance of copper. For equal sections the conductivity of aluminum is 62% that of copper.
Magnetic Properties	Aluminum is nonmagnetic. Thus electrical losses and disturbances are reduced in applications such as cable shielding and electronic equipment.
Heat Conduction	Because it transmits heat rapidly and efficiently, aluminum is widely used for kitchenware, automotive pistons, industrial equipment, and similar products.
Reflective Properties	Aluminum reflects both light and heat with high efficiency.
Low Elasticity	A low modulus of elasticity gives aluminum the ability to withstand considerable impact without deforming permanently.
Miscellaneous	Aluminum is nontoxic and odor-free; thus it is widely used in food-processing plants and home kitchens. It is non-sparking, so it is safe near inflammable and explosive substances.

SOURCE: Aluminum Company of America.

Magnesium has a relatively low tensile strength (14,000 psi [980 kg/cm^2]), but it is the lightest of all industrial metals. It is mostly used in the alloy form with aluminum, zinc, and manganese. As an alloy, magnesium has an excellent castability, especially for die casting relatively intricate parts.

Tin is produced from a tin oxide (mineral) called *cassiterite*, which is heated above the melting point in an inclined hearth furnace. The melted tin moves downward on the inclined hearth, leaving the unmelted impurities behind it. Among the most unique properties of tin are exceptional corrosion resistance, coating ability, low melting point, and castability. Because of its excellent corrosion resistance and coating ability, most tin is used to coat other metals, such as "tin cans," which are actually made of tin-coated steel strip. Due to its low melting point, tin is extensively used in the soldering of metals as alloys of tin and lead.

Zinc is produced from zinc oxide and zinc sulfide ores by melting (pyrometallurgical method) and by electrolytic processes. Zinc has a relatively low melting point, excellent electroplating characteristics, low shrinkage, good fluidity in casting, and good corrosion resistance. Its excellent casting characteristics make it desirable for die casting of automotive parts, household utensils, toys, and novelties. Zinc is available in slabs, strips, sheets, rod, and wire. Because of its corrosion resistance, it is used as the coating for galvanized steels and steel products.

Lead is produced from the lead sulfide (mineral) called *galena*. It is one of the oldest metals used. Among the unique properties of lead are density and weight, softness and malleability, low melting point, electrical conductivity, and corrosion resistance. Lead is of relatively low strength and has the lowest-strength-to-weight ratio of all metals. Because of its high density, it is extensively used as a shield against radiation. Because it is soft and has good self-lubricating characteristics, it is used as bearing metal in the automotive industry. It is also used in the chemical industry and in the soldering of metals. Caution is necessary in handling lead because it is highly toxic.

Among the precious metals used industrially are *gold*, *silver*, and *platinum*. Much of the quantity of these metals for industrial use are recovered during the production of such other nonferrous metals as copper, tin, zinc, and lead. Most of these metals are refined to increase their purity by the electrolytic process.

Gold and silver have been used for thousands of years and have played an important role in many civilizations. Unique properties of gold are corrosion resistance, workability, castability, softness, malleability, and ductility. Silver's unique properties are thermal and electrical conductivity, high reflectivity, good coating characteristics, and photosensitivity. Platinum is soft, ductile, and highly corrosion resistant. It is used as a catalyst in chemical processes. All these metals are used in jewelry, the petroleum and chemical industries, electrical industries, and the metal-finishing industry.

Production of Plastics

Plastics are among the fastest growing industrial materials in modern industry. The wide variety of plastics and their properties make them the most flexible of all materials in terms of application.

The basic plastic (polymeric) molecule is based on carbon. The raw materials for the production of plastics are petroleum and carbon-based gases. The basic resins are produced by chemically reacting *monomers*

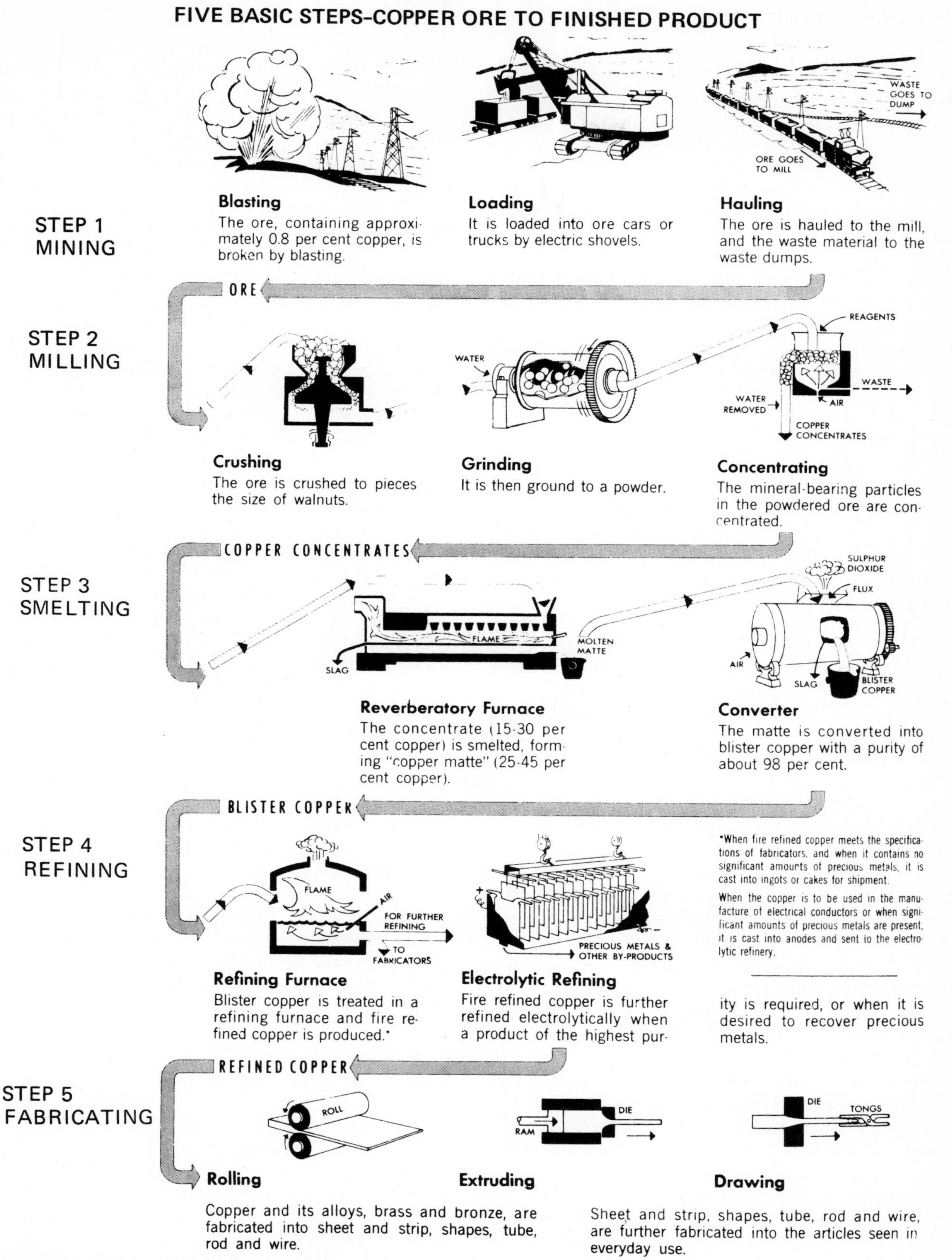
FIVE BASIC STEPS–COPPER ORE TO FINISHED PRODUCT
STEP 1
MINING
Blasting
The ore, containing approximately 0.8 per cent copper, is broken by blasting.
Loading
It is loaded into ore cars or trucks by electric shovels.
WASTE GOES TO DUMP
ORE GOES TO MILL
Hauling
The ore is hauled to the mill, and the waste material to the waste dumps.
ORE
STEP 2
MILLING
WATER
REAGENTS
WASTE
WATER REMOVED
AIR
COPPER CONCENTRATES
Crushing
The ore is crushed to pieces the size of walnuts.
Grinding
It is then ground to a powder.
Concentrating
The mineral-bearing particles in the powdered ore are concentrated.
COPPER CONCENTRATES
STEP 3
SMELTING
FLAME
SLAG
MOLTEN MATTE
SULPHUR DIOXIDE
FLUX
AIR
SLAG
BLISTER COPPER
Reverberatory Furnace
The concentrate (15-30 per cent copper) is smelted, forming "copper matte" (25-45 per cent copper).
Converter
The matte is converted into blister copper with a purity of about 98 per cent.
BLISTER COPPER
STEP 4
REFINING
FLAME
AIR
FOR FURTHER REFINING
TO FABRICATORS
PRECIOUS METALS & OTHER BY-PRODUCTS
*When fire refined copper meets the specifications of fabricators, and when it contains no significant amounts of precious metals, it is cast into ingots or cakes for shipment.
When the copper is to be used in the manufacture of electrical conductors or when significant amounts of precious metals are present, it is cast into anodes and sent to the electrolytic refinery.
Refining Furnace
Blister copper is treated in a refining furnace and fire refined copper is produced.*
Electrolytic Refining
Fire refined copper is further refined electrolytically when a product of the highest purity is required, or when it is desired to recover precious metals.
REFINED COPPER
STEP 5
FABRICATING
ROLL
DIE
RAM
DIE
TONGS
Rolling
Extruding
Drawing
Copper and its alloys, brass and bronze, are fabricated into sheet and strip, shapes, tube, rod and wire.
Sheet and strip, shapes, tube, rod and wire, are further fabricated into the articles seen in everyday use.

Fig. 3-33 Basic steps involved in the production of copper: mining to fabrication of copper mill products. (*Kennecott Copper Co.*)

to form long-chain molecules called *polymers*. This process is called *polymerization*. Polymerization is accomplished by two methods: *addition polymerization*, where two or more similar monomers directly react to form long-chain molecules, and *condensation polymerization*, where two or more dissimilar (different) monomers react to form long-chain molecules plus a byproduct (i.e., water).

Properties of Plastics. Plastics behave differently under load than any other industrial material. The principal reason is that the behavior of most thermoplastics under load is *viscoelastic* in nature; these materials have a combination of viscous and elastic responses to applied loads. Unlike metallic materials, which fail primarily by plastic deformation (slip) under load, plastics fail due to viscoelastic deformation. When a load is applied on a plastic, there is a combination of rapid elastic change (elastic response) and slow viscous change (viscous response). This viscoelastic deformation is due primarily to the long-chain molecular structure of plastics. Under load, the long chains move past one another, and the amount of movement is determined by the type of bond; plastics with weak bonds deform more easily than plastics with strong bonds. Therefore, most of the mechanical properties of plastics such as tensile, compressive, impact, and flexural strengths are determined by the viscoelastic behavior of plastics.

An understanding of the characteristic properties of plastics is important. Among the most important characteristic properties of plastics are low density, high corrosion resistance, and low thermal and electrical conductivity. Unlike metallic materials, which have relatively high densities, plastics are considered low-density materials. Therefore, plastics are extensively used in applications requiring low weight. In general, all plastics are of high corrosion resistance and are used in relatively corrosive environments. Plastics are used extensively as heat and electrical insulators because of their low thermal and electrical conductivity, and they are limited as to service temperature.

Classification of Plastics. Plastics are classified in two categories, thermosets and thermoplastics. *Thermoplastics* are those which can be remelted or softened by heat and can therefore be reused like metallic materials, which are remelted and recast. *Thermosets* are those plastics which cannot be remelted or softened after they have been solidified.

Commercially available plastics in the thermoset group are epoxy, melamine, phenolics, and urethane. *Epoxy* plastics are resistant to moisture, acids, and solvents. They are used for coatings, adhesives, electrical insulation, and many other applications. *Melamines* (amino plasts) are hard and strong, resistant to oils, and can be colored. They are used for buttons, laminates, tableware, and electrical wiring devices. *Phenolics* are hard and elastic, have good resistance to chemicals and good insulating properties. They are used for jewelry, toys, electrical insulators, and cabinets for radios and TV sets. *Urethanes* are relatively strong and tough, are resistant to moisture, and are good thermal and electrical insulators, and they have good weather resistance. They are used for sponges, toys, industrial tires, insulation and padding for furniture, and excellent varnishes.

Commercially available plastics in the thermoplastic group are many, but among the most commonly used are ABS (acrilonitrile-butadiene-styrene), acrylics, cellulosics, nylon, polystyrene, and vinyls, which include PVC (polyvinyl chloride).

The *ABS* plastics are relatively strong and resistant to weather. Typical applications of ABS are boats, automotive fenders, luggage, telephone components and helmets.

Acrylics are strong and rigid and have exceptionally good optical properties and good electrical and wear resistance. Typical applications are covers for lights, transparent shields and windows in place of glass, illuminated signs, and paint.

Cellulosics are tough, have good insulating properties, and can be easily colored. Typical applications are lamp shades, automotive accessories, pipe and tubing, drafting equipment, telephone sets, and toothbrushes.

Nylons are tough and elastic but have low resistance to certain acids and other chemicals. Typical applications are bushings, gears, washers, and mechanical components.

Polystyrenes are relatively rigid and hard, have low resistance to hydrocarbon solvents, good resistance to most inorganic chemicals, and are easily colored. Typical applications are food containers, utensils, toys, and furniture.

Vinyls are either rigid or flexible, are strong, and have good resistance to chemicals. Typical applications are pipes, insulation, food packaging, coatings, and tiles.

Processing of Plastics. The plastics industry consists of the plastic resin manufacturers, the processors, and the fabricators. The plastic materials manufacturers usually produce the plastic raw material in powders, granules, liquids, standard forms such as sheets, tubes, structural shapes, and laminates. The processors usually convert the raw material into complete products, and the fabricators usually finish the plastic products.

The most commonly used fabrication processes are injection molding, extrusion, compression molding, transfer molding, blow molding, thermoforming, rotational molding, calendering, reinforcing, laminating, foaming, coating, machining, joining, and casting (Fig. 3-34). Most of these processes are common to other types of materials and will be discussed in detail in later chapters of this book. Therefore, only a brief description of these processes will be offered in this section.

Injection molding is used mainly with thermoplastics although some thermosets are also used with this process. The plastic material, usually in granular or powder form, is heated in an injection chamber and forced by means of a reciprocating plunger or a screw. The semiliquid plastic material is injected into the cavity of the mold, producing the desired shape of the part. Figure 3-35 shows the principle of an injection-molding process. The part configuration is produced by the cavity of the mold or die. Therefore, the making of the mold or die is the most important phase of injection molding.

Extrusion is used mostly with thermoplastics. This process is also used with metallic and ceramic materials. In extrusion, the plastic material, usually in a powder or granular form, is heated in the extrusion chamber to a semiliquid form and forced in a continuous manner through a preshaped extrusion die. The cross-sectional configuration of the extruded part is determined by the shape of this die. Extrusion is

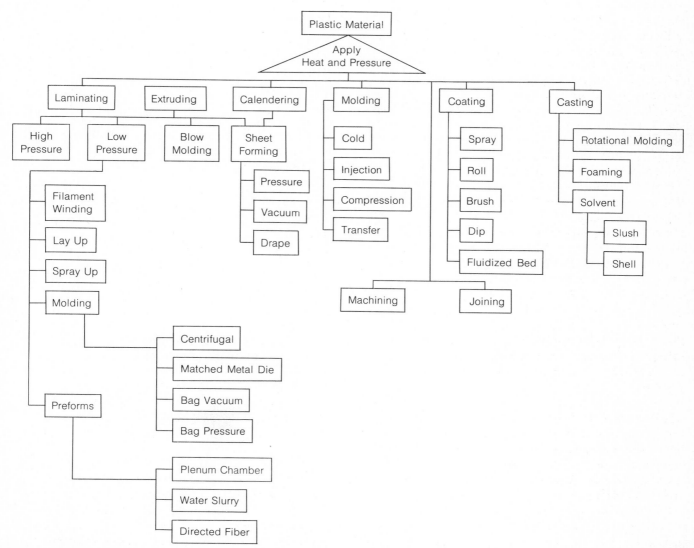

Fig. 3-34 Classification of common plastic processes.

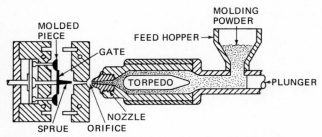

Fig. 3-35 Injection molding of plastics. (*The Society of the Plastics Industries, Inc.*)

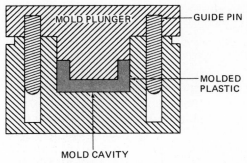

Fig. 3-37 Compression molding of plastic materials. (*The Society of the Plastics Industries, Inc.*)

used to produce such products as tubes, rods, sheets, films, pipe, and rope. Actually, extrusion is used to produce almost an unlimited variety of profiles. Extrusion is basically the same process as injection molding. The difference is that in extrusion the part configuration is generated by the extrusion die and not by the mold, as in the case of injection molding. Figure 3-36 shows the basic elements of an extrusion process.

Compression molding is used mostly for thermosets. The thermosetting plastic, usually in powder or granular form, is placed in a heated die, and the upper half of the die compresses the material which melts and fills the die cavity. After compression, the part solidifies (polymerizes or "cures"), the die is opened, and the part is removed. Compression pressures vary from 2000 to 10,000 psi [140 to 700 kg/cm²] depending on the size of the part and its configuration. Figure 3-37 shows the basic elements of a compression-molding process.

Transfer molding is the process used mostly with thermosets. Like compression molding, transfer molding employs both heat and pressure to form the part. The difference between compression and transfer molding is in the place where the plastic material is compressed. In the case of transfer molding, the plastic material in powder or granular form is heated and compressed outside of the die cavity and is then forced (transferred) as a semiliquid into the die cavity

through a sprue. After curing the plastic in the mold, the mold is opened and the part is removed. Transfer molding is used with parts of intricate design or multiple parts which require the material in the liquid form to be compressed to fill the cavities. Figure 3-38 shows the basic elements of transfer molding.

Blow molding is used mostly with thermoplastics. The heated plastic in the form of a parison is enclosed in the mold and air pressure is applied at the center of the parison forcing the hot plastic to fill the mold and assume its configuration. Then the mold is opened and the part is removed. Blow molding is extensively used to produce plastic containers of all sizes and configuration (Fig. 3-39).

Thermoforming is applied only to thermoplastics. The plastic material, usually in sheet form, is heated to a formable state and air pressure in the form of a vacuum (or other form) forces the hot sheet to cover the cavity of the mold and assume its configuration. Fig. 3-40 shows the basic elements of a thermoforming process using vacuum as the required force.

Rotational molding is applied to plastics to form hollow parts such as balls, squeeze bulbs, and hollow toys. The material is placed in the mold, which is rotated

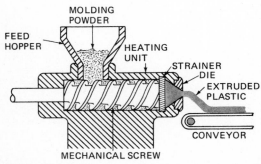

Fig. 3-36 Extrusion molding of plastics. (*The Society of the Plastics Industries, Inc.*)

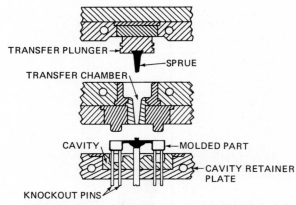

Fig. 3-38 Transfer molding of plastics. (*The Society of the Plastics Industries, Inc.*)

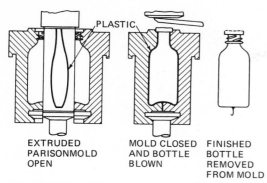

EXTRUDED PARISONMOLD OPEN

MOLD CLOSED AND BOTTLE BLOWN

FINISHED BOTTLE REMOVED FROM MOLD

Fig. 3-39 Blow molding of plastics. (*The Society of the Plastics Industries, Inc.*)

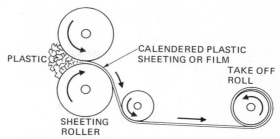

PLASTIC

CALENDERED PLASTIC SHEETING OR FILM

TAKE OFF ROLL

SHEETING ROLLER

Fig. 3-41 Calendering of plastic materials. (*The Society of the Plastics Industries, Inc.*)

both vertically and horizontally in an oven. When the material melts, it is forced by centrifugal force to take the shape of the mold. The mold is then cooled before it is opened, and the part is removed.

Calendering is used to coat textiles or paper with plastic materials and also to produce plastic sheets and films. The plastic is passed through revolving rollers and forced on the surface of the textile or paper sheet or forced to form a sheet. The thickness of the film is controlled by the opening between the revolving rollers. The rollers are heated to facilitate the easy flow of the plastic material. Figure 3-41 illustrates the calendering process.

Reinforcing and laminating are used to reinforce the strength of plastic sheets and/or to cover the surfaces of reinforcing materials such as cloth, glass fibers, wood, paper, and metals. The plastic material is forced by high pressure until the reinforcing material and the plastic adhere and form a strong laminate. The number of layers (sheets or plies) forming the laminate depends on the type of laminate required and the type of material used. Usually two or more layers of lamina or plies are used in laminates. The process

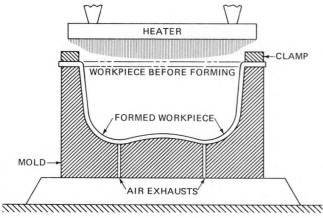

HEATER

CLAMP

WORKPIECE BEFORE FORMING

FORMED WORKPIECE

MOLD

AIR EXHAUSTS

Fig. 3-40 A schematic of thermoforming vacuum of plastic material.

is widely used to laminate table tops with plastic material, such as formica tops.

Casting of plastics is done by melting the material in a suitable container and pouring the molten plastic in a preshaped mold. After the plastic has partially cooled, the mold is opened and the part is removed. Casting of plastics differs from the molding processes discussed in the preceding paragraphs in that pressure is not applied. Casting of certain plastics can also be accomplished without heat; in those cases the plastic material is liquefied by solvents rather than heat.

Machining and *welding* of plastics will be discussed in detail in later chapters of this book.

Production of Ceramics

Ceramics have been used for centuries in a variety of applications. However, their industrial importance is of recent origin. Ceramics are used industrially because of their unique characteristics of high electrical resistance and high temperature resistance. It is their high heat resistance which makes ceramics the most important material for high temperature applications.

Ceramic materials consist of a wide variety of clay-based materials: glass, cements, concrete, and abrasives, to name only a few. In the manufacture of ceramic products, two basic steps are involved: first, the ceramic raw materials are produced and, second, the raw ceramic material is used to manufacture the finished ceramic products.

Clays. Most ceramic raw materials are produced from clay. *Clay* is a natural material extracted from the earth's crust and improved before it is used. The most common types of clays are kaolin, fireclay, feldspar, flint clay, grog, glacial clay, ball clay, and slip clay. *Kaolin* is produced as a white, fine clay and is used extensively for the manufacture of chinaware. *Flint clay* is produced from coal deposits and used as a fireclay. *Fireclay,* like other clays, is produced from natural deposits and used to manufacture fire-resist-

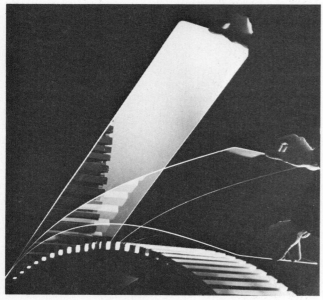

Fig. 3-42 Chemically treated and heat-treated glasses can be produced with desirable properties. (*Corning Glass Works.*)

ant products, such as firebricks used to line furnaces. *Ball clay* is produced from natural deposits and used to make whiteware. *Glacial clay* is produced from rocks of glacial deposits and used to manufacture

structural bricks. *Slip clay* has a moderate melting temperature and is therefore used for glazing of ceramic products. All clays are extracted from natural deposits, their impurities removed, ground into desirable powder form, grated, and packed for market use.

In most cases, the production of ceramic products requires three basic steps: (1) the clay is mixed with appropriate additives and water to a desirable consistency and the product is shaped, (2) the product is dried to increase its rigidity and strength for safe handling, and (3) the product in its dried state is fired in a firing furnace called the *kiln*. During firing, the clay reacts thermochemically to bond its particles, producing a strong, hard, and permanently formed product. Among the products produced by this method are structural bricks, firebricks or refractories, and whiteware.

Glass. Glass is classified as a ceramic even though its structure differs significantly from the normally crystalline structure of ceramics. Glass is produced by mixing *silica* with other minerals, milling them together and melting this mixture. The resulting liquid cools into a noncrystalline mass called glass.

Production of glass is a continuous process. The raw materials include silica (silicon dioxide), soda,

Fig. 3-43 Various glass products produced from different kinds of glasses. (*Corning Glass Works.*)

lime, manganese, alumina, boric oxide, lead oxide, and potash, depending on the type of glass. The mix is fed into the rear of a melting furnace called the *refractory tank,* then melted and gradually moved forward to the exit end of the furnace. The melted glass is moved from the furnace and processed manually or by machines into the desired products.

By varying the mix and the process, different glass having distinct characteristics can be produced. Glass varies in its properties more than any other material. There are soft and hard glasses (Fig. 3-42), strong and fragile glasses. However, glasses can be classified into six groups. Table 3-6 summarizes the six types of glasses. Glass products are produced by blowing, cast-

ing, drawing, pressing, and finishing processes that yield a great variety of glass products (Fig. 3-43).

Cement and Concrete. Cement and concrete are classified as ceramic materials even though their structure is quite different from the normally crystalline structure of ceramics. While most ceramics require heat to solidify them into hard and strong material, cements do not require heat. Based on the process of hardening, cements are classified into two groups: *hydraulic,* those which are hardened by water through the process of hydration, and *nonhydraulic,* those which harden in air through setting.

Cements are produced from siliceous material

Table 3-6 **Common Types of Glasses, Their Composition, Characteristics and Typical Applications**

Glass Type	Composition	Characteristics	Typical Applications
1. Soda-lime-silica	silica, soda, lime, magnesia, alumina	Most common of all glasses, accounting for nearly 90% of glass used in the world. Easy to melt and shape; inexpensive; poor thermal and chemical resistance.	Window glass, plate glass, containers, building blocks, light bulbs.
2. Borosilicate	silica, boric oxide, soda, alumina	Good resistance to thermal shock and corrosion, low thermal expansion. Hard to work because of high softening temperature.	Laboratory apparatus, piping, sealed beam headlamps, telescope mirror blanks, ovenware.
3. Lead-alkali-silicate	silica, lead oxide, soda, potash, lime	High index of refraction, good infrared transmission and electrical properties.	Crystal glass for tableware; thermometer and neon sign tubes; optical lenses.
4. Fused silica	silica	Simplest glass chemically; made directly from silica. Exceptional radiation resistance, optical qualities, and thermal shock resistance. Low thermal expansion, at a rate of 5 as compared to 90 for soda-lime-silica glass.	Lightweight, spaceborne telescope mirrors; space vehicle windows; laser beam reflectors.
5. 96% silica	silica, boric oxide, oxides	Made by removing the non-silicate ingredients from borosilicate glass. Expensive. Exceptional thermal properties and chemical resistance (can be heated cherry red and plunged into ice water without damage).	Missile nose cones, space vehicle windows, optical lenses, chemical glassware.
6. Aluminosilicate	silica, alumina, lime, magnesia, boric oxide	Outstanding thermal shock resistance, good electrical and chemical properties, very high softening temperatures.	High temperature thermometers, stove top cookware, laboratory apparatus.

SOURCE: Corning Glass Works.

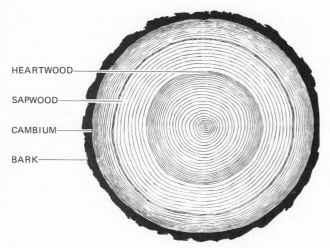

HEARTWOOD

SAPWOOD

CAMBIUM

BARK

Fig. 3-44 Cross section of a tree trunk showing the various types of wood structure.

(clays) which are burned together in a rotary kiln. After burning, the material is ground to a powder form which has the capacity to harden when mixed with water into a noncrystalline, hard, strong, and durable product. Although cements are used for many different applications, most cement is consumed in the production of concrete.

Concrete is a mixture of cement, which forms the binder, fine and course aggregates of sand and gravel, and water. These materials are proportioned,

mixed, and cast into an infinite variety of structural elements used in construction. Based on application and production, several types of concrete are available. Among these are reinforced concrete, prestressed concrete, air-entrained concrete, precast concrete, and asphalt concrete.

Production of Wood

Wood is the oldest building material known, and it has played an important role in industrial development. Wood is an organic material of plant origin, produced from trees which grow in the wild in almost every part of the world or are cultivated in certain parts of the world.

The structure of wood consists of long, hollow cells held together by a natural resin called *lignin.* A cross section of a tree reveals its structure. It consists of the outer corklike dry layer called the *outer bark,* the *inner bark* which is just under the outer bark, the *cambium,* the living layer of wood where growth takes place (Fig. 3-44). The wood cells arrange themselves into two distinct patterns, resulting in two different types of wood: the *hardwood* characteristic of the broadleaved trees (oaks and maples) and *softwood* characteristics of the cone-bearing trees (pines and firs).

Properties of Wood. Because of its unique properties, wood is considered among the most versatile ma-

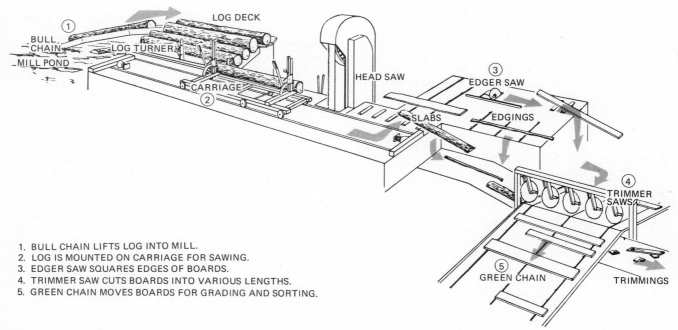

1. BULL CHAIN LIFTS LOG INTO MILL.
2. LOG IS MOUNTED ON CARRIAGE FOR SAWING.
3. EDGER SAW SQUARES EDGES OF BOARDS.
4. TRIMMER SAW CUTS BOARDS INTO VARIOUS LENGTHS.
5. GREEN CHAIN MOVES BOARDS FOR GRADING AND SORTING.

Fig. 3-45 Manufacture of lumber in a typical saw mill. (*American Forest Products Industries, Inc.*)

terials. One of the most unique characteristics of wood is its affinity for water which provides the moisture in wood. The moisture content of fresh cut wood can be up to 60 percent of its weight. Moisture affects the behavior of wood and its dimensional accuracy. Wood shrinks as its moisture is reduced or swells as its moisture is increased. These changes may result in warpage and distortion. Wood is relatively soft, is easy to work, and has good decorative features, but it has relatively low strength in certain directions and has relatively high strength-to-weight ratios. Although wood burns easily, its thermal insulating capacity is very good. It should be remembered, however, that all properties of wood are affected by moisture content and by the direction of the grain.

Classification of Wood. After the tree is cut, the log is processed in the lumber mill into marketable lumber (Fig. 3-45). The lumber is then graded in standard grades. Grading is based on the amount of defects present in the wood, with best grades having no defects. The hardwood is graded into *Firsts,* which is the highest grade; into *Seconds,* which combine the Firsts and Seconds; and into *Selects,* followed by *No. 1 Common, No. 2 Common, No. 3A Common,* and *No. 3B Common.* Softwood is graded in different ways by the various national associations. However, there is a general simple classification known as the American Lumber Standards. This classification includes three main classes: *yard* lumber, *structural* lumber, and *factory* or *shop* lumber.

In addition to lumber, wood is produced in *fiberboards, plywoods,* and *modified wood* which have certain unique characteristics not found in natural lumber. Among these characteristics are increased strength, resistance to splitting, dimensional stability, appearance, and workability. Wood is processed into finished products by machining, forming, joining, fastening, and finishing.

MISCELLANEOUS MATERIALS

In addition to the basic materials discussed in the preceding paragraphs, there is a host of other materials which can be classified as miscellaneous. These materials are extensively used in modern industry for many unique applications. Among the most commonly used miscellaneous materials are fibers and fabrics, leather, composites, lubricants and fuels.*

REVIEW QUESTIONS

3-1. Why is it important that we know about the nature and properties of materials in manufacturing?

3-2. Explain the electron shell theory as applied to materials.

3-3. Why are valence electrons the most important electrons in the study of materials?

3-4. Explain the various types of bonding of materials.

3-5. Define and explain the crystalline structure of metallic materials.

3-6. Explain the process and need for alloying metallic materials.

3-7. Why are equilibrium diagrams important in the study of metallic materials?

3-8. Outline the steps used to produce steel from iron ore.

3-9. Describe and compare the various steel-making methods.

3-10. Why is aluminum considered the most versatile of metallic materials?

3-11. Why do plastics behave differently under load than any other material?

3-12. Describe the various processes used to manufacture plastic products where heat is not used.

3-13. Describe the three basic steps normally used in the production of ceramic products.

3-14. What is the most unique industrial characteristic of ceramic material?

3-15. Compare wood and plywood in terms of strength, cost, and extent of application.

* For a detailed description of the production processes of these materials, the reader is referred to Chap. 7 of the textbook by Kazanas, Klien, and Lindbeck, *Technology of Industrial Materials,* Charles A. Bennett Co., Inc., Peoria, Ill.

CHAPTER 4

INTRODUCTION TO MATERIALS TESTING

Almost all known materials are used in some way by modern manufacturing technology. But to use the right material to improve the performance of equipment, machines and structures, and to identify defects, a knowledge of the way various materials behave under certain conditions, then, is necessary. The main purposes of *materials testing* are (1) to study the behavior of materials under specified conditions and (2) to identify defects in materials and products. It is the purpose of this chapter to present some of the *tests* and testing procedures most commonly used in modern manufacturing technology.

PREPARATION FOR TESTING

In general, tests (and testing procedures) used in manufacturing may be classified as either *destructive* or *nondestructive* tests. This classification is based on the condition of the specimen after testing. In destructive tests, the specimen is destroyed and cannot be used again. The main purpose of destructive tests is to determine the ultimate mechanical properties of materials under loads. Nondestructive tests do not destroy the specimen, and are used primarily to determine whether or not there are defects in materials and finished parts. These nondestructive tests are used in production lines for inspection purposes.

Specimens

In testing materials in manufacturing, the size, shape, and composition of the specimens used should be given special consideration. In many testing applications, unfinished *as-cast, as-cut,* or *as-machined* specimens are used. Other testing applications may require highly finished and carefully prepared specimens. For reliable results, a number of specimens are required for each test performed.

Testing Machines

Most tests are performed on commercially available testing equipment. However, there are cases where special tools and fixtures must be built in the testing laboratory if commercially made tools and fixtures are not available.

Materials testing requires that standard procedures be followed as much as possible to assure the reliability and validity of the tests. The American Society for Testing and Materials (ASTM) describes standardized tests and testing procedures which are published in *ASTM Standards.* Many manufacturers, however, may complement the ASTM standards with standards which are unique to their particular products.

Economic Efficiency of Testing

The economic efficiency of tests is especially critical in manufacturing where tests must be repeated many times in the course of production. In such cases, among the factors to be considered in the selection and adoption of a particular test are the following: (1) volume of production over which the test would be repeated, (2) rate at which the test must be repeated, (3) testing cost, and (4) capital investment in testing tools and equipment. Many tests can be applied in large-volume production, but others may require certain modifications to make them applicable. In some cases, standardized tests which are time-consuming to set up and perform cannot be applied in production situations requiring a high rate of test repetition.

Testing cost, labor, maintenance, capital investment in testing tools and equipment, and the means of preparing specimens should all be taken into account in selecting a particular test. In many production situations where automated or semiautomated testing equipment is needed, the cost is high and must be justified in terms of production output and product quality.

Testing and Inspection

Although the terms *testing* and *inspection* are sometimes used interchangeably, this book differentiates between them. *Testing* refers to the performance of tests for the purpose of collecting data to determine, among other things, the behavior of materials. *Inspection,* on the other hand, refers to the examination of materials (or products) to identify flaws, defects, or undesirable qualities present.

Both testing and inspection are used in manufacturing. Inspection usually makes use of nondestructive tests based on the analysis of visual, electrical, radiographic, magnetic, ultrasonic, and laser-beam phenomena. Testing requires such destructive tests as creep, torsion, shear, fatigue, impact, bending, hardness, and compression or tension.

Nature and Scope of Failure in Materials

To predict the behavior of materials under specified conditions, an understanding of the general ways materials fail is necessary. A material (or a product) has failed if it no longer performs the service functions for which it was intended.

Materials and products may fail under adverse conditions of *excessive loads or forces,* resulting in cracking, breaking, splitting, crushing, buckling, tearing, spalling, abrading, or wearing. Materials may fail under adverse environmental conditions such as chemical attack, resulting in rusting and corroding. They may also fail because of biological attacks or a combination of the factors named above.

One of the most important types of failures of materials and products is that induced by excessive load or force, resulting in *deformation* (dimensional change). Before a material fails under excessive load, it resists deformation. The resistance of a material to failure by deformation is its *strength,* and is due primarily to the interatomic forces of the material as explained in the preceding chapter.

The principal process of deformation, particularly in metallic materials, is called *slip.* Slip is the "sliding" or "gliding" of atomic planes in the space lattice of the material due to excessive force. The fracture induced by slip may occur suddenly, without an appreciable deformation, as in the case of such *brittle* materials as cast iron or concrete, or it can occur slowly, *preceded by a noticeable deformation,* as in the case of such *ductile* materials as rubber and soft steel.

DESTRUCTIVE TESTING

Destructive testing is the most commonly used testing of materials. The main purpose of destructive testing is to collect data relative to the behavior of materials under specified conditions. Among the common tests used in destructive testing are tensile-compression, bend, impact, fatigue, shear, and hardness.

Tensile-Compression Testing

Tensile and compression tests require that a specimen be subjected to uniaxial tension or compression loading until it fractures. In the case of tensile tests, the specimen is gripped at its two ends and pulled apart. In compression tests, the specimen is compressed perpendicularly until it fails (Fig. 4-1). In other words, in compression testing the specimen is subjected to exactly opposite conditions than in tensile testing. Therefore, tensile-compression testing is basically the same in terms of the behavior of the specimen and the procedures used in testing except that in tensile the specimen is pulled, whereas in compression it is compressed.

Most materials used in manufacturing may be tested in tensile and compression. Among the materials for which the ASTM has standardized tensile-compression tests are steel products, cement, concrete, rubber, wood, plastics, and insulating materials.

When performed according to recommended procedures, tensile and compression tests can provide data on basic properties of materials for the determination of their behavior under stress-strain loading conditions. *Stress* is the force per unit of area. It is measured in *pounds per square inch* (psi) in the U.S. customary system of measurements. In the SI metric

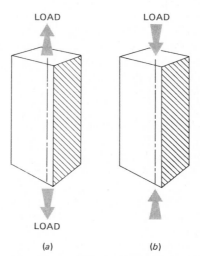

Fig. 4-1 Uniaxial tensile (a) and compression (b) tests.

system, stress is measured in *pascals,* units equivalent to *newtons per square meter. Strain* is the change in the original gauge length of a specimen due to stress and is measured in inches of change per inch of the original dimension (in/in), or millimeters per millimeter (mm/m) in the SI metric system.

A useful method for recording the behavior of a specimen under tensile load is to use a stress-strain curve or diagram.

The diagram can be plotted for any specimen under tensile testing. Figure 4-2 shows a stress-strain diagram for mild steel. The stress in 1000 psi [70 kg/cm²] is indicated on the vertical axis (ordinate) of the graph. Strain at each 1000 psi [70 kg/cm²] stress point is indicated on the horizontal axis (abscissa) of the graph.

Among the properties directly observed (and those that can be estimated) in tensile and compression testing are (1) proportional limit, (2) yield point and yield strength, (3) tensile (or ultimate) strength, (4) fracture (or breaking) strength, (5) ductility, elongation, and reduction of area, and (6) modulus of elasticity.

Proportional limit is the point at which the strain (deformation) is no longer proportional to the load. Up to the yield point, materials behave "elastically," meaning that there is no observable permanent deformation. Therefore, when the load is removed, theoretically the material returns to its original condi-

tion. This is often referred to as *Hooke's law of elasticity.* Beyond the yield point, materials behave "plastically" and will not return to the original condition when the load is removed.

Yield point is the point at which the strain increases without a corresponding increase in the load. Yield point is one of the properties that can be measured (observed) directly under tensile-compression testing. For ductile materials such as hot-rolled, low-carbon steels, the yield point is about 60 to 80 percent of the tensile strength of that material. There is no observable yield point for many brittle materials.

Offset yield strength under tensile testing is the stress at which a specified percent elongation occurs, point *c,* Fig. 4-2.

Tensile or *ultimate strength* is the maximum load a material can withstand when its tension (maximum load) is divided by its original cross-sectional area and expressed in U.S. customary units as:

$$\text{Tensile strength (psi)} = \frac{\text{maximum load (lb)}}{\text{original area (in}^2)}$$

Tensile strength is measured in *pounds per square inch* (psi) and should not be confused with load, which is measured in pounds. Tensile strength is applied to many design problems in manufacturing as one of the most important properties. For some materials, mostly the brittle ones, tensile strength coincides with the fracture strength, but for ductile materials it is different.

Fracture or *breaking strength* is the breaking load divided by the original area and expressed in U.S. customary units as:

$$\text{Breaking strength (psi)} = \frac{\text{breaking load (lb)}}{\text{original c.s. area (in}^2)}$$

A material is *ductile* if it can be deformed plastically more than 6 percent (permanent deformation) without fracture. During tensile testing the ductility of many materials can be determined in terms of percent of area reduction of the specimen. The percent of elongation can be computed if the original gauge length of a specimen is subtracted from the final length, divided by the original length, and multiplied by 100:

Elongation (percent)
$$= \frac{\text{final length (in)} - \text{original length (in)}}{\text{original length (in)}} \times 100$$

The percent of reduction of area can be estimated if the final cross-sectional area of the specimen is subtracted from the original, divided by the original area, and multiplied by 100:

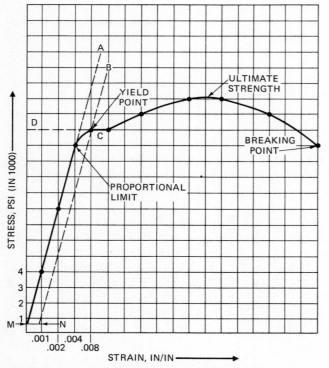

Fig. 4-2 A typical stress-strain curve for mild steel.

Reduction of area (percent)

$$= \frac{\text{original area (in}^2) - \text{final area (in}^2)}{\text{original area (in}^2)} \times 100$$

It is stated by Hooke's law that below the proportional limit the stress and strain are proportional, and the slope of this line is the *modulus of elasticity* or *Young's modulus.* Young's modulus is unique for each material and is measured in pounds per square inch (psi). Below the proportional limit, Young's modulus can be computed by dividing the unit stress by its corresponding unit strain. This is often used as a measure of the "stiffness" of a material.

Specimens in Tensile Testing. The accuracy of a test depends, among other things, on the correct selection, preparation, and use of the specimen. Specimens for some materials are specially prepared, while others may be used in the unfinished condition as cast, rolled, forged, extruded, and so on.

Specimens for many materials have been standardized by ASTM. For tensile testing of materials, the design of the *ends* and the *gauge length* are critical factors of the specimen (Fig. 4-3). For round specimens, three end designs are recommended: (1) shouldered end, (2) threaded end, and (3) plain end. Figure 4-4 shows a shouldered-end and a threaded-end specimen. For flat specimens, two designs are recommended: (1) plain end and (2) pin end. The most commonly employed gauge lengths are 2 in ±0.005° and 8 in ±0.010°.

In preparing compression specimens, the ratio of length to diameter (L/D) should be considered. Based on this ratio, the ASTM recommends three types of specimens for metallic materials: *short, medium,* and *long.* Short length specimens are those with length equal to $9/10$ the diameter (L = 0.9D). This type of specimen is recommended for bearing metals and those with similar applications. Medium length specimens are those with length equal to three diameters

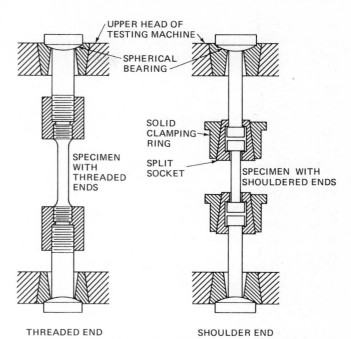

Fig. 4-4 Typical setups for tensile tests. *Left,* for threaded-end specimens; *right,* for shouldered-end specimens. *(ASTM.)*

(L = 3D). These specimens are usually employed to determine the strength of metallic material under compressive loads. Long length specimens are those with length equal to eight to ten diameters (L = 8D to 10D). Long specimens are used to determine the modulus of elasticity in tension of metallic materials. For concrete, the length of specimens recommended is equal to two diameters (L = 2D). Rectangular specimens of $2 \times 2 \times 8$ in [$5.08 \times 5.08 \times 20.3$ cm], parallel to the grain, are recommended for wood.

Testing Machines. Many different tensile testing machines have been developed and are commercially available. Some machines are light, are bench mounted, and are used for relatively low-strength materials such as plastics, wood, paper, textiles, and rubbers (Fig. 4-5). Other machines are heavy, are floor-mounted, and are used for high-strength materials such as steels and concrete. If the machine has been designed to perform more than one type of test, such as tensile, compression, and bending tests, it is referred to as a *universal testing machine* (Fig. 4-6).

The tensile-compression testing machine consists of the frame, which comprises the upper head, the lower (adjustable) head, the table with hydraulic ram, the hydraulic loading system, the measuring indicators, the grips, and the various controls. In tensile testing, the specimen is gripped between the upper head and the lower head. In compression and bend testing, the specimen is placed between the lower

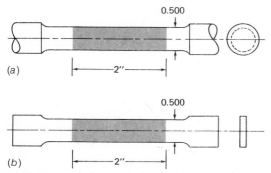

Fig. 4-3 Tensile-testing specimens for metallic materials: (a) round with 2-in gauge length, and (b) flat with 2-in gauge length.

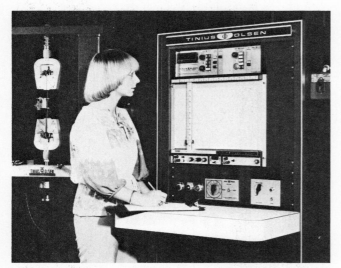

Fig. 4-5 Bench-mounted tensile-testing machine with recording system for testing plastics, ceramics, leather, fabrics, and other low-strength materials (*Tinius Olsen Testing Machine Co.*)

Fig. 4-6 A setup for a tensile test on a universal testing machine completely hydraulically operated. (*Tinius Olsen Testing Machine Co.*)

head and the table. The lower head can be adjusted in relation to either the upper head, for tensile tests, or the table, for compression and bending tests.

A testing machine must be carefully calibrated to assure accuracy of the test. Although all machines are factory calibrated, a check upon installation at the place of use and periodic checks thereafter are recommended. Several methods are recommended by the ASTM to calibrate testing machines. A practical (although not precise) method is comparison with specimens of known tensile or compressive strength.

Most testing machines are equipped with certain attachments designed to increase their versatility and accuracy. Some of these attachments are strain measuring extensometers; stress-strain recording systems

for plotting the stress-strain curve; automatic program attachments for automatic testing; center punches for setting off the gauge length marks on the specimen; percent gauge for direct indication of percentage of elongation; reduction of area gauge; attachments for compression, bending, shear, and hardness tests; and special grips. Some attachments come as standard equipment with the machine; others must be built or purchased independently.

Testing Procedures. Before a tensile-compression test begins, the procedure should be outlined. The ASTM recommends specific procedures for most standardized tests. These may be adapted to suit specific situations. To record data as accurately as possible, it is desirable that appropriate tables, graphs, and charts be made before the test begins.

Bending Tests

In manufacturing, sections of machines (shafts, machine bases, chassis, and so on) are subjected to bending. Bending refers to a type of loading which induces tensile stresses on one side of the part and compressive stresses on the other side. Figure 4-7 illustrates these stresses in bending. Bending tests are applied to brittle and ductile materials such as metals, concrete, wood, plastics, brick, building stones, and gypsum. Bending-test setups for several types of materials are shown in Fig. 4-8.

Bending tests are more frequently used to study the characteristics (cross-sectional design, length, and so on) of a specimen than to study its properties. However, among those properties studied by bending tests are modulus of fracture (breaking), modulus of elasticity, deflection, and ductility. The *modulus of fracture* refers to the maximum load-carrying capacity of a specimen before fracture.

Bending tests are performed on the universal tensile-testing machine. However, certain bending attachments are required to perform the tests accurately. Most bending tests have been standardized by ASTM, and dimensions of specimens and testing procedures are outlined in *ASTM Standards*.

Failures in Bending. Most failures in bending depend on the material, cross section, size and shape of

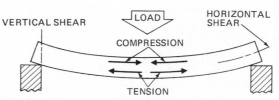

Fig. 4-7 Stresses in a simple structural member.

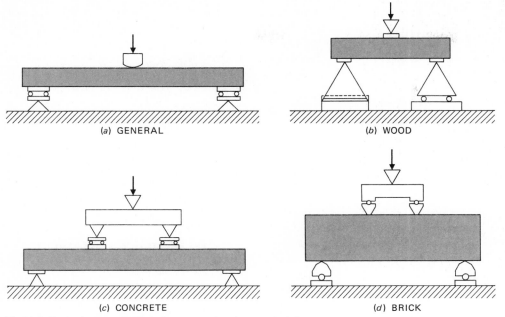

Fig. 4-8 Typical setups for bending tests of various materials.

the specimen, and on the type of loading. Three characteristic failures are yielding, buckling, and shearing. Yielding failures of the outermost fibers under tension are characteristic of massive solid specimens. Buckling of the fibers under compression is characteristic of either slender solid specimens or tubular specimens. Failures associated with shearing occur primarily at the points of support where a high concentration of force is present.

Specimens in Bending. On the basis of material, shape, and type of loading, a variety of bending specimens have been standardized by ASTM. Bricks are of $2 \times 2 \times 8$ in and are tested flat with a 7-in span, the distance between the two supports. Specimens for clear wood are made $2 \times 2 \times 30$ in and tested in a 28-in span. Concrete specimens are based on the size of aggregate used. For aggregates up to $1^1/_2$-in size, specimens are made $6 \times 6 \times 18$ in and for aggregates up to $2^1/_2$-in size, specimens are made $6 \times 6 \times 24$ in. Dimensions of wood and concrete specimens are kept to ± 0.01-in accuracy.

Bending tests for metals are used only as indicators of the extent to which metals resist fracture under bending. These tests are performed to determine changes in ductility due to manufacturing processes such as heat treating, rolling, forging, and welding.

Four basic bending tests for metallic materials have been standardized by ASTM. The *free bend test* is performed by using a plunger to force the specimen into an attachment of predetermined size. This test is used

extensively by the welding industry to check interior conditions of welds. The *semiguided bend test* is performed by applying force on the specimen at the point of bend. The *guided bend test* is performed by applying force to a specimen, which is shaped around a mandrel. The *wrap-around bend test* is performed by wrapping a specimen around a pin of predetermined diameter. Figure 4-9 illustrates these four tests.

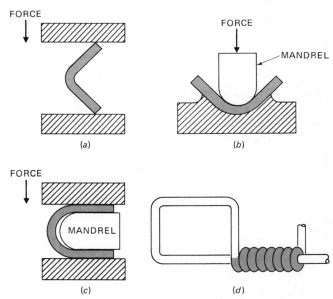

Fig. 4-9 Metal bending tests. (*a*) Free bend test; (*b*) guided bend test; (*c*) semiguided bend test; and (*d*) wraparound bend test.

Impact Testing

It has been observed that while certain materials demonstrate considerable resistance to static loads, they fracture easily under dynamic loads. A good example is glass, which can withstand considerable tensile and compressive stresses but shatters with a minor blow of a hammer. Other materials which behave in the same way are cast irons, rigid plastics, ceramics, and high-carbon steels. *Impact tests* are among the tests developed to study the behavior of materials under impact conditions. The main purpose of impact testing is to determine the *energy absorbed* by a specimen to cause fracture. The property associated with impact testing is *toughness,* which is defined as the capacity of a material to resist fracture under impact loads. Many mechanical parts and structures in manufacturing, such as bolts, shafts, hammers, anvils, and forging dies, are subjected to dynamic (impact) loads.

Impact tests are of three types: *torsion, tension,* and *beam,* based on the way the specimen is loaded. Torsion impact tests are rarely used (principally for tool steels). Tension impact tests are used to some extent but have not been standardized. Beam impact tests are the most common and are of two kinds, *Izod* and *Charpy* tests, which differ from each other in the method employed for loading the specimen. In the Izod test, the specimen is loaded as a cantilever-beam, supported on one end, and struck on the opposite end (Fig. 4-10*a*). In the Charpy test, the specimen is loaded as a simple-beam, supported on both ends, and struck on the side opposite of the notch (Fig. 4-10*b*). Both tests use notched specimens which are then fractured in flexure. Izod and Charpy tests have been standardized by ASTM for metallic materials and plastics. The Hatt-Turner impact test has been used for wood; however, the use of this test is not very common in manufacturing.

Various types of impact testing machines have been developed and are commercially available. However, the universal impact-testing machine, the machine

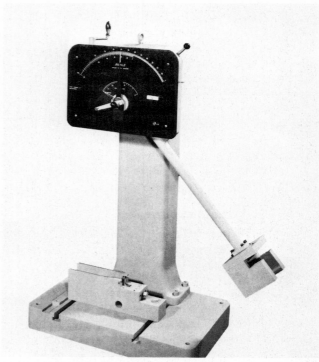

Fig. 4-11 A universal-impact testing machine. (*Courtesy of Riehle Testing Machines, Div. of Amelek, Inc.*)

which performs more than one type of impact test, is the most common (Fig. 4-11). Universal impact-testing machines are designed to fracture the specimen with a single blow and indicate directly the energy absorbed by the specimen in foot-pounds (ft-lb). Results of notched impact testing are rarely applied directly to solve design problems in manufacturing, unless the specimen and its testing condition exactly duplicate the actual part and its service condition. Figure 4-12 shows the testing of an automobile body as it may occur in actual service conditions. Impact tests are mainly used as control or acceptance tests of materials for certain applications in manufacturing. In the plastics industry, falling dart tests are used to determine impact properties.

Fatigue Testing

Machine parts sometimes fail because they are subjected to *repeated loads.* These failures are caused primarily by fatigue. *Fatigue* has been defined by ASTM as the process of progressive localized structural change occurring in a specimen subjected to conditions of fluctuating stresses and strains, which may result in cracks or complete fracture after a sufficient number of fluctuations. There is a maximum stress,

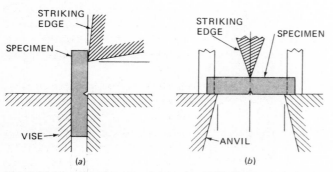

Fig. 4-10 Impact-tests setup. (*a*) Izod and (*b*) Charpy.

Fig. 4-12 Testing of an automobile body for roof integrity using a specialized roof crush testing machine which electronically records data on a graph. (*Fisher Body Division, General Motors.*)

Fig. 4-13 Universal-fatigue testing machine for materials and components testing. (*Schenck.*)

however, which can be applied for the design life of the part without causing fracture. This stress is called the *endurance limit*. The *fatigue strength* of a material is the stress at which the material fractures by fatigue. The repeated loading of a specimen is *cyclical* in nature; that is, in each repetition (cycle) the stress reaches its minimum and maximum limits.

Most fatigue fractures originate as microscopic cracks, causing high concentration of stresses, and gradually spread to the point of complete fracture of the part. Many of the microscopic cracks originate as a result of slip lines formed in the part from repeated stresses. Other sources which can start small cracks are imperfections such as surface scratches, notches, and inclusions. Surface defects are among the most critical imperfections because stresses on the surface are the highest.

Fatigue testing is employed to study the behavior of materials under repeated loads. Therefore, specimens are usually subjected to one or to a combination of the following types of repeated loads: *flexural, torsional,* and *axial.* On the basis of the type of loading, several kinds of fatigue-testing methods and machines have been developed for testing specimens or components. Figure 4-13 shows a universal fatigue-testing machine.

Shear Testing

In *shear testing*, the shear forces are acting parallel to an area but in opposite direction, producing "sliding" of one portion of the material past another, as in the case of a pair of scissors cutting a piece of paper. The purpose of shear testing is to determine the shear stress which may occur because of tension, compression, or torsion. There are many machine parts in manufacturing subjected partly or entirely to shear stresses. Some typical examples are riveted structures, bolts, pins, and shafts. Shear tests are usually performed with the universal tensile-testing machine, using special attachments.

Hardness Testing

Hardness is the result of a combination of various properties of a material, and cannot be considered a basic property. Thus, the hardness numbers derived by hardness testing depend on many of the properties of the material and on the conditions under which the testing is performed. None of the known hardness tests today can isolate the "hardness" of a material from such properties as its elasticity, ductility, brittleness, or toughness.

On the basis of the inability to isolate the hardness of a material from its physical properties and the need for establishing some means to measure the material's resistance to certain conditions, various tests have been developed. Some of the hardness tests (such as Brinell and Rockwell) measure the resistance of materials to penetration and others (such as the scleroscope test) measure the rebound or resilience energy.

To measure the resistance of materials to penetration (permanent deformation), the Brinell, Rockwell, Vickers, and microhardness tests are among those most commonly used. The basic principle of these tests is that a penetrator of known size is forced into the surface of a specimen by a predetermined load. The size of the penetration is determined by measuring its diameter (or its depth), and the hardness number is derived by dividing the load over the corresponding area of penetration.

Rebound hardness tests, such as the Scleroscope, are based on the amount of energy absorbed by a material under testing when a "hammer" of known weight is dropped on its surface from a predetermined height. The hardness numbers in these tests are based on the amount (height) of rebound of the hammer.

Scratch tests are based on the concept that a hard material will scratch a soft one when forced over its surface. Among the scratch tests are the Mohs and file tests.

Brinell Hardness Test. Various types of machines have been developed to perform the Brinell test. During the test, a hardened steel ball, usually 10 millimeters (mm) in diameter, is hydraulically forced with a load of 3000 kilograms (kg) into the surface of the specimen. Two other standard loads, 1500 kg and 500 kg, are available for softer materials, and sometimes a 5-mm ball is used for thin materials.

Brinell hardness numbers depend on the applied load, the length of time that the load is applied, and the size of the penetrator. The hardness numbers vary from 16 to 600 for the 10-mm ball and standard loads of 3000 kg, 1500 kg, and 500 kg—the smaller the numbers, the softer the material.

Because of the large area covered by the impression, Brinell tests are particularly suited for "porous" materials such as castings, forgings, and structural steel members. The tests are relatively easy to perform but time consuming.

Rockwell Hardness Test. The Rockwell hardness test is probably the most widely used hardness test in this country. These tests are performed on testing machines which are manually or automatically operated. Figure 4-14 shows a Rockwell testing machine in operation.

Rockwell testing machines have fifteen different scales (Table 4-1), with B and C being the most common. Two types of penetrators are used: the diamond spheroconical penetrator (Brale), with an angle

Fig. 4-14 A Rockwell testing-machine setup to test the hardness of a gear-tooth face. (*Wilson Mechanical Instruments Division, American Chain and Cable Co.*)

of 120°, and four steel-ball penetrators of $1/16$-, $1/8$-, $1/4$-, and $1/2$-in diameters, the $1/16$-in-diameter ball being the most commonly used.

One of the advantages of the Rockwell tests is that a wide range of hardness can be measured by using the different loads, scales, and penetrators. Also, the hardness numbers are read directly on a dial, eliminating possible errors in measurement but requiring proper test procedures and constant calibration. Employment of the minor load allows good contact of penetrator with specimen, thus eliminating errors in reading due to surface imperfection. Because a very small penetrator is used, tests can be performed on

Table 4-1 **Rockwell Hardness Scales, Loads, Penetrators, and Applications**

Common Scales	Major Loads, in kg	Dial Figures	Penetrators Employed	Applications on Commonly Used Materials
A	60	Black	Diamond-Brale	Very hard materials, such as cemented carbides, thin steel, and shallow case-hardened steel.
B	100	Red	$1/16$-in ball	Copper and copper alloys, soft steels, aluminum and aluminum alloys, malleable iron, etc.
C	150	Black	Diamond-Brale	Medium-hard steels, hard cast irons, pearlitic malleable iron, titanium, deep case-hardened steel, and other materials harder than B 100.
D	100	Black	Diamond-Brale	Thin steel, medium case-hardened steel, and pearlitic malleable iron.
E	100	Red	$1/8$-in ball	Cast iron, aluminum, and magnesium alloys; bearing metals.
F	60	Red	$1/16$-in ball	Annealed copper alloys, thin-soft sheet metals.
G	150	Red	$1/16$-in ball	Malleable irons, copper-nickel-zinc and cupro-nickel alloys. Upper limit G 92 to avoid possible flattening of ball.
H	60	Red	$1/8$-in ball	Aluminum, zinc, lead.
K	150	Red	$1/8$-in ball	
L	60	Red	$1/4$-in ball	
M	100	Red	$1/4$-in ball	Bearing metals and other very soft or
P	150	Red	$1/4$-in ball	thin materials. Use smallest ball and
R	60	Red	$1/2$-in ball	heaviest load that does not give
S	100	Red	$1/2$-in ball	anvil effect.
V	150	Red	$1/2$-in ball	

finished products where a small indentation on the surface is not objectionable.

In performing a hardness test, it is recommended that at least three readings be taken and averaged for better accuracy of results. Before a test is made, the testing machine is verified by employing standard testing blocks provided with the machine or materials of known hardness.

Rockwell Superficial-Hardness Test. Superficial-hardness testing has been developed to test *hard-thin* materials, such as razor blades and "case-hardened" parts which cannot be adequately tested by any of the regular tests described above. The Rockwell tester previously described is used for superficial tests, but different loads are applied. Three standard loads,

15 kg, 30 kg, and 45 kg, and a 3-kg minor load, are used with the superficial test. Five scales have been standardized, but only two, N and T, are extensively used (Table 4-2).

Vickers Hardness Test. The Vickers hardness test is a penetration test. A square pyramid penetrator is forced by a predetermined load into the surface of the specimen. The impression is measured optically, using a microscope, and the hardness numbers are derived by dividing the load over the area of penetration.

Unlike other penetration tests, Vickers tests provide accurate hardness measurements for both small and large parts. The load can be varied from 1 gram (g) to 120 kg. The load is applied for 15 seconds and

Table 4-2 **Rockwell Superficial-Hardness Scales and Major Loads**

Major Loads, in kg	Penetrators and Scales				
	Diamond Penetrator (N Scale)	$1/16$-in Ball (T Scale)	$1/8$-in Ball (W Scale)	$1/4$-in Ball (X Scale)	$1/2$-in Ball (Y Scale)
15	15N	15T	15W	15X	15Y
30	30N	30T	30W	30X	30Y
45	45N	45T	45W	45X	45Y

released automatically. Because of the small size of impressions, the use of diamond penetrators and varied loads, and the precise measurement of the impression, Vickers tests can be used for hard and thin materials. The test is relatively sensitive, and is considered mostly a surface test. It is used extensively in manufacturing.

Microhardness Tests. Although microhardness tests have not yet been standardized in this country, they are growing in importance and application. Microhardness tests are basically penetration tests, and the hardness numbers are derived by dividing the load over the area of the impression. A diamond pyramid penetrator (Knoop) with an included angle of 136° is used. The *long diagonal* of the impression is optically measured with a microscope.

Microhardness tests are still considered laboratory tests and are used primarily for research purposes. They can be used for testing extremely thin materials

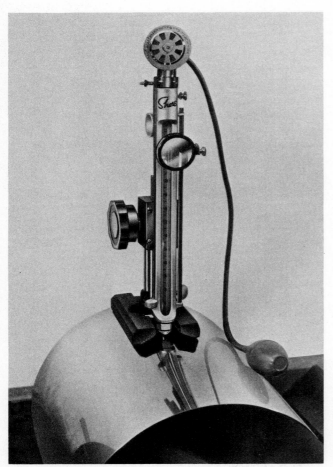

Fig. 4-15 A Shore scleroscope used to test the hardness of a machined shaft. (*Shore Manufacturing Company.*)

such as coatings, films, and foils and for testing individual grains or constituents in metallic materials.

Rebound (Dynamic) Hardness Test. In the rebound hardness test, a hammer (load) is dropped from a fixed height onto the surface of the specimen, making a small impression (Fig. 4-15). After the hammer hits the surface of the specimen, it rebounds to only a portion of its original height, because certain amounts of energy have been absorbed to make the small impression. The property measured is elastic resistance to penetration. Therefore, the hardness of a material under testing is proportional to the amount the hammer rebounds; the higher the hammer rebounds, the harder the material is. Table 4-3 provides a comparative summary of basic characteristics of hardness tests discussed in this chapter.

NONDESTRUCTIVE TESTING

Any test which does not destroy the specimen under testing can be classified as nondestructive. Nondestructive testing has been used to identify defects in materials and products and to determine properties and other characteristics of materials.

On the basis of the principle of operation, most nondestructive tests can be classified as: (1) visual, (2) radiographic, (3) ultrasonic, (4) magnetic, and (5) electrical tests.

Visual-Examination Test

Visual-examination testing is one of the oldest and most widely employed inspection methods. These tests are used primarily to identify surface defects (cracks, porosity, and so on) induced by such manufacturing processes as casting, welding, rolling, forging, machining, and heat treating. Visual tests are relatively inexpensive, simple, easy, and fast to perform and are very flexible.

To improve vision, the surface under inspection must be clean and well illuminated. Sometimes liquid penetrants are used which penetrate the defect and makes it more visible. Among the most common optical aids employed are various types of portable magnifying lenses, optical comparators, and microscopes. Some microscopes are equipped with polaroid cameras for instant photography.

Liquid-Penetrant Test. Liquid-penetrant testing is basically visual, using various types of high-visibility (fluorescent or red) liquids over the inspected area to increase the visibility of any defects (Fig. 4-16). The liquids may be applied to the specimen by dipping,

Table 4-3 **Summary of Basic Characteristics of Common Hardness Tests**

Characteristics	Brinell	Rockwell	Vickers	Microhardness	Scleroscope
Penetrators used	10-mm ball (5 mm for thin mats)	120° Brale and $1/16$-, $1/8$-, $1/4$-, and $1/2$-in balls	136° Pyramid	136° Knoop	A small hammer dropped from a predetermined height
Applied loads	3000 kg, 1500 kg, and 500 kg	For regular—10 kg minor load; 60 kg, 100 kg, and 150 kg major loads. For superficial—3 kg minor; 15 kg, 30 kg, and 45 kg major.	1 gm to 120 kg— most common 5 kg to 30 kg	1 gm to 1 kg	
Hardness scales used	One scale	For regular—15 scales are used, but B and C are the most common. For superficial—5 scales, but N and T are the most common.	One scale	One scale	One scale
Hardness numbers	Derived by dividing load by size of impression	Derived by dividing load by size of impression	Derived by dividing load by size of impression	Derived by dividing load by size of impression	Derived on the basis of rebound on heat-treated steel
Major applications	Usually for large heavy parts such as structural steel, heavy castings, large forgings, etc.	Usually for small finished metallic parts, such as sheet metal, wire, forming tools, cutting tools, gears, valves, etc. Also plastic parts or sheets.	Mostly the same applications as Rockwell plus for highly finished parts and thin sections down to 0.005 in	Usually for extremely thin and hard materials such as glasses, plated surfaces, coatings, foils, individual constituents and grains of mats.	Used for almost any application for small and large parts, for thin and thick parts, for metallic and nonmetallic mats.
Surface condition of specimen	Relatively clean and flat	Clean and smooth surface required	Clean and smooth surface required	Very clean, smooth, flat, and often polished surface required	Clean and smooth surface required

Fig. 4-16 Fluorescent Penetrant Test indicating porosity on aluminum missile component. (*Magnaflux Corp.*)

spraying, or brushing. Because of capillary action of the surface defect, the liquid is pulled into the cavity of the defect. Excess liquid is washed off, but the liquid in the cavity is not removed and thus makes it more apparent for inspection purposes.

One of the oldest liquid-penetrant methods is the *oil-whiting test* employed to locate surface cracks otherwise difficult to detect. The most recent liquid-testing aids are the commercially available liquid penetrants sold under such trade names as *Zyglo, Spotcheck, Met-L-Check, Dy-Check,* and *Dyeline.* These tests can be applied to any nonporous material, such as plastics, metals, and certain kinds of ceramics. However, the limitation imposed by these tests is that only surface defects can be located, and there is no means of determining the depth of the defects, and the surface must be *absolutely clean.*

Radiographic Testing

Industrial radiographic tests are basically photographic processes that employ penetrating radiation instead of visible light. These tests are based on short-wave radiation which penetrates solid materials such as metals, plastics, and ceramics. As radiation passes through the tested specimen, part of it is absorbed by the specimen and the rest passes through to make the photograph on the film. The "amount" absorbed depends, among other things, on the density of the specimen. The photograph produced on the film is the result of differences in radiation absorption, which is due to differences in the density of various portions of the specimen. If a defective steel specimen is tested, a different "amount" of radiation will be absorbed by the defect from that absorbed by the rest of the specimen. This will show on the film as a darker or lighter area, depending on the comparative densities of the defect and the specimen.

The two most common sources of radiation employed in industrial radiography are X-ray and gamma-ray radiations. The difference between the two is in the source of radiation. In the X-ray source, the radiation is supplied by an X-ray tube which is housed in the X-ray machine. In the gamma-ray source, the radiation is supplied by a radioisotope generated in a nuclear reactor and housed in the gamma-ray camera.

The X-ray tests are employed extensively by the welding, foundry, and construction industries. In welding, X-ray tests are considered among the most reliable means of locating defective welds. The most common welding defects tested for are inadequate penetration in welding, porosity, incomplete fusion, undercutting, and cracks. In the foundry industry, X-ray methods are used to test castings for such defects as inclusions, porosity, blowholes, cracks, and shrinkage. In the construction industry, these tests are used primarily to check the welded steel structure of buildings, bridges, and so forth. Radiography also provides a permanent record of the quality of a part in service. Suppliers to the military and commercial aircraft fields must maintain these records for the life of the part.

Gamma rays result from disintegration of natural or manufactured radioactive materials. Among the most important radioactive materials used are radium, iridium, and cobalt. Gamma-ray cameras are small, portable, low-cost, and can be used much more conveniently than X-ray machines. Both X-ray and gamma-ray methods are used for the same applications.

One of the problems associated with gamma rays, and to a lesser extent with X rays, is the safety hazard. Like other materials, the human body is penetrated by and absorbs radiation. Beyond a certain level, the amount absorbed becomes fatal. Among the various protective means developed against radiation are shielding with lead and controlling the length of exposure and distance from the source.

In addition to the basic radiographic methods discussed above, certain related methods have been developed in recent years. Among those methods are *fluoroscopy, xeroradiography,* and *electron-neutron radiography.*

Ultrasonic Testing

High-frequency sound waves (vibrational waves) having frequencies above the range of normal human hearing are called *ultrasonic.* The most outstanding properties of these waves, which make them desirable for testing purposes are (1) their essentially straight-line travel and (2) their capability of being transmitted by all materials. Ultrasonic testing is based on use of these waves.

In testing a specimen, a beam of ultrasonic waves (vibrations) is generated by a source (usually a piezoelectric crystal) and sent through the specimen. A defect (discontinuity) present in the specimen serves as a barrier to the ultrasonic waves and reflects some or all of them back to the source. The reflected waves are received by the testing instrument and are converted into electric energy. Electrical pulsations are then used to measure the amplitude of the waves and the time of travel through the specimen. Some of the waves, however, are reflected from the opposite side of the specimen and are used as reference location to measure the total length of the specimen.

Ultrasonic tests are employed to determine the size, shape, and location of such surface and subsurface defects as cracks, inclusions, and blow-holes. They are used to measure physical characteristics of materials such as thickness. They are also used to determine differences in the structure and properties of materials. Among the materials ultrasonically tested are metals, plastics, and ceramics.

Magnetic-Analysis Testing

Magnetic-analysis testing is based on the principle that the magnetic characteristics of a material are related to its structural composition. Therefore, changes in the magnetic characteristics of a material would result from changes in its composition. By observing the variations in the magnetic characteristics, any variations in composition due to defects can be detected. Iron filings are used for observing the discontinuities in the magnetic flux of a part.

When a specimen with no defects is magnetized, the iron filings are not crowded at any particular spot but are dispersed along the path of the magnetic field. If the specimen has any such surface or subsurface defects as blowholes, cracks, or inclusions, the path of the magnetic field is distorted because the defect has a different composition, and therefore, different magnetic characteristics. When the magnetic field is distorted, some of the particles of the magnetic flux are crowded at the point of the defect, and the defect acts as a local magnet that attracts and holds the iron filings at that spot.

Most of the particles used are made from black magnetic iron oxide ground to fine particles (100 mesh sieve), then coated with fluorescent material to increase its visibility under black light. The particles are applied by spraying or dipping, or automatically, either as dry powder (dry method) or as water-solid suspension (wet method).

Various types of equipment to perform magnetic testing have been developed and are commercially available as production-line equipment, laboratory-testing equipment, and portable units for field testing. Figure 4-17 shows a typical Magnaflux machine used for magnetic-analysis testing.

Some of the advantages of magnetic-analysis tests are that they can be used for locating surface and subsurface defects, are relatively simple and of low cost, and, with portable units, can be used almost anywhere. The basic limitation is that they can be used *only* for testing magnetic materials such as ferrous metals and alloys and parts must be demagnetized.

Electrical-Analysis Testing

Eddy-current testing is the most notable development of electrical-analysis testing. Unlike other nondestructive tests, which are used primarily for detection of defects, eddy-current tests are employed to: (1) detect such defects as cracks, voids, and inclusions, (2) evaluate and measure such properties as electrical conductivity and modulus of elasticity, and (3) measure such characteristics as hardness, alloy composition, and thickness of materials.

Eddy-current tests are based on the principle that a coil carrying alternating current induces an alternating current—an eddy current—of similar characteristics in the specimen. The induced eddy current is affected by variations in homogeneity, mass, and conductivity of the specimen. As the induced eddy current "flows" through the specimen and encounters a defect, such as a crack or a void, it is changed because it is forced to detour around the defect. As it detours around the defect, it is compressed, delayed, and weakened. The change due to the defect is re-

Fig. 4-17 A typical Magnaflux unit for magnetic-analysis testing featuring infinitely variable and self-regulating current control. Also features timer to control duration of magnetizing cycle. (*Magnaflux Corp.*)

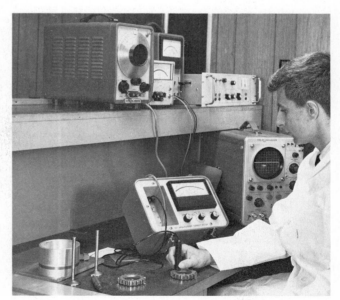

Fig. 4-18 A general purpose eddy-current testing instrument used for testing the composition of metallic materials, heat-treatment condition, surface and subsurface defects, sorting of alloys, thickness measurement, and coating measurement. (*F. W. Bell, Inc.*)

flected in the electrical characteristics of the test coil. By careful study and measurement of the changes caused by differences in the properties of the specimen, the defects can be located.

Eddy-current testing equipment has been developed, and units are commercially available for various applications. Among the most widely used types of equipment are those for locating defects such as cracks, voids, inclusions, seams, and laps; those for sorting parts according to alloy composition, temper, and electrical conductivity during manufacture in automated production lines; and those for measuring thickness and other characteristics of materials in manufacturing (see Fig. 4-18 on page 91).

OTHER TESTING METHODS

Many other testing methods and testing machines have been developed and are used mostly for materials research purposes. One of these methods is the spectroanalysis technique used to determine the composition of materials. Figure 4-19 illustrates a modern spectrometer used for advanced materials research

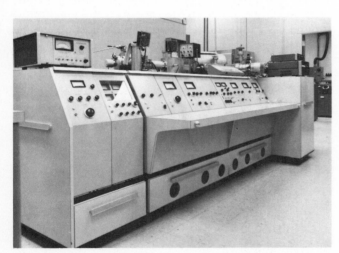

Fig. 4-19 A modern spectrometer used by NASA for testing the composition of materials. (*NASA.*)

and development. However, most of these methods are extremely complicated and of limited direct use in manufacturing and, therefore, beyond the scope of this text.

REVIEW QUESTIONS

4-1. Explain the difference between inspection and testing.

4-2. Differentiate between destructive and nondestructive testing.

4-3. What properties of materials can be measured by testing?

4-4. Explain why materials in service should not be loaded beyond their proportional limit.

4-5. Explain the difference between stress and strain.

4-6. What properties can be measured by using the tensile test?

4-7. Explain the mechanism of metallic materials exposed to tensile stresses.

4-8. Define and explain impact strength, endurance limit, and flexural strength.

4-9. Define hardness of materials and describe some of the most commonly used methods to test this property.

4-10. Define nondestructive testing and state some of the advantages and limitations.

4-11. Compare magnetic analysis testing with electrical analysis testing in terms of cost, extent of application, and importance.

4-12. Explain the principle of eddy current used in electrical-analysis testing.

4-13. Compare the Brinell with the Rockwell hardness testing in terms of cost and extent of application.

CHAPTER 5

PRINCIPLES OF CASTING AND MOLDING TECHNOLOGY

Casting is one of the oldest metalworking processes known. Metals have been cast for over 5000 years. However, the ability to cast iron and steel has been limited, and the oldest known casting of an iron object is an ornamental lion cast in China about 500 years A.D.

Casting is a process of giving a shape to an object by forcing liquid material into a formed hole or cavity called the *mold* and allowing the liquid to solidify. As the material solidifies in the shaped cavity, it retains the desired shape. The mold is then removed, leaving the solid shaped object.

Molding is similar to casting. A shaped mold is used to impart the shape. However, the material is not a liquid but is only in a softened or "plastic" state. The material is forced into the mold under pressure and allowed to solidify. The mold is then removed to expose the shaped object.

Almost every home and office has numerous objects made by casting or molding. The average automobile employs a great variety of castings of different materials made by different casting processes. Many different materials may be cast and the casting processes take many different forms, sizes, and variations. Examples range from the towering industrial machines cast from iron and steel to tiny toys made of molded plastics. In between these extremes are such examples as ceramic vases, glass bottles, electronic components, and a multitude of metallic items from gold rings and silverware to automobile parts in an endless variety of metal alloys, ceramics, plastics, glass, concrete, and other materials.

CLASSIFICATION OF CASTINGS

To describe a complex variety of processes, some system of description is necessary. As shown in Tables 5-1 and 5-2, some processes can be used for different types of materials, while several different processes may often be used to produce the same object. To provide a basis for understanding the various processes, a knowledge of classification and categories of casting processes is essential.

Casting processes are first categorized by the manner in which the materials are forced into the mold cavity. The two basic forces are gravity and pressure systems.

The second classification of casting processes is by the mold material. The mold can be made of sand and is destroyed when the object is removed. This is called *sand casting* and is normally used only for metals. However, there are many variations of the sand processes. Each variation has a certain advantage and disadvantage relating to accuracy, cost, and the types of metals that can be cast. Other types of materials such as plaster and ceramics are also used for the destructible molds similar to those used for sand casting. Molds can also be made of permanent materials so that the mold may be reused.

Plaster molds used for making small statues and figurines in ceramics and porcelain are examples of reusable molds using the force of gravity. Complex pressure-fed permanent molds made of metal are commonly used for making such items as automobile carburetors, plastic toys, or lead pipe. The molding material is forced into a preshaped permanent mold under pressure. After the object has been cast, the mold is opened and the object is removed, allowing the mold to be reclosed for immediate reuse.

Casting processes are also categorized by the material being cast. A *foundry,* for example, is a shop or factory where metal castings are made. Figure 5-1 shows an artist's concept of a completely automated foundry. An identical molding process used to cast plastics would occur in a *casting shop* and not a foundry. Shops specializing in casting iron objects

Table 5-1 **Classification of Casting Processes**

Category	Process	Product Material	Mold Material	Mold Class (Unit or Permanent)	Typical Product
Gravity	Lost wax (investment)	Metals	Sand or plaster	Unit	Statues, turbine parts, transmission impellers
	Pit molding	Iron and steel	Green sand	Unit	Large industrial machines, turbines, and generators
	Floor molding	Iron and steel	Green sand	Unit	Medium to large industrial equipment, pumps, engines
	Bench molding	Iron and steel nonferrous metals	Green sand	Unit	Engine parts, plaques, hardware
	Shell molding	Ferrous and nonferrous metals	Cured sand	Unit	Semiprecision cast parts—gears, engine parts
	Ceramics molding	Ferrous and nonferrous metals	Cured sand	Unit	Precision parts, dies, rotor blades
	Rotational molding	Plastics, concrete, metals	Metal	Permanent	Pipe, trash containers, balls, toys
	Pouring	Concrete	Steel or wood	(1) Unit (2) Permanent	(1) Building construction (2) Pipe, blocks, steps
	Slush and slip molding	(1) Nonferrous metals (2) Ceramics	(1) Iron and steel (2) Plaster	Permanent	(1) Toys, small figures, and statuary, (2) Vases, trays, dishes, etc.
	Dip molding	Plastic, metal, rubber	Metal	Permanent	Insulated tool handles, toy ballons, housings
Pressure	(1) Centrifugal investment	Metal	Plaster	Unit	(1) Custom jewelry, intricate machine parts
	(2) Sand	Metal	Sand	Unit or permanent	(2) Intricate machine parts, pipe, housings
	(3) Centrifuge	Metal	Sand	Unit	(3) Small machine parts
	(1) Die casting	Nonferrous metals, glass	Iron and steel	Permanent	Sewing machine cases, door handles, carburetors
	(2) Injection molding	Plastics, rubber	Metal	Permanent	Toys, molded rubber handles, eating utensils, auto interior trim, "foamed" styrene products
	Blow molding	Plastics, glass	Iron and steel	Permanent	Glass containers and plastic containers
	Continuous	Metal (steel, copper, and aluminum alloys)	None or graphite "guides"	Permanent	Wire, rod, and intricately shaped rods
	Compression molding	Thermosetting plastics	Iron and steel	Permanent	Electrical construction hardware, gears, insulators
	Transfer molding	Thermosetting plastics	Iron and steel	Permanent	Small intricate parts, transistor sockets, etc.
	Lay-up molding	Fiberglass plastics	Iron and steel	Permanent	Boat hulls, machine covers, chairs
	Vacuum forming	Thermoplastics	Metal	Permanent	Packaging "blisters," small machine housings (electric drills, etc.), auto trim
	Straight-pressure molding	Thermoplastics	Metal	Permanent	Packaging "blisters," small machine housings (electric drills, etc.), auto trim
	Free blowing	Thermoplastic	None	None	Airplane canopies, skylight domes

would be called *iron foundries,* where a shop utilizing similar processes for making specific items from plastics might be called a *plastics speciality shop.* Almost any material that can be liquified may be shaped by the casting process. Also included are materials which, although not liquid, are given some degree of fluidity, such as granular plastic materials that can be later bonded or solidified by heat or chemical action. Ma-

Table 5-2 **Comparison of Casting Methods*** (approximate)

	Green Sand Casting	Permanent Mold Casting	Die Casting	Sand–Shell CO₂–Core Casting	Ceramic & Investment Casting
Relative cost in quantity	Low	Low	Lowest	Medium high	Highest
Relative cost for small number	Lowest	High	Highest	Medium high	Medium
Permissible weight of casting	Unlimited	100 lb	30 lb	$\dfrac{\text{Shell oz–250 lb}}{CO_2 \ ^1/_2 \text{ lb to tons}}$	Oz to 100 lb
Thinnest section castable, inches	$^1/_{10}$	$^1/_8$	$^1/_{32}$	$^1/_{10}$	$^1/_{16}$
Typical dimensional tolerance, inches	0.012	0.03	0.01	0.010	0.01
Relative surface finish	Fair to good	Good	Best	$\dfrac{\text{Shell good}}{CO_2 \text{ fair}}$	Very good
Relative mechanical properties	Good	Good	Very good	Good	Fair
Relative ease of casting complex design	Fair to good	Fair	Good	Good	Best
Relative ease of changing design in production	Best	Poor	Poorest	Fair	Fair
Range of alloys that can be cast	Unlimited	Copper base and lower melting preferable	Aluminum base and lower melting preferable	Unlimited	Limited to unlimited

* Some metals cast easier than others.
SOURCE: Courtesy American Colloid Company.

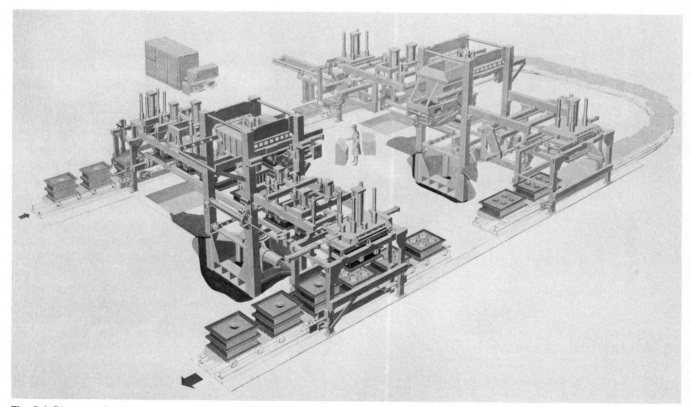

Fig. 5-1 Diagram of a high-production automated foundry. (*The Osborn Mfg. Co.*)

terial commonly shaped by casting processes include plastics, glass, ceramics and porcelain, metals, and even paper products.

Molds can also be characterized by whether they are reused or whether they are destroyed during removal of the cast object. Both types of processes are extensively used in industry. When the mold is regarded as permanent, it is often referred to as a *die*. When the mold is a unit that is destroyed in the removal of the object, it is referred to as a *mold*. Permanent molds are generally made from durable materials such as iron or steel. Permanent molds for plastics may be made of softer, more easily worked metals such as aluminum or brass.

Molding processes are also used in manufacturing material shapes for the construction industry. Concrete, for example, is a material that must be formed by a mold. Concrete molds may be constructed on the site for such things as basements, foundations, supports, sidewalks, and other similar items. However, permanent molds are used in casting various standardized concrete objects such as precast concrete construction beams, precast concrete brick and paving blocks, steps, large-diameter drain pipes, and other similar items used in the construction industry.

A third form of concrete molding is called *slip forming* and is typically used in the construction of highways and drainage ditches. A machine known as a *slip-forming machine* lays the concrete in place and smooths it into the desired shape as it lays the concrete. This is like a continuous-forming process in which the form is moved or slipped forward at a slow rate so that the machine leaves behind a fully formed section of concrete. As long as concrete is fed into the machine and is guided, it lays a smooth, continuous strip of highway or drainage ditch.

To illustrate the extensive use of casting processes, iron- and steel-cast products alone make up the sixth largest industrial grouping. This does not include the casting of nonferrous metals nor does it include casting of other materials such as glass, ceramics, plastics, or concrete. The casting of iron and steel products by various sand-casting processes form the largest portion of all castings produced. This is closely followed by casting nonferrous metals by die-casting and sand-casting processes and the casting or molding of plastics.

SELECTION AND ALLOYING OF METALS FOR CASTING

A great many variations in properties exist in metals having the same generic name, so that it is insufficient to specify simply "iron," "steel," "aluminum," and so

on. The proper selection and specification of metals, including the alloying elements, is necessary in order to ensure that the castings will have the particular properties needed in finished parts depending on their function in assembled products. Also, pure metals are often difficult to cast. Alloying elements improves fluidity and solidification characteristics for casting.

Iron and Steel

Iron is the most commonly cast metal. However, there are three main types of cast irons that are commonly used, and each has different properties.

Gray Iron. Gray iron is the type of cast iron most commonly used. However, gray iron is high in carbon content and impurities and has tiny gas bubbles in it.

Because of the free graphite in its grain structure, gray iron can be easily machined. It is hard and rigid and will absorb shock. Because of these features, it is used for the bodies of most machine tools such as lathes, mills, and so forth. It is also used for engine blocks in automobiles, iron pipe and fittings, ornamental castings, hydraulic cylinders, camshafts and bathtubs. This is but a small segment of a wide range of uses.

However, because gray iron is hard and rigid, it is also brittle. This means that it can be machined (cut) easily, but it is hard to shape by other processes.

White Iron. White iron is hard and brittle. Unlike gray iron, white iron is difficult to machine. This cast iron is chilled rapidly so that the iron and carbon cannot separate. The result is a much harder state due to the carbon. White cast iron is used where durability to abrasion and hardness are the desired qualities. White iron can be used for crusher and mixer jaws or rollers.

Malleable cast iron begins as white cast iron. To change white iron to malleable iron, it is subjected to prolonged heat treatment. This changes the carbon into spherical rosettes rather than the usual flakes. As a result, the iron becomes a little softer and more ductile. Because of this ductility, shaping operations other than cutting may be performed. Common operations include coining for dimensional control, moderate forging and bending, and press fitting. Common malleable iron products include brake pedal arms and supports, gear cases, rocker arms, sprockets and gears, and other engine and implement parts for automobiles, and maritime and agricultural equipment.

Nodular or Ductile Cast Iron. Nodular cast iron has been alloyed with magnesium or cerium. These alloys reduce the silicon content and change the carbon into round nodules instead of the usual flakes. No heat treating need be done to improve the machineability or to reduce the brittleness. The qualities of ductile iron include high strength and toughness. Common uses include crankshafts, cylinder heads, drive gears and pulleys, cams and camshafts, and other similar items.

Other alloys. Other elements may be added to cast iron to provide particular qualities. Corrosion resistance can be attained by the addition of nickel, chromium, or silicon. Chromium and silicon alloys are also more heat resistant. Strength at high temperatures can be improved by the addition of nickel and molybdenum. However, intensive alloying usually reduces either the ductility or the machineability of the cast iron.

Steel. Steel may be cast also. Although steel is iron alloyed with carbon, most modern steels contain other alloy elements as well. These elements include tungsten, molybdenum, chromium, nickel, and several others. These elements are all used to add specific qualities such as making the steel weather or corrosion resistant, stainless, harder, or more wear resistant. The alloys for iron and steel are added during the melting process and become a soluble part of the iron or steel before it is poured into the mold.

Nonferrous Metals

Nonferrous metals are also commonly alloyed. One of the most commonly used nonferrous metals is *pot metal*. Pot metal is a zinc alloy commonly used in die casting processes which produce such items as automobile carburetors, door handles, hardware for cabinets, and similar items.

Most nonferrous metals are cast as alloys to improve various characteristics of the metals. Most pure metals are either too soft, too brittle, or they are not easily machined. The alloying of nonferrous metals generally improves these characteristics. *Aluminum* is commonly alloyed with copper to increase both its strength, shock resistance, and its ability to be machined smoothly and quickly. *Zinc* is commonly alloyed with metals such as bismuth and antimony to control shrinkage. Most metals shrink as they cool; however, bismuth and antimony expand, and the alloying of these metals to zinc allows the alloy to maintain its dimensions as it cools. *Bronze* is a copper alloy but may have other elements added to it, such as

graphite, to decrease its coefficient of friction. *Gold* and *silver* are alloyed with nickel, copper, or tin to increase the hardness because pure gold or silver are too soft to withstand wear and use.

Melting Metals for Casting

Metal is melted in several ways. Blast furnaces such as those in Fig. 5-2 are used to melt iron ore and make iron. Open-hearth furnaces are used to convert iron to steel. The open-hearth processes, developed in England in the 1860s, have been used throughout the world as the main process for making carbon steels but are gradually being replaced by the much more efficient basic-oxygen process.

The modern basic-oxygen process is replacing the open-hearth process for production of carbon steels. The word *basic* is used in the sense of alkaline. The alkaline is absorbed from the firebrick lining used in the basic-oxygen furnace. Instead of compressed air, high-purity oxygen is forced into the molten iron at about 150 psi [10.5 kg/cm²] pressure through a retractable oxygen lance. The pure oxygen rapidly combines with the carbon in the pig iron, producing carbon monoxide, which then escapes through the top of the furnace.

During the oxidation period, a certain amount of limestone is added through a retractable chute. As it melts, the limestone combines with other impurities forming a slag that floats on top of the molten steel and is afterwards removed. All factors can be closely controlled in this process. A complete heat, up to about 200 tons, can be processed in less than an hour. This is far more production than in the open hearth, where each heat requires from 8 to 10 hours.

The electric-arc furnace is used extensively to produce fine alloy steels, particularly those used in tool making. However, the electric furnace is also used for nonferrous metals as well. The advantages of the electric furnace is that it does not depend upon the burning of fuels which contaminate the melted metals with impurities. Another type of furnace that uses electricity for power is the induction furnace. The induction furnace, as shown in Fig. 5-3, uses high-frequency alternating current. This induces eddy currents into the metal to be melted, and the heat from these induced currents then melts the metal. This type of furnace produces the highest quality steel. There are no burning elements to mix with the steel at any point of the process. The induction process also stirs the melt and gives a more evenly mixed alloy.

Reverberatory furnaces are also used for nonferrous metals. The reverberatory furnace features a mixture

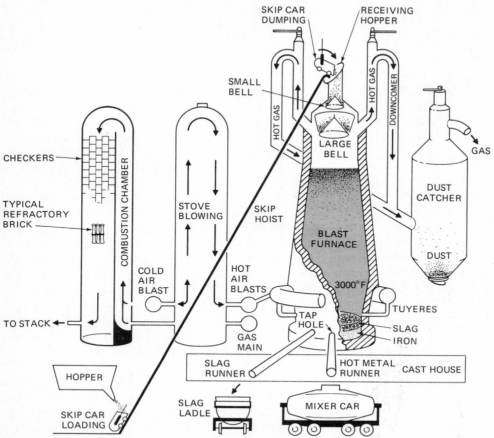

Fig. 5-2 Blast-furnace diagram. Alternate layers of coke, ore, and limestone are charged into the top of the furnace; the iron is separated and melted; and the limestone mixes with impurities to form a slag. The slag is lighter than the iron and floats to the top where it is drawn off through the cinder notch, leaving the iron to be drawn off through the tap hole. The iron is either taken to a steelmaking furnace by the mixer car or cast in the cast house, as needed.

of air and fuel gas made to swirl around a crucible as the fuel-air mixture is burned. The swirling action produces even and quick heating but also adds impurities and gases which must be flushed from the melted metal before it is used. The melted metal is cleaned by adding a flux which liberates the gas and combines with the impurities forming a sludge which then floats to the top and is skimmed off. The reverberatory furnace is not suitable for ferrous metals because it does not normally produce high enough temperatures. However, it is very appropriate for low-temperature metals such as aluminum, lead, or brass when high-purity states are not required. The cupola furnace is extensively used to melt cast irons.

CASTING AND MOLDING OF NONMETALLIC MATERIALS

Materials shaped by casting processes are traditionally referred to as *cast* rather than *molded*. In the case

of nonmetallic materials, some of these (most plastics, glass, and porcelain) are molded rather than cast, whereas concrete is more often cast rather than molded. Because the objective of this book is not to reform terminology, both terms are used for processes that are logically identical or very similar.

As in the case of metals, nonmetallic materials with the same generic name (for instance *glass*) are made in a variety of types with different properties. The main processes for molding or casting these are described in paragraphs that follow.

Ceramics and Porcelain

Ceramic and porcelain materials are prepared by mixing special clays with water much like mixing fine concrete. These materials are set by removing the water from the mixture rather than by cooling. Excess water is initially removed simply by being absorbed into the porous plaster molds. This process is

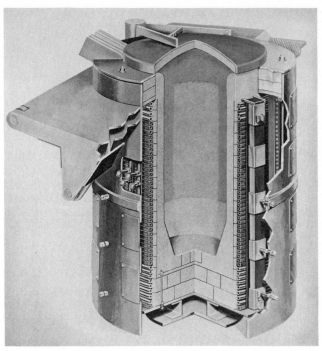

Fig. 5-3 An induction furnace melts the metal without exposing it to air, allowing close control of alloys and freedom from oxides and other impurities. (*Inducto Therm Corp.*)

called *slip casting*. Actually, only a thin shell of material is hardened. The rest is poured back out of the mold. The hollow object is later removed from the mold and fired to harden the object through solid-state diffusion.

Colors and finishes are added by painting with a "glaze" compound before the final firing. The glazes are absorbed into the outer layers of the cast materials and all fuse into one continuous film during firing.

Plastics

Plastics can be separated into two basic categories. *Thermoplastic* materials become pliable, or soft, when heated. *Thermosetting* materials become hard when exposed to heat. Molded plastics are of either variety; however, both categories of plastics can be effectively molded using either a solid plastic or a liquid plastic that is set or hardened in the mold. Variations also use chemical action to set the plastic.

Before molding, foamed polystyrene plastics are small, granular spheres called *beads*. Each bead has a bubble of air or gas trapped within it. The beads are poured into a mold and the mold is then sealed. The mold is heated which causes each particle to soften and expand to many times its original size. The gas or air inside the beads forces the softened particles

against the sides of the mold and against the other beads. The beads cohere and take the shape of the mold. The mold is then cooled and the object is removed.

Glass

Glass has been manufactured since ancient times. Its basic composition throughout these times has been sand, soda, and limestone or chalk. The primary compound of glass is silica, which is approximately 75 percent of the raw material and frequently is quartz sand. The remaining 25 percent is approximately an equal mixture of soda and various limestone compounds.

In modern glass manufacturing, approximately 30 to 50 percent of the manufactured glass is made of reprocessed glass. Reprocessed glass is glass salvaged from garbage and dumps that has been cleaned and crushed into small grains called *cullet*. The raw materials and cullet are mixed together and then melted in furnaces which hold up to 1500 tons of the melted glass. The glass may also be melted in pot furnaces which contain a number of small pots or crucibles, each holding a desired amount of glass.

The basic melting process is carried out at approximately 2500°F [1370°C]. Various materials are added at the lower temperatures to remove undesirable discoloration, to change the color of the glass, or to add particular qualities desired to make special types of glass.

Special types of glass are normally manufactured in fairly small pots. The pots used for this purpose are usually constructed of a high melting point material such as fused graphite or aluminum oxide. Discoloration is reduced by the addition of materials such as manganese dioxide or arsenic. Colored glass is produced by the addition of metallic oxides such as the oxides of iron, copper, manganese, or chromium. An example of a special type of glass is optical glass, to which lead has been added. The addition of lead improves the achromatic qualities of the glass so that the light rays are not distorted.

Glass used in molding is normally cast at the lower temperatures at which the glass is held during the secondary heating process. The molten glass is a thick liquid which may be shaped by several processes including pressure molding, die casting, and blow molding.

Concrete

Concrete is not only a widely used construction material but is a widely used manufacturing material as well. Generally, the difference between "construction" and "manufacturing" is that "construction" is the assembling of materials on a building site, whereas "manufacturing" is the fabricating of cast pieces in a factory. These cast pieces of cement are later transported to a specific location to be used.

Concrete is a combination of Portland cement, sand, and gravel particles called *aggregate*. The ingredients are mixed together with water to form a semiliquid material that is easily molded. Portland cement is named after the city of Portland, England, where the process was discovered. Although Portland cement, sand, and gravel are the basic ingredients for concrete, a wide variety of various ingredients, as well as various sizes of the aggregates, are used in the manufacture of concrete products.

The Portland cement acts as a binder or glue to hold together the various particles which give the desired characteristics. Generally, concrete consists of a larger proportion of aggregates and sand than of cement. This mixture provides stronger structural qualities but leaves poor surface finishes due to the large aggregate sizes. Concrete that contains a greater proportion of sand or cement is used for finishing surfaces or providing smooth surfaces.

Various types of gravel may be added to the surface of concrete as it hardens. Different colors and textures of gravel can be used to provide several decoratived effects. In addition, concrete may be colored for decorative purposes by the addition of various coloring agents, such as lamp black and various oxides. Colored concrete is used in the manufacturing of patio tiles, roofing tiles, decorative panels used for sides of buildings, and various other manufactured products. Special aggregates can be used with colored concrete to form products that have a granite appearance and are then finished by grinding and polishing. The most commonly used colors in concrete are black, green, and red.

Concrete hardens very slowly by the process of *hydration*. During the initial hardening processes, concrete should be kept from drying too rapidly. For large sections, such as a poured slab for a house, the surface can be moistened periodically, or it can be covered with plastic film. Smaller objects may simply be kept enclosed in the mold. Concrete fully cures and hardens after several days. The initial setting will usually be completed in two to five days. Factors that affect the time required include the types and amounts of additives used, the size of the casting and the type of weather to which the casting is exposed.

Concrete is a versatile material that can be molded into a variety of shapes and uses. It is quite literally artificial stone which can be molded quickly to the desired shape instead of being laboriously chipped from solid stone. "Molded stone" products are used extensively in building construction to achieve special corner effects and carved effects, as in Fig. 5-4. Such materials, although used in the construction industry, are manufactured articles.

In addition to being a vital construction material, a multitude of concrete products are manufactured for use in ornamental building construction, which includes such products as railings, flower and plant boxes, splash pans, tiles, pipes, septic tanks, water troughs for livestock, paving and terrace tiles, masonary, and construction blocks manufactured from a mixture of concrete and cinder, as well as decorative garden furnishings. Additional products include lamp posts, street markers, curbing, and paving.

Entire buildings may be cast in one spot and moved to the desired location. A recent innovation in hotel and hospital construction features the casting of a complete room on the ground, where the room is furnished and made livable. Once the room is finished, it is then hoisted into place. Water, sewage, and electrical connections are built directly into the concrete walls with various plastic gasket materials around the pipes. Electrical cables enable each unit to be simply "plugged in" to previously installed units. This allows

Fig. 5-4 Such building construction features exclusive use of manufactured concrete products in corner stones, ornamental and decorative effects, window casements, and drains.

buildings to be constructed in much the same way that a child builds a toy castle from small blocks.

TYPES OF SHOPS

As explained previously, shops that cast metals are called *foundries*. Shops that cast other materials such as plastics and ceramics are generally called either *casting* or *molding shops*. However, foundry, casting, or molding shops may also be categorized by their products.

Production shops normally restrict their output to one type of product which they mass produce in extremely large quantities. A production shop, for example, might produce only handles for an automobile, but it might produce millions of these handles during a year. Jobbing shops, on the other hand, produce a variety of products. However, they are not normally equipped for high-production rates and produce only small quantities of each of the items that they make.

Foundry and other casting shops are described by how they market their products. A captive shop is a foundry or casting shop owned by a larger company. The larger company uses all or most of the output of the casting shop. For example, a large automotive corporation might own a captive iron foundry to produce motor blocks and a separate captive casting shop to produce trim and interior accessories. The automotive corporation that owns the shops would use all the products of the two shops and no products of these shops would be marketed elsewhere.

An independent shop, on the other hand, is usually privately owned and produces items by contractual agreement. Many independent shops are small nonferrous foundries that specialized in one or two materials and one or two types of products. For example, an independent jobbing shop might do only brass or aluminum medallions or plaques and other similar items. Many famous ceramic shops produce only a few types of figurines and perhaps a special Christmas plate, using only one or two processes.

MAKING THE MOLD

A key element in the casting of any object is the making of the mold. An inaccurate or poorly formed mold will result in an inaccurate and poorly formed object. Although sand molds are the most widely used, permanent molds are also in wide use. These molds are usually made of steel although other metals are also used. The mold cavity in permanent and die-

casting molds (dies) is painstakingly formed from blocks of metal by highly skilled tool and die makers. The molds may be reused thousands of times and still produce accurate, quality products.

The major exception to the great reusability of "permanent" molds is the plaster molds used for molding ceramic figurines. These permanent molds are made of porous plaster to absorb moisture, as previously described. These molds, however, are not as durable because plaster is not very hard. The mold life is further reduced by the repeated absorption of water, which softens the plaster. The molds can produce extremely high quality products only a few times.

Unit molds, those that are destroyed when the castings are removed, are made by embedding a pattern (Fig. 5-5) in the mold material. After forming the mold, the pattern is removed and the mold is then filled with the casting material. Several different methods are employed for removing the pattern, but the most common is the *split mold* process in which the mold comes apart for the easy removal of the pattern and the inspection of the mold cavity. This process is used in both sand casting for ferrous and nonferrous metals. The patterns for such molds may be mounted on permanent plates for high-production rates, or they may be split into sections for easy removal from complicated molds.

The earliest types of patterns were made of wax, and the mold was not disassembled. The mold was heated to melt the wax which then drained from the mold. This is the *lost wax* process and is still used. This process is now used for casting precision parts. Variations are used for some steels including jet-turbine blades and golf clubs. The process is still used for art objects, statues, dental bridges, and jewelry; and it is known today as *investment casting*.

Patterns can be made from wax, clay, plaster, plastic, wood, or metals. Patterns are made by highly skilled craftspeople known as *pattern makers* whose skills includes the ability to work to exact dimensions

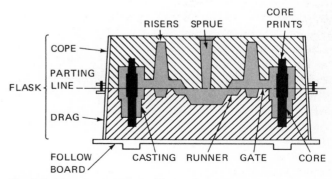

Fig. 5-5 A two-impression mold complete with cores.

in all the common pattern materials. When patterns are to be reused many times in a production shop, the pattern is normally made of metal so that in use it will not deform or wear down. When the pattern is to be used only a few times, it may be made of plaster or wood or a combination of several of these materials.

MAKING HOLLOW CASTINGS

Hollow castings are often necessary to reduce the weight, the amount of material, or the cost of an object. Two methods may be employed for producing a hollow casting. The oldest is the *slush casting* process, used extensively in toy making and for making ceramic vases and figurines.

The slush process is one in which the hot fluid material is poured into the mold. The side of the mold causes the outer edges of the fluid to harden while the inside remains melted. The inside is then quickly poured out of the mold, leaving a thin, solid shell.

The second process, used extensively in sand casting, features the use of sand *cores*. These cores are

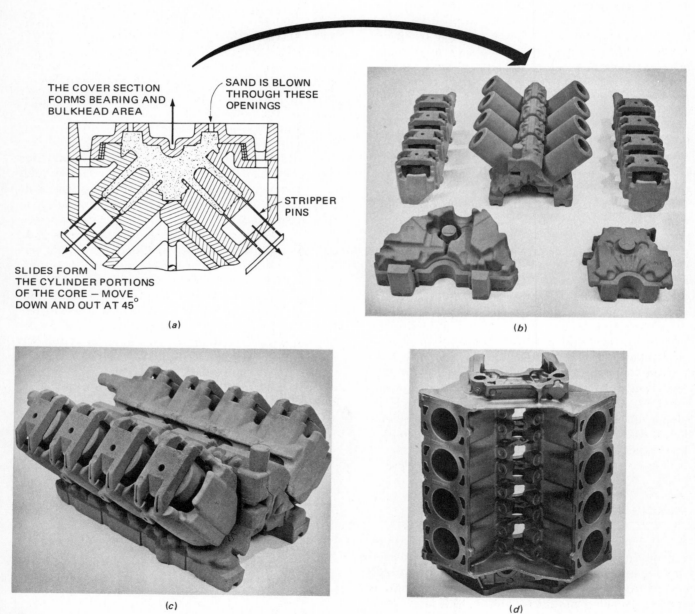

Fig. 5-6 An advanced method in using cores for casting an automotive block of an eight-cylinder engine.

made of treated and hardened sand inserted into the mold. When the metal is poured into the mold, it forms against the side of the mold and around the core. When the metal has hardened, the mold and the core sand are shaken out, leaving the hollow space as desired. Figure 5-6 shows the use of cores in casting.

QUALITY AND COST OF CASTING

Defects in cast materials reduce the quality and usability of the castings. For example, bells with defect in casting do not have clear tones, and they often crack or shatter when used. However, many common casting defects may be repaired through modern welding and repairing techniques. Casting defects may be caused by improper mixing of the material, pouring of the material at an improper temperature, conditions of mixing, obstructions in the mold, defects in the mold, or gas pockets. Casting defects may also be caused by improper removal of the mold or removing the product from the mold too quickly before the product has fully solidified.

Castings must be continuously inspected for proper dimensions and proper quality. Inspections maintain quality of the casting and they also quickly pinpoint types of consistent defects indicating a fault in the mold or the molding process. Sagging, for example, would indicate that the object has been removed from the mold too quickly. Other defects such as cracking or warping of metals would indicate improper removal before the metal cooled. Figure 5-7 shows a typical inspection process.

Surface mars and blemishes which do not affect the integrity and strength of the object can often be quickly repaired. It is also important to remove defective castings from the production line quickly. Doing so prevents operator and machining costs from being expended on unusable products. Defective castings should be removed and sorted into categories of defects that can be repaired with minor operations, defects which can be repaired only with complicated operations, and defects that cannot be repaired.

Modern industry generally operates on a low-profit margin. Keen competition prevents any one producer from doing otherwise. Producing unsatisfactory products can quickly lead to both loss of profit and loss of customers to competitors. It is also important to produce the products only to the desired accuracy and quantity needed. To make any item better than needed in great quantity would increase costs; to make any item more accurately than is desired increases costs as well. Some costs are necessary and fixed, such as the making of the pattern and the preparation of a basic quantity of casting material.

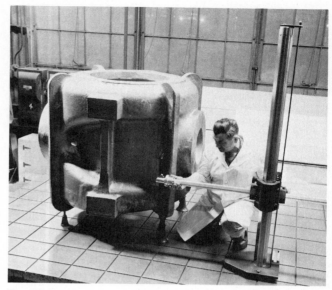

Fig. 5-7 Dimensional checking using a height gauge on a layout table.

The price of one casting might be considerable due to these fixed factors. Similarly, the fixed cost per casting would be greater for 10 castings than for 10,000. However, the material cost is greater for 10,000 castings, whatever the initial cost of the pattern. A factory that obtains a contract for 10,000 identical items can afford to make less profit per item than on the production of 10 items, due to labor and to pattern making and other fixed costs.

The need for closer dimensional tolerance also increases cost of castings. Very accurate castings can be made by the die casting process or using shell molding. However, these more accurate processes are also more expensive, as they require greater fixed costs and operational costs than do the basic sand-casting processes. The casting for a linkage rod for an automobile steering system does not require a great degree of accuracy. Therefore, it could be easily cast by the sand-casting process. The primary considerations of such an item would be strength and length. These characteristics can be easily controlled during the sand-casting process, and subsequent machining operations would ensure the necessary length and hole size. The manufacture of this item by a more expensive die-casting or shell-casting process would not improve the strength or the final dimensions but would only increase cost of production considerably.

The type of material cast would also effect cost consideration. In the steering link example, high-tensile strength would be required. The material should also be ductile to assure that it could withstand impact loads and not fracture from sudden vibration or shock. Steel or certain iron alloys would be desirable for such an item. Because of the higher melting costs and higher-temperature mold materials, it is more expensive to produce steel for this item than iron. However, if steel is the only material that would produce the desired strength and quality within the desired size, it must be used. Low-strength metals such as zinc or aluminum alloys, while not very expensive, would not be satisfactory. This fact would rule out consideration of die casting, which is almost entirely a process for casting low-temperature and low-strength metals.

Four major characteristics must be considered in the production of any cast product. These are quantity, accuracy or precision, fixed costs in mold preparation, and finishing and cleaning operations.

The quantity required affects the decision on the type of molding process to be used. This, in turn, affects the operational costs, handling processes, subcontracting decisions, and use and scheduling of equipment. The accuracy or precision specified also affects the decision relating to the type of process and subsequent handling and operational costs. Accuracy and precision requirements also relate to establishing and maintaining quality controls and inspection processes. Fixed costs, or operational costs, include the costs to prepare the molds and patterns; operate the mold curing, coring, and material preparation equipment; labor costs; material costs; and similar factors.

Finishing and cleaning factors include the surface smoothness or dimensional accuracy required, the amount and precision of welding, bending or machining needed to convert the casting into its final shape, and the method and accuracy needed to clean, descale, and paint or plate the casting.

REVIEW QUESTIONS

5-1. What is casting?

5-2. What products are normally found in a home that are castings or moldings?

5-3. What materials may be cast or molded?

5-4. How may casting processes be classified?

5-5. Why are there so many types of casting processes?

5-6. Why are castings considered important to industry?

5-7. Why is the preparation of the casting material important?

5-8. How do the following materials differ? *a.* Gray iron, *b.* White iron, *c.* Nodular iron, *d.* Steel, *e.* Pot metal.

5-9. What types of furnaces are used to melt metals? What are the advantages and disadvantages of each?

5-10. What is a captive shop? A jobbing shop?

5-11. What is a tool and die maker? A pattern maker? Why are their jobs so important?

5-12. What factors determine the material selected for making a pattern?

5-13. How are hollow areas made in castings?

5-14. What factors affect quality and costs of castings?

5-15. What are fixed costs? Operational costs?

CHAPTER 6

PATTERNS, CORES, MOLDS, AND DIES

Factors that determine casting quality include the quality of the materials, suitability of equipment, and the skill of the craftspeople and operators during all phases of the manufacturing processes. These processes include drawing the specifications and design of the cast product, constructing the pattern (or mold for permanent molding), preparing the mold material, preparing the mold and coring, filling the mold, removing the cast object, and conducting the quality-control operations.

In addition to these processes, castings must be finished. Finishing includes the removal of flashing and minor surface imperfections, and the cleaning and machining operations.

Each of these operations plays an important role in achieving and maintaining the overall quality of the product. Technical factors include the flow of materials into the mold cavity, the type of molds, the pressure under which the materials are introduced into the mold cavity, cleaning and finishing of the mold, and the particular molding techniques and equipment utilized.

CASTING DESIGN

The first step in manufacturing a quality casting is to determine the exact function to be fulfilled by the product. Service requirements determine the design requirements and limitations, and must be carefully drawn. Service requirements describe functions or roles of the casting (either alone or in conjunction with other parts) and also the weights, loads, pressures, or other conditions to be met in service. Sometimes the weight limitations of the casting itself must be considered. The specific size and mass of the casting needed to fulfill these qualifications becomes a basis for the design of the shape, finish and accuracy specified. Tolerances also play a vital role in deter-

mining the finishing and machining operations as well as the shape and type of casting to be made.

The type of material is determined once the function and limitations have been identified. Where weight of the material from which the casting is made is critical, such as in aircraft components or space capsules, material selection may be limited. Other physical limitations include heat resistance, chemical reaction, and weather resistance.

The type of material also affects the cost of the finished product. Some materials, such as gray cast iron, are inexpensive. Other cast materials such as high alloyed steels, are much more expensive. The degree of accuracy required in making the part also affects the design. Some casting processes are extremely accurate and produce surfaces that are smooth enough to require no machining. However, these processes are more expensive and more complex than processes that produce rougher, cheaper castings. Therefore, the amount of machining and the amount of accuracy becomes an important factor in the design of the casting. The finished part cost rather than the cost of the raw casting must be the guiding consideration.

The casting process to be used thus depends upon the type of material needed, the type of machining required, the amount of tolerance required, and the number of parts to be produced. It is less expensive to produce one or two parts and machine them extensively than it would be to construct the mold and molding equipment to make an accurate casting. The permanent mold-casting processes require greater expense for construction of molds. Eliminating this tooling cost for small production quantities would more than offset the machining costs required to produce the same product by more accurate casting processes. However, production of large quantities reduces the cost per casting for the product and makes initial tooling expense justifiable. Cost for com-

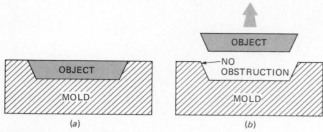

Fig. 6-1 *Draft* is the taper or slope that allows a pattern to be removed from a unit mold. Draft must also be included for permanent molds to allow the casting to be removed from the mold. All objects must have draft, but different mold types and materials require different allowances.

Table 6-1 **Cast-Metal Shrink and Contraction Allowances**

Metal	Contraction, in/ft
Gray cast iron	$1/10 - 5/32$
White cast iron	$1/4$
Malleable cast iron	$1/8 - 3/32$
Meehanite metals	$1/8 - 1/10$
Aluminum alloys	$5/32$
Magnesium alloys	$5/32$
Yellow brass	$5/32 - 3/16$
Gun-metal bronze	$1/8 - 3/16$
Phosphor bronze	$1/8 - 3/16$
Aluminum bronze	$1/4$
Manganese bronze	$1/4$
Open-hearth steel	$3/16$
Electric steel	$1/4$
High-manganese steel	$5/16$

SOURCE: American Colloid Company.

plex permanent molds can often exceed $100,000. Such costs are only justifiable when many items will be produced. For this reason some processes are inherently high volume processes. An example is the die-casting process. Other examples are transfer molding and injection molding of plastics.

Physical limitations in casting design include the amount of coring needed, the removal of the pattern or the product from the mold, the shrinkage inherent in the type of material, and the strength and melting temperature of the material to be cast. Products that must be removed from permanent molds, and patterns that must be removed from unit molds, must both be tapered for easy removal. This taper, or slope, is called *draft*. A reverse draft would prevent the object from being removed from the mold (see Figs. 6-1 and 6-2). In addition, where hollow spaces are desired, cores must be used. Mold design must consider methods for the insertion and support of these cores during the pouring of the material.

Since nearly all materials shrink during solidification and coding, allowances must be made for the shrinkage. Patterns and molds must be made larger so that the product will shrink to the desired size as it solidifies. Table 6-1 shows typical contraction and shrinkage allowances per linear foot for commonly

cast metals. Allowances for shrinkage are made in the pattern for items produced in unit mold processes, such as sand casting. The shrinkage allowance must be in the mold, however, when the product is to be produced in a permanent mold. In addition, the larger the item, the more shrinkage that occurs.

Most of the shrinkage occurs as the metal changes from a liquid to a solid. This is termed *solidification* shrinkage. Uncontrolled, it can cause casting defects as well as poor dimensional tolerance. Casting defects due to shrinkage are controlled by adding *risers* which store extra liquid material. As the casting shrinks, a vacuum is formed drawing extra material from the riser into the mold (see Fig. 6-3).

An additional factor in the physical limitations of the mold includes the strength of the mold itself, particularly in molds that are made of plaster and sand. Sand molds, as well as plaster molds, have less strength where the mold has small sections. If metal is allowed to rush into the mold across a thin section, it can cause the mold to break so the casting will have missing segments, as shown in Fig. 6-4. In addition, the segment of the mold that is broken away forms a

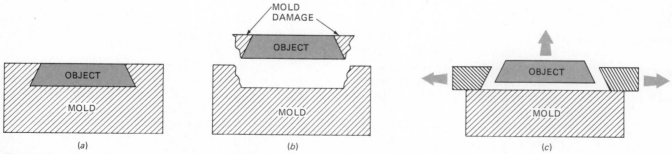

Fig. 6-2 *Reverse draft* prevents an object from being withdrawn from a mold without damage to the mold. Some permanent molds incorporate reverse draft into the mold but overcome removal problems by using movable mold segments.

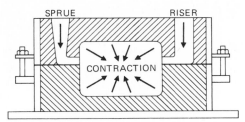

Fig. 6-3 Risers and sprues help compensate for contraction as the casting solidifies.

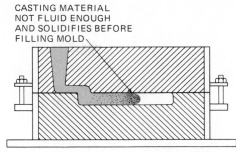

Fig. 6-5 Thin molds require casting material to be more fluid. Metals must be hotter, ceramics must contain more moisture, and plastics must contain more solvents or reactants.

cavity within some other spot of the mold, causing an additional imperfection. Although these imperfections due to breakage of small sections within the mold can frequently be repaired, they do require extra time, operations, equipment, and expense in the repair of the product. Obviously, consideration of the mold strength and how the material is allowed to enter the mold can remedy many such problems.

The fluidity of the material as it enters the mold is also a consideration. Greater fluidity is required in casting thin pieces to allow the fluid material to flow all through the thin mold areas, as indicated in Fig. 6-5. However, this increased fluidity also increases problems within the mold from trapped gases which can cause products to be defective. Thin sections may also warp or break during solidification and coding.

Green sand molds, which derive much of their mold strength from moisture, pose particular problems when casting thin sections. The fluidity of metal for sand casting is increased by alloying and by heating the metal to higher temperatures. This increased heat creates more steam from the moisture in the molding sand. This, in turn, has less area to escape through the thin sections. In addition, thin and thick segments of a mold harden at different rates. This difference in rates of solidification can cause tearing and cracking. A common technique used in casting metals with thick sections is to insert pieces of metals called *chills* into the mold. These chills assist in cooling the mass of metal in the mold, thus preventing defects caused by cooling too slowly, or uneven cooling rates at different portions of the casting (see Fig. 6-6).

Many products tear or crack due to uneven solidification processes. Plastics can tear or crack in the mold due to rapid setting or curing. The too-rapid cooling of metals will cause similar defects. Ceramics that have the moisture absorbed too rapidly from the slip also suffer similar damage. The most common type of defect is creating a tear or a crack partially through adjoining thick and thin sections. The changing of the product from a liquid state to a solid state at uneven rates causes most of the tearing and cracking problems. Also, when materials solidify too slowly, they may be removed from the mold too soon. This causes damage and distortion because the material is either still liquid or is too soft and does not have sufficient support when removed from the mold.

PATTERNS

A pattern is an object that is made to provide a shape to the mold cavity. The mold cavity may be used in

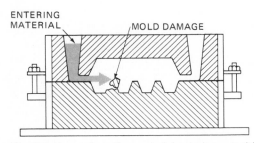

Fig. 6-4 Force of casting material entering a mold can cause mold damage that can cause casting defects.

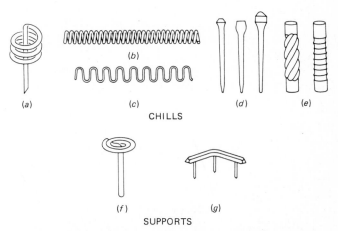

Fig. 6-6 Chills prevent porosity from uneven cooling at thick, isolated sections and reduce gas bubbles by reducing temperatures of metal in mold sections. Chills may also be applied externally. Internal chills are usually made from approximately the same alloy being cast. NOTE: *b* and *g* are chaplets (supports); the rest are chills.

Fig. 6-7 Various foundry patterns. (*REN Plastics, Inc.*)

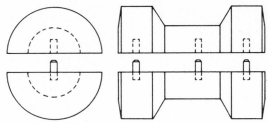

Fig. 6-9 A split pattern.

producing castings or it may be used to produce a permanent mold which, in turn, produces the cast product. Patterns are more frequently used to make unit molds which are destroyed as they are used. In cases where the pattern is reused many times for high-production rates, the pattern is made differently than one that is to be used only for a few times. When a pattern is to be used continuously to produce large numbers of objects, it should be made of hard, durable materials to withstand repeated uses. When a pattern is to be used only for a limited number of times, it may be constructed of softer materials that are easier to work but are less durable. Figure 6-7 illustrates different patterns.

In addition, if a mold is to be used only once but the mold is to be made of plaster, the pattern is made differently than if the mold is to be made of sand. Different processes require different pressures and different finishing techniques. When the pattern is used repetitively, it is frequently made into what is known as a *matchplate*. Matchplates allow several steps to be combined into one and reduce the time required to make the individual mold.

There are several types of patterns. The simplest to construct is the *solid pattern*. The pattern is the shape of the object desired and is a one-piece unit (Fig. 6-8). Draft, or taper, is needed to allow the pattern to be removed from the mold. It is in one direction only. In two-piece *split patterns,* the parting line, where the top half meets the bottom half, may be even and

straight. However, some objects may have uneven parting lines that require special operations performed by the molders. For round objects which have draft in both directions, the pattern is frequently split to allow easy removal from both top and bottom portions of the mold. Split patterns may have only two sections, as in Fig. 6-9, or may consist of several layers. The patterns for an engine block, for example, often have as many as five or six different layers. They are all classed as split patterns even though they have many sections.

The matchplate (Fig. 6-10) is a pattern that has been made from a solid piece of metal so that it provides the surface upon which the pattern is made and the pattern itself as one solid object. This eliminates the need for careful positioning of both halves of the pattern and various other manipulations by operators. Matchplates greatly increase the speed of the molding operation. The making of the matchplate from metal, although an exact and expensive process, also increases the useful life of the pattern as well as increasing the speed of production. For long production runs, the matchplate is ultimately the most economical way of producing patterns for small objects.

Another detail in the design and construction of patterns is providing certain projections on the patterns that will form cavities in the mold in order to support the core when it is inserted after the pattern

Fig. 6-8 A solid pattern.

Fig. 6-10 A matchplate pattern.

is removed. These projections on the pattern are called *core prints* (Fig. 6-11) and the corresponding hollow areas to support the cores are called *core seats.* Core prints are designed so the core seats will be completely filled by the core and form no part of the casting itself.

Patterns must also be made slightly oversize to allow for the shrinking and contraction of the material after it solidifies. Different pattern allowances must be made for each type of metal, as shown in Table 6-1. Aluminum, for example, contracts a different amount than cast-iron alloys. Other materials such as concrete, ceramics, and plastics have different shrinkage values; therefore, the dimensions of the pattern must compensate for the shrinkage of the specific material from which the casting is to be made.

Patterns are made in several manners. Small patterns are frequently made of solid materials such as wood or plaster, or a combination of materials. Wax, wood, leather, plaster and metal are all used. Outside corners are usually rounded by cutting the corner away. Inside corners are rounded by filling in the corner. These are called *fillets* and are made from wax or leather. Rounded corners in the mold do not break as easily and thus help produce cleaner and better castings.

Metal pattern letters are also used to make the patterns for objects with printing cast into them. These metal pattern letters are usually glued or attached to wooden bases which may have strips around the edges made of either wood or leather.

Patterns for smaller objects that are round or cylindrical in shape are frequently made of plaster. They are made by taking a rod and putting quick hardening plaster or plastic onto the rod and turning the blob of material against a sheet metal profile. This metal profile acts as a cutting tool and gives the revolved blob the desired shape.

Larger patterns, such as patterns for massive dies used to stamp automobile bodies or other similar objects, are made in hollow segments. The basic shape is built up of wooden or metal strips with plaster or clay

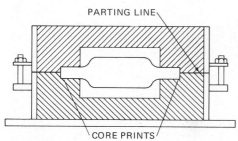

Fig. 6-11 Core prints are used to support the core in the mold cavity.

used to smooth over the frame evenly and to give the final shape to the pattern. Such patterns are generally used for production of dies and machine cases comprising parts of gigantic industrial tools and machines.

The pattern may include the making of a prepattern (preliminary pattern) used to cast a production mold from plaster, plastic or even metals so that patterns may be produced in sufficient number to be used with large-scale production. In such cases extra allowance must be made for shrinkage and the control of both the prepattern and final pattern as well. Pattern makers are greatly concerned with shrink allowances and frequently use metals such as bismuth in the construction of these molds.

Bismuth, as previously mentioned, expands as it solidifies rather than shrinking as do most metals. The careful use of such metals as alloys with other metals provides stable materials that neither contract or shrink as they solidify. Although they are soft and not suitable for production line use, they are very suitable for the making of prepatterns.

GATES AND RISERS

Gates and risers are necessary in all types of casting and for all types of materials. They are used in permanent and unit molds alike. Gating is the process of providing a path for the fluid material to enter the mold cavity itself. Most casting operations do not allow pouring the fluid directly into the mold cavity. Direct pouring increases the chance of forming gas pockets and frequently causes damage to the mold, particularly in sand casting. Therefore, a complex system is used to reduce possible defects.

Generally, the fluid material is poured into a receiving hole called a *sprue.* From the sprue the fluid material goes into a horizontal tunnel called a *gate,* which conducts it to the mold cavity. Where multiple parts are cast simultaneously in a single pour, additional tunnels called *runners* lead from the gate to each of the mold cavities. Figure 6-12 depicts a typical gating system.

The sprue is molded into the top half or cope portion by means of a tapered sprue pin, and the sprue hole generally extends to the lower or drag portion of the mold. Usually the gate or gate and runners are cut in the top of the drag by means of a gate cutter. One of the advantages of matchplates is that they generally include patterns for the entire gating system, saving a great deal of time for the molders.

Several runners and gates may be used when pouring thin sections where high fluidity is necessary to enable the metal to enter quickly before it cools. In

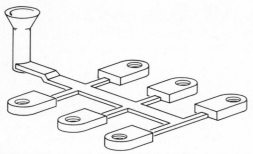

Fig. 6-12 "Gang" or multiple casting allows each molding cycle to produce several parts.

casting smaller objects usually only one runner and gate are necessary. Gang casting, shown in Fig. 6-12 or "pigging" is the practice of using one mold, gate, and riser to produce several parts. This is particularly true in the casting of plastic and the die casting of small metal parts.

Each mold cavity must be properly designed for its runner and its gate to feed the liquid material into the mold. In addition to runner design, air or gases that would create voids in the product should not be trapped in the mold. The air or gas must be allowed to escape from the mold. Small holes called *vents* are made in the cope by means of a vent rod (Fig. 6-13) very near the mold cavity to allow gases to escape. In permanent molds made of metals or plaster, these vents must actually enter the mold cavity. For sand casting the vents need only be placed near the cavity because the porous sand allows some of the gases to escape through the sand for short distances.

Risers, as previously mentioned, are vertical holes similar to sprues. They provide reservoirs to compensate for the contraction of the bulk of the material to prevent deformation during the solidification process. Risers can also serve to allow air and gases to escape from the mold. Also, they provide places for slag or impurities to rise from the mold cavity into a noncritical mass.

The placement of the riser is very important for two reasons. The first reason is that the reservoir of

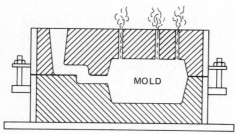

Fig. 6-13 Vents are small passages in molds that allow trapped gasses (air, steam, etc.) to escape from the mold and to prevent air pockets in the casting.

hot fluid must be placed where there is no obstruction to the return of fluid material from the riser to the mold during contraction. The second consideration concerns removal of the riser in subsequent operations. Risers are generally attached to the mold by a gate very similar to the gate that connects the sprue to the mold cavity. Although risers may be connected directly to the mold cavity, this often complicates their removal and increases the possibility of surface defects. Therefore, castings often have risers that are connected by a thin gate that is easily broken or cut from the casting without damage.

BASIC MOLD CONSTRUCTION

To better understand the general casting processes, one basic process will be explained in detail. This process is the *bench-molding* process for sand casting and contains all the basic considerations for almost any casting process. Once the making of a sand-bench mold is understood, the concepts and principles vital to other molding processes are easily understood. The procedure for making a bench mold of sand is outlined in the paragraphs that follow.

The mold is constructed in a container called a *flask* (Fig. 6-14). The flask has a top half and a bottom half; the top half is the *cope* and the bottom is the *drag*. A follow board is placed between the cope and the drag so that only one half of the mold may be constructed at a time. This board supports the pattern during the construction of the drag. The drag is generally prepared first. The pattern is placed in the desired position on the follow board, allowing for gates, risers, and runners, and is dusted with a parting compound to prevent the sand from sticking to the pattern.

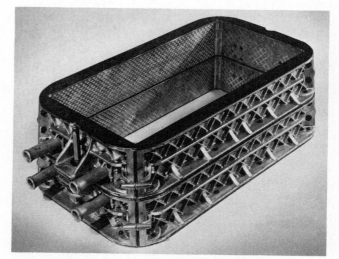

Fig. 6-14 Heavy-duty "slip" flask equipped with "pop-off" handles which permit easy flask removal. (*The Hines Flask Co.*)

Sand is then shaken onto the pattern through a screen called a *riddle*. This is done so that finer sand without lumps or air pockets are placed next to the pattern. This gives the mold the smoothest possible shape. A thin layer of sand has been riddled onto the pattern. The remainder of the drag is filled with sand and rammed solid. The sand must not be packed too firmly, however, or it will not allow gases to escape. It must be packed firmly enough to provide the necessary strength and rigidity for the mold.

Because the rammed sand is uneven in height after the ramming, it must be leveled to provide an even surface for support. The rammed sand is leveled by "striking" it. To do this, a rod is moved across the drag portion of the flask. A bottom board is then placed on top of this leveled sand and the mold is turned over to allow the construction of the cope. The follow board, which is now on top of the drag, is lifted off and placed to one side. The cope section of the flask is then placed directly onto the drag, which still holds the bottom part of the split pattern, and the mold face is dusted with parting compound. The molder then joins the upper part of the pattern to the bottom part of the pattern and repeats the same procedure as for constructing the drag portion of the mold.

After ramming the sand, the molder may cut sprues or may insert sprue pins for sprues and risers during the ramming operation. These sprue pins are tapered dowels and serve as patterns for the sprue and riser cavities after they are removed. After ramming and striking the cope portion of the mold, the molder pierces vents, using a vent rod (essentially a stiff wire of proper diameter and length).

At this point the mold is nearly complete but still contains the pattern which, of course, has to be removed. The cope is then lifted carefully from the drag and placed near it. The molder then taps the drag pattern to shake it slightly to loosen it, and finally withdraws the pattern from the drag, using special lift pins inserted into the pattern. The same procedure is followed to withdraw the cope pattern.

At this point the molder cuts the necessary gates and runners leading from the mold cavities to the sprues and risers, and removes the sprue pins from the cope.

Before the pattern is removed, the sand around the edges of the pattern are moistened slightly. A brush or mist spray is used. This moisture provides added strength to the edges of the mold cavity so that the pattern may be removed with less damage to the mold. Care must be taken to prevent soaking the edges too much as this would create gas problems when the metal is poured.

A final inspection is made, and any loose particles of sand are removed from the mold. The cope is then placed back upon the drag. The mold is now ready for pouring.

The flask may be used as an integral part of the mold. However, where a molder produces molds for small objects continuously, most foundry workers use a flask known as a *slip flask*. The slip flask unbuckles and expands away from the sand mold leaving it complete and allowing the flask to be reused. This allows the manufacturer to use less equipment for continuous production and reduces flask handling and storage expenses.

TRIMMING AND CLEANING CASTINGS

Once a casting is removed from a mold, certain finishing operations must be done. Gates and risers for castings are removed by chipping, grinding and/or other similar processes. Because most molds do not close perfectly, small amounts of material flow around the parting lines and create thin sections around the parting line of the object. This is known as *flashing*, as in Fig. 6-15, and it must be removed before the casting is converted into the finished product. Flashing may be removed by chipping, cutting, grinding, or the casting may be placed on a punch-press die and the flashing sheared off cleanly and accurately. Die castings are frequently cleaned by this type of punching operation. However, this punching operation is seldom used for cleaning sand castings or

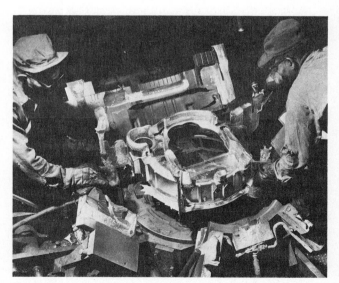

Fig. 6-15 An aluminum-flywheel housing for a diesel engine emerges from a permanent mold. The permanent mold has a collapsible permanent core. Note the "flashing," or untrimmed, thin overflow around the edges of the casting. (*Alcoa.*)

for materials such as ceramics, pottery, and most plastics.

Flashing from ceramics and porcelain products is removed before the firing which hardens the products. The flashing is removed quite simply from the soft clay by using a stiff brush. A finer brush is then used when a smooth, even texture is desired that will make the parting line virtually impossible to detect.

Most castings, particularly metal castings, are given additional cleaning or polishing operations to remove sand, mold particles, scale, or to produce better smoothness or even texture on the product. However, the more accurate and precise the casting process, the less polishing and cleaning is required. The sand-casting process featuring the use of green sand requires the most cleaning and polishing operations, with casting processes such as investment molding and shell molding requiring only minor cleaning and polishing operations.

Products made by green sand casting processes are often cleaned by tumbling the object in rotating bins. These castings are trimmed of gates and risers and then placed into the bins. Small particles called *jack stars* may also be added to increase the cleaning action. These jack stars may be small, hard, pointed objects or they may be lead pellets. The bins are then rotated and the action of the castings against each other and the small particles against the castings provide a cleaning, polishing, and smoothing action to the castings. Cleaning and polishing by tumbling must be done before machining as this process would affect any machined surface.

Sand castings are also cleaned by air blasting with either lead shot or sand. Blasting produces a finer texture than the tumbling process but is more expensive due to initial equipment costs and wear and tear of equipment. Brushing with a wire steel brush or portable grinder may also be done to metal castings.

Large surface blemishes may be removed by grinding, by chipping or by sawing or machining. However, most chipping operations are done to repair surface defects rather than to polish or clean the casting. The exception is the removal of gate or riser stubs from metallic castings. Flashing is generally removed by grinding or sawing.

REVIEW QUESTIONS

6-1. What is used to create a mold cavity?

6-2. What factors should be considered in determining which molding process to use?

6-3. Why must mold cavities be oversized?

6-4. What does a riser do?

6-5. Define gate, riser, sprue, and runner.

6-6. What is a split pattern?

6-7. Why must molds be vented?

6-8. Define cope, drag, matchplate, riddle, flashing, and jack star.

CHAPTER 7

GRAVITY-CASTING PROCESSES

The gravity-casting processes are the oldest of the casting processes. Perhaps the oldest types are the lost-wax casting and the sand-casting processes. The sand-casting process is still the most commonly used casting process today and comprises a large segment of modern industry. Gravity-casting processes are used extensively for all metals, for most of the porcelain and ceramic materials, most concrete applications, and many processes for plastics and rubber products. Gravity-casting processes are used with both permanent and unit nonpermanent molds.

LOST-WAX AND INVESTMENT-CASTING PROCESSES

Lost-Wax Process

The oldest form of casting, which has been used for over 5000 years, is the lost-wax process. The early lost-wax process consisted of making a clay core over which wax was coated and then modeled or sculpted to the desired shape and smoothness. The pattern was placed over a large platform and sand and clay were then packed around it. After the sand was packed in place except for an opening at the top, the base platform was removed and a fire was kindled under the platform opening. The heat from the fire melted the wax which then drained from the mold cavity. The wax used was carefully weighed so that the molder could determine when all the wax was exhausted from the mold cavity.

Once the wax had drained, the platform was coated with sand, clay, and plaster. This platform then sealed off the bottom of the mold. The mold was then filled with molten metal and allowed to harden for a time—sometimes several days for large metal castings such as cannons and statues. After the metal was allowed to harden, the mold was dug away, leaving the finished casting.

Precision Investment Casting

Modern versions of the lost-wax process are known as *precision investment casting*. The accuracy and finish are greatly improved by the modern processes which use plaster and ceramics to form the molds (see Fig. 7-1). Plaster contains moisture which limits the melting point of the metals that can be used with it. Unlike the moisture in green sand, the moisture in plaster cannot escape from the mold easily. Plaster investment molds are limited to use with metals with lower (less than 1500°F [815°C]) melting points. For metals with higher temperatures two types of investment materials are used, a ceramic one and a plaster one. During the wax burnout, the ceramic shell is hardened, loses its water content and also becomes impervious to water. The wax pattern (see Fig. 7-2a) is dipped into a mixture of extremely fine ceramic mold material called *slurry*. This coats the wax with a hard, refractory material which imparts great dimensional stability and smoothness to the casting. However, the slurry only forms a thin shell and must be reinforced with plaster or sand (Fig. 7-2b). After the plaster mold is finished, the wax is melted out. This is done by placing the mold in an oven. This investment-casting process results in great accuracy and is extensively used for statuary, art objects, and complex industrial shapes, such as compression blades for turbines (Fig. 7-2c). Generally, investment casting is expensive and is used where the saving in machining, or the inability to machine the complex shape, justifies the expense of the mold preparation. A strong advantage of investment casting is that it combines the capability to produce fine detail with the capability make castings of high temperature metals.

For production runs of investment or lost-wax castings, molds must be made in which to produce the wax patterns. These molds are individually finished, and allowance must be made both for the shrinkage

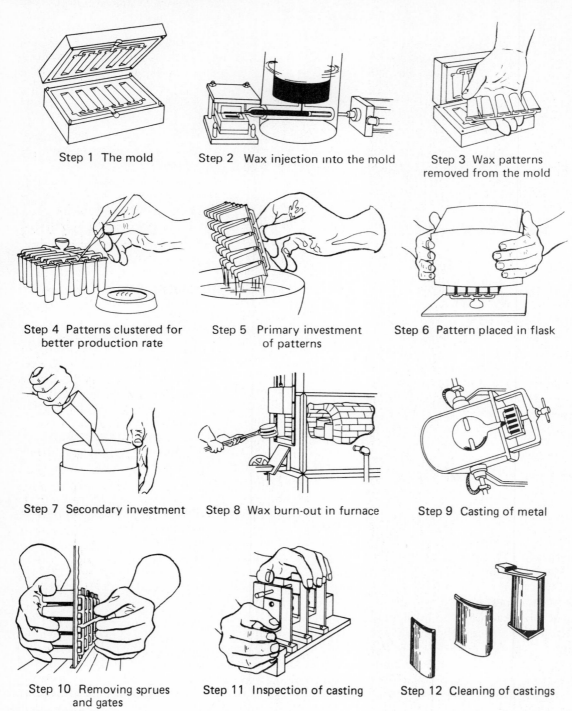

Step 1 The mold

Step 2 Wax injection into the mold

Step 3 Wax patterns removed from the mold

Step 4 Patterns clustered for better production rate

Step 5 Primary investment of patterns

Step 6 Pattern placed in flask

Step 7 Secondary investment

Step 8 Wax burn-out in furnace

Step 9 Casting of metal

Step 10 Removing sprues and gates

Step 11 Inspection of casting

Step 12 Cleaning of castings

Fig. 7-1 Basic steps involved in the lost wax (investment) method.

of the wax as it solidifies as well as for the shrinkage of the material to be cast.

Full-Mold Process. The full-mold process is another variation of the investment-casting process. The major distinction between investment casting and the full-mold process is that in the full-mold process, the pattern is not withdrawn from the mold. The pattern is not made from wax; it is made from a styrene plastic foam which is comprised mostly of large air bubbles in the combustable plastic. When a hot molten metal touches the plastic, the plastic is rapidly burned

(a)

(b)

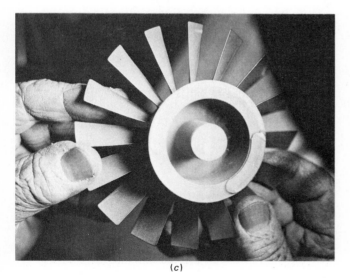

(c)

Fig. 7-2 (a) The modern investment casting process begins with the making of a wax pattern in a complex permanent mold; (b) the slurry-coated pattern is then coated with a strong, coarse shell. The shell is cured and the heat melts the wax which drains from the mold; (c) after curing and draining, the mold is poured. The finished casting with its fine texture and smooth

away. Because there is a relatively small amount of plastic material used in forming the bubbles within the foam, little ash or residue is produced by burning it. The residue that is produced is very light in weight and will float on the surface of the metal so that the ash or residue readily rises to the top of the sprue and thus does not cause defects in the casting. Care must be taken so that all gases from the burned plastic are allowed to escape from the mold.

As with the precision investment casting processes, a pattern or mold must first be made to prepare the individual patterns. The degree of finish or accuracy that can be attained is similar to that for the investment casting process. However, the use of the full mold process eliminates the need for melting and withdrawing the wax from the mold. In addition, less care is needed for handling the patterns because the plastic foam patterns will not melt or sag at warm temperature as would wax patterns.

Other Related Processes. There are several other variations of investment casting. Most are highly specialized and are used for intricate, detailed parts or for casting metals in controlled atmospheres.

High-vacuum casting is an example. It is an expensive process for metals that must be poured within a controlled atmosphere. A vacuum chamber is connected, through a valve, to one end of the mold. The metal is then pulled directly into the mold by the vacuum. A high degree of finish, accuracy, and detail is possible.

Frozen mercury patterns are sometimes used instead of wax to form intricate patterns. The advantages of using frozen mercury are that it produces superior surface finish and detail. Its main advantage lies in the ability to weld frozen mercury simply by pressing it together. This allows patterns to be "stacked" or attached together quickly and without fastening or melting processes that could cause distortion. By stacking patterns, extremely complex and intricate patterns can be built with great accuracy.

SAND CASTINGS

Sand castings, used extensively in metals, are the quickest and least expensive of the casting processes. The greatest disadvantages of using sand-casting processes are the relatively poor surface finish and lack of dimensional accuracy. The metal shrinks a great deal in cooling, the molds are not extremely accurate, and

shape requires only a minmum of finishing. (*The Stellite Division of Colso Corp.*)

the sand of the mold leaves a grainy texture on the surface.

Most sand processes are unit (destructible) molds. Some permanent molds are used with gravity processes, however. Some gravity processes employ variations of the sand-casting principles and thus have unit molds. Special processes can be used to improve surface texture, degree of finish, and dimensional accuracy. However, to increase the dimensional accuracy and finish usually requires a greater degree of care in the preparation, handling, and storage of casting patterns and molds.

Special additives such as cement, concrete, plaster, and cereal binders are sometimes added to the sand to impart better smoothness and reduce shrinkage. These additives upgrade the process but make the molding operations longer and more expensive for two reasons. The first reason is the greater care and longer time needed to prepare the mold. The second reason is that there is no easy way to reprocess the sand; it cannot be recycled or reused without special cleaning and purification processes which also add to the production cost.

Green-Sand Casting

Sand-casting processes use fine sand as the material for making the mold; however, several variations are used extensively. The most common is *green-sand* molding. Green sand is simply sand that has not been *cured* or hardened by baking or heating. The natural color of the sand ranges from white to tan, but the sand may turn black from use. The sand does not normally have enough strength to hold its shape. It is mixed with a "binder" (often clay) to give it strength. The mixture is then "tempered" by adding a small amount of water to make it stick together. The sand is reusable, and only small amounts of binders need be added to the sand over a period of several years of use.

Green-sand molding is generally classified by the position or place at which the molding is done. Although the processes and factors involved in the successful casting of an object are virtually identical, small to medium molds are generally prepared at bench-molding "stations." The molds for larger castings weighing from several hundred pounds to one or two tons are done on the foundry floor and are called *floor moldings*. Castings of several tons or more are cast in a pit built into the floor of the foundry. Floor and pit molding processes avoid handling and movement of the heavy and cumbersome molds for large objects. Any movement must be done with large machinery such as cranes and derricks, as the molds

are virtually impossible to prepare in the normal bench-molding fashion.

Bench Molding. *Bench molding* is a means of making a mold by hand or by machine on a workbench. The basic process has been covered in Chap. 6. However, the use of automated machinery, as in Fig. 7-3, has eliminated much of the slow, tedious handwork required for making small to medium size sand molds. Molding machines are also used to compress the sand by jolting or by squeezing. These machines are also used to turn over (Fig. 7-3, bottom right) and to open the mold and thus eliminate much of the heavy lifting and moving from the process as well.

For the typical bench-molding process where large numbers are produced, *matchplates* are also used with *slip flasks* (Fig. 7-3, top left). The matchplate is a specially prepared one-piece metal pattern. The slip flask is a special flask which allows easy removal of the pattern from the sand mold. It is reusable. This allows one flask and one pattern to be used rapidly and reused over and over again.

The molding process begins when a small amount of sand is *riddled* (screened) next to the pattern. Instead of ramming, the sand is slung (using a special machine) or allowed to fall with force to fill the drag (bottom part of flask). The mold is then flipped over, either by the machine (Fig. 7-3, bottom right) or by the operator, and the drag is filled. The sand is compressed by squeezing the mold between the bench top and a pressure plate. Then the mold is rolled over again. The mold is then disassembled to remove the pattern and finish the mold.

During the removal of the matchplate pattern, the matchplate is vibrated to "rap" the pattern for easier withdrawal and to reduce the chance of damaging the mold cavity. After the inspection and finishing of the mold, the mold is reassembled and the slip flask is removed. The mold on the bottom board is then moved to a conveyer for positioning in the pouring area. A good operator can complete a small mold of this type in only one or two minutes.

The controls for molding machines are usually placed so that slight pressures from the operator's foot or knee against control peddles will operate the various features of the machine such as the matchplate vibrator, the jolt mechanism or the squeeze mechanism. The machines are usually powered by air or air-hydraulic combinations.

Typical machines include three basic actions. The first is the *jolting action,* and machines which use only the jolting actions are called *jolt* machines. The second action is the *squeeze* done by forcing a pressure plate against the bench surface, thus squeezing the

Fig. 7-3 The "Rota-Lift" molding machine for making matchplate moldings. (a) Operator begins sequence by positioning the matchplate pattern between the cope (*top*) and drag (*bottom*). The two parts together are called the *slip flask*. (b) The operator raises the lower table to close the matchplate between the cope and drag. (c) The cope is filled with molding sand and a top board is placed over it. The mold is then squeezed and vibrated to pack the cope. (d) After striking the sand in the cope level, the bottom board is placed over the cope and the flask is turned over with the drag on top. The drag is then filled and rammed and the mold is rotated again for finishing. Notice the knee control pedals. (*The Osborn Mfg. Co.*)

mold in between. Machines incorporating these principles are called *squeeze* machines. Third are machines that position the molds and *rotate* them, enabling the operator to work on the cope and the drag without lifting the mold, are called *rollover* machines.

Machines that incorporate the jolt and squeeze features are sometimes called *jolt-squeeze machines* and these are quite common in production work. A machine that incorporates all three functions is called a *jolt-squeeze-rollover* machine and is also quite common for production of castings weighing from about 80 to 200 lb. [36.3 to 90.8 kg].

Floor Molding. Larger sand castings that are both difficult to move and require equipment to move them are often made directly on the foundry floor. Floor molding as shown in Fig. 7-4 helps to eliminate the need for heavy, bulk molds and also eliminates the possibility of mold damage from moving and handling. Floor molds are made individually, take considerable time to construct and are generally expensive. However, large, heavy objects must be cast this way. The mold may be several feet across, and the operators and molders may even walk over the mold surface while they are slinging sand into place and ramming it with portable pneumatic rammers. The mold is filled by a mobile sand slinger. Such molds are extensively used in the casting of parts for large industrial machines such as housings for turbines and generators, forges, large industrial engines for ships and factories, and other similar items. The basic molding principles and processes are identical to those used in bench molding except for the size and some equipment used in moving cores, sections, sand ramming, and disassembling of the mold.

Pit Molding. Pit molding, as shown in Fig. 7-5a and b, is similar to floor molding except that it is used on even larger castings. Pit molding also utilizes the floor of the foundry, but the mold is assembled in a pit dug in the foundry floor. The pit is normally lined with concrete and the walls are padded or lined with green sand to form the mold walls. The pit is the flask. The cope or top portion is usually constructed from cores or from cope sections made separately and assembled to form the entire top. Pit molding is normally done on extremely large shapes weighing several tons or more, such as industrial machinery, hydroelectric generator parts, and large industrial engines.

Special Molding Sands. Where smoother finish or greater accuracy than can be obtained by normal sand casting is required, large floor and pit molds can be made using several variations of the basic sand casting processes. A *loam mold* is constructed of approximately 50 percent molding sand and 50 percent clay. The mold is usually built up around a supporting structure of common brick.

Fig. 7-4 A mobile sandslinger rams these large molds for the production of locomotive castings.

(a)

(b)

Fig. 7-5 (a) A pit mold under construction. The finished casting will become the crosshead for a 50,000 ton hydraulic press; (b) the pit mold partly assembled with the cores. (*Mesta Machine Co.*)

Cement-bonded molds are similar to loam molds except that 8 to 12 percent of a high-early-strength cement is used instead of clay to bond the sand together. Permeability of the mold is maintained by a low-cement content that allows great strength with a relatively small amount of concrete. Molds may also be built up from sections made from previously cured sand and cemented together to form the complete mold.

CONTROLLING GREEN-SAND CHARACTERISTICS

Green molding sand is carefully selected sand mixed with a fire-resistant clay such as bentonite which gives it the strength to make it stick together. As dry clay and dry sand do not naturally stick together, water is normally used to mix the sand and clay together to provide the bonding of the sand. Green sand must be properly mixed to provide enough water for adequate bonding strength. It should not contain large amounts of water which could cause gas problems as a result of hot metal striking moist sand and creating steam. Steam creates pockets and pressure within the mold that could cause casting defects. In production work, great care is maintained in the mixing of the sand so that it contains sufficient water for bonding and also considers other factors such as the size, thickness or thinness, shape of the object to be poured,

and the pouring temperature required for the metal to be used.

Variables to consider in the proper mixture of the sand are the ingredients of the sand* itself, the methods of mixing the sand (Fig. 7-6), and how the sand is prepared for molding, which includes the length of time needed to convey it to the molding stations, the methods used for making the mold, and the pressures used in molding to cause the sand and clay to bond. Casting variables include the heat at which the metal will be poured, the weight of the metal, the shape of the mold, the type of gating, and the force of entry of the metals into the mold cavity.

Two basic types of sands are *natural sand* and *synthetic sand*. Natural sand is sand that already has sufficient clay and needs only tempering (the adding of water) and conditioning (the even mixing of the sand) to be ready for use. Synthetic sands are simply natural sands without sufficient clay content and are compounded by mixing them with clay and other ingredients to provide the desired bonding characteristics.

Certain additives may be used to impart special qualities to sand as they are desired. Fluids such as alcohol may be added to speed the drying of the sand. In other cases, cereal binders much like wheat paste are added to the water to provide extra strength for

---

* Particle shapes and size distribution of sand grains are important also.

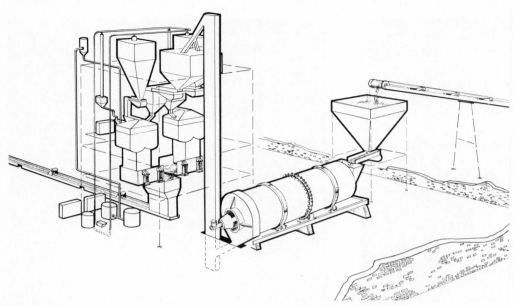

Fig. 7-6 Foundry sand is continually reprocessed and remulled. Here the sand is conveyed to a rotary furnace where impurities are reduced. The sand is then elevated to hoppers and mulls for mixing. After mixing, the sand is conveyed to various molding stations.

the bonding of the sand. The sand is frequently treated with cereal binders made of corn starch to give the thin letters and figures of the mold enough strength to withstand the force of the molten metal pouring into the mold cavity.

The mold is also sprayed with dryer and various mold preparations and allowed to dry thoroughly before it is closed. Mold coatings can be mixtures of minerals or cereals to add strength, stability, or added surface smoothness to the finished casting for special applications such as plaques and other decorative pieces, usually made in small quantities for special orders. Additives and driers are not normally used for the production of large quantity items.

Large bronze plaques, for instance, are thin and must be poured at hotter than normal pouring temperatures so that the metal will remain fluid long enough to flow throughout all parts of the thin and intricately shaped mold. If the metal is not poured at a sufficiently hot temperature, the heat absorbed by the walls of the mold will chill the metal and cause it to solidify before it reaches all parts of the mold. Such a casting is generally poured very hot and quickly so that the metal rushes swiftly into the mold. In doing so, it usually exerts a strong force as it strikes the very thin parts of the mold. The molding sand for such a mold would contain cereal binders for added strength and would be carefully dried to prevent gas pockets from forming.

Other additives commonly used include *pitch* and *asphalt,* which improve the hot strength of the sand and *sea coal,* which is finely ground soft coal that improves the surface finish and makes the casting easier to clean. *Graphite* may be added to improve the stability of the mold and increase the surface smoothness. *Gilsonite,* which is an asphaltive material, is used in a similar manner to sea coal. *Fuel oil* is also sometimes added to the sand to improve its moldability. *Wood flour,* which is actually made from corn cobs and cereal husks, improves the collapsibility and expansion qualities of the sand. *Vermiculite* improves the hot stability of sand and is also used to insulate risers so that they remain molten longer to better compensate for shrinkage in bulky castings. *Dextrenes* are used in molding sands to increase the dry strength and to improve the edge hardness of the molds. The various properties that are important to green sand include the following:

A. *Green strength*   Green strength refers to the plasticity needed to make the mold of sand. It is the ability of the molding sand to clot or stock together to retain the necessary shape.

B. *Dry strength*   Dry strength refers to the sand strength needed to retain the shape of the mold cavity as it dries. As the metal touches the walls of the cavity, it evaporates the water, leaving a dry sand.

C. *Hot strength* and *thermostability*   Hot strength refers to the strength of the sand as metals heat it after it has been dried. Hot-strength characteristics and thermostability characteristics enable the sand not to deteriorate or to change in dimension at the temperatures of approximately 212°F [100°C]. This is the boiling point of water and the molding sand rarely gets hotter before the metals solidify. Once the metal solidifies and dries the edges of the mold, the sand may get much hotter. However, at that point the metal has already solidified and the dimensional stability of the sand is no longer quite as critical.

D. *Permeability* Because the water in the molding sand causes steam and gases to form as the metal touches the sand, some means for allowing these gases to escape is essential. Although a large portion of the gas escapes through risers and vents, a great deal must also escape through the sand. Permeability refers to the porosity which enables the steam to escape through the sand. Adding binders or clay can cause reduction of the permeability that can cause defects in castings due to gas pockets and pressures.

E. *Refractory quality* Refractory qualities refer to the sand's ability to maintain its integrity. In other words, the molding sand does not melt, bend, or otherwise deform in the presence of high heat. Bentonite clay, the same clay as used to make fire brick, has high refractory qualities. If the sand were to melt or to otherwise deform in the casting process, the sand could fuse with the casting. These defects would result in poor quality products and nonreusable sand.

F. *Flowability* Flowability refers to the responsiveness of the sand to the mold process and the ease by which it takes the desired shape. Sands with greater flowability take less pressure to mold them. Pressure is a vital factor in molding because great pressure reduces the permeability of sand and increases the possibility of defects due to the trapping of gases within the mold.

G. *Collapsibility* Collapsibility is a desirable quality in sand and refers to the ability of sand to be shaken loose or removed after the casting has solidified. If the sand contains binders or clays that become extremely hard when they are heated and dried, the sand is difficult to remove from the casting.

Testing Sand

Because desirable qualities of the sand vary with the type of job to be performed, and because qualities of the sand are an essential factor in determining the quality of the casting, great care is taken to control the nature of the sand used. Tests are conducted to determine the initial qualities of the sand used and the necessary ingredients to change the sand as desired. In addition, sand testing is done continuously during the molding process to make sure that the sand meets the molding requirements throughout the molding process. Continued inspection is necessary because certain additives may be exhausted through use and may need to be replenished to compensate for such losses. In addition, almost all sand deteriorates and looses certain binding qualities from extended use. Small particles (dust and fines) are formed in use. These reduce the porosity (permeability) of the sand, and must be periodically removed.

Various tests are performed to maintain the proper moisture content, to determine the various strengths of the sand and to test the binder content and permeability. Figure 7-7a–e shows the testing equipment used in controlling sand qualities in the laboratory of a foundry.

Moisture content is tested with a *moisture teller* (Fig. 7-7a). The moisture teller drives heated air through a sample of sand at a constant rate. Moisture content may be determined by the length of time needed to dry the sample.

Strengths are tested on a universal sand strength testing machine (Fig. 7-7b). The strength of a given sample of material may be tested for green and dry strength and may be exposed to compression testing, shear testing, tension testing, and loading testing, on the same machine. This machine uses a cylindrical plug of sand approximately 2 inches in diameter and 2 inches high. The test plugs are prepared in a special sand rammer (Fig. 7-7c) so that the plugs are prepared with uniform impact and size. The rammer uses a special tube to control sample size. The number of times the dead weight drops and tamps the sand determines core strength.

Permeability is tested by a permeability meter (Fig. 7-7d) measuring the amount of time necessary to force a given quantity of air through a sand sample. Under a constant air pressure, sand with less permeability will pass less air than sand that is more porous or permeable. Usually, the amount of air is relatively small (measured in cubic millimeters or centimeters of air). The test measures the amount of air passing in a specified time of one or two minutes.

Mold hardness, clay content, and sieve (Fig. 7-7e) analysis are also performed on the sand. These determine fineness and grain size of the sand, the exact clay or binder content, and how hard the surface of the mold becomes in the molding process.

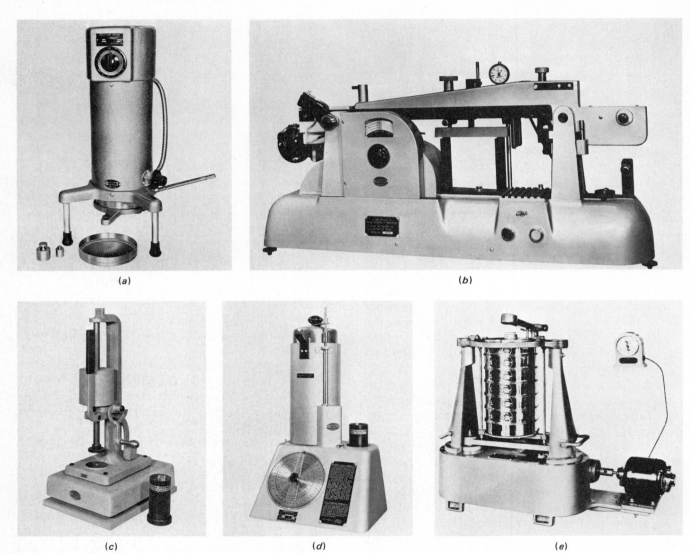

Fig. 7-7 (a) Moisture teller used to determine moisture content of molding sands; (b) universal sand strength machine for green and dry testing of foundry sands; (c) sand rammer employed to ram sand specimens for sand testing; (d) direct-reading universal permeability meter for measuring the green and dry A.F.S. permeability of foundry sands, cores, washes, investment materials, refractories, and other porous substances; (e) sieve shaker for determining fineness of foundry sands. (*Courtesy Harry W. Dietert Co.*)

CURED SAND MOLDING

Shell Molding

The process of forming a thin shell of sand bonded by a thermosetting plastic resin is known as *shell molding*. Figure 7-8 illustrates the process by which the shell is formed and cured over the pattern. The heat of the pattern melts and bonds the plastic and sand mixture to form a thin shell over the pattern. The remainder of the sand and resin mixture not affected by the heat of the pattern is returned for later use.

The thin shell of plastic and sand is then moved into an area where it is baked to cure and harden into a sturdy but thin shell. After curing, each shell is removed from the pattern and the shells are assembled for use.

Shell molding was originally developed for making cores but is now extensively used for production casting of parts (Fig. 7-9). Shell molds produce castings which have a greater smoothness and dimensional accuracy than parts cast in the normal green-sand-casting process. Although the shell molding process is

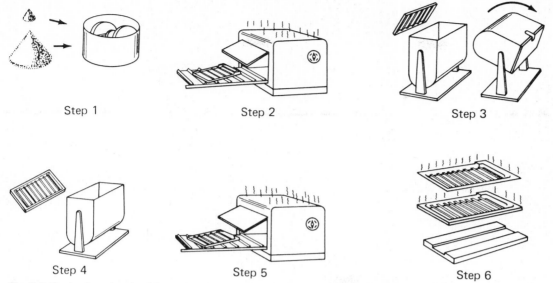

Step 1   Step 2   Step 3

Step 4   Step 5   Step 6

Fig. 7-8 Basic steps involved in shell molding: (1) mixing the resin and sand; (2) heating the pattern; (3) investing; (4) removing the invested pattern; (5) curing; and (6) stripping the cured shell from the pattern. (*Caterpillar Co.*)

somewhat more expensive than green-sand molding, it is much less expensive than plaster and investment molding and is much more adaptable to production processes.

The shell-molding process is used extensively to cast iron and steel products as well as those of nonferrous metals. This is a great advantage because many of the casting processes that give great dimensional accuracy cannot be easily used with iron and steel. The use of shell molding, however, is generally re-

stricted to small- and medium-size castings. The cost of preparing the shell molding pattern is more expensive than for the patterns used for bench molding. However, the greatest expense in the cost of the shell-molding process is the cost of the aldehyde and phenolic resins used as binders. The fact that the sand used in shell-molding processes cannot be recovered nor can the binders and resin be recovered also increases production costs. These cost factors restrict shell molding to high-volume production of accurate castings of medium size.

The shells formed for the shell-molding process are usually made in halves and assembled to constitute the finished mold (Fig. 7-10). Cores, chills, and chaplets are also inserted as needed before the closure of the mold. The gates, sprues, runners, and risers are formed in the basic shell so that the amount of handwork and finishing is greatly reduced. The cured shell sections are assembled by clamping or gluing.

PLASTER MOLDING

Plaster molding is particularly useful in casting copper and aluminum-based alloys which have relatively low melting points. Plaster molding features the use of molds constructed of a mixture of sand and plaster or gypsum. Ingredients such as talc, asbestos, silica flour, and other additives may be mixed with the sand and gypsum to produce characteristics pertaining to shrinkage, heat resistance, and moisture retention.

Fig. 7-9 Partial view of shell-molding foundry making cores and molds. (*Foundry Equipment Division, The National Acme Co.*)

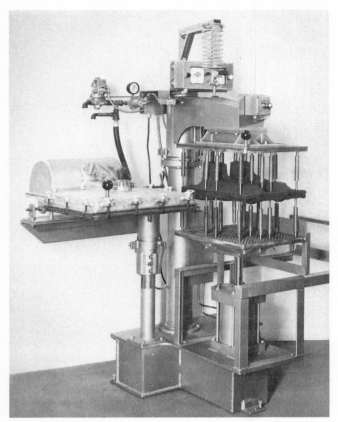

Fig. 7-10 A hydraulic shell-mold bonding press used to assemble the two mold shells to form the completed mold.

Figure 7-11 shows examples of plaster molds and products.

The plaster molding process gives great accuracy ranging from ±0.005 to ±0.015 in [±0.127 to ±0.381 mm], depending upon the size of the casting. Extremely fine surface textures can also be produced using this process. The great disadvantage is the long curing time needed to dehydrate the molds to prevent the formation of steam pockets. However, this type of molding is particularly useful in the preparation of industrial tools such as matchplates, forming dies and permanent molds, particularly for use in plastic- and rubber-molding processes. Such molds need not be refinished for use with many of the products such as tread molds for rubber tires.

Molds are constructed from sand and plaster powder mixed with water to form a slurry. The slurry is then poured around a pattern which has been treated with silicone or petroleum products so that the plaster will not bond to the pattern. The pattern is removed when the plaster has set. The mold is then cured and the casting is produced. Plaster molding is normally a unit-mold process and each object requires a new mold.

CO_2 Molding

The CO_2 process is similar to shell molding in the cost and quality of castings. Large CO_2 shells are also reinforced with plaster, as shown in Fig. 7-12. The process is used to form a mold made of cured sand and

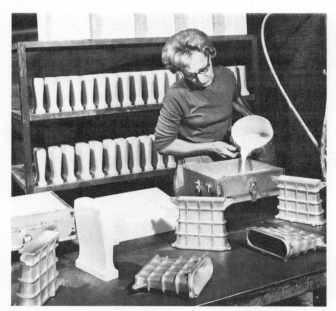

Fig. 7-11 Pouring plaster mold for one of 14 different premium-engineered cast parts being made for an Air Force supersonic fighter. (*Alcoa.*)

Fig. 7-12 Workers are assembling a CO_2 mold to produce a transmission housing weighing 186 lb. The shape in the foreground is part of mold assembly. (*Alcoa.*)

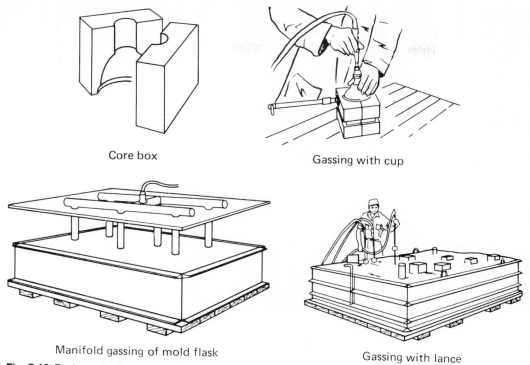

Core box

Gassing with cup

Manifold gassing of mold flask

Gassing with lance

Fig. 7-13 Basic methods used for gassing in the CO_2 process. (*Alcoa.*)

resin. Unlike the shell-molding process, however, thicker sections can be easily formed and cured without the use of heat which can cause dimensional variations in the mold itself. The special resin is cured by exposure to carbon dioxide (CO_2) rather than by heat (Fig. 7-13).

The process was originally developed to manufacture cores. It was found that intricate cores could be made and cemented together successfully. This core assembly was then extended to use for the mold as well. Although somewhat more expensive than shell molding, the CO_2 process is widely used, particularly for larger castings that require complex shapes which are hard to mold due to the need for several parting lines.

PERMANENT MOLDS

Permanent molds have extended reusability and are made of metal (Fig. 7-14). Metals that require high casting temperatures, such as iron (2300 to 2700°F [1259 to 1481°C]) and steel (3000°F [1647°C]), shorten the life of permanent molds, which are mainly used for high-volume production in molding metal products made of zinc-, copper-, and aluminum-based alloys. They are also extensively used in molding plastics, pottery, and rubber products.

Many of the unit molding processes that provide good accuracy and finish, such as the cement bond casting and plaster casting methods, are extensively used to produce permanent metal molds for making such things as rubber tires, plastic toys, glassware, and numerous other nonmetallic items. Permanent molds may often be used without finishing the mold cavity. However, permanent molds are frequently machined and finished to provide for better surface finish and dimensional accuracy.

It is sometimes necessary to cool the permanent molds as they are being used. This is done by rapidly circulating water or air. The molds may also be moved from the molding area to the storage area to cool naturally.

Permanent molds are very expensive. However, where the number of castings required is sufficiently large, great initial expense is justified to increase the possible production rate and to decrease machining and handling expense (Fig. 7-15).

SLUSH AND SLIP MOLDING

Slush molding is used for production of metal goods where wall thickness and accuracy of the interior are not critical factors. A process similar to slush molding (called *slip casting*) is also used extensively in the ce-

Fig. 7-14 A permanent mold casting for tandem axle beam saddle is being removed from the mold. Note the ladle at top left which was used to pour the casting. (*Alcoa.*)

ramics industries for making plaster and pottery objects.

Slush molding with metals is done by pouring molten metal into a cold mold cavity. The metal next to the cold mold surface is chilled and solidified rapidly. The remaining molten metal in the mold is quickly poured out of the mold again, leaving the thin solidified shell. The mold is then opened and the thin shell is removed. The solidification of the casting shell

Fig. 7-15 Shows the casted part removed from the permanent mold. (*Alcoa.*)

mold occurs through absorbtion of heat from the molten metal.

The slip-casting process in ceramics industries features solidification of the fluid-casting materials, called *slip*, from hydration. The molds are normally made of plaster or ceramic materials which absorb water. The slip is poured into the mold cavity, as in Fig. 7-16*a*, and the slip near the mold surface is solidified by the rapid loss of its moisture into the walls of the mold. The excess slip is rapidly repoured from the mold, leaving the thin, dehydrated shell next to the mold surfaces. Figure 7-16*b* shows a broken section. The mold is then opened and the shell removed for later setting and firing. Because of the absorbtion of the water into the mold itself, such molds cannot be used more than a few times before they must be dried and cured. Figure 7-17 shows an example of the mold and casting, also some other typical slip-molded products.

Pottery and ceramic products produced by such materials are not very accurate due to great shrinkage of the materials. This shrinkage is due to the amount of water that must be rapidly absorbed and due to further loss of moisture from in settling and in firing and hardening processes. Thus shrinkage occurs in two places: first, in the mold and secondly, during the hardening and curing processes after the product has been removed.

Two methods are commonly employed to reduce the shrinkage of plaster products. First, less water is

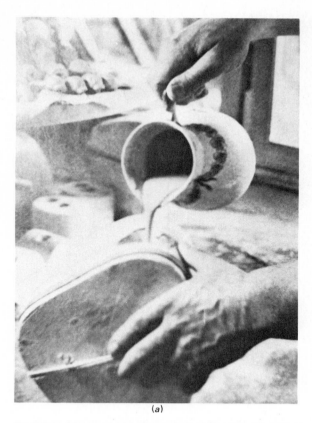

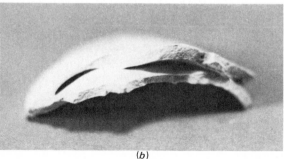

Fig. 7-16 The slip casting process and products. (a) The slip is poured into a plaster mold (*Courtesy Bing and Grondahl of Copenhagen*); (b) broken fragment shows how thin shells are formed by the rapid drying of portions near the edge of the mold.

used. However, if less water is used, the fluidity of the slip is not as good. To keep a high fluidity without adding materials *deflocculents* are used. A *deflocculent,* such as water glass, increases the fluidity without adding the water. The second method requires the addition of small particles called *grog.* Grog is composed of *bisque,* which is ground material that has been previously fired and is thus impervious to water. Bisque may be made by reprocessing fired ceramics or by processing materials such as fired *bentonite* (clay similar to firebrick) and other materials. Because these materials do not absorb water and have already been

fired and shrunk on previous processes, their addition reduces the total shrinkage.

DIP MOLDING

Dip molding is used in metal, rubber, and plastic industries. It essentially consists of dipping a shaped pattern into fluid material and withdrawing it. The temperature of the pattern causes the metal or plastic to solidify much as in the slush-molding process. The pattern is quickly withdrawn, leaving a thin but somewhat irregular coating on the pattern, as illustrated in Fig. 7-18.

For metal or plastics, a chilled pattern is inserted. For ceramic products the pattern is drier and will absorb moisture. For some plastics, a highly volatile solvent is used to liquify the plastic. A heated pattern is used; the heat quickly evaporates the solvent leaving a solid film over the pattern.

Dip molding is used extensively to form thin metal cases in which plastic parts may be mounted. Dip molding is also frequently used to put a plastic coating over metal objects such as tool handles and small coin purses, and to give weather-resistant coatings. The toy balloon is an example of a dip-molded rubber product.

Fluidized coating is a variation of the dip-molding process. Heated metal objects are immersed into powdered plastic resins. The coatings are used for decorative coloring, weatherproofing, and electrical insulation. However, the heat required for the fluidizing process makes it unsuitable where the heat of the process could distort the shape of the object.

The advantages of the dip-molding process are that it is fast, inexpensive, and requires very little equipment. The disadvantages are inaccuracy, poor control of wall thickness, and poor strength. Dip molding for metals is generally restricted to the zinc- and aluminum-based alloys.

ROTATIONAL MOLDING

Rotational molding is a process where a mold, usually round or cylindrical, turns slowly, allowing the force of gravity to coat the inside surface of the mold with the desired shape, as in Fig. 7-19. The process is used extensively to produce hollow rubber and plastic balls and items such as round plastic garbage containers.

Rotational molding allows a small amount of material to be poured into a mold, as opposed to completely filling the mold, as in slush molding. Better surface qualities and more accurate wall thicknesses are usually obtained by rotational molding than by slush molding. The process is well suited for produc-

(a)   (b)   (c)

Fig. 7-17 (<i>a</i>) A typical mold and product of the slip-casting process. Note the hollow section in the bottom part of the casting. (<i>b</i> and <i>c</i>) Other typical slip castings.

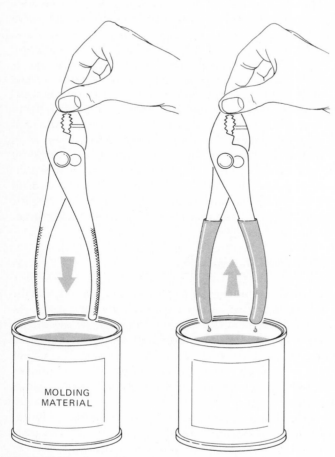

Fig. 7-18 The dip molding process.

tion of round and cylindrical shapes where wall thickness is not critical. The equipment for rotational molding, however, can be expensive both to purchase and to maintain. However, large items can be produced with relative ease, and it is possible to mold hollow products without complicated pressurized molds. The primary disadvantage is the long cycle for producing a single item.

Molds for plastic and rubber products may be rather simple in construction, but they must be well polished to produce a good surface finish and must be carefully joined to minimize flashing or visible seams at the parting line.

FLEXIBLE RUBBER MOLDS

Flexible molds made of rubber or flexible plastics play an increasing role in the production of cast plastic and plaster products. Flexible molds have several advantages over traditional molds. The master pattern for a flexible mold is shown in Fig. 7-20<i>a</i>, and Fig. 7-20<i>b</i> through 7-20<i>h</i> (see page 130) illustrates various steps in the flexible-mold process.

First, the master pattern is simply coated with a liquid that solidifies to form the mold directly over the pattern. The flexible mold can then be stripped from the master without regard for irregular shapes, pattern draft, and similar conditions because the mold will stretch during removal and snap back to its original shape when removed. The second major advantage of flexible molds is that pieces can be duplicated

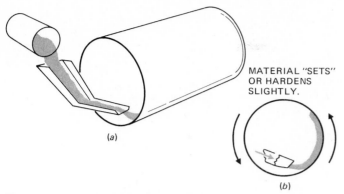

MATERIAL "SETS" OR HARDENS SLIGHTLY.

(a)

(b)

Fig. 7-19 Rotational molding. (a) Material is poured into a rotating mold; (b) material coats the inside of the mold as the mold turns.

from an original without preparing a special master pattern.

Modern silicone plastics allow molds to be made that will give perfect reproduction of items with deep undercuts (great reverse draft) and yet provide durability for several hundred castings. Epoxy and urethane plastics as well as plaster products are commonly cast using such molds. Products cast by this process include special carrying trays, "carved" panels for furniture, picture frames, and decorative items.

Casting may also be prefinished by combining stain or paint with the mold-release compounds used to prevent the castings from sticking to the mold. The inside of the mold is coated with the combined release and finish material. When the casting is poured, the surface of the casting absorbs the finish so that the casting is completely stained or painted when it is stripped from the mold.

CASTING CONCRETE

Concrete is cast in both unit and permanent molds. Although the most widely known use of concrete is for floors and foundations of buildings, concrete is a cast material and manufactured concrete products are extensively cast by a variety of methods.

Concrete hardens very slowly through chemical action. Once the hardening process begins, additional water will not prevent the completion of the hardening process except when the water is present in such large quantities that it actually washes or "leeches" away the cement from the sand and gravel mixture.

When hardening, the chemical action generates heat that can prevent proper hardening rates throughout the casting. Large, massive concrete sections often are constructed so that pipes embedded within them can circulate cooling water for proper

hardening. Hoover Dam near Las Vegas, Nevada, for example, used hundreds of miles of water pipe for this purpose. Concrete that dries in hot sunlight must often be sprinkled with water and sometimes must be covered with damp cloth to prevent surface scaling due to the uneven hardening rates between the surface and the mass underneath.

When concrete must be poured for underwater structural supports, temporary dams called *coffer dams* are used to keep out the water. The coffer dams are erected from wood or sheet metal sections and are continuously pumped dry until the concrete is poured. While concrete will set under water, moving water will cause damage. After the concrete has set, mold forms and coffer dams can be removed.

Concrete first hardens in a setting process. After the initial setting, concrete continues to harden for a long period of time. Complete hardening continues from 28 to 30 days under normal conditions. However, hardening can continue for much longer periods.

Concrete can be reinforced by adding steel rods before the pouring of the concrete, as in Fig. 7-21. Steel rods add strength and toughness and prevent the loss of strength through cracking. Concrete has a large expansion and contraction rate from changes in temperature or weather, and the addition of steel to the concrete reduces cracking and separation of the pieces.

The use of concrete molding in building construction provides monolithic, or "one stone," structure and strength. Concrete is commonly cast for houses, apartments, offices, and farm buildings, and it can be made into masonry blocks.

Concrete may be cast on the building site or it may be cast in permanent metal molds or it may be cast in sectional plaster molds in the same manner that porcelain is cast to produce vases and figurines. Concrete

Fig. 7-20 The basic steps in flexible rubber molds. (*a*) Casting epoxy plastics in flexible or silicone molds. A brass master pattern is carefully machined to size; (*b*) the liquid silicone is forced carefully into all parts of the master pattern; (*c*) the flexible, solidified silicone is peeled from the master pattern; (*d*) the silicone mold is coated with special agents to prevent the epoxy plastic from sticking to the mold. (*General Electric.*)

Fig. 7-21 Concrete reinforcement.

forms take many shapes and sizes. Concrete cast on local site is usually made into wall sections, using forms built on the construction site. The forms are erected to form the mold for the shapes desired for the particular building. Structural steel is added as needed when the concrete is poured. Once the concrete is set, the forms are removed, leaving a finished concrete section in the shape desired.

The forms are generally coated with a material to prevent the concrete from sticking to them. Although the form may be constructed of wood, plywood, or metal, the most common practice in construction industries today is to use sectional units that can be assembled into the mold or form desired, as shown in Fig. 7-22. These forms provide stronger, more easily assembled features with great reusability. The concrete forms can be easily constructed and easily disassembled.

Another variation of concrete can be achieved through *prestressing*. Prestressed concrete is concrete that has been compressed under certain conditions during the setting process. This is achieved by an-

choring the reinforcement cables or rods to powerful hydraulic jacks at each end of the mold form. These rods or cables are then stretched under extremely high tension and the concrete is poured around the tensioned cables.

After the concrete sets and hardens to form a bond with the roughened rods or cables, the tension is slowly released, squeezing the concrete particles tightly together. The rates for the release of tension are very carefully computed and controlled to maintain the desired quality of the concrete product.

As illustrated in Fig. 7-23, prestressing can produce concrete forms that are actually strengthened by the addition of weight or pressure in one direction. The compression of the force on the form will tend to make the product stronger. Prestressed concrete forms are normally used in manufacturing beams of concrete for construction purposes. These are used for the construction of industrial and office buildings and the construction of bridges and highway over-

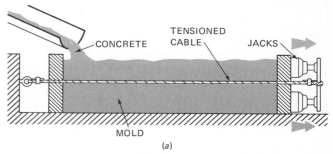

(a)

THE CABLE OR ROD IS TENSED (STRETCHED) AND THE CONCRETE POURED.

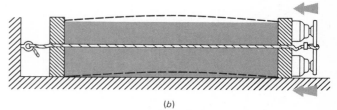

(b)

RELEASING THE PRESSURE CREATES INTERNAL PRESSURES ON THE SOLIDIFYING CONCRETE PARTICLES CAUSING A SLIGHT BOWING.

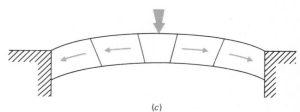

(c)

A KEYSTONE EFFECT IS CREATED THAT ACTUALLY STRENGTHENS THE BEAM WHEN FORCE IS APPLIED.

Fig. 7-23 Prestressing.

Fig. 7-22 Sectional units.

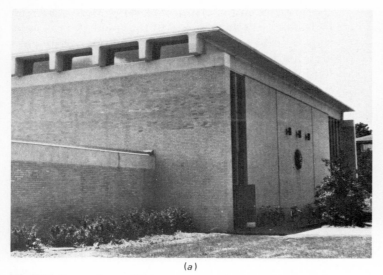

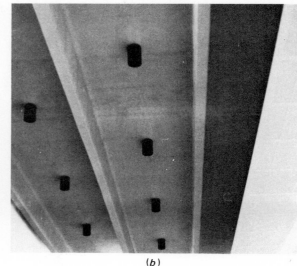

(a)　　　　　　　　　　　　　　　　(b)

Fig. 7-24 Spans of over 100 ft.

passes. Spans of over 100 ft [31 m] without supports have been achieved, as shown in Fig. 7-24.

Concrete castings form materials for mosaic flooring and building panels, and may also be used in other processes for sand casting. Concrete products can also be combined with asbestos fibers to produce hard, durable, fireproof roofing materials and panels.

When concrete is to be combined with other materials such as colored rocks to make panels where the faces of the aggregates materials are to show, a mold form is floored with a paper containing water-soluble cement. The stones that are to be visible are glued to the flooring material by the water-soluable cement. The concrete is then poured onto the stone-covered paper. The water from the concrete dissolves the cement holding the stones in place at the same time that the concrete begins the hardening and setting process. Thus, the stones are held in the desired position at the same time. Final finishing is achieved

Fig. 7-25 Building panels.

simply by washing the face of the panel to remove the paper from the stones. The building in Fig. 7-25 features outer walls and a courtyard made with these panels.

Concrete products such as the round sections for use as concrete railings and banisters are made in permanent molds. These molds are sectional molds much like those used for making plaster and porcelain products. They are made by casting concrete or plaster around clay or plaster forms. The walls are formed for the removable sections by inserting divider pieces of metals, modeling clay or wood. The molds are then assembled and wrapped with a strap to keep the mold together and the concrete then poured into the cavity as indicated. When the concrete has set, the molds are then removed, leaving a casting that can be trimmed and finished into an attractive product.

Concrete is also used to manufacture products from open permanent molds. These molds may be made from wood or metal. The open mold is simply a mold form that has only one piece. The concrete is poured into the open mold, leveled, or struck with a rod and allowed to set. Once set, the concrete product is taken from the mold simply by turning the mold upside down.

REVIEW QUESTIONS

7-1. How are the lost-wax and full-mold processes similar?

7-2. What casting process is used most extensively? What are its advantages and limitations?

7-3. What process can be used to cast large objects such as parts for heavy industrial machines?

7-4. What process could be used to cast a smooth, intricate shape that would be virtually impossible to machine to shape?

7-5. What products can be made using rotational molding?

7-6. What is slush molding?

7-7. Why is the conditioning of molding sand important?

7-8. What qualities are factors in controlling molding sand?

7-9. What are the advantages of shell molding over green-sand castings?

7-10. What are the advantages and disadvantages of using permanent molds?

CHAPTER 8

PRESSURE-CASTING AND MOLDING PROCESSES

Pressure casting is almost exclusively used with permanent molds. Exceptions are made for some variations of investment- and full-mold castings where centrifugal force is used to force metal into intricate molds for products made in small numbers. In many cases, pressure-casting processes are almost identical with gravity-casting permanent mold processes. Equipment is very similar and the mold may be identical. The difference is in the method of forcing the molten or fluid material into the mold cavity.

Pressure-molding processes are like pressure-casting processes in that a material is forced against the contours of a mold or die. However, the material in molding is not in the fluid state when it enters the mold. The material may be placed in the mold and then softened, or it may be forced into the mold in a pliable, plastic condition.

DIE CASTING

Die casting is perhaps the most widely used of the pressure-casting processes. Die casting is used for such materials as glass, rubber, plastic, and metals. Versions of it are also known as injection molding (specifically for nonmetallic material such as plastic and rubber). Die-casting metals are restricted to nonferrous metals such as the zinc and aluminum alloys. The die-casting processes feature extreme accuracy, dimensional control, surface finish, and high-production rates. Typical products include hardware such as the handles shown in Fig. 8-1, for business office machines; cases for portable power tools; plumbing fixtures such as spouts, handles, and controls for water faucets; carburetor parts for engines; toys; and many others. The major disadvantages in die casting are (1) the limitation which excludes the use of ferrous metals, (2) the extreme expense in the initial tooling operation and (3) porosity from oxides trapped during the injection. The dies can be very complex and may contain large numbers of moving parts to allow for the withdrawal of complicated shapes from the molds. Small parts are frequently ganged together so that one molding cycle is used to form several pieces. This complicates the preparation of the molds and several identical shapes must be made into the mold.

In the die-casting process, fluid material is forced into the mold cavity with sufficient pressure to fill the entire cavity evenly and completely. The pressurization enables the process to be used to cast small pieces, complex shapes, and extremely thin sections.

Two methods of pressurizing metals for die casting are used. The *cold chamber process* requires an operator to pour a proper amount of molten metal from a ladle into a chamber where a plunger applies the pressure to force the metal from the chamber into the mold. The metal is called a *shot* and the pressure chamber is called a *shot chamber*. Figure 8-2a shows the cold chamber principle.

The *hot chamber process* is more automatic in that the operator does not load the shot chamber for each casting cycle. As can be seen in Fig. 8-2b, the hot chamber is submerged in the molten metal so that each time the pressure piston withdraws, the shot chamber fills from the pool of molten metal. The hot-chamber process is faster than the cold-chamber process, but the equipment-installation cost is also greater.

INJECTION MOLDING

Injection molds are much the same as die-casting molds. The major difference between die-casting and injection-molding processes is the manner in which the materials are loaded and enter the mold cavity. There are great differences in the fluidity and castability of the materials which dictate differences in

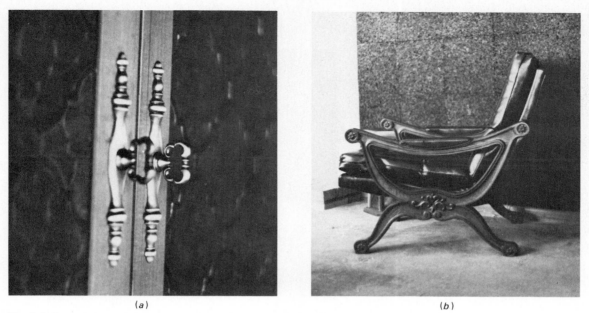

(a)   (b)

Fig. 8-1 Typical die-cast products include the metal door knobs in (a) and the solid cycolac plastic sides of the chair in (b) which were "injection" molded, a version of die casting.

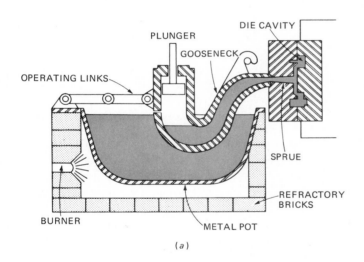

(a)

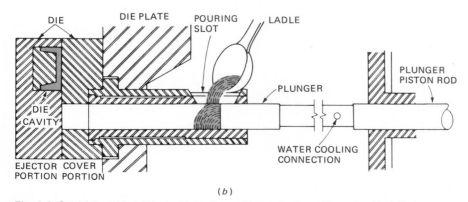

(b)

Fig. 8-2 Cold (a) and hot (b) chamber die casting processes. (*Greenlee Tool Co.*)

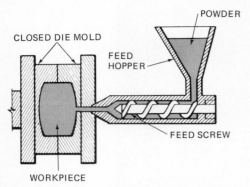

Fig. 8-3 Showing the injection molding process.

Fig. 8-5 Styrene-foam products are manufactured using injection molding techniques.

loading. Glass and rubber products are plasticized by heat and are injected as a hot, viscous material from a shot chamber similar to that of a metal die-casting machine.

Plastics are loaded as small solid particles. The feed augur creates enough friction heat to soften the particles into a viscous mass. The feed augur can then force the softened material into the mold cavity where it is shaped (Fig. 8-3).

After the material has been injected into the cavity, the molds may be chilled by cold water causing the plastic to become hard enough to be removed from the mold. The object is then ejected from the mold by knockout pins.

Typical products made by the injection-molding process include plastic toys, telephone cases, rubber handles and knobs, glass door knobs, and hardware such as cabinet doors, handles, and many other products (Fig. 8-4). One needs only to glance around the home or office to find numerous examples of products made by die-casting or injection-molding processes.

Foamed plastic products are made using one of three variations to the injection-molding process. The small cooler chest in Fig. 8-5 is a typical product. Such

styrene plastics are made into many products. Because of their ability to absorb shock and to act as insulation, they are used to package fragile dishes and intricate machines such as typewriters and cameras and to act as insulation and flotation devices.

Foamed plastic products are made by one of these three variations of injection molding:

1. Small beads which have a bubble of gas in the center are blown into a mold. Steam or heated air is blown through the mold. The heat softens the beads and allows the gas inside the bead to expand. This expansion forces the beads against the mold wall and against the other beads so that they will fuse.

2. Chemicals are added to the resin in the barrel of the injection machine. It generates a gas which is dissolved in the softened resin. Remember that the resin is softened in the barrel by friction heat. On injection into the mold, a metered shot delivers less resin than the mold requires. At the reduced pressure in the mold, the dissolved gas comes out of the resin and forces the resin against the mold walls.

3. Inert gas (usually nitrogen) is injected into the resin at the nozzle. The gas injection causes the resin to foam.

Fig. 8-4 The plastic case for a small transistor radio shows the dimensional control and surface finish possible with injection molding.

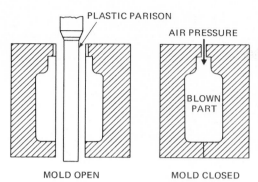

Fig. 8-6 Blow molding is used to make plastic bottles. A plastic tube is extruded between an open mold. The mold is closed and the tube is blown up to take the shape of the mold.

BLOW MOLDING

Blow molding is extensively used for glass, rubber, and plastic products. It is used to form glass jars and bottles, drinking glasses, jugs and bottles made of plastic, rubber nipples for milking machines, toys, and many other products from many materials. The blow-molding process begins by forming a lump of molten material into a tube-shaped section. This section is generally called a *pre-form* or a *parison*. The parisons are transferred to a blow-molding machine where air pressure is applied to the inside of the parison (see Fig. 8-6). The air pressure blows, or forces, the soft fluid material against the walls of the mold to give it the desired shape. A typical machine and products are shown in Fig. 8-7*a* and *b*.

The advantages of blow molding are that it may be used to make products that are dimensionally accurate and have a high-surface finish, and, in addition, the process is comparatively fast. Although the tool-

ing costs are quite high, similar to those for die casting and injection molding, the production rate is also high. The major disadvantage is that the initial installation cost is high, and the process must be used to manufacture items in large quantity. One additional problem sometimes occurs where chemically setting plastics are used. Once the chemicals are mixed, the blow-molding operation must continue until the supply of mixed plastic is exhausted. Otherwise, the plastic will set and harden outside the mold and not be usable.

COMPRESSION AND TRANSFER MOLDING

Compression Molding

Compression molding is a process where a measured quantity of thermosetting plastic resin is placed into a mold cavity. The mold is heated and pressure is applied to the mold. The molten resin is forced by compression to completely fill the cavity where it undergoes a chemical reaction to harden into the permanent molded shape, as in Fig. 8-8. Commonly used compression-molding plastics include phenolic and alkyd resins and the aldehyde and urea products. The resins may be used in the form of powder, granules, flakes, and rope- or rod-shaped briquettes.

The length of the molding cycle varies depending upon the plastic used. There may be a variation of from 3 to 20 minutes before the mold may be opened and the product removed. The mold temperatures must be maintained throughout the process and the temperature ranges—depending upon the materials—from around 250 to 400°F [121 to 204°C]. The mold may be either a single- or multiple-cavity mold, and the mold sections are usually made of tool steel

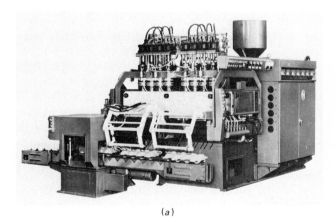

(a)

(b)

Fig. 8-7 (a) A blow molding machine tooled to produce plastic half-gallon containers. (*Unilog Division of Hoover Ball and Bearing.*) (b) These plastic containers illustrate the wide variety of shapes applicable to blow-molding processes. The process is readily adaptable to a variety of plastics, glass, and rubber products. (*Beloit Corp.*)

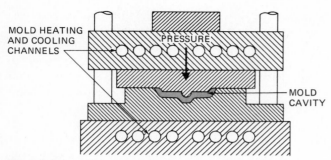

Fig. 8-8 Schematic cross section of a typical single-cavity compression mold. (*U.S. Industrial Chemicals.*)

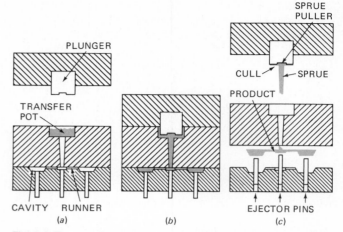

Fig. 8-9 The transfer molding cycle. (*a*) Mold opens and resin is placed in transfer pot; (*b*) plunger forces resin through the sprue into the mold cavities; (*c*) the cull and sprue break loose as the mold opens and the part is removed by ejector pins.

that is highly polished to produce the desired surface finishes on the products. The pressure for the molding process is usually applied hydraulically. Typical products include plastic parts for automotive electrical systems, plastic gears, and the plastic panels used for automobile control panels and interior trim. Because compression forces are used to force the plastic into all parts of the mold cavity, a small space must always be left around the edges of the mold sections for runout of excess resin. This allows the material to fill the mold completely. However, it also forms flashing, which must be removed in an added operation.

Transfer Molding

Transfer molding involves the same principles and materials as in compression molding. The primary difference between the two processes is in the method in which the molding resin is placed or fed into the cavity. In compression molding, the resin is placed directly into the mold cavity and pressure from part of the mold forces the material throughout the mold cavity much as in a forging operation. In transfer molding, the pressure to force the material into the mold cavity is applied, as shown in Fig. 8-9*a–c,* from a different cavity called a *transfer pot.* A mold plunger squeezes the material in the transfer pot into a sprue and through gates into the mold cavity. The resin is heated in the transfer pot under pressure and then a higher pressure is exerted to force the softened resin into the mold cavities. Reinforcing fibers or fabric may be added to the part. Pressures from 6000 to 12,000 psi [420 to 840 kg/cm²] are used during the molding process. Transfer molding requires less finishing than does compression molding, and the use of runners makes the production of multiple parts easily done. A small disadvantage is incurred in that the sprue and runners must be removed from the mold before the next charge is added because the thermosetting plastic cannot be scrapped or reused. The transfer molding process is particularly suited to the

manufacturing of small intricate parts such as those used in manufacture of electrical products (circuit breakers, housings, and insulation segments). Other products include the production of pumps and impeller cavities, switches, and electronic devices.

CENTRIFUGAL CASTING

Centrifugal casting features the use of a rapidly spinning mold to create high centrifugal forces (up to 100 gravities) which force the heavy casting material into the spinning mold cavity. The heavy casting material is forced into the mold, and lighter impurities and gases are forced to the center outside the mold cavity where imperfections are easily removed by finishing and machining, as in Fig. 8-10. Centrifugal-casting processes may be used to produce intricate shapes with high surface qualities, close dimensional tolerances, and shapes that would be difficult to cast under gravity processes due to gating and risering requirements. Normally, castings produced by this process do not require risers and only require a very minimum of gating.

Centrifugal-molding processes are used extensively for casting plastic, concrete, ceramic products, and almost all metals. Centrifugal molding can be used to cast completely enclosed one-piece hollow shapes of relatively uniform thickness. No other casting process is capable of producing such hollow shapes. Disadvantages of centrifugal-casting processes include the equipment cost and maintenance for the rotating machinery and the relatively slow production rate due to the care needed in loading and to the low cycle time.

Two variations of the centrifugal-casting processes

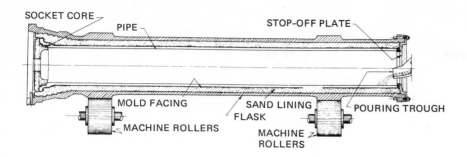

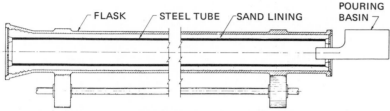

"TRUE CENTRIFUGAL" CASTING-CROSS-SECTION OF A MOLD FOR MAKING A STEEL TUBE AS MOUNTED ON ROLLERS OF THE CASTING MACHINE.

Fig. 8-10 Schematic showing the centrifugal casting process for casting cast iron pipes. (*American Cast Iron Pipe Company.*)

are widely used. The "centrifugal" casting process is used to make one product in one mold. The mold for that product is spun rapidly and the metal flows from the sprue in the center to the outer edges of the mold cavity where it solidifies to form the product. The second variation is called *centrifuge* casting. It differs from centrifugal casting in that more than one mold cavity is utilized. Centrifuge casting is used to produce several distinct and separate parts with each pouring of molten material.

Generally, centrifugal castings, which are produced one at a time, are symmetrical. The insides are usually round although the outside of the casting may take a variety of shapes including square, hexagonal, or other symmetrical shapes. Centrifuge casting may be used to produce objects of irregular size and shape, however.

The molds used for centrifuge and centrifugal castings may be either permanent or unit molds. The molds may be made of metal, plaster, graphite, or either green or cured sand. Typical products of centrifugal casting include rolls, sleeves, rings, bushings, cylinders, and tubes, pressure vessels, pump rotors, and pipe (Fig. 8-11). Typical products of the centrifuge process include turbine parts, jewelry, and other intricate shapes.

CONTINUOUS CASTING

Continuous casting is used extensively for the manufacture of steel-, copper-, and aluminum-alloy prod-

ucts. It is used to produce round rod, wire, and shaped rods similar to those shapes that are produced by extrusion processes.

Continuous casting uses gravity as a pressure to force the material through dies or guides much as a stream of molten metal dropped in a cool chamber would solidify before it reaches ground level, as in Fig. 8-12. The newly formed solid shape remains soft and is allowed to be pulled downward from the die by gravity through a cooled space where the shaped form continues to solidify and harden. The formed material is then sliced to length or coiled for further handling.

The advantages of the continuous-casting process is that material is produced directly from a molten to a solid state to the desired shape without complicated

Fig. 8-11 Centrifugal casting is also used to produce products made from nonmetallic materials such as these concrete pipe sections.

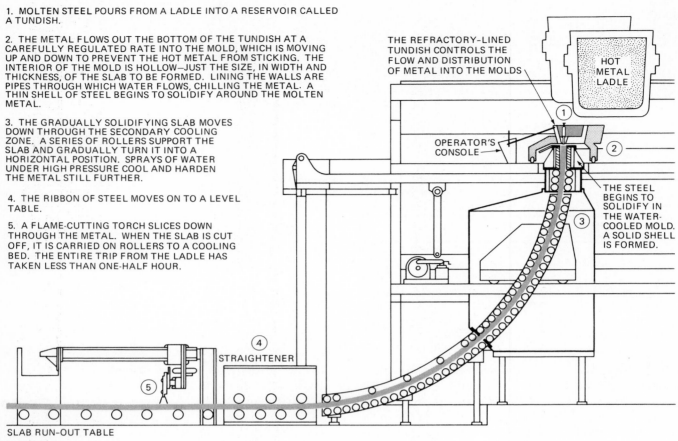

1. MOLTEN STEEL POURS FROM A LADLE INTO A RESERVOIR CALLED A TUNDISH.

2. THE METAL FLOWS OUT THE BOTTOM OF THE TUNDISH AT A CAREFULLY REGULATED RATE INTO THE MOLD, WHICH IS MOVING UP AND DOWN TO PREVENT THE HOT METAL FROM STICKING. THE INTERIOR OF THE MOLD IS HOLLOW—JUST THE SIZE, IN WIDTH AND THICKNESS, OF THE SLAB TO BE FORMED. LINING THE WALLS ARE PIPES THROUGH WHICH WATER FLOWS, CHILLING THE METAL. A THIN SHELL OF STEEL BEGINS TO SOLIDIFY AROUND THE MOLTEN METAL.

3. THE GRADUALLY SOLIDIFYING SLAB MOVES DOWN THROUGH THE SECONDARY COOLING ZONE. A SERIES OF ROLLERS SUPPORT THE SLAB AND GRADUALLY TURN IT INTO A HORIZONTAL POSITION. SPRAYS OF WATER UNDER HIGH PRESSURE COOL AND HARDEN THE METAL STILL FURTHER.

4. THE RIBBON OF STEEL MOVES ON TO A LEVEL TABLE.

5. A FLAME-CUTTING TORCH SLICES DOWN THROUGH THE METAL. WHEN THE SLAB IS CUT OFF, IT IS CARRIED ON ROLLERS TO A COOLING BED. THE ENTIRE TRIP FROM THE LADLE HAS TAKEN LESS THAN ONE-HALF HOUR.

THE REFRACTORY-LINED TUNDISH CONTROLS THE FLOW AND DISTRIBUTION OF METAL INTO THE MOLDS

HOT METAL LADLE

OPERATOR'S CONSOLE

THE STEEL BEGINS TO SOLIDIFY IN THE WATER-COOLED MOLD. A SOLID SHELL IS FORMED.

STRAIGHTENER

SLAB RUN-OUT TABLE

Fig. 8-12 A schematic of continuous-casting installation to cast blooms, slabs, and billets.

or costly operations such as slabbing, rolling, drawing, or extrusion. Wire is normally produced from rods that were slabbed, rolled, and then formed into thick rods, which were then further reduced to size by repeated drawing through dies. The continuous-casting process eliminates most of the intermediate operations from ingot casting to initial drawing. Almost any shape that can be rolled, drawn, or extruded can be produced with great savings in the cost of operation and equipment. The process further reduces the product cost and the operating cost as well as the cost of maintaining special equipment for the intermediate processes.

The equipment is costly to install and many manufacturers prefer to operate existing machinery rather than to purchase new machinery. The continuous-casting process is also used to produce rough rod that is later drawn by regular processes to reduce it to the desired size (Fig. 8-13). This eliminates the disadvantage pertaining to uniform thickness and density of the continuous process while reducing the drawing operations to a minimum.

LAY-UP MOLDING

Lay-up molding is used in the manufacturing of fiberglass products. Fiberglass molding involves the cementing together of glass fibers with chemically setting resin around a mold. The glass fibers may be chopped into short lengths and blown onto the resin-coated mold to form a feltlike mat, or the fibers may be woven into a coarse fabric much like burlap and placed on the resin-coated mold. Lay-up molding involves pressing the glass fibers and resin against the form to provide the desired shape; the pressure must be maintained until the resins have completely set. The mold form must be coated with a chemical to prevent the resin from adhering to the mold.

The lay-up process may use gravity, but pressure systems have certain advantages. Complicated shapes may be molded, and more uniformed wall thicknesses may be achieved with pressurized processes and better wetting of the fibers. Vertical sections may also be more easily molded without sags and uneven wall thickness. The chief disadvantage of this process is in

Fig. 8-13 Small diameter rod produced in a continuous casting machine. (*Nichols Wire and Aluminum Co.*)

the long cycle time required to clean, coat, and prepare the mold and then to apply the fiberglass mixture and to keep it pressurized until setting occurs. Pressure is normally applied to a rubber diaphragm that will press evenly against any mold contour. Air pressure is applied to the outside of the diaphragm, as in Fig. 8-14. A variation of this process is shown in Fig. 8-15, where the air beneath the diaphragm may be exhausted, creating a vacuum. Normal atmospheric pressure (14.7 psi) then presses the diaphragm firmly against the fiberglass mixture and presses this against the mold shape.

Typical products include hulls for small boats, one-piece plastic chairs, the "hard hats" used in electrical construction as protection from head injury and electrical shocks, briefcases, luggage, and similar products.

VACUUM FORMING, PRESSURE MOLDING, AND BLOWN FILM

Molding of industrial plastics began with the use of thermoplastic sheets. As explained in Chap. 2, thermoplastic materials are materials that soften with the application of heat. Most thermoplastic materials may be heated and shaped repeatedly. Thermoplastic

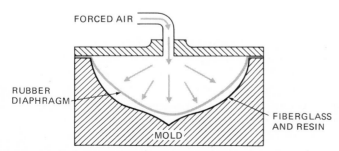

Fig. 8-14 Principle of air pressure used for lay-up molding.

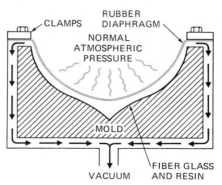

Fig. 8-15 Using a vacuum for pressurized lay-up molding.

molding dates from the early 1900s with attempts at shaping cellulose sheets. Recent improvements in plastic materials have extended both the quantity and quality of possible products. Pressure-molding techniques were developed for forming thermoplastic sheets. Today, however, they are used on both thermoplastic and thermosetting materials.

Partially polymerized and wetted-fiber woven-set strips called *pre-preg* can be formed into final shapes by thermoplastic-molding methods. The formed pieces are then finish cured in autoclaves under pressure to complete "monocoque" designs for large objects. Examples of this process include helicopter bodies and truck bodies (driver's cabs and engine hoods).

Vacuum Forming

Vacuum forming, depicted in Fig. 8-15, is very similar to the process used in pressure-lay-up molding. The major difference is that no flexible rubber diaphragm is used or needed. A thermoplastic sheet or film is heated and placed over a mold. The air between the mold and the plastic is exhausted. Heat may be applied to the plastic from the mold or from a separate heater, or the plastic may be heated prior to insertion in the mold. As the heated plastic softens, the atmospheric pressure on the outside forces the plastic to take the shape of the mold. The process is extensively used in blister packaging products for protection and easy display. The packaging shown in Fig. 8-16 is typical of vacuum-forming products. Vacuum forming is restricted to shapes with single draft, that is, that slope in one direction, and is more effective with thin-wall sections. The process does not produce uniform wall thicknesses, particularly in molding thicker materials that have sharp corners. Products include the production of plastic toys, win-

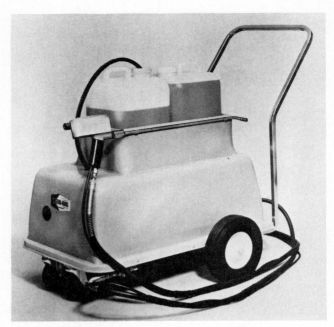

Fig. 8-17 A sweeper and scrubber machine featuring a vacuum-formed cycolac case. (*Marbou Division of Borg-Warner.*)

dow shutters, and, to a limited extent, special custom automobile bodies. Figure 8-17 shows vacuum-formed products used to form machinery housings and other industrial products.

Pressure Molding

Pressure molding is the opposite of vacuum molding, in that air is pumped into a cavity forcing the heated material to expand against the mold shape. A shaped mold may be used or the pressure may be used to expand the sheet to a domelike bubble. When the pressure is used to force the sheet against a shaped mold, the process is called *straight-pressure* forming (Fig. 8-18). When the material is expanded like a bubble, it is called *free blowing* (Fig. 8-19). Free-blown shapes may

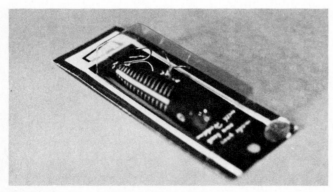

Fig. 8-16 A clear plastic "bubble" or "blister" is vacuum formed to produce attractive packaging and protection for small products.

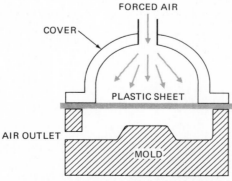

Fig. 8-18 Principle of operation for straight-pressure molding.

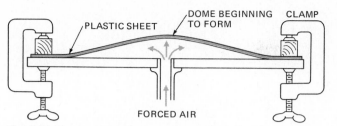

Fig. 8-19 The free-blowing process is used to form skylights, airplane canopies, and other large dome-shaped products.

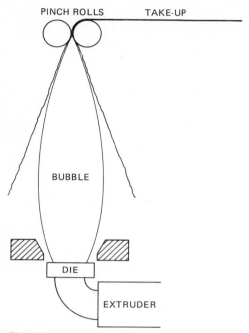

Fig. 8-20 A blown film setup. The plastic tube is extruded by air injected to blow the tube into a large bubble.

have square, rectangular, round, or oval bases and may also be several feet across. Typical free-blown products include aircraft canopies and skylights used on the roofs of buildings. Advantages of free-blown acrylic plastics include good optical quality and even, regular contour. A minimum of equipment is needed for free blowing and almost any base shape can be quickly set up with a minimum of expense. No special materials are needed to construct the mold board and

base. Relatively low-production rates are the primary disadvantage of free blowing. The primary disadvantages of straight-pressure molding are the cost of making the dies and of purchasing the equipment. Straight-pressure molding is very adaptable to high-production rates, however.

Blown Film

The process of blown film is similar to pressure or vacuum molding in some respects. A tube of thin plastic is extruded through a die, as in Fig. 8-20. The tube is pulled up through a die to pinch rolls on a tower some distance above the die. Air is then injected through the die and blows the plastic tube into a bubble. The plastic cools and solidifies by the time it reaches the pinch rolls. The thin film can then be slit, folded, or sealed into containers. Plastic sheets, bags, and wraps are typical products.

REVIEW QUESTIONS

8-1. What are the common methods of molding thermoplastic materials?

8-2. What forces are used in pressure-casting processes?

8-3. How does die casting differ from injection molding?

8-4. What materials may be molded using centrifugal-molding processes?

8-5. How is centrifuge molding different from centrifugal molding?

8-6. What products are commonly made by the blow-molding process?

8-7. How does compression molding differ from transfer molding?

8-8. What is continuous molding?

8-9. What is lay-up molding?

8-10. What is a "parison"?

CHAPTER 9

MATERIALS–FORMING TECHNOLOGY

Materials-forming processes utilize pressure forces to change the shape and/or size of the material being worked. The use of pressure to shape the material distinguishes this broad grouping of processes as opposed to cutting (material removal) or casting processes. There is a wide variety of both processes and equipment used, and the selection of each depends upon many factors including size, material, use, cost, and availability. Materials most commonly formed by these varied processes include most of the ferrous and nonferrous metals, plastics, metal powders, and a variety of other substances including wood, cloth, rubber, food products, and some ceramic products before they are fired or hardened. Some plastics are worked by processes that are the same or similar to forming processes. However, in the plastics industry, as well as with rubber, ceramics, and some other substances, some of the processes are more often referred to as "molding" and as such are specifically discussed in other chapters relating to casting and molding processes.

Figure 9-1 lists a variety of processes and some of their various subgroupings. From the figure it can be seen that there are four broad families of operations. These are pressing, drawing, bending, and shearing. There are also some special processes such as high-energy-rate forming that are not normally categorized into any of these; however, the special processes do utilize the basic principle of pressure in the forming processes. These processes are characterized by their use of pressure and lack of cutting or removing material to form the objects.

As can be seen in Fig. 9-2a, the grain structure of an object formed by cutting (removal of material) has been severed and interrupted. The removal of material in this manner can potentially weaken the structure of the material. As can be seen from Fig. 9-2b, the grain structure of the formed metal actually flows around the shaped areas and is compressed or stretched at appropriate places. The continuous grain structure enhances the overall strength of the formed piece. The forming processes also provide this same advantage over cast products, because the grain structure of cast products is generally uniform and neither adds to nor detracts from the strength.

Two other major classifications are made in forming processes. In dealing with metals these are often referred to as *hot* and *cold* forming processes. However, with other materials such as plastics, cloth, and ceramics "hot" is not a valid classification, as heat may not be the factor that causes the material to assume a pliable crystalline structure.

Most materials have a range of temperatures between a melting point and a solid, cold point.

Most materials may be permanently deformed—shaped or formed, that is—between a range of temperatures. The material in question may be completely solid as in steels, or a semisolid where a mixture of solid and liquid material exists. An analogy can exist for soluble materials where the amount of solvent is more (hotter) or less (colder).

The cold point in any case is that point where the material will break rather than assume a new shape.

The most obvious advantage of working the material in the more hot (or more solvent) state is the ease with which the materials are formed. Less force is required, less time is required, and no strain hardening occurs when the material is in the more pliable states. In addition, there is no increase in hardness or decrease in other desirable qualities as the forming occurs. Since most materials contain minute spaces between the crystalline structure or pockets of impurities within the material, working during the hot state also helps to refine the grain structure.

To be most effective, the hot forming of metals should be above the *recrystallization* point. This is a

Fig. 9-1 Forming processes. H = hot state; C = cold state.

Category	Process	Variations	Predominant Working State
Pressing	Forging	Smith forging	H
		Drop forging	H
		Upset forging	H and C
		Press forging	H and C
		Coining	C
		Hobbing	C
		Swaging	H and C
	Rolling	Rolling mill	H and C
		Roll milling	H and C
		Shape forming	H and C
	Surfacing	Knurling	C
		Burnishing	C
Drawing (pushing or pulling)	Drawing	Wire	H and C
		Bar	C
		Fluid rubber	C
		Kirksite drawing	C
		Tubular	H and C
		Spinning	C
		Stretch forming	C
		Shell drawing	C
	Extruding	Solid	H and C
		Tubular	H and C
Bending	Braking	Angular bending	H and C
		Folding	C
		Seaming	C
	Rolling	Straightening	H and C
		Cylindrical forming	C
		Conical forming	C
	Roll forming	Forming	H and C
		Flanging	C
	Bending		H and C
Shearing	Shearing	Shearing	H and C
		Notching and nibbling	C
		Blanking and dinking	C
		Piercing	H and C
		Trimming	H
		Shaving	H
	Slitting	Slitting	H and C
		Truing	C
Special	High energy rate	Explosive	C
		Impactive	C
	Magnetic forming		C
	Powder metal Compacting		Granular-F

specific temperature at which the grain of the metal will change shape. For example, if a metal has been cold formed and the grain has become elongated and great stressing has occurred, it must be treated. It is heated to a temperature high enough for the grain to reform back into its normal, unstressed shape. To form the metal at the high temperature first reduces the energy and force needed for shaping as well as the subsequent heat treatment. The recrystallization point will vary greatly upon conditions and materials. Figure 9-3 shows some typical recrystallization temperatures.

All of this means that less energy is required during the forming process so that less-massive equipment and hence less-expensive equipment is required and the inital cost and cost of power for operation are also reduced.

To prevent sagging, special handling processes are required to prevent distortion occurring to the softened, finally shaped material. The potential for wear and degeneration of the machinery and forming dies due to the heat or chemical action is increased.

However, the choice of hot or cold working allows the processor to capitalize upon the peculiar advan-

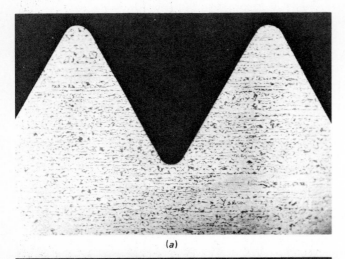

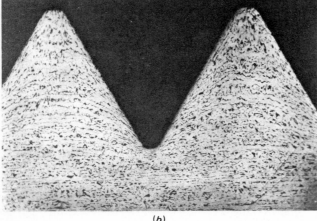

Fig. 9-2 (a) Grain structure of threads formed by cutting part of the material away; (b) grain structure of threads formed by pressing (rolling). Note that the grain has been formed to the configuration of the threads increasing the ability to withstand shear forces. (*Reed Thread and Die Co.*)

tage of each forming technique. These not only include cost and equipment factors but include the refinement of the grain structure to improve the physical characteristics of the material in regard to strength, durability, rigidity, toughness, and other

Fig. 9-3 Recrystallization temperatures.

| Metal | Approximate Temperature | |
	°F	°C
Aluminum	300°	148°
Copper	390°	198°
Iron	840°	449°
Lead	Below 75°	25°
Magnesium	300°	148°
Nickel	1110°	600°
Tin	Below 75°	25°
Zinc	75°	25°

similar aspects. In addition, "cold working" generally increases the hardness, but this is accompanied by a proportionate increase in brittleness. This may be a desirable quality but also increases the difficulty in working with the materials and can add the requirement of further treatment or processing to remove undesirable brittleness or hardness.

Factors to be considered when engineering the product include (1) the operational stresses that the formed object is expected to withstand, (2) the weight of the object and the type of material that will meet conditions such as strength, resistance to corrosion, and other factors, and (3) the feasibility of constructing the object considering the complexity of shape and the operational cost required to produce that shape (this includes the denseness and hardness of materials, machine time, worker time, and other similar cost aspects). Weight and materials are closely related to each other and include the use of standard strength and weight and volume tables. These factors are particularly critical in aerospace industries.

Stresses to be considered include those illustrated in Fig. 9-4. These are compression and buckling, tension or tensile pull and necking, combined multiaxial compression and/or tension resulting in fractures, torsion or twisting, and biaxial compression or shearing. The part geometry for a given material must be designed to withstand these particular forces. Construction and design considerations include an analysis of the intended use of the product to provide shapes that would be resistant to these stresses without adding bulk, materials, and costs to the manufacturing process.

One other stress factor must be considered. In hot working, little or no internal grain stress is formed. However, when materials are worked in their cold state, internal stresses are formed after the *loading* or forming process occurs. This causes the material to *unform* slightly after the forming pressure has been released. This is generally referred to as *springback*. Since most cold-worked materials exhibit some springback tendencies, "overforming," or forming a little too much and allowing the material to "unform" back to the desired shape is generally done to compensate for this characteristic. Generally, the harder materials accrue more springback than the softer materials, especially in bending operations. Two major factors involved in selecting a process include: (1) machine expediency and (2) physical requirements. *Machine expediency* means doing the job with the machines and equipment at hand, particularly when small production runs are involved. Using a less desirable process for producing a few pieces can prevent major retooling expense and/or purchase of

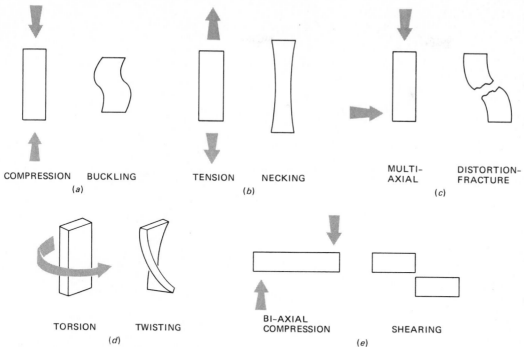

COMPRESSION   BUCKLING
(a)

TENSION   NECKING
(b)

MULTI-
AXIAL   DISTORTION-
FRACTURE
(c)

TORSION   TWISTING
(d)

BI-AXIAL
COMPRESSION
(e)   SHEARING

Fig. 9-4 Stresses upon formed parts.

unnecessary, expensive, specialized equipment. The second factor, *physical requirements,* refers to the type of crystalline structure desired, including the various characteristics of the product such as hardness, ductility, toughness, and dimensional accuracy.

PRESSES

As previously mentioned, four basic forming process categories are readily identified. To exert the pressures for these four basic categories, machines known as *presses* are utilized. Although not all the possible process variations require presses, they are used for forging, shearing, drawing, blanking, piercing, bending, and a variety of different subgroupings. Because of the importance of presses to the entire field of forming processes and their widespread use, a broad understanding of these machines is vital.

The basic press is the machine age's answer to the village blacksmith of days gone by. Machines have been used to replace the large hammer and the sweating muscles. In addition, greater forces and more complex operations are performed. However, the principles and the basic operations that the old village blacksmith performed are still utilized in industry today.

Two basic types of presses are extensively utilized. The first is the press that exerts its forces slowly with a squeezing action. The second are presses that provide both sudden impact as well as pressure. The machines that provide only pressing action are called *presses,* while machines that include impact are classed as *hammers.* The mechanics of a press used in bending may be very similar to the machine used in press forging. However, some specialized adaptation will be necessary. Further, not all machines are variations of a press. Although some type of pressure-forming process is involved, not all may be the actual press operation. These specialized equipment situations will be taken up in later sections where they are appropriate to the process being studied.

The Basic Press

The basic press is utilized primarily to squeeze materials to rough shape before being formed in subsequent operations. The shape and the accuracy depend primarily upon the skill of the operator in machine operating and upon the manipulation of the workpiece. Heavy pieces and pieces that are too dangerous to touch due to temperature or chemical hazards are moved using mechanical manipulators that are operated either by the press operator or by helpers working under the operator's direction.

The basic working sections of a press are illustrated in Fig. 9-5. A stationary bed or support is provided upon which a hardened work surface called an *anvil* rests. The material to be formed is placed on the

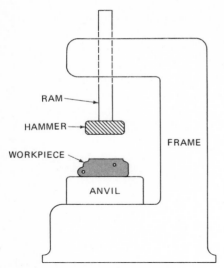

Fig. 9-5 Press parts.

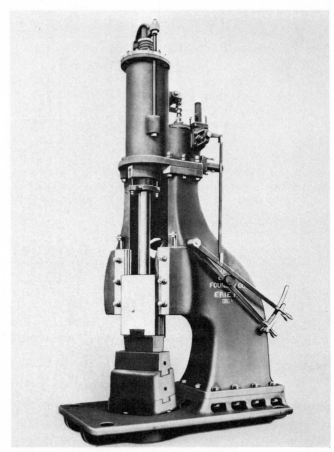

Fig. 9-6 An open-frame press. (*Erie Foundry Co.*)

anvil, and force is exerted upon the material by a hammer which is driven by the ram. The power of the press is applied to the ram and not necessarily to the hammer itself. Separate hammer and anvil pieces are provided so that when these become unusable due to wear and normal usage, they may be replaced or reconditioned without completely rebuilding an expensive piece of machinery.

Figure 9-6 illustrates a single- or open-frame press. Because of the openness of construction, the frames of these presses provide less support and therefore are used primarily in comparatively lighter operations dealing with pressures up to 15,000 or 20,000 lb. Open-frame presses are generally used for lighter operations and can be used for such operations as stamping, hobbing, coining operations, and other similar processes on light stock.

Most of the basic operations performed on the press are done with flat hammer and anvils. Shaped-hammer attachments are provided for working with round shapes, forming holes, and for cutting operations. Some presses may be rigged for simple-press operation or may be converted for both pressing and impact usage combined. Heavier machines, such as the double-frame press shown in Fig. 9-7, are utilized for forming heavy shapes and massive pieces and may range up to capabilities of thousands of tons of pressure. Needless to say, such machines are extremely expensive.

Drop Hammers

The features of the drop hammer are similar to those of the simple press, except that some arrangement is made to provide impact as well as the basic pressure

Fig. 9-7 A double-frame press. (*Erie Foundry Co.*)

process. Impact creates extremely high pressure forces for a brief instant; this allows a relatively small press—in terms of "squeeze" pressure—to provide larger forces. The machines are called *drop* hammers even though the ram may be powered. The term is derived from the early machines which raised the hammer and ram to a height and allowed it to drop to provide the impact. Drop hammers which utilize a gravity drop are still used extensively and are classified according to the manner in which the hammer is raised. Ropes, belts, chains, cogs, or gears may be utilized to raise the hammer. A very common style which pulls up a shaft is the *board* drop hammer. This is from the early mechanism shown in Fig. 9-8(*a* and *b*). The hammer was attached to a board which was pulled up to a height. Cams were used to hold the board once it was raised. To release the hammer,

these cams were pulled away from the board allowing the hammer to fall.

Drop hammers fall into two broad categories. The first is the category of *gravity* hammers such as the *board, rope,* and *trip* hammers. The latter utilize cam actions to raise the hammer before gravity is allowed to pull it down. The major disadvantage of gravity hammers is the limitation of applied force from the size and weight of the falling hammer. The second grouping is the *power* drop hammer. Steam or hydraulic pressure is utilized to both raise the hammer and also to provide additional power for the downward thrust. These power hammers provide greater accuracy and control as well as greater force and impact.

Drop-hammer operations also differ from press operations in that a closed die is ordinarily utilized.

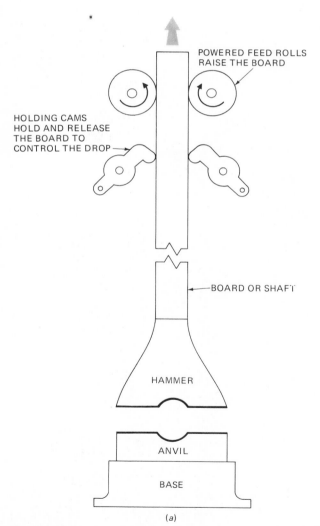

POWERED FEED ROLLS RAISE THE BOARD

HOLDING CAMS HOLD AND RELEASE THE BOARD TO CONTROL THE DROP

BOARD OR SHAFT

HAMMER

ANVIL

BASE

(*a*)

(*b*)

Fig. 9-8 (*a*) Board drop-hammer mechanism; (*b*) a typical board drop hammer of current usage. (*Erie Foundry Co.*)

Fig. 9-9 The closed die of a drop hammer serves to mold and shape an object. This bottom die is matched with a similarly shaped top die. (*Ajax Co.*)

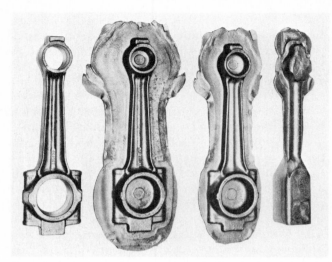

Fig. 9-10 A connecting rod for an engine is formed in a progressive die of a drop hammer. (*Ajax Co.*)

As can be seen in Fig. 9-9, the exposed die provides a mold for the material.

Dies may be single units as in Fig. 9-9, or they may contain several cavities. When several cavities are involved, these cavities advance from only a vague, roughly formed shape to more finished appearances. Such dies are termed *progressive* dies. When progressive die forging is done, the blank is placed upon an anvil containing successively more complex cavities. The first stroke of the hammer provides only a very general shape to the object. The object is then removed and placed in successive cavities, each of which cause a more refined shape to be given to the object. The impact provides greater pressure than a similarly sized press could provide, as well as speed and accuracy. After the last forging stroke, the object may be trimmed. As can be seen in Fig. 9-10, excess material has been squeezed out around the edges of the forming cavities. This material is known as *flashing* and may be sheared off in the last operation known as *blanking.* Both the hammer and the anvil will contain shaped cavities for such operations.

Press Position

The working elements of a press may be constructed to facilitate the forming process. Since presses are often constructed to produce only one specific item, it is common to find the press built with major consider-

ation given to ease, expense, or safety of moving or handling the formed object. The position of a press will fall into one of three basic patterns depending upon the moving axis of the ram. These are vertical, inclined, and horizontal. Figure 9-11 shows a large

Fig. 9-11 A large vertical press capable of performing many processes. (*Erie Foundry Co.*)

Fig. 9-12 This horizontal extrusion press forms aluminum billets by forcing preheated aluminum through dies under pressure. The process can be likened to squeezing toothpaste from a tube. (*Alcoa.*)

vertical press capable of being adapted to a variety of processes. The press shown in Fig. 9-12, however, is a horizontal press designed to only perform one process.

Another variation of the horizontal press is the impact hammer. Impact hammers, such as the one illustrated in Fig. 9-13, utilize mechanical, steam, or hydraulic pressure to drive opposing hammers to impact with the object across horizontal guides. These machines have less vibration and less noise. Although they are generally not capable of pressures as great as those of the power hammers, they are more adaptable to production lines where constant, high rates of output are required. They may also be combined with automatic controls more readily than other types of presses and hammers. The impact hammers are frequently used, particularly for smaller pieces, in conjunction with automatic production line sequences and provide successive, rapid forming operations.

Press Drives

Several press mechanisms are used. All are used extensively to provide the forces for press work and each has specific advantages and limitations. Naturally enough, the more complex and versatile the mechanism, the greater is its expense for both purchase and operation.

The *crank*-driven press is commonly used for blanking and piercing operations in light weight sheet and plate materials. *Knuckle* drives, also called *knee action* drives, are used where speed of operation is combined with higher mechanical advantages. Typical process applications include coining, hobbing, punching, sizing, and some fluid-rubber processes. Knuckle presses are commonly used in forming sheet and plate and for some forging types of operations for stock of light and medium weight and sizes.

The *eccentric* press is normally used for situations where a short ram stroke is required. Eccentric presses are relatively fast acting but are generally restricted to lighter operations. The *toggle* drive is used for a wide range of weights and sizes of material. Its principal advantage is in providing dual actions in drawing operations where blank holders act at different times than the ram.

Hydraulic presses and *steam* presses operate on similar principles. Figure 9-11 shows a large double frame hydraulic press, while Fig. 9-14 shows a smaller hydraulic press. Both steam and hydraulic presses are

Fig. 9-13 Impact hammers are horizontal hammers featuring two moving hammers. They are used for high production rates. (*Courtesy Erie Foundry Co.*)

Fig. 9-14 A small open-throat hydraulic straightening press. Compare this press with Fig. 9-11. (*Courtesy Greenerd Press Co.*)

actuated in much the same manner. However, newer presses tend to be hydraulic rather than steam driven. This is due to the early use of steam for power and the more recent advances in hydraulic controls which provide greater versatility, easier control, and less complicated power generating systems. Hydraulic presses range in size from small to an almost unlimited capacity. They can be made for impact, press work, or for both. Several power cylinders, or *slides* as they are termed, may be grouped in one frame to provide varied timing strokes or for added pressures.

Presses may be controlled from remote control panels or from controls on the machine. The operator may activate the mechanism by electrical switches or by mechanical controls. Most larger presses are controlled by electrical switches, however. The controls may be *push button* types or foot-operated. The latter are called *kick presses,* regardless of the actual drive mechanism used.

It is also common to find double switching devices for safety purposes. Because many presses, particularly the fast action hammers, involve considerable hazards to the hands of the operators, a common situation is to require the operator to press two different switches at the same time. Thus, the operator must remove both hands from the danger zone in order to press the switches that activate the machine.

Variations of presses include rolling presses which provide continuous pressure to reduce thick slabs of material into thin sheets. The rolling mills, as in Fig. 9-15, are used in making sheet or plate stock in plastics, metals, glass, and other materials.

SPECIAL FORMING PROCESSES

The special forming processes consist of three major forming techniques. These are (1) magnetic forming, (2) high-energy-rate forming, and (3) powder metal-

Fig. 9-15 A rolling mill is a type of press that forces thick slabs between rollers that squeeze the material into thin sheets. (*Courtesy Mesta Machine Co.*)

lurgy. The magnetic-forming process induces sudden magnetic forces about the object to be shaped, causing it to assume the shape of walls or molds surrounding it. This type of forming may be used to form extremely large surfaces of thin materials and is primarily used in the aerospace industries. It is also extensively used to form complex tubing shapes.

High-energy-rate forming is used primarily to form relatively small objects, up to three or four feet in size, of relatively thin material. This process is actually the rediscovery of a process used in the nineteenth century to form brass cuspidors for the saloon trade. The forgotten process was rediscovered by a group of aerospace engineers who were posed with a seemingly insoluble problem in shaping some sections to be used in the aerospace industry. The process uses an explosive charge to generate a sudden force upon a liquid. The liquid then exerts even pressures in all directions upon a material within the liquid which then presses the material evenly against a mold shape. The rapidly forced movement of the liquid stretches and forms the material tightly against the walls of the mold, thus allowing intricate shapes to be formed quickly and without complex drawing and forming operations.

Powder metallurgy is a widely used series of forming processes that have been known for many years. Several variations within the usage of powdered products include the forming of intricate gears and springs for light-duty machinery, wire, and cores for magnetic devices where strength is not critical; for imparting magnetic properties to metals otherwise difficult to magnetize due to extreme grain densities; for filters for liquids; and for including lubricant in impregnated bearings.

REVIEW QUESTIONS

9-1. What is the recrystallization temperature?

9-2. What is the difference between hot and cold forming?

9-3. What are the advantages of hot and cold forming?

9-4. What are the disadvantages of hot and cold forming?

9-5. What is "springback"?

9-6. What four broad categories are included in the forming processes?

9-7. How do forming processes differ from machining processes?

9-8. What materials are commonly shaped by the forming processes?

9-9. What factors should be considered when engineering a product?

9-10. What factors should be considered when selecting a forming process?

9-11. How are presses classified, categorized, or described?

9-12. How is a "hammer" different from a "press"?

9-13. What is an impact hammer?

9-14. What is a progressive die?

9-15. What are the advantages or features of: *a.* Toggle drives, *b.* Eccentric (cam) drives, *c.* Crank drives, *d.* Knuckle (knee) drives, *e.* Hydraulic drives.

9-16. What are the special forming processes?

CHAPTER 10

HOT- AND COLD-FORMING PROCESSES

This chapter deals with those processes in which materials are formed or forced into a desired shape by squeezing, pulling, bending, or shearing. These processes form a class of operations different from the shaping of materials by casting or molding, or by the cutting or chip-removal processes.

As explained in the previous chapter, two working states exist. These are hot and cold and affect the grain structure of the particular material. The advantages and disadvantages of each general state were described.

Before materials can be worked or shaped, three conditions of preparation must be met in order to achieve accuracy and economy.

1. The material must be formed to a uniform size and thickness within certain limitations to assure the ease of working, the accuracy of the machines, and the ability to control the quality of the product developed by the process.

2. The material must be free of surface scale and defects and must be relatively smooth and even. Without these characteristics, product spoilage and machine damage are more probable.

3. The material must be prepared to the proper consistency considering the strength, hardness, toughness, resiliency, and other factors. Material that is too hard will not be properly reformed or shaped and can cause machine damage. Material that is too soft will not be hardened properly and the hardness and toughness specifications will not be up to desired standards. In addition, consideration must be given to the hardness of the material before forming because cold-working processes normally add hardness to the material. Care must be taken that the material is not too

hard for customer requirements after it has been formed and shaped

PRESSING

The *pressure forming* or *pressing processes* feature both squeezing and impact types of forces. *Impact* provides a variety of momentary pressures, particularly when distributed over small areas. Impact provides the advantages of focusing an extremely great force upon a very small area using relatively light equipment. Simple *squeezing* operations, while generally requiring greater pressures, have advantages in that the slower application of pressure gives the material more time to change shape and flow into all areas of the shaped-die cavities. Included in these pressure classifications are other pressure processes, including *rolling,* which is the process of using a continuously rotating "hammer," and *surfacing,* another variation of rolling.

Figure 10-1 shows another advantage of using pressing processes for forming materials—that of reducing material waste. However, material waste is not always the main consideration. A primary consideration is that of labor and machine costs. From Fig. 10-1 one can see that the machining process took several operations and may have required transportation to different machines, handling, and a number of different processes. However, the pressure-forming process for that same piece resulted in an item produced in a few strokes of one machine, required approximately one-third as much material, produced 75 percent less waste, and was also 18 percent stronger.

FORGING

Forging implies the use of impact and pressure to form objects. The impact processes include smith forging, drop forging, and upset forging. Black-

Fig. 10-1 Forging saves 68 percent more material than machining in making this coupling. (*Courtesy National Machinery Co.*)

smiths since Roman times used a tool called a *nail header* to manually form the heads of nails—this is perhaps the oldest upset forging process. Blacksmithing and other metal-smithing processes formed the basis for improving human technology and the shaping of materials through pressure processes. Forging dates from the old processes of tool flaking and the making of stone implements, and has evolved from the arm and hammer to the water-powered drop hammer of the sixteenth and seventeenth centuries to the modern drop hammer. As pointed out in the previous chapter, power forges include combination drop and power hammers as well as ram-type and impact-type systems. All of these contribute to the formation of products ranging from hypodermic needles to machines weighing thousands of tons and which are vital to maintaining modern society.

Smith Forging

Smith forging is perhaps the oldest type of metal working process and has influenced the formation of other materials through the ages. Smith forging was formerly the process envisioned when we think of the village blacksmith wielding a hammer against a piece of hot metal placed upon a rigid anvil. The modern process is essentially the same except that the craftsperson uses a mechanical hammer and he or she uses mechanical manipulators to move heavy pieces. Smith forging is normally done with a flat hammer and a flat anvil, as indicated in Fig. 10-2. The press is usually an impact press but can be a squeezing type of press. The applications of smith forging in industry today are operations done predominantly in the hot state to provide a rough shape or size reduction to an object. For example, a large billet of steel that is to have a bearing surface machined on one end may be roughly reduced to the correct size and formed into a

roughly round section before the bearing surface is machined.

Drop Forging

Drop Forging is similar to smith forging except that shaped dies on both the hammer and anvil are normally utilized. Figure 10-3a and b illustrates the types of dies and products from these dies that typify drop forging.

Fig. 10-2 Smith forging. (*Courtesy National Machinery Co.*)

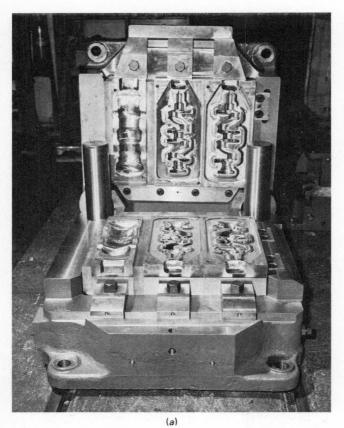

Fig. 10-3 (*a*) The progressive dies used in drop forging an engine crankshaft; (*b*) the product of these dies before the flashing is trimmed. (*National Machine Co.*)

Drop forging uses successive dies to first roughly form the product, provide a semifinished form to the product, and then to form and "blank" or trim the flashing from the formed object. One, two, or more blows may be given by the drop-forge operator to the object in each of the die cavities. When the operator deems the product formed well enough, depending upon size and shape, he will flip the forging into the next cavity and continue with the forging operation. The grain structure in a forging operation is greatly enhanced, and the characteristics of hardness, toughness, resiliency, and other similar characteristics are improved. Figure 10-4*a* and *b* illustrates how the grain structure of an object has been squeezed together to provide reinforcement, increased strength, and greater material density.

The dies used in drop forging must be made from extremely strong and tough materials in order to withstand the impact loads and temperatures of the process. The dies are usually made from various types of tool steel and are usually designed so that

they may be refinished when the surfaces of the forming cavity become worn or pitted through use. The design of these dies has several considerations: (1) The parting line should be in a single plane and near the center of the piece. This enables both dies to easily free themselves from the forging and also facilitates trimming or blanking the flashing from the finished piece. (2) Draft must be provided to both top and bottom mold cavities. On simple pieces a draft of seven degrees may be used, but on more complex shapes greater draft must be allowed. (3) Ribs and vents should be low and wide, and fillets, rounds, and radii should be included for all corners. These should be of larger size than those used in molds for casting processes. (4) Sections of the forging should be balanced and near symmetrical to avoid extreme variations in the rates of material flow into parts of the die cavities.

If these considerations are followed, drop forging generally can provide good dimensional tolerance in a plane perpendicular to the parting line. However,

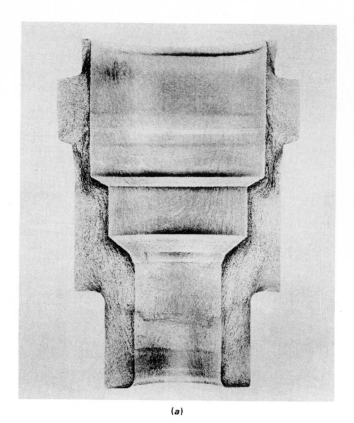

(a)

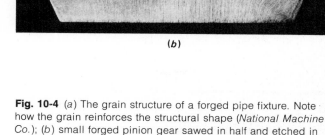

(b)

Fig. 10-4 (a) The grain structure of a forged pipe fixture. Note how the grain reinforces the structural shape (*National Machine Co.*); (b) small forged pinion gear sawed in half and etched in acid to show grain structure. *Forgings are stronger.*

due to draft considerations and to parting-line considerations, less dimensional accuracy can be obtained in a plane parallel to the parting line.

Impact Forging

Impact forging is a process normally done in a plastic state. It is similar to drop forging except that two powered horizontal rams drive opposing hammers against the workpiece. A comparison of impact forging with conventional drop forging is illustrated in Fig. 10-5. The impact-forging process may also be done hot or cold. Either state is extremely appropriate for mass producing small parts and can easily be connected into an assembly-line situation.

Upset Forging

Upset forging is similar to impact forging in that the ram that drives the hammer usually moves in a horizontal direction. Upset forging is done in either the hot or cold states as with impact forging. However, two features distinguish upset forging from impact forging: (1) there is usually only one ram and hammer in upset forging and (2) the object is formed by displacing or deforming the material, which is usually

cylindrical or rod-shaped, into a formed cavity similar to drop forging. Both the ram hammer and the die may be shaped, although the ram hammer in many situations is not shaped and is flat. Particular applications of upset forging include the formation of bolt and rivet heads, engine valves, and other types of products where a wide head is formed upon a smaller stem. Figure 10-6 shows the process for making bolts, which is called *cold heading*. Most bolts are cold formed to improve the grain and strength characteristics. Upset forging may be used to increase the

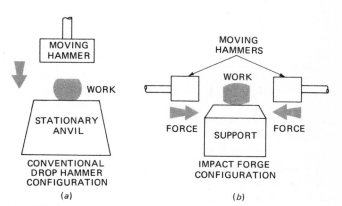

Fig. 10-5 Comparison of impact forging.

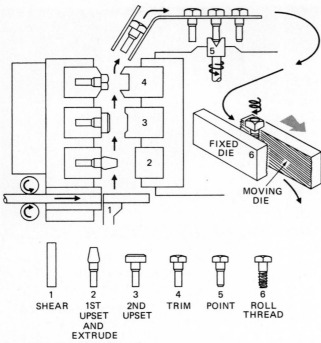

Fig. 10-6 The process of boltmaking by upset forging. (*Courtesy National Machine Co.*)

thickness of a round shaft prior to machining or to form head sections as in bolt making.

Three general considerations must be made in the design of upset forging dies. These are:

1. Length of the upset portion must be no more than 3 times the diameter of the object.
2. The diameter of the upset portion should be no more than $1\frac{1}{2}$ times the diameter of the bar.
3. When material to be upset projects beyond the carry of the die, it should not project more than the diameter of the material being formed in the cavity.

Swaging

Swaging is a process used to reduce the diameter of round objects, both solid rod and cylindrical shapes, by repeated impact. Figure 10-7 illustrates how revolving dies are forced against the stock as rollers are quickly rotated against the die housings. The swaging process may be used to form cone shapes and to reduce the diameter of rounded sections. The machine may be constructed so that the work axis may be either horizontal or vertical. Further, the swaging process can be done to reduce both interior and exterior sections. Figure 10-8 illustrates some typical swaging applications.

Press Forging

Press forging is a process similar to drop forging except that a slow squeeze is applied without impact. The advantage of press forging includes allowing time for the material to flow to shape and allows careful formation of a shape such as the bending of a large piece into a shape with careful, measured applications of force as illustrated in Fig. 10-9. Press forging is normally done without shaped dies and both hammer and anvil are normally flat. However, some shape of a very restricted nature is used to provide general squareness, general roundness, or to do cutting operations prior to more accurate operations with other machines and equipment.

Typical applications of press forging include the rough forming of larger pieces of sheet or plate into curved areas, the rough reduction of diameters of

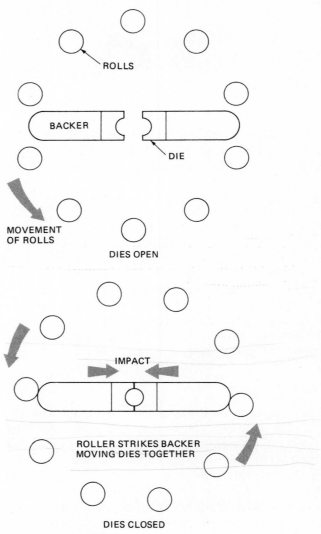

Fig. 10-7 Swaging process.

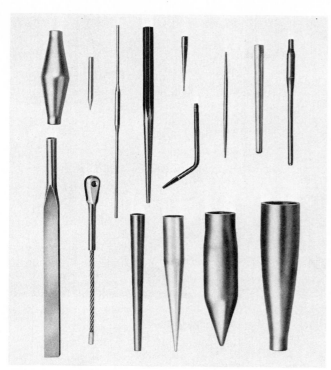

Fig. 10-8 Typical swaging applications. (*Courtesy Torrington Co.*)

shafts before machining to critical dimensions, the shaping of aircraft fuselages and structural members, and the formation of car bodies. Shaped dies made to finished dimensions are used for extremely large pressing operations, such as those that produce car bodies and aircraft-fuselage members.

The advantages of forging such large surfaces are that by capitalizing upon the grain reformation and the accompanying grain strength per weight ratio, considerable bulk and weight are eliminated. In addition, the elimination of this bulk and weight also elim-

inates or reduces greatly the time required for costly (including labor and machine costs) work from other slower processes.

Coining

Coining is an operation done primarily in the cold state to small pieces. The stock is displaced by force and impact into the cavities of the dies. Because the cavity is completely enclosed within tightly fitting die shapes, the volume of the blank material must be very carefully controlled. It is not possible to do coining in which flashing or excess material is formed; therefore, to overfill the die cavity would either cause damage to the machine or would result in faulty products. Coining is especially appropriate for the production of small pieces that require extremely fine surface detail and finish. The name of the operation is derived from its principal application, making coins, medallions, and other similar pieces.

Hobbing

Hobbing is similar in some respects to the coining process. Hobbing is used in the preparation of dies except that the impression is made in a larger piece of metal so that the displaced metal is pushed out into an open area around the pattern impressed into the stock. Figure 10-10 illustrates this situation. Hobbing is appropriate for the shaping of small parts where fine surface requirements must be met. It is frequently used in making small molds which in turn are used in making stampings in softer materials or in casting plastics and other materials. It is generally easier to make the male-shaped object and press the de-

Fig. 10-9 Press forging a slight curve into a thick slab of cold steel.

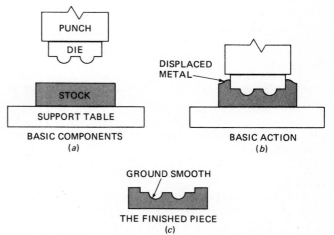

Fig. 10-10 Hobbing is used to make a formed cavity into a free block.

sign into a softer material than it is to machine the female image in the die material. When making dies, the die surface is generally ground flat after the hobbing process. It is used for making a number of identical mold cavities for multicompression molds.

ROLLING

Rolling Mill

Rolling-mill operations, also called *reduction mills,* account for an important segment of modern industry. These mills are mammoth machines designed to reduce thick slabs into shimmering strips, as in Fig. 10-11. The thick slabs enter at a speed of a few feet per second and emerge from the final end of the mills as thin, flexible strips streaking past the operator at speeds of 25 or 30 mph [40 or 48 km/h].

Rolling-mill operations may be done to a variety of materials and may be done in either hot or cold states. The mills operate upon the principle of continuously squeezing the slab into thinner and thinner sections as in Fig. 10-12. Greater reduction is achieved simply by passing the material through greater numbers of rolls. Variations occur in these mill operations which are used to work most metals, glass, rubber, and many plastics.

Shape Rolling and Roll Forging

Shape rolling and roll forging utilize formed rolls to impress deep patterns upon billets or pieces of blank stock as they are passed between the rolls. Shape rolling is normally done to small parts such as small gears and the threads upon bolts and screws. Roll forging

Fig. 10-11 A slabbing mill rolls an ingot into a giant strip of metal. (*Courtesy Mesta Mfg. Co.*)

Fig. 10-12 A hot strip of aluminum as it enters the continuous hot strip stands. (*Reynolds Aluminum Co.*)

employs shaped rollers and is normally done in the hot state to larger pieces. Shape rolling is normally a cold-working process. The advantages of both operations are that they form objects of greater strength, there is lower production cost, and the production rates are greater than with cutting processes. However, the processes are most adaptable to high-production rates and less adaptable to shaping extremely heavy equipment or extremely long pieces. Figure 10-13a, b, and c illustrates several variations of the thread-rolled process.

Figure 10-14 illustrates the roll-forging process. The roll-forging process is much more adaptable to larger pieces than is shape rolling. Shaped pieces are extremely common; and similar operations may be applied to produce hot- or cold-rolled rod and various construction shapes such as angle, channel, and H- and I-beams (see Fig. 10-15). Roll forging provides the advantage of enabling a machine smaller than a rolling mill to produce products of this nature. In addition, the machine is more versatile in that rolls may be changed so that different shapes can be formed in the same machine. Disadvantage of the roll-forging process include the ultimate size of the object and the length of the object that can be produced.

Roll Forming

Roll forming is a variation of shape rolling used to form thin sections such as strip sheet metal into various shapes as illustrated in Fig. 10-16. The process is

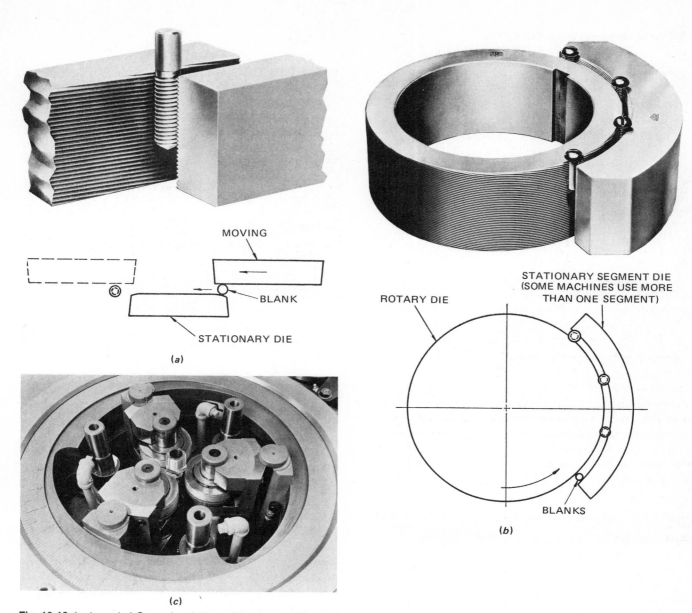

MOVING

BLANK

STATIONARY DIE

(a)

STATIONARY SEGMENT DIE
(SOME MACHINES USE MORE
THAN ONE SEGMENT)

ROTARY DIE

BLANKS

(b)

(c)

Fig. 10-13 (a, b, and c) Several variations of the thread-rolling process.

Fig. 10-14 The roll-forging process. (*Ajax Co.*)

Fig. 10-15 Several parts produced by roll forging. (*Ajax Co.*)

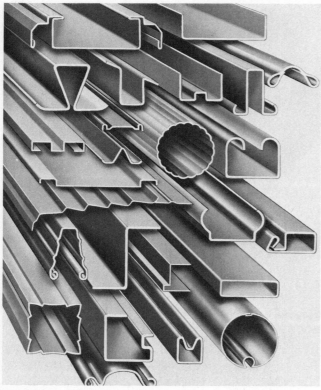

Fig. 10-16 Many shapes attained by roll forming flat stock. (*Courtesy Yoder Co.*)

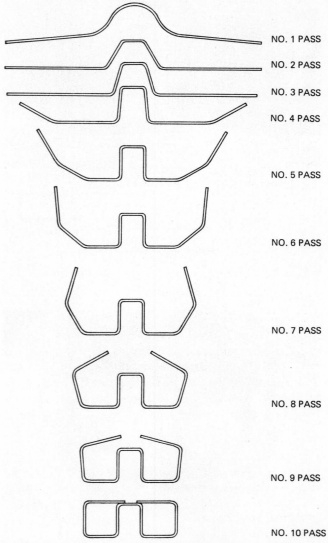

NO. 1 PASS

NO. 2 PASS

NO. 3 PASS

NO. 4 PASS

NO. 5 PASS

NO. 6 PASS

NO. 7 PASS

NO. 8 PASS

NO. 9 PASS

NO. 10 PASS

Fig. 10-17 How successive "passes" through roll-forming dies can develop a complex shape. (*Courtesy Yoder Co.*)

particularly appropriate for making sheet metal pipe, strips, gutters, shaped sheet metal products, and material for use in molding and trim production. The roll forming process is normally done in a cold state to smaller pieces, pieces of thinner cross section, or softer metallic materials such as aluminum and copper. However, the roll forming process is done to heavier pieces such as steel pipe in a hot state. Figure 10-17 shows the multistage forming steps for complex sheet formations. The same multistage process is used to give formed shapes large surface areas (see Fig. 10-18).

Pipe-Making Processes

There are three basic processes for forming *pipe* and *tubing:* (1) tubing, which rolls a hot steel billet against a shaped piercing rod to produce a seamless tube (Fig. 10-19), (2) butt-welded pipe is made by rolling the strips of sheet stock into a round shape so that the two ends meet where they are welded together, and (3) by drawing a strip of metal called a *skelp* through a mandrel where the strip is bent into a general round shape and welded. The pipe formed by both welding processes is given final sizing and shaping operations.

Another type of pipe-making process is very similar to the rolled butt-welded pipe except that the metal at the edges does not butt together. Instead, the edges overlap slightly, and when the formed pipe is drawn through the final sizing and shaping stages, the force of the material being drawn over the reducing mandrel welds the lapped joint together. This type of process is primarily used in making nonferrous, metallic pipe such as copper or brass pipe.

Surfacing

Surfacing operations are normally done cold to impart some desired finish or shape to a surface. Three basic processes are considered. Figure 10-20 shows how

Fig. 10-18 Successive stages are used to form a large surface area. (*Courtesy Yoder Co.*)

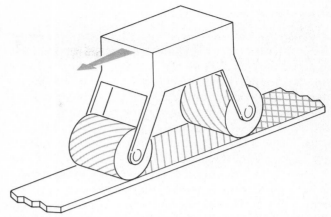

Fig. 10-20 Knurling of flat surfaces can be formed by rolling the impression into the surface.

shaped rollers are pressed against flat stock and moved to impress into the surface the knurled shapes. This is a process similar to thread rolling but is normally done to provide surface decorations or "nonslip" surfaces.

The *burnishing* operation by drawing is done pri-

marily to reduce surface blemishes, to increase surface smoothness and appearance and to provide critical accuracies of very close tolerances.

The third surfacing operation is that of embossing. Figure 10-21 shows how rollers with a design carved into them press a raised surface, usually in some form of design, onto a workpiece. This slightly raised or depressed image is primarily for decorative purposes. Variations of the embossing process can be done on presses that are almost like printing presses.

DRAWING AND EXTRUDING

In both drawing and extruding processes, material is forced against shaped dies in such a manner that the material is forced to assume the same configuration or shape as the die. For extrusion, the material may be pushed or squeezed through shaped openings. The great advantage of extrusion is in the wide variety of complex cross sections that can be formed. Drawing normally "pulls" sheet, rods, or tubular stock into a conforming die or mold cavity.

Fig. 10-19 Seamless steel pipe being pierced.

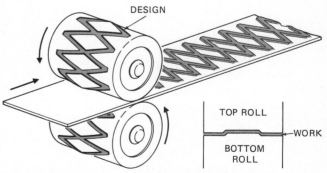

Fig. 10-21 Embossing a slightly raised design onto a sheet by rolling.

Drawing

Drawing operations may be used to form cylindrical shapes, hollow rectangular shapes, solid shaped rod, and angular stock. Drawing may also be used to form tubular shapes from sheet stock as in a previously mentioned process for making pipe. Drawing operations may be done in either hot or cold states. There are a great variety of process variations, including the drawing of bolts and wire, tubular shapes, spinning, and a number of others. Drawing is used to form walls of cylinders, compressed gas tanks, ammunition casings for large weapons, ammunition casings for common rifle shapes, walls of curved storage tanks, and other similar products. Drawing offers the advantages of providing relatively uniform, thick-walled sections over large curvatures. Most of these operations are done in a plastic state. Cold drawing of metal parts, however, forms a large segment of the drawing operations in modern industry. They are used to produce a wide variety of shapes and sizes from wire to automobile bodies.

Bar Drawing. *Bar drawing* is most frequently performed in the cold state. As illustrated in Fig. 10-22, it is commonly used in making rods, primarily of round configuration. The process may, however, be used to produce rods of complex cross sections and thus provide many of the advantages of extrusion. Drawing such shapes may be advantageous in producing small runs rather than considering the purchase of expensive extrusion equipment. Where great reductions in size or shape must be made, multiple passes or stages are required.

Bar drawing is basically a pulling process, a pulling head grips the material and pulls it through the die. Where the pulling head grips the material, a piece of scrap or waste is created. In hot-bar drawing operations, tolerances of $1/64$th of an inch are considered good. However, cold-drawing operations provide greater accuracy and surface finish. Great dimen-

sional accuracy, as well as increased toughness and hardness qualities, are characteristic of cold-drawing processes. Drawing is used to shape both soft metals and hard metals and is almost exclusively used on metals. This is because a high tensile strength is needed for the material to be pulled through the forming dies.

Wire Drawing. *Wire drawing* is essentially bar drawing on a smaller scale. However, wire drawing is the oldest of the drawing operations which dates from ancient Egypt where a process was used that is virtually the same process used in wire drawing today. Wire drawing generally starts with rolled quarter-inch bars which are then pulled through successive dies. Drawn wire develops cold-worked hardness characteristics which necessitate annealing between draws to restore pliability. As in most drawing operations, high quality lubricants must be utilized to lubricate and cool the dies, and to reduce the force and energy required. These lubricants may be natural, but modern chemistry has developed a number of synthetic lubricants.

Tubular Drawing. *Tubular drawing* is done in both hot and cold states, but is done primarily upon metals due to the tensile strengths required to pull the materials through the dies. Tubular drawing provides a seamless tube or pipe which has certain advantages over welded sections. Before welding techniques had achieved modern quality, welding generally created a weakness in the pipe which reduced its effectiveness for use in high pressure applications. Most applications of pipe for high pressure work such as for steam and hydraulics specify tubular, seamless pipe.

Shell Drawing. *Shell drawing* is a process of making contoured shapes or containers where one piece forms sides and bottom. Hot-shell drawing of tubular materials is usually done by stages. Products of shell drawing range from the formation of oxygen tanks, as in Fig. 10-23, through hypodermic needles, to contrast the size and accuracy requirements possible. The material may be *stretched,* where the finished diameter or size is greater than the initial diameter of the stock, or the material may be classed as *shrink* drawing, where the final diameter is decreased from the diameter of the original workpiece.

Two additional classes may also be specified. The drawing may be considered *shallow* drawing or *deep* drawing. Shallow-drawing operations are those in which the depth of the draw is less than the initial diameter of the material. Deep-drawing operations are those in which the depth is greater than the initial diameter. Deep-drawing operations are usually done in

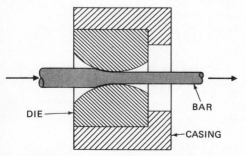

Fig. 10-22 Basic components in bar drawing.

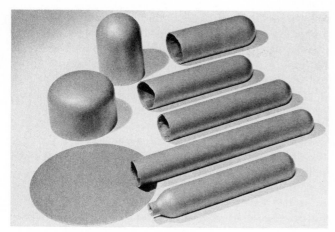

Fig. 10-23 The stages of formation by the hot-draw process for making the oxygen tank. (*Courtesy Pressed Steel Tank Co.*)

several successive stages as in the deep drawing of a tubular oxygen tank to allow annealing for pliability to be restored.

In most drawing operations of this nature, lubricants are used to enable the dies to be matched, thus pressing the material closely between them. Without lubricants the material frequently does not flow smoothly which causes loss of accuracy and control and the increase of wrinkles or deformities in the material itself.

Presses used for shell-drawing operations may be either single or double action. A single-action press is simply a press where only one member of the press moves. A double-action press provides movement to both the top and the bottom (ram and anvil) of the forming press. The motion may be sequential or may be simultaneous.

Kirksite Drawing. *Kirksite* is a plasterlike cast material of medium hardness and density. It is used to make the dies used for production of limited num-

bers of products where the dies will not be subjected to continued and extensive use. The material is useful for making dies used for deep drawing operations on stock done in the cold state where the depth of draw is no more than two or three times the diameter. Kirksite drawing is done mostly in softer metals on a drop hammer or a press using a shim device called a *book.* Figure 10-24 illustrates how the initial draw is made and one leaf of the shim book is removed and the press is allowed to make another drop. The successive drops cause successive formations until the final drop makes the completed shape.

Fluid Rubber Drawing. *Rubber* is normally not considered a fluid. However, when completely confined and subjected to extremely high pressures, it exhibits all the characteristics of fluid. Because the pressure, as in a confined fluid, is exerted equally in all directions, certain operations may be performed without the creation of expensive dies. This means that for the production of small numbers of products where die expense is prohibitive, fluid-rubber processes are used effectively to reduce costs. The fluid-rubber processes may be used for both drop and squeeze operations.

Figure 10-25 illustrates the *Guerin process,* in which the rubber section becomes a pad upon the moving punch part of the die. The blanks are simply laid over shaped dies upon a work table, and the press ram is forced down upon the material. The rubber forces the material to take the shape of the forming blocks placed upon the table.

Figure 10-26 shows the *marforming process,* which is very similar to the Guerin process except that the workpiece or blank is laid upon a stationary rubber platform and the shaped punch or die is moved rapidly against the blank. The pressure of the rubber, when confined, forces the workpiece evenly against the shaped punch or die thus forming the material into the desired shape.

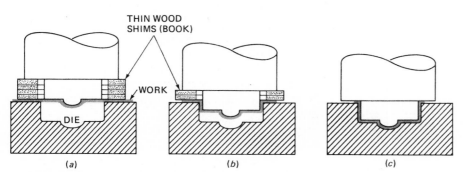

Fig. 10-24 Kirksite deep-drawing method for use on a drop hammer using removable shims from a "book."

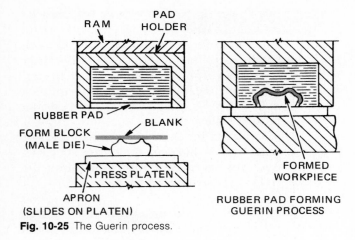

Fig. 10-25 The Guerin process.

The third variation of the fluid-rubber process is that of *bulging*. The bulging process is particularly appropriate to forming sections where the material bulges out from the side of a basic cylinder. A cylindrical blank is placed inside a shaped steel mold and a rubber plug is fitted into the blank. A punch then exerts force upon the rubber which in turn transmits the force equally in all directions forcing the metal to bulge out and assume the mold shape. This process is adaptable and frequently used for large production runs. Its main advantage is that it provides a method of removing the interior forming pieces (the rubber in this case) from the formed object.

Stretch Forming. *Stretch forming* is a process developed primarily from aerospace applications in which large sections of sheet stock are pulled over forms to develop curved sections. The process is similar to drawing of large forms but is usually done on sections larger than those capable of being done in most pressing applications. The process is confined to the

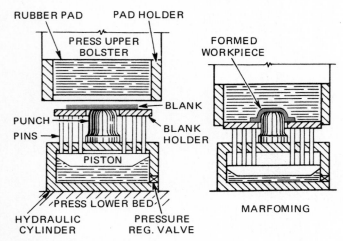

Fig. 10-26 The marforming process.

stretch forming of sheet and plate materials in the cold state. As indicated in Fig. 10-27, the process consists of placing a sheet of flat material over the form, gripping it tightly by mechanical means on two ends, and forcing it around the form shape. The force can be applied either vertically or horizontally depending on size, space, and equipment factors.

Spinning. *Spinning* is a forming process that may be done both in hot and cold states. The process is normally done in the cold state. However, extremely large pieces of hard, tough materials may be worked in the hot state to facilitate ease of forming. Spinning is a process of forcing the material to a shape over a rotating male die. The material and the die rotate while a stationary tool forces the material to shape. Spinning provides the advantages of using relatively inexpensive dies made of wood or metal and uncomplicated forming tools of wood or metal. Spinning may be done on common production machinery, such as an engine lathe, which adds versatility and adaptability to its list of advantages. Spinning shrinks the diameter of the material when compared with the original diameter of the blank, but the shrink in diameter is accomplished through the stretching of the thickness of the material.

One of the disadvantages of spinning is that it is an operation that requires some machine time and when done by an operator, requires a great degree of skill. However, the process may be mechanized for production purposes to reduce the degree of skill required and the machine time involved. However, the adaptation of the spinning process to production requires additional tooling expenses, making the mechanized process most adaptable for higher-production requirements.

When it is desirable to spin reverse-curve situations that would normally prevent the removal of the object from the pattern, two processes may be utilized. Offset chucks may be used to spin the double-curved

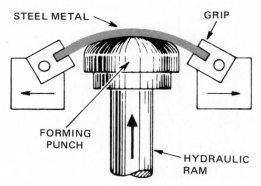

Fig. 10-27 The stretch-forming process.

sections or a die may be specially built to collapse into a smaller diameter, allowing it to be withdrawn from the formed shape. However, most spinning production processes do not involve the double-curved sections.

Recently developed variations of the spinning process include the *shear-forming* process. The shear-forming process is one in which a hardened metal tool is forced against a comparatively thick and semi-formed piece of blank stock. Great pressure is exerted upon the forming tool which literally shears or pushes the material out of the way and spreads it into thinner sections and greater areas. The shear-forming process involves frequently a reduction of 50 percent or more in thickness of the wall sections.

Extruding

As illustrated in Fig. 10-28, *extruding* is a process in which material is forced through a die made in the form of a given shape. As the material is forced through the shaped opening, it assumes the shape of the opening. The process could be compared to the squeezing of toothpaste from a tube or to decorating a cake with a frosting gun. Extruding is done in both hot and cold states. However, extrusion is done in the hot state in the production of large metallic pieces and almost all aluminum extrusions. Extrusion is done in both states with softer materials including rubber, ceramics, plastics, and other materials. The primary advantage of extrusion is the variety of shapes that can be formed in an almost unlimited array of cross sections. The process may also be used to form materials of relatively low strengths that could not be shaped by drawing processes. Typical shapes are illustrated in Fig. 10-29.

Extrusion processes provide extremely good dimensional accuracies, finished surfaces, grain refinement, working strengths, and toughness properties. Figure 10-30 shows the basic grain deformation process of extrusion. There are two basic variations of the extrusion process. The *direct extrusion* process

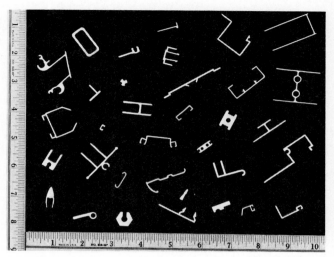

Fig. 10-29 An extrusion is a form achieved by ramming a hot billet of aluminum through a die configuration, much like squeezing toothpaste from a tube. (*Alcoa.*)

utilizes the principle of a moving force behind the material which is confined in a die cavity. As force is applied to the rear of the material in the confined space, the material is squeezed through the shaped opening. In the *indirect process,* the material is confined in a closed cavity. The die opening is made into the moving ram that applies the force. As the force is applied, the material is squeezed out through the die opening in the moving piston or ram. The die opening may be either in the center or the die opening may be around the edges of the moving ram. *Impact extrusion* frequently utilizes the latter process.

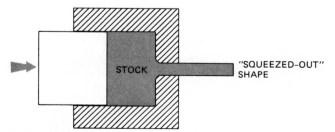

Fig. 10-28 The basic extrusion principle.

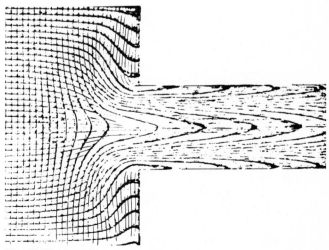

Fig. 10-30 Metal-flow pattern in extrusion.

Fig. 10-31 The cross-shaped solid extrusions, typical of the complex designs required by the aircraft industry, will help to attach and secure the 747's giant wings, spanning 196 ft, to the aircraft's 225-ft-long fuselage. The high-strength aluminum parts depicted here are known as main upper-rib chords, and are 26 ft long. Each weighs slightly more than a ton. (*Alcoa.*)

Solid Extrusion. *Solid extrusion* is normally done to materials in the hot state when shaping large pieces such as metallic billets from which other objects such as pipe will be made. Solid extrusion is used to produce a large variety of shapes for large structural processes, as illustrated in Fig. 10-31. It is a process applicable to most metals and plastics and is also used for a variety of other applications including the production of glass rods and producing decoratively shaped plastic rods from which thin slices are cut and applied as surface decoration to other objects.

Hollow Extrusion. *Hollow extrusion* is the variation of the extrusion process used to produce hollow and tubular shapes. For larger, harder materials the process is done in the hot state. However, with smaller, softer materials the process is frequently done to materials cold. Figure 10-32 illustrates two variations for producing hollow extrusions. Both a moving mandrel and a stationary mandrel may be used to extrude hollow shapes in a variety of cross-sectional configurations. The advantage of a moving mandrel is in the better forcing surface provided. However, the disadvantage of using a moving mandrel is that the length of the extruded material is limited to some extent by the length of the moving force. However, it should be noted that the length of the extruded material will greatly exceed the length of the mandrel due to the amount of material displaced. In both cases, a hollow billet is used.

Figure 10-33 illustrates another variation for producing hollow extruded shapes. The mandrel may be supported within the die cavity by thin sections which literally cut or split the material as it is extruded through part of the die opening. The mandrel and its supports are called a *spider.* However, the force of the material moving through the die opening and the pressure of the die and the mandrel reweld or recombine the material into one solid piece. For this reason, this process is normally done to hot materials. Spiders may be used to form both round and other tubular shapes.

Impact extrusion is used primarily upon softer materials in their cold state to form closed containers, cylinders, and other shapes. As illustrated in Fig. 10-34, the shapes may be hollow or may have a variety of combinations. The process is frequently used on materials such as plastic, lead, tin, and aluminum. It is basically a single-impact process used to produce small objects such as containers for food, beverages, and cases for items such as lipstick, cigarette lighters, and other devices. The process is also used to produce toothpaste tubes and small bearings. The material used in impact extrusion must be very carefully

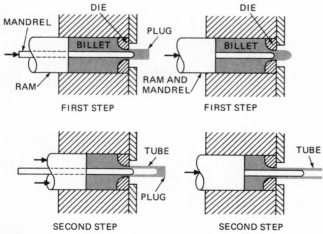

Fig. 10-32 Two methods of extruding hollow shapes.

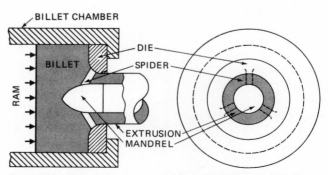

Fig. 10-33 Method of using a spider mandrel to extrude hollow shapes.

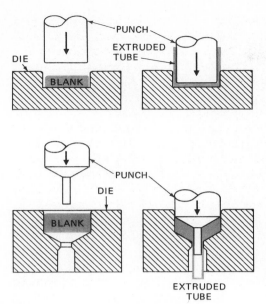

Fig. 10-34 The two types of impact extrusion.

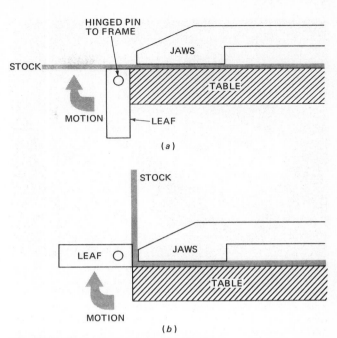

Fig. 10-35 The basic hand-operated brake.

measured to ensure proper flow and forming as well as to prevent damage to the dies.

BRAKING, BENDING, ROLLING, AND ROLL FORMING

Braking

Braking is a process of bending sheet stock into various angular shapes while the sheet stock is in a cold working state. Figure 10-35 illustrates the basic machine process used in most hand-forming operations. These machines are extensively used in the sheet-metal industries for custom-production pieces but are less adaptable to the high-cyclical rates used for mass-produced goods. The machine operates by rotating a leaf against the work which is clamped against a table by movable jaws. Several varieties of these brakes include a box and pan brake used for the formation of sheet-metal boxes and pans; a cornice brake used for the formation of sheet-metal ducting, cornices, and other long pieces; and the bar folder. The bar folder features an adjustable leaf which will move closely against the stock, forming a sharp angular bend, or which can be moved away from the jaw and table slightly so that the resulting bend is somewhat rounded. This gives an advantage of providing a smoother, rounder area or of providing enough "round" so that wire may be inserted to form wired rims around rectangular objects.

Much of the braking that is done for production, however, is done on a press brake, such as the one seen in Fig. 10-36. The press brake uses a fixed die

upon which the stock is placed and a moving die which is forced against the workpiece, clamping the material between the two shaped jaws and thus bending it to the desired shape. The advantages of this process are that it is quicker, it can be adapted to

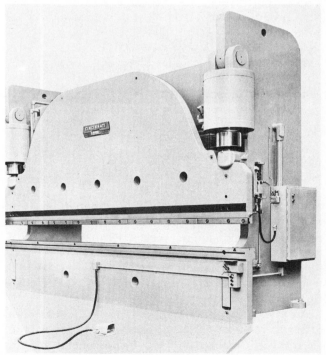

Fig. 10-36 A powered press brake. (*Courtesy Cincinnati Machine Co.*)

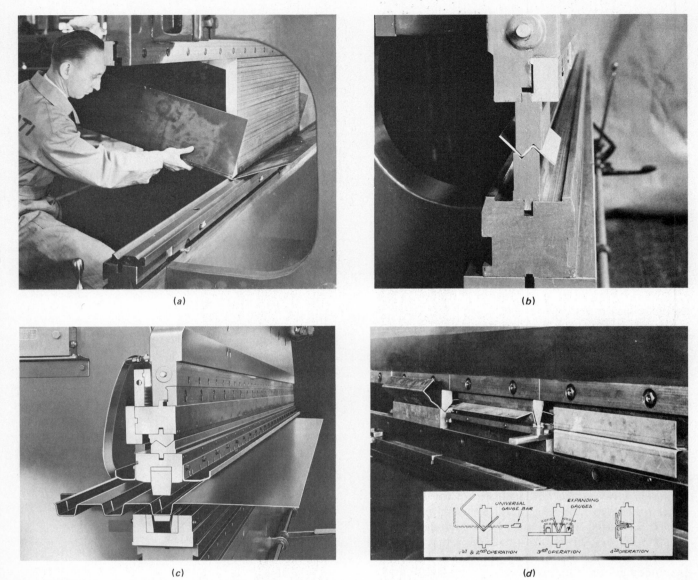

(a)

(b)

(c)

(d)

Fig. 10-37 (a–d) Applications of angular bends. (*Courtesy Cincinnati Machine Co.*)

quick use of jigs and fixtures thus eliminating set-up time for individual pieces, and it can be used on thicker stock and longer pieces. Figure 10-37a–d illustrates the variety of angular shapes that may be formed in press-braking operations and Fig. 10-38 illustrates how the press brake may be used to form rounded sections.

Another use of braking operations is the formation of seams in sheet-metal sections such as downspout, sheet-metal ducting, and other sheet-metal applications. Figure 10-39 illustrates commonly used sheet-metal seams.

Bending

Bending is the process of forming bar stock, tubing, and pipe into a variety of shapes. Bending is normally done cold, but for larger and heavier pieces it may be done hot. One of the basic bending variations forces the workpiece around a shaped die either by a rotating knuckle or by moving slides. The work may also be simply pulled around the shape, either by hand or by machine manipulation.

Special forms are needed when bending pipe and hollow shapes as well as when special other shapes

Fig. 10-38 Application of the press brake for current bending. (*Courtesy Cincinnati Machine Co.*)

such as angle or channel are bent. These forms must include special support features such as a round groove for bending pipe to prevent the stock from buckling or becoming distorted as it is bent into shape.

Multiple bending stages may be utilized for complex operations. Such bending operations are used for automatically making chain, wire and link fencing, as well as for bending individual pieces. For the production of such goods as wire and link fencing, the bending operations are "ganged" so that many different bends are produced in different strands of wire at the same time and are connected by cross strands automatically. Such machines operate faster than the eye can easily follow and are only appropri-

ate for the production of large volumes of objects due to the initial expense of building or purchasing such specialized equipment.

Rolling

The basic rolling operation consists of *rolling* bar, wire, tubing, or flat sheet stock between two feed rollers which force the material against a raised third roller causing the material to bend at a constant rate as the stock is fed against it, as illustrated in Fig. 10-40. Rolling operations may be done to hot materials when they are large or extremely hard and tough. Normally, rolling is done to materials cold, particularly in sheet-metal applications. The machines used in rolling range in size from the small bench-mounted model in the making of sheet-metal products by individual craftspeople to extremely large machines capa-

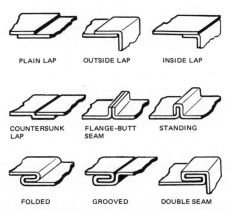

Fig. 10-39 The most commonly used sheet-metal seams.

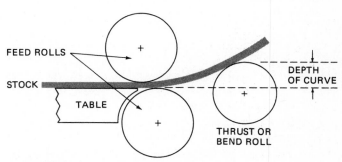

Fig. 10-40 Basic rolling process to curve stock.

ble of rolling pieces of plate metal approximately 8 by 16 ft [2.4 by 4.9 m] in length into curved sections used in the construction of oil storage tanks. Another variation of the rolling process is setting the bending roller at a nonparallel setting to roll conical shapes.

Roll Forming

Roll forming is a shaping process normally done in the cold state to sheet materials, including both metals and plastics. It differs from the rolling operation in that the rollers used in roll forming are comparatively small, thin shaped wheels that are used to bend relatively small areas of flat stock into various shapes. This is much like die forming, except that the material is passed through continuously rotating dies. Forming and flanging are two general classifications of these types of operations. The flanging operations are basically roll-forming operations done at or near the edges of sheet or plate stock. These specific operations include flanging, seaming, beads and grooves, crimping, and other similar operations. Figure 10-41 illustrates a rotary machine used in making these types of operations. Rotary machines range in size from small, bench-mounted machines that are hand operated to large, stationary machines designed for automated production work. Such machines are extensively used in canning and packaging industries.

Roll-forming operations are basically the same as flanging types of operations, except that they are performed upon other areas of the stock besides the edges. They are used in the construction of shaped, sheet metal tubing and other types of products. Some types of embossing operations could be considered to be similar types of operations. Figure 10-42 illustrates the roll-forming process. Roll forming may also be done to hot materials; a typical example is the forming of pipe by the butt-welded construction method.

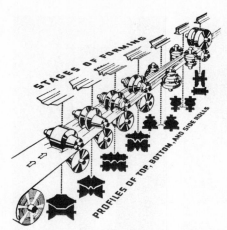

Fig. 10-42 The roll-forming process used for shaping objects.

SHEARING

Shearing is a cutting process extensively used on sheet and plate materials. The process features cutting without chip removal, heat, or chemical reaction. The process is clean, quick, and accurate but is limited in the thickness that can be cut by practical machine limitations and by the hardness and density of the material. Shearing is the term generally given to straight-cutting situations while cutting to other shapes, including angular, round, oval, and irregular shapes are given other names such as notching and piercing. Shearing is primarily done cold, particularly when dealing with thin materials, and is done to a wide range of materials including paper, fibers, cloth, ceramics, plastics, rubber, wood products, and most metals. Shearing principles are used to cut crankshaft blanks from heavy forged-steel pieces and are also used to cut blanks of cloth used in mass producing clothing items.

Shearing is a process normally done in the cold state for most materials. Shearing, as a term, is generally classified as performing straight cuts, usually across the entire width or length of the material. This cut may be perpendicular to the width or length or may be at any angle. The basic shearing action, shown in Fig. 10-43, involves moving a knife or blade in opposition to a table or bed; the process is as much a controlled fracture or break as a cutting type of operation. Most shears have a very small relief angle to the blades. For some specific machine operations, such as blanking and piercing, no such relief angles may be utilized.

The basic shearing operation may be done to a variety of materials, including paper shears used in bookbinding and the squaring shear used in sheet-metal work, which is illustrated in Fig. 10-44.

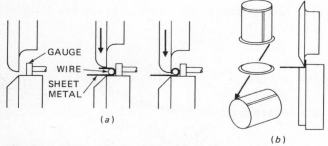

Fig. 10-41 The roll-forming process for Flanging Operations. (*a*) Flanging process adapted to enclose wire in an edge of a sheet-metal object; (*b*) turning a burr or flange to prepare the burr for bottom, preparatory to setting down and double seaming.

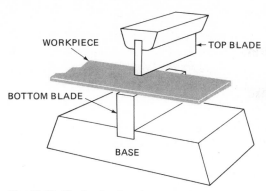

Fig. 10-43 The basic shearing process.

Notching

Notching is a shearing process normally done cold to sheet and plate stock and differs from shearing primarily in that the shape of the cutter is of some angular configuration. The cutter may actually be almost any shape provided that the intent of the forming or shearing operation is to cut out small portions of the stock.

Nibbling

Nibbling is a notching process that simply features successive bites or cuts until a larger shape or cut-out portion is produced. Interior shapes can be started from drilled holes quite easily and used to produce large pierced sections. Nibbling is used instead of punching and piercing operations when the production rate is relatively low in either numbers or specific rate. Machines used in nibbling are extremely versatile, inexpensive and relatively simple to operate and maintain. However, the production rates are comparatively slow.

Fig. 10-44 A squaring shear. (*Courtesy Cincinnati Machine Co.*)

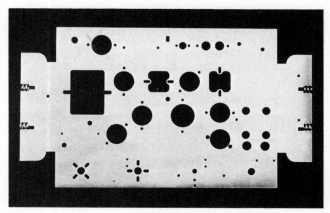

Fig. 10-45 Typical piercing operation. (*Courtesy S. B. Whistler & Sons.*)

Piercing

Piercing is a process of shearing a shaped hole in a sheet of plate or sheet stock. It is normally done cold and results in a shaped hole of almost any desired shape. Figure 10-45 illustrates some typical principles and applications of the piercing process. Such applications include making holes in washers, rivet holes for structural steel members, openings in panels that will be formed by other processes so that instruments or other equipment may be mounted, and similar operations.

Blanking

Blanking is basically a piercing operation except that the desired outcome is not a hole, but is rather a trimmed or sized product. As in piercing, the process is generally done to cold materials. Figure 10-46 compares the blanking and piercing processes in the manufacture of cut washers. As in the illustration, the first

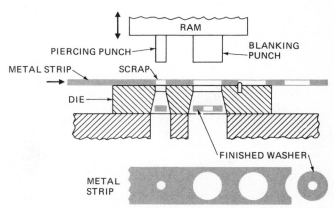

Fig. 10-46 Progressive die for piercing and blanking a simple washer. Notice that making the hole is "piercing," while shearing the body from the strip is "blanking."

operation is to produce the hole in the washer. The production of the hole is classified as a piercing operation. The second step of the process is to cut the washer from the stock with the hole already formed. In principle, the same process as piercing is employed except that the product is a formed object and is thus classified as a blanking operation.

To produce such items as cut washers, and other similar devices, a compound operation may be developed. A compound die or compound operation simply combines two steps into one action.

The blanking operation is also used to trim many other shapes in a variety of applications. Figure 10-47 shows a trimmed forging.

In most blanking operations, three essential parts are included in a blanking die. The punch must be shaped and sized to mate with the die cavity. In addition, some device must be included so that when the punch is withdrawn, the material, either the flashing (the trimmed-off excess) or the formed object, will not cling to the withdrawn punch. As illustrated, these are usually projections, or edges of projecting metal that catch the piece as it comes up ensuring its separation. These devices are known as *strippers*.

Dinking

Dinking is the same basic operation as blanking except that the term "blanking" is used primarily in reference to metals. "Dinking" is used in reference to most materials of lesser strengths or toughness such as cloth, rubber, fiber, and plastics.

Trimming

Trimming is a shearing operation, the same in principle as blanking. "Trimming" is for all practical purposes an alternative term to blanking. For example, a forged part may have its flashing "trimmed" with a blanking-type die as illustrated in Fig. 10-47.

Perforating

Perforating is an identical operation to piercing, except that more than one hole is produced in the same operation. Perforating is in essence ganged piercing or punching. The term "ganged" simply indicates that there is assembled into one machine operation a group of dies or punches that makes several holes with each motion. Perforating is used extensively and is well illustrated in making pegboard from the hardboard perforating operation shown in Fig. 10-48.

Slitting

Slitting is a cold shearing operation. In principle it is a continuous shearing process done by rotary knives rotating at the same pace as a moving material. The process principle is illustrated in Figure 10-49. This shows a typical industrial slitting machine in operation. The process is normally used to produce a number of strips of material from a wider sheet or plate of stock. These cut or sheared strips are then used in subsequent operations and are formed to produce other products.

Truing

Truing is a shearing or slitting operation. Its principle is exactly the same as slitting except that it is usually done to the edges of sheet materials, usually just after

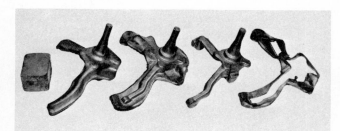

Fig. 10-47 Automobile steering knuckle produced in three operations on forging press. Bust, block, and finish. Note the blanking process of the last operation. (*Courtesy Ajax Co.*)

Fig. 10-48 Perforating is "ganged" for multistage piercing as illustrated on this section of perforated hardboard. (*Courtesy Cincinnati Machine Co.*)

Fig. 10-49 A typical slitting machine in operation. (*Courtesy Yoder Co.*)

the stock has been rolled to the desired thickness, and serves to straighten and align the edges of the rolled material. Truing operations are used primarily for nonmetallic products manufactured from plastic and rubber.

Shaving

Shaving is a cold shearing operation done to small, shaped parts. Many parts, particularly those parts that have been pressed or forged into round shapes, often have minor irregularities around their edges. These edges may frequently be designed to fit evenly against other surfaces but are unable to be fitted properly due to surface irregularities. Shaving is simply a process of shearing off a very small surface area to produce a surface area free of minor irregularities. Shaving may be done to either the edge or to the flat surface. Frequently, shaving is done to formed gears for small sewing machines and adding machines to remove burrs and rounded edges of the gears from the forming process.

REVIEW QUESTIONS

10-1. What distinguishes hot-working situations from cold-working situations?

10-2. What are the reasons for working materials in either hot or cold states?

10-3. What are the advantages of forging, shearing, drawing or extruding, and rolling processes?

10-4. Compare the tolerances generally obtained on hot and cold working processes.

10-5. What is cold heading?

10-6. What are the main considerations in deciding to produce a product by either drawing or extruding?

10-7. What factors determine whether an item will be drawn or whether it will be extruded?

10-8. How do bending and forming differ?

10-9. How do the hammer and anvils differ in press forging and in drop forging?

10-10. What is the basic difference between a progressive die and a compound die?

10-11. What is the major difference between shear forming and spinning?

10-12. What are the two major classifications of presses?

10-13. What process can be used to form deep drawing processes using a drop hammer?

10-14. What are the advantages of the fluid rubber processes?

10-15. How does notching differ from piercing?

10-16. How does blanking differ from piercing?

10-17. What factors should be considered when deciding whether or not to produce an item by either notching or by piercing?

10-18. How are the following items made? *a.* Toothpaste tubes, *b.* Half-dollar pieces, *c.* Beverage cans, *d.* Engine crankshafts, *e.* Nails.

CHAPTER 11

SPECIAL FORMING PROCESSES

The special forming processes include a variety of processes that utilize pressure to form materials to shape. However, because of the unique features of each process, the processes are not classified with those discussed in Chap. 10. The special forming processes differ from the normal forming processes in either the method of applying the pressure, the nature of the material, or both. The special pressures utilized include the use of explosions, expanding gases, compressed liquids, and magnetism. The processes are restricted to metals in the cold state and powdered metals.

There are a number of unusual variations of the pressure-forming processes used in the aerospace industries for forming sheet and plate metals. However, only those processes which employ either a distinct and different method of applying pressure, or a different type of material are considered in this chapter. Some exotic shaping processes are omitted because they are relatively rare and normally restricted to one segment of the broad spectrum of industrial classifications. For example, some processes are used exclusively in the aerospace industry in forming thin sheet sections of nonferrous metals. While these processes are impressive, they are of such limited scope that they have little impact upon the industry as a whole at this time.

POWDER METALLURGY

Powder-metallurgy (P/M) processes consist of pressing metallic powders into some shape (Fig. 11-1). The pressing process is done upon presses similar to those used in normal forming processes. However, the P/M processes use more complex dies, and the materials that are pressed to shape require baking to harden them to a usable condition. The process has been "discovered" at various points in history ranging from an-

cient Egypt, to the South American Inca Empire, to Victorian England. The first modern industrial application was that of forming wire from powdered materials that were too hard to work or melt by any of the then current processes.

Powder metallurgy is used extensively in industry to form a variety of small parts. Typical parts are shown in Fig. 11-2. The advantages of using powder metallurgical processes include:

1. *Shaping to close tolerances.* Tolerances of ±0.002 in [±0.0508 mm] are considered normal. The close tolerance of the shaped product reduces the machining required even though the shape may be complex.

2. *Versatility of use.* The type of products made from metallic powders include magnets, cutting tools of extremely tough and hard materials, filters for gasoline and other liquids, gears, machine parts, and other applications.

3. *Economy of production.* With the close tolerances and complex shapes that require little or no machining, low cost of production is a major feature. In the production of impregnated bearings, no machining is required and the cost of production is approximately one-fifth of the cost of bearings produced in conventional cutting processes.

4. *Hardness utilization.* Materials of great hardness may be shaped in a powdered form. Tungsten carbide is extensively used for industrial cutting tools. However, in its metallic form it is almost impossible to shape and to cut effectively into the desired shapes. However, when tungsten carbide is reduced into its powdered form, it can be pressed into the desired shape and used without further shaping.

176

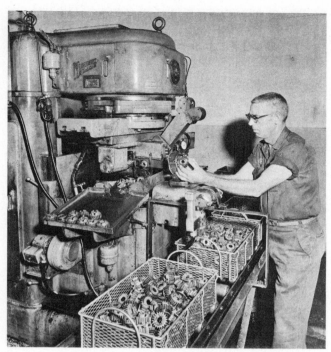

Fig. 11-1 A typical press operation for forming powdered metal parts. (*Courtesy Haller Division.*)

5. *Range of forming operations.* Secondary operations can be performed to increase the accuracy, finish, or other desired physical characteristics of the shaped powdered material. Secondary operations include impregnation with oil, graphite, wax, or other lubricants for use in bearings. Secondary operations include impregnation, infiltration, plating, heat treatment, pressing, and machining.

Impregnation

Cold-metallic bearings may be impregnated with oil, graphite, wax, or other lubricants. Impregnation provides a maintenance-free sealed bearing with the lubricant built into it so that exterior lubrication is not necessary. Such products find extensive use in bushing-type bearings for use in automotive products such as water pumps, alternators, starters, and other similar equipment.

Infiltration

Infiltration is a process of strengthening the powdered product and of making it more dense. Infiltration involves placing a solid metal piece upon the formed, sintered, powdered piece and resintering, or baking, both pieces. The second piece melts and is literally

Fig. 11-2 Multilevel parts made of powdered metal compacted in a 20-ton press. (*Courtesy Penwalt Chemical Corp.*)

soaked up by the porous powdered object. This process imparts qualities of toughness, durability, strength, and density to the powdered-metallic product.

Heat Treatment

Powdered-metallic products may be *heat treated* by conventional methods. The heat-treatment processes are done for the same purposes for which other metals are heat treated, to impart qualities of toughness, hardness and other desired metallurgical characteristics. However, in heat treating powdered metals, the heating must be done carefully because the powdered metal products conduct heat much more slowly and less evenly than do solid metals. The heating period for heat treatment must, therefore, be of longer duration than for similar pieces of solid material. The cooling, however, should be quicker. Care must be taken during the heat treatment processes to prevent oxidation within the porous structure. This is usually done by heating within inert gases. Oxidation can reduce strength and create weaknesses and impurities within the object. Oxidation problems are more acute in powdered metals than in solid metals.

Plating

Powdered-metal products may be *plated* with most of the metals normally used for any other plating process. These include copper, gold, silver, chromium, and other metals. The plating process is virtually the same and conventional plating equipment may be utilized. However, the porosity of the powdered-metal piece presents some problems that are normally not present with solid metal products. Care must be taken to prevent the electrolytic solution used in plating from becoming entrapped within the porous structure of the powdered-metal piece. Should the electrolytic solution become entrapped, a galvanic, or battery-type, action would occur resulting in corrosion which would weaken the structure or ruin the surface. To offset this problem, powdered-metallic objects are normally impregnated with resinous or plastic substances to prevent the electrolytic solution from being absorbed within the piece.

Pressing

Powdered-metallic objects that have been shaped by *pressing* and sintered to give them shape and physical strength may be resized in operations similar to blanking and coining. Pieces may be trimmed in blanking-type operations or they may be slightly shaped in operations similar to coining. The two operations are often combined for powdered-metal products, which are normally heat treated after such operations to increase hardness and to make the grain density and structure more uniform.

Machining

Although a primary purpose of using powdered-metal processes is the reduction of *machining,* limited machining processes can be successfully undertaken. The processes are, in most cases, conventional machining processes with no initial tooling or machine adaptations required. Cutting lubricants are not normally used due to the porous nature of the formed objects. Because no cooling action occurs, great heat is generated in the cutting processes. This heat can decrease the useful life of the cutting tools. In some situations where lubricants must be used, volatile fluids—such as carbon tetrachloride—are used; such fluids will evaporate quickly and easily from within the structure without stains or residue. Machining processes are normally confined to surface smoothing or fine finish cuts and are not normally employed for removing major portions of the shaped object.

DISADVANTAGES

Powder metallurgy has a number of disadvantages. These include relatively low tensile and compression strengths, uneven material flow during forming which results in uneven strengths, uneven curing and heat distribution rates, and meticulous handling requirements due to the relative fragility (before sintering) of the parts. Also included is the cost of producing the dies used in shaping or pressing the powders to shape. The dies are very expensive because they must be very smooth with highly polished (and often plated) and hardened surfaces. The dies, to fulfill these requirements, are normally made of high-quality, hardened, tool steel. Tolerances are close for dies with more than one part and the movement of the mated parts of the dies is normally held to a minimum due to the abrasive action of the powder. Because of the expense of the dies, powder-metallurgy processes are normally limited to products that can be mass produced so that the large numbers that are produced will offset the initial tooling expense. The great care required in making the metallic powders creates another disadvantage in working with powdered-metal products. Great care must be taken so that the particle size will be sufficiently small to assure the ability of the particles to flow uniformly throughout the object during forming and treatment. Uneven particle size reduces the ability of the material to be formed, as well as reducing its strength and toughness. There are also optimal size and shape factors. Further, care must be taken relative to the purity of the powdered product during the processes of manufacturing the particles and storage and transportation of the particles and during the forming and treating processes.

Because the powder particles do not flow easily, the size of the piece that can be formed is limited. Normal pressures for shaping powdered products run from 10 to 30 tons per square inch (tons/in²). As most presses in most industries are relatively small, this limits the size of the piece that can normally be done. Pieces can be formed up to approximately five inches square and up to six inches thick where the available presses have sufficient force. However, extremely large presses are required for items of this size, and the process is more readily adaptable to smaller and thinner pieces that may be fabricated using presses of the size normally found in industrial concerns.

Production of Metallic Powders

There are three main processes for producing the metals and then reducing them to metallic powders. Figure 11-3 illustrates the three main processes for reducing metal to powder of working specifications.

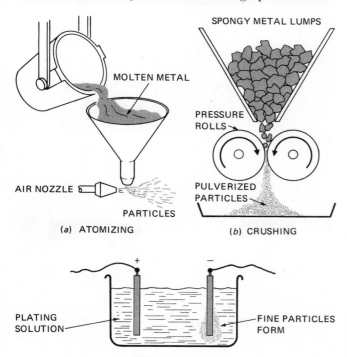

(a) ATOMIZING (b) CRUSHING

(c) ELECTROLYSIS

Fig. 11-3 Three main processes in metallic-powder production.

Normally, the metallic powders are made from metal that has been processed specifically for the purpose of being reduced to powder. The care required for purity and the specific metallic alloys desired dictate the particular type of process that might be used to produce the metallic ore. Some metals destined to be powdered are produced by electrolysis. These metals include iron, silver, tantalum, and copper. Some alloys of iron, nickel, cobalt, molybdenum, and tungsten have higher levels of impurities and are normally produced by reducing ores in furnaces. Furnace operations, due to the gases and heats generated and the oxidation that is always present in such processes, generally develop higher proportions of impurities than those metals produced by either electrolysis or induction furnaces.

Shaping the Part

Shaping the part is a sequential process involving up to six specific operations. The powder must be mixed to specifications detailing purity, alloys, and size of particle. After being mixed, the powdered material is pressed into shape. This process is called *briquetting*. Briquetting dies are one of three specific types. A simple pressing operation may be done with a compacting single-action die. Compacting double-action dies produce better grain structures as illustrated; for parts of differing cross-sectional size, compound-action dies may be preferred. After the part has been formed it may be presintered to increase its green strength for ease in handling and storage operations. The final hardness is attained through a sintering operation in which the material is baked for a determined length of time at temperatures ranging from approximately 70 to 80 percent of the melting temperature for that particular metal or alloy. In sintering, an initial "burn out" is required to burn or evaporate lubricants used in shaping and forming. Gradual increasing heating is done to prevent explosion. As in heat treating, controlled atmospheres are required to prevent oxidation. When required, secondary operations may be performed to develop finished, machined surfaces or dimensional tolerance requirements on the piece as previously mentioned. Whether secondary operations are performed or not, heat treatment of the product is common.

HIGH-ENERGY-RATE FORMING

High-energy-rate forming (HERF) is in principle simply another system of forging metals. The process is principally done cold to sheet and plate metals. However, instead of dropping a hammer, an instantaneous force of great magnitude (such as an explosion) is created and transmitted to the workpiece through a liquid medium. High-energy-rate forming processes are used to make both simple and complex shapes. Because pressure applied to liquids is exerted evenly in all directions and because the extreme magnitude and quickness of the force combined with the property of most metals in which they deform readily and evenly under these conditions, the processes may be used to form shaped sections of great quality and uniform wall thickness. Further, metals that are difficult to form in the traditional forging or pressing processes may be readily formed by the extremely large pressures generated in these processes. Stainless steel and titanium, in particular, are metals easily formed by this process which are difficult to form by conventional pressing processes. As in conventional pressing, large or deep sections may be formed by a progression of several successive applications of force.

The inexpensiveness of the dies is yet another significant advantage of the process. The dies may be formed from cast Kirksite, from shaped wood, or other inexpensive and relatively soft materials. Normally, only the female die is required as the force itself provides the shaping part. Further, the pressures that can be generated by such processes are in excess of one million pounds per square inch over large areas. These pressures are almost impossible to obtain even in the largest of conventional presses when forming relatively large areas. Complex shapes may be easily formed as the pressures are distributed evenly to all areas regardless of their complexity of shape. There is no springback in forming with this process. In conventional forming processes, when metals are formed cold, they must be overformed because of the characteristic of metals to literally spring back into shape. The high-energy-rate forming system does not have this characteristic. It is not known why; however, it is sufficient to know that there is little or no springback in the process. Finally, the amount of forming that can be done in one step or stage is greater in high-energy-rate forming than in conventional processes. Extremely complex and deep shapes may be formed in one or two steps compared with four or five in conventional processes. The forming also improves the grain structure and physical properties of the shaped part.

The process is believed to have originally been used in forming brass cuspidors common to the late nineteenth century. However, the process was lost as the era passed. The process was not rediscovered or revived until the needs of the aerospace industry called

for complex forming of exotic metallic alloys which were difficult and expensive to do by conventional means.

Explosive Forming

Figure 11-4 illustrates the three variations of the *explosive-forming process,* and Fig. 11-5 shows a die container and some typical products formed in this particular set of equipment. The explosive forming process uses high explosives, such as dynamite, which are exploded within a closed container of water. The equipment and area requirements are small. The process has a relatively low rate of production and is not suitable for the production of great numbers of pieces. There is some danger involved due to the nature of the high explosive igniting under confined conditions.

Explosion or Gas Forming

The *explosion- or gas-forming process* differs from the explosive-forming process in two respects. First of all, the explosive is a slower acting charge, such as black

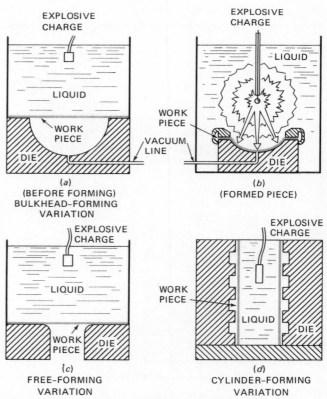

Fig. 11-4 Three variations of the explosive-forming process.

Fig. 11-5 Typical products formed by explosive forming. (*Courtesy Vought Aeronautics.*)

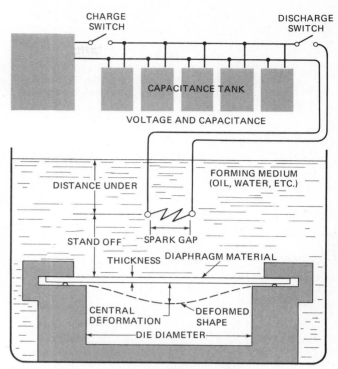

Fig. 11-6 Schema of the spark-forming process.

powder, and the force generated is from the rapid expansion of the gas rather than the force of the explosion. The process could be likened to that of forcing a bullet from a rifle by the expanding gas rather than the explosive force of the powder. The second distinction is that the force of the expanding gas is transferred to a piston rather than directly to the confined liquid. The piston impacts against the confined liquid which in turn transmits the force equally to all parts of the shaped cavity. This process is much quicker than the free-formed explosive forming operation. It is readily adaptable for use on small parts, and permanent dies may be used to ensure high uniformity of products. Because of the low explosive force of the material, it is safer, and it has fewer requirements concerning bulk and safety shielding.

Spark Forming

Spark forming is similar to explosive forming (Fig. 11-6) in that the force is generated within a liquid, such as water, when the spark generates gases from the liquid. It has been found that a high voltage electrical spark when ignited in water will generate up to 6000 horsepower (hp) in as little time as 40 millionths of a second. Because of almost instantaneous generation of great force and because of the characteristics of liquids that transmit such forces equally in all directions, the use of electrical sparks is becoming common in place of explosives. Spark forming is used essentially for the same purposes for which explosive forming is used. However, greater safety (no explosives to handle or store), better control, and lower cost are desirable features when compared with the explosive-forming process.

HYDROFORMING

Hydroforming uses fluid pressure as in the high-energy-rate forming processes. However, the hydroforming process features a slow pressing operation and can be done on adapted conventional presses. There is no impact or instantaneous force as in explosive or explosion forming. As can be seen in Fig. 11-7, the hydroforming process is best compared to the fluid rubber processes. A moving punch moves against a blank that is backed by a diaphragm which contains the liquid. As the ram of the press moves further into the liquid, the force of compression creates greater pressures, which are distributed evenly throughout the liquid, thus helping to form the part evenly and uniformly. The primary advantage of hydroforming is that there is little if any thinning of the metal as it is formed. Of course, as in any drawing operation, a certain reduction of thickness is experi-

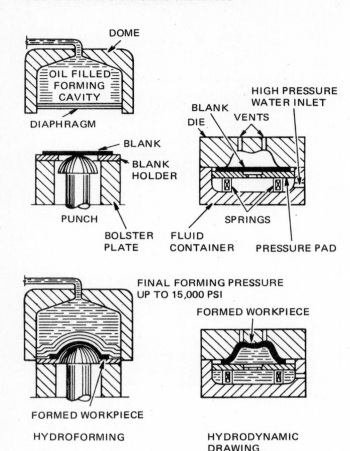

Fig. 11-7 Shows two variations of the hydroforming process.

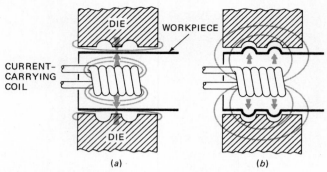

Fig. 11-8 Principle of magnetic forming. (a) Strong magnetic fields are formed on both the inside and outside of the workpiece to create a balance of force. (b) The sudden collapse of one side ends the balance. This causes the other field to expand toward the permeable die with impact force that presses the workpiece against the die.

enced, but the thinning is uniform without thin spots or wrinkling. Hydroforming can also be used to accomplish greater depth of forming than can be experienced in normal pressing or drawing operations.

MAGNETIC FORMING

The basic *magnetic-forming process* is illustrated in Fig. 11-8. The process has certain similarities to the explosion-forming process in that the force field of the magnet is expended in an extremely brief period of time, thus attaining the effect of instantaneous impact. The basic process features the transferral of magnetic lines of force from one place to another. As the magnetic forces can be transferred through the metallic objects which are being formed and which also have certain degrees of magnetic permeability, the force of the passage shapes the part. Pressures of one-half million pounds per square inch for durations of up to six millionths of a second are common.

The process may be used to expand or to compress materials into desired shapes. Split or segmented dies and mandrels may be utilized to allow the formed die

to be dismantled and withdrawn or removed from the finished, shaped piece.

In the magnetic-forming processes, an immense electrical current creates the dense magnetic flux and is turned on in a few millionths of a second, and the force is thus generated. However, far more complex operations with other variations are also utilized. The flux may be built up in both halves of a matched set of dies. The flux is turned off of one die by suddenly turning off the current that creates the magnetic flux. This creates an imbalance which causes the flux to be attracted by the die and in transferring creates the force that shapes the object.

To control the rapid start and stop of electrical current, special switching devices, called *ignitrons,* are used. These are large pieces of electronic equipment, and the operation is not visible. A further requirement of such equipment is that extremely large currents and voltages be present. These are normally generated and stored in large capacitor banks just before the process is triggered. The process can be utilized for sizes ranging from quite small to moderately large and is, particularly for small parts, readily adaptable for mass production. The entire triggering and generating equipment for small parts can easily fit upon an average workbench.

REVIEW QUESTIONS

11-1. Why is a force generated within a liquid for high-energy-rate forming?

11-2. What are the advantages of HERF?

11-3. What are the different processes in HERF, and what are the advantages of each?

11-4. What are the main disadvantages of HERF?

11-5. What products are made by HERF?

11-6. What are some typical products made by the powder-metallurgy process?

11-7. What are the advantages of the powder-metallurgy process?

11-8. What are the disadvantages of the powder-metallurgy process?

11-9. What supplemental processes may be done to powder-metallurgy objects?

11-10. What special considerations must be made in shaping a powdered-metal part?

11-11. What is magnetic forming?

11-12. Is magnetic forming suitable for: *a*. High production rates? *b*. Large parts? *c*. Interior work? *d*. Exterior work?

CHAPTER 12

TECHNOLOGY OF MATERIAL REMOVAL

Modern manufacturing employs a variety of techniques and engineering principles to shape or form materials into usable products. Industrial processes such as joining, forming, molding, casting, and cutting are employed in manufacturing. The process of cutting a material in order to provide a desired shape within specifications is called *machining*. Machining processes are associated with material removal by producing a chip or separating the material. Plastic, wood, and metal parts are commonly machined during some stage in their manufacture. In addition to chip-producing and separation processes, other machining and/or contemporary processes include: laser, electrodischarge, and electrochemical methods. However, the majority of material-removal processes utilize chip-producing methods and are therefore the concern of this chapter.

MACHINABILITY

The ease with which a material can be removed in planning, shaping, turning, drilling, sawing, boring, threading, broaching, and grinding is referred to as *machinability*. In addition to the ease of material removal, the quality of the resultant surface finish and life of the cutting tool are of prime importance in the concept of machinability. A number of factors influence the resultant surface finish, tool life, and ease of material removal.

Factors Influencing Machinability

Factors related to machinability can be classified by material considerations, tool design, tool type, and lubrication. Material considerations include the physical properties, internal structure, and heat treatment of the material.

Physical properties that affect machinability include hardness, tensile strength, and compressive strength. Soft materials may cause considerable drag on the tool, which causes heat generation, decreased tool life, and the potential for poor surface finish. As the material's hardness increases, the heat being produced is reduced, yielding an improved surface finish and increased tool life. However, when the material reaches a very high degree of hardness, the machinability decreases, due to abrasive wear on the tool.

Internal structure of the work material such as the grain structure, abrasive inclusions, and alloys affect the machinability of a material. A uniform microstructure, small amounts of alloying elements, such as sulfur, manganese, and carbon increase the machinability of certain steels, while small amounts of abrasives or inclusions, large gain structure, and very high alloy content tend to reduce the machinability. A very low (0.03) or very high (0.70) percent of carbon will decrease the machinability.

Heat treatment of the work material is the final material consideration directly affecting the machinability of the material. Hot and cold working of plain carbon steel tends to increase the machinability of high- and low-carbon steels respectively. However, tempering, annealing, and normalizing of work-hardened material will increase the machinability. Various methods of quenching and cooling heat-treated materials also has a direct bearing on the machinability of the work.

Tool design and cutting-tool materials will have a direct affect on the machinability of the work. Design of the tool must allow a clearance between the work and the noncutting edges of the tool. In addition, the tool design controls the type of chip generated and aids in removing the chip from the work. Improper design will result in heat generation, rough surface finish, reduction in tool life, and uncontrollable chip formation. Cutting-tool materials must be hard enough to maintain a reasonable cutting edge when machining hard materials. The tool material should also be capable of maintaining strength and a sharp cutting edge during heat generation in the machining operation. Rigidity of the tool on the machine must be considered when selecting the tool material. Related machining factors also have a direct influence on the machinability of the material. Depth of the cut and the amount of feed affect the quality of the surface finish. However, the properties and size of the work piece dictates the depth of cut and feed that should be used. Cutting speed, also dependent on the type of workpiece material, is another variable affecting the ease of material removal. The final variable affecting machinability is the type of lubrication employed. Lubrication aids in cooling the workpiece and cutting tool while reducing friction during machining.

Machinability Ratings

It is apparent from the previous discussion that machinability of a material is dependent upon a number of variables. When the machinability of a material is determined for a specific process, such as drilling, the information should not be applied to other processes such as turning or milling. As an aid in determining the cutting speed and feed in machining processes, machinability ratings are applied to various materials. The rating is based on the machinability of AISI-B1112 carbon steel which is given the value of 100. If a material has a machinability rating of 75, the overall machinability is approximately 75 percent that of B1112 steel. The power required for machining is increased while tool life and surface finish are decreased with lower machinability ratings.

CHIP FORMATION

Material removed is usually associated with the formation of a chip. Through the machining process, chips are formed, then discarded, yielding the desired shape and size of the product. Formation of the chip is important in machining therefore, a brief discussion is warranted. Basically chips are the result of a cutting tool moving under or into and along the work material. As the tool moves through the work, material failure results producing a chip. For ease in understanding the concept involved in material removal, an example using orthogonal cutting principles is used. *Orthogonal* cutting refers to two dimensional cutting where the tool is perpendicular to the work and direction of feed. The width of the tool is the same as the width of the material being machined.

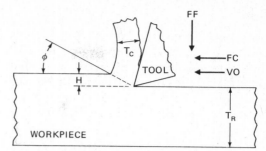

Fig. 12-1 Principle of orthogonal cutting.

Figure 12-1 illustrates the principle of orthogonal cutting. The tool moves at a velocity VO through the work material with the force of cutting action FC and the force of the feed FF. As the tool moves into the work with a depth of cut H, forces FC and FF cause the work material ahead of the tool to be compressed. The area under compression is in a state of plastic deformation, within this area is a section referred to as the *shear zone*. Material failure within the shear zone occurs at a plane called the *shear plane*. The shear plane occurs at the *shear angle* ϕ, approximately a 45° angle. The work material moves up and over the face of the cutting tool and breaks off the parent material. Cutting action causes the work material in the shear plane to be compressed and move up and over the face of the tool. The deformation occurs in both the chip and the workpiece below the cutting tool (Fig. 12-2). Shear deformation can be compared to a deck of playing cards when pushed, causing a sliding motion. Deformation occurs at different stages in machining, depending on specific materials. Failure or separation from the workpiece occurs when a critical value is reached for any material. Cast iron, a brittle material, will fracture easily because the material's resistance to plastic deformation is high. Soft ductile materials have lower resistance to deformation and fracture. Materials with a low resistance to plastic deformation will form a continuous type chip. Brittle materials have a high resistance to deformation and will have small and easily fractured or segmented chips.

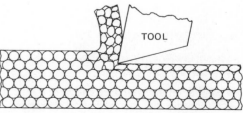

Fig. 12-2 Deformation in the chip and workpiece resulting from machining.

Type of Chips

During the machining process, various types of chips are produced. The type of chip depends on the cutting speed, tool design, properties of the work material, temperature in the cutting area, and feed and depth of cut. Chips are classified as either discontinuous, continuous, or continuous with a built-up edge (Fig. 12-3a, b, and c).

Discontinuous or Segmented Chips. *Discontinuous* or *segmented chips,* usually result when machining brittle materials such as gray cast iron. This type of chip is formed because the work material has a high resistance to compression and plastic deformation. Therefore, the material breaks up into small segmented chips. Surface finish is usually of a high quality when discontinuous chips are formed. In addition, the power required to machine a material is usually lower when discontinuous chips are formed. When machining hard materials, the cutting tool edge tends to wear down becoming rounded. When the tool wears, additional pressure on the tool is necessary. The additional pressure required causes increased friction between the work and tool, which generates additional heat. Dependent on the cutting-tool material, an increase in heat can reduce the tool life. Surface finish of the work material becomes rough as the cutting tool edge wears down. Discontinuous chips can be formed when cutting a ductile material if a slow cutting speed and heavy cutting feed is used. Cutting tools can be produced with chip breakers (an obstruction on the tools face) to aid in formation of a segmented chip. Discontinuous chips have an advantage over other chip forms in that they are easily disposed of.

Continuous Chips. *Continuous chips* are caused by a continuous deformation of the material directly ahead of the cutting tool. The chip flows up along the cutting tool without fracturing. Usually a continuous chip is formed when cutting a ductile material at relatively high speeds. This form of chip has a smooth back surface (cutting-tool side) and a rough deformed front. When continuous chips are produced, they tend to snarl around the cutting tool rather than curl and break off as in discontinuous chips. In addition to snarling, the continuous chip is difficult to dispose of.

Continuous Chips with Built-Up Edge. *Continuous chips with built-up edge* are very similar to continuous chips, however, a built-up edge or mass of material forms at the edge of the tool. During machining, small particles of material weld or fuse themselves to the tool. During cutting operations, these particles remain on the tool a short period of time then slide off and remain on the machined surface. The build up of particles starts again and the process is repeated. When a continuous chip having a built-up edge forms, the cutting action changes to a rubbing rather than a cutting action. Rubbing increases the friction between the tool and work thus generating higher temperatures. Higher temperatures cause increased distortion in the workpiece and reduce the tool life. In addition to the built-up edge being formed by machining ductile materials, other factors influence the formation of the type of chip. Friction between the tool and the workpiece can cause a built-up edge to be formed on the tool. Friction can be reduced by using an appropriate lubricant. Lubrication reduces friction and in turn aids in reducing the heat generated, it also reduces the probability of a built-up edge forming. Another factor influencing the cutting action and type of chip is a rough ground cutting tool. A rough surface on the tool face can cause the material being cut to adhere to the tool's surface causing a built-up edge. A heavy feed, producing large-size chips can also cause a continuous chip with a built-up edge.

CUTTING TOOL MATERIALS

Cutting tools are made of various materials each having specific properties and characteristics. Common cutting tool materials include: high-carbon steels, high-speed steels, cast alloys, cemented carbides, ceramic materials, and diamonds.

Carbon Steels

Carbon-steel cutting tools were the first type of tool commercially available. Carbon content in the cutting tool ranges from 0.90 to 1.10 percent. The tools are forged or machined into the desired shape then hardened. Tempering of the tool reduces internal stress in the shaped tool which enables the use of the tool to

(a)   (b)   (c)

Fig. 12-3 Sectional view of (a) discontinuous chip, (b) continuous chip, and (c) continuous chip with a built-up edge
NOTE: The three illustrations should be combined to form one figure with three separate sections. See Fig. 12-5.

temperatures up to 400°F [205°C]. Carbon-steel tools possess good strength and hardness. The strength enables the tool to maintain a sharp cutting edge. In addition to the ability to keep an edge, the tool has good shock resistance. Shock resistance enables the tool to be used for intermittent cutting or rough cutting. Carbon-steel tools are used for low or short production runs and are low cost. The most common use of carbon steel is for reamers, taps, dies, and tools requiring hardness, strength, and wear resistance.

High-Speed Steels

High-speed-steel tools are composed of various alloys to enhance the properties of the tool. The most common tool steel available consists of 18 percent chromium, 4 percent nickel, and 1 percent vanadium alloys. Alloy high-speed-steels are superior to carbon tool steel in that the tools retain a cutting edge at higher temperatures and cutting speeds. High-speed steels are identified according to their alloy content such as cobalt, molybdenum, and tungsten. Tungsten, as an alloy, increases the range of the cutting tool relative to heat resistance. Carbon-steel tools soften at 350 to 400°F [177 to 205°C] and loose the cutting edge, while tungsten-steel tools maintain their edge to 1000°F [538°C]. With the increased heat resistance, the cutting speed can be increased proportionately. Tool steels are further classified by their heat treatment (water or oil quenched), and by physical properties (heat resistance, wear resistance, toughness, and deformation). Impact resistance is increased as the carbon content is increased, however, abrasion resistance is decreased. High-speed steel is used for drills, reamers, mills, punches and dies, also lathe and milling machine tools, and so on.

Cast Alloys

Cast alloys as the name implies, are cast into shape rather than forged or machined. Cast alloys are a combination of chromium, tungsten, carbon, and cobalt. The material is very hard and difficult to machine, therefore the mixture is cast into shape. The material can be polished to a very smooth surface thus reducing friction and the possibility of a built-up edge. Cast-alloy tools exhibit high resistance to abrasion and corrosion. Abrasion resistance prevents craters and wear spots from forming on the tool. Cast-alloy tools will soften at temperatures above 1200°F [649°C] but will regain their hardness when the tools cool. Higher cutting speeds are possible with cast alloys than with high-speed steels. The cast-alloy tools find application in machining most materials.

Cemented Carbides

Cemented carbide tools are composed of carbon, a binder, and some alloy usually tungsten, tantalum, or titanium. The tool is made through the powder-metallurgy process where the ingredients are pressed into shape. Two manufacturing methods are used in the powdered metallurgy process, hot and cold pressed. Ingredients are mixed and compressed with high pressures, then heated to 1500 to 3000°F [816 to 1650°C] in the cold-pressing methods. Hot-pressing methods apply heat and pressure to the mixture simultaneously. Heating to a high temperature causes the particles in the mixture to fuse together. Binders are used to form the bonds between the particles. Cobalt is the most common binder used in cemented carbide tools. Varying amounts of alloys and cobalt are mixed to achieve varying degrees of hardness and toughness. Increasing the cobalt increases the toughness of the tool but reduces the material's hardness. Cemented carbides are classified by the type of cut, surface wear, and the material's impact resistance.

Four types of cuts in the classification system include: rough, general purpose, light finishing, and precision. Classifying by the type of cut is further identified by the type of material the tool is used on, either ferrous or nonferrous. The second classification system includes three degrees of surface wear on the tool. Finally, cemented carbides are classified into three groups based upon their impact resistance. Application of cemented carbides includes roughing and finishing cuts on ferrous and nonferrous metals, machinery parts, dies, and gauge construction. Carbides provide a very hard surface for cutting tools. Tungsten carbide is probably the most common type of cemented carbide tool. Carbides cost considerably more than tool steel. To reduce the cost of the tool, carbide cutting edges are produced and brazed on the shank of steel tools or used as replaceable inserts. Cemented carbides are very hard and have a high heat resistance. When carbides are used, the cutting speed can be from two to five times the speed of high-speed-steel tools. The tools are able to withstand high heat but the temperature increase should be gradual.

Ceramic Materials

Ceramic materials (cemented oxides) are another type of cutting tool material. Actually, the ceramic materials used in tools are composed of aluminum, silicon, or magnesium oxides. Ceramic tools are manufactured by the powder-metallurgy process. The oxides are compressed and sintered by either cold or hot pressing. Cold pressing compacts the material in the mold using pressures in excess of 20 tons per square inch, then the shaped tool is removed and sintered. Hot pressing utilizes pressure and heat together to form and partially sinter the material. Ceramic tools are very hard with high degrees of abrasion resistance. In addition to their ability to keep an edge, ceramic tools are noncorrosive, nonmagnetic, nonconductive, and possess high compressive strength. During machining operations, ceramics are capable of withstanding high temperatures without failure. Ceramic tools are capable of high-speed cutting; in fact, they are designed for cutting speeds two to four times higher than carbides.

Ceramic tools are very brittle and will break apart or chip if subjected to vibrations or intermittent cutting. Ceramic inserts are supported on a rigid holder and discarded when the edge deteriorates or chips. Inserts come in various shapes, sizes, and cutting angles. A locking device on the tool holder keeps the insert in place. Unlike carbides, ceramics cannot be brazed to the tool holder.

Diamonds

Diamonds, the hardest known material, are highly desirable as cutting tools. Needless to say, diamonds are very expensive and their use in cutting tools make the investment very costly. Diamonds as a natural material have been used for many years in cutting tools. To reduce the cost, General Electric Company developed an artificial diamond possessing the same qualities at a lower cost. Diamonds possess hardness but are very brittle. Therefore, vibrations and intermittent operations should be avoided. Because of the cost the diamonds are usually crushed and the diamond chips brazed to a tool holder. Diamonds are used to achieve a high quality finish in machining. When using diamonds, high cutting speeds and light cuts should be used. Diamonds are used to true standard grinding wheels or are impregnated in special wheels as the abrasives.

CUTTING-TOOL GEOMETRY

Cutting tools are classified basically as *single-point* or *multiple-point* tools. Both types of tools have a cutting edge and components common among themselves. In order to understand basic tool geometry a single-point tool is used for descriptive purposes. Single-point tools are available in various shapes and sizes which are designed for specific purposes. A single-point-lathe cutting tool is used for this discussion. Common characteristics of cutting tools include: cutting edges, clearance angles, rake angles, and cutting-tool nose (Fig. 12-4).

Cutting Edges

Cutting edges of the single-point tool are located on the side and end of the tool's shank. The *side cutting edge* is the primary cutting edge of the tool. The side cutting edge is ground along the side of the tool away from the nose of the tool (Fig. 12-4). The angle (lead angle) at which the side edge is ground ranges from 0 to more than 20°. Primarily, the side cutting edge allows the load or force of the cutting action to be distributed along the shank of the cutting tool. By distributing the load the point of the cutting tool does not take the initial shock of the cutting edge and provides a means to gradually remove material from the workpiece. The *end cutting edge* is located at the end of the tool shank. An angle (end-cutting-edge angle) is ground from the nose of the tool to the side opposite the side cutting edge angle. End cutting edges provide a means for the tool's nose to contact the workpiece. In addition, the cutting edge and angle allows the tool to move into the workpiece for a deeper cut.

Rake Angles

Rake angles primarily used to direct the flow of the chip include the back-rake and side-rake angles. Rake angles are located on the face of the tool sloping back and away from the cutting edges. The inclination of the rake angle affects the compression of the work material ahead of the cutting tool. Since it affects the compression of the material, it also affects the shear zone and formation of the chip. Figure 12-5 shows a single-point tool having positive rake (slopes away from cutting edges). A small rake angle compresses the work material to a greater degree than a large rake angle. Increased compression causes increased

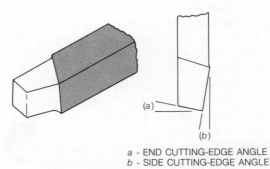

a - END CUTTING-EDGE ANGLE
b - SIDE CUTTING-EDGE ANGLE

Fig. 12-4 Side- and end-cutting edges. (*a*) End-cutting-edge angle; (*b*) side-cutting-edge angle.

Fig. 12-5 Side-rake angle.

friction, therefore, small rake angles cause higher temperatures in the work material and the cutting tool. Chips produced while using a small rake angle will be hot and highly deformed. By increasing the rake angle the friction and compression forces are reduced. With increased rake angles, the tool's strength and ability of the tool to conduct heat away from the work is reduced.

Side-Rake Angle

Side-rake angle is the angle on the face of the tool sloping back from the side cutting edge. The side-rake angle is measured on a plane which is parallel to the direction of the feed or perpendicular to the end of the tool. The side-rake angle determines how the chip leaves the workpiece and how it is directed away from the cutting edge. As the side-rake angle increases, the cutting edge is more susceptible to breaking since its support is reduced. Larger side-rake angles are desirable for cutting ductile materials. Characteristics of ductile materials allows the rake angle to be larger with greater cutting speeds possible. The side-rake angle is reduced for harder materials since the cutting edge requires greater support. Negative side-rake angles have the face of the tool slope toward the cutting edge. Negative side rake strengthens the cutting edge by increased support. However, greater cutting force is required when negative side-rake angles are used.

Back-Rake Angles

Back-rake angle is the slope of the tool face from the nose toward the back of the tool. The back rake provides an additional means of guiding the chip. As with the side rake, the back-rake angle should be kept to a minimum adding support to the cutting edge. In addition to the added support of the cutting edge, a small rake angle enables heat to be conducted away from the cutting area. Five to ten degrees is usually satisfactory for the back rake. Brittle materials require a very small back-rake angle. Irrespective of the material being machined, the rake angle should be kept as small as possible. Negative back rake is where the face slopes toward the nose of the tool. A negative rake forces the chip to bend and eventually break.

Figure 12-6 illustrates cutting tools with positive and negative back-rake angles.

Relief Angles

Relief angles, also called *clearance angles,* are provided below the cutting edges to allow the tool to penetrate the workpiece. The relief angles prevent the cutting tool from rubbing the work during the cutting operation. Relief angles should be kept as small as possible to allow support for the cutting edge. Angles from 6 to 8° are usually sufficient to reduce friction. However, the tool design and work material dictate the angles that should be used. Increased relief angles reduce heat conduction and may cause chatter. *Chatter* is a form of resonant vibration in the tool which causes tool failure and rough surface finishes. Relief angles are increased for soft and ductile materials. Large relief angles requires frequent sharpening of the cutting edge due to minimum edge support.

End Relief

End relief is provided on the end of the cutting tool where the surface lopes down and away from the end cutting edge and tool nose. A relief angle is provided below the side cutting edge and is called the *side-relief angle.*

Radius or Nose

Radius refers to the tip of the cutting tool which is slightly rounded. The nose or radius of the tool is where the end and side cutting edges intersect. Primarily, the nose radius is provided to improve the surface finish and strengthen the cutting tool. A large radius increases the strength of the tool because of larger contact area. With a larger radius the pressure on the tool is reduced because pressure on the tool is directly proportionate to the contact area. As the cutting tool and workpiece contact area increases the pressure on any one point is reduced. A nose radius

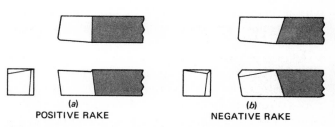

Fig. 12-6 Positive and negative back-rake angles.

of from $1/4$ to $11/32$ in is usually sufficient for cutting tools. Greater cutting speeds are possible with a smaller radius.

LUBRICATION

Lubrication in the cutting area serves a multitude of purposes: primarily the purpose of lubrication is to reduce the friction between the tool and the workpiece. Lubrication cools the workpiece and the cutting tool, therefore, prolonging tool life. Also the time period between tool sharpening is increased because lubrication reduces wear on the tool. As a result of lubrication the power required for machining is reduced. Lubrication aids in improving the surface finish and reducing the possibility of a built-up edge. In addition to the primary reasons for lubrication it also aids in removing chips from the cutting area by flushing away the chips. Secondly, a lubricant allows machining to a close tolerance because distortion from heat is reduced. Finally metal is removed faster because cutting speeds can be increased when a lubricant is employed. Various types of lubrication used in machining include: soluble oils, mineral oil, lard oil, chemical additive oils, and synthetic oils.

Straight-Cutting Oils

Straight-cutting oils are classified into two major categories of vegetable and animal oils. Vegetable oils such as straight mineral oil is commonly used in light machining operations. The advantage of mineral oil is its ability to cool and lubricate the work and tool. In addition the mineral oil provides lubrication of the machine parts. Mineral oil is commonly used on automatic machines because of this ability to lubricate the machine sliding parts. Animal oils are derived from animal fats with lard oil being the most common. Other animal oils include: fish, seal, and whale oil. Lard oil provides ample lubrication for heavy machining operations. Regardless of its lubrication ability lard oil presents a number of disadvantages. Lard oil and other animal-based oils are very expensive. Since they come from animals, bacteria breeds in the oil. Bacteria causes the oil to spoil and become rancid with a distinct objectional odor. Bacterial growth may also cause skin inflammation. A mixture of mineral oil and lard oil provides an acceptable and widely used combination. The blend provides for very good lubricating ability while maintaining lower costs. The proportion of lard oil and mineral oil may vary depending on the material being cut. Higher concentrations of lard oil are desired for heavy cutting or cut-

ting hard materials. Mineral lard oil provides for reduction of friction between the work and the tool yielding a high-quality surface finish.

Soluble Oil

Soluble oils are a combination of oil and water. Since water and oil do not mix into a soluble solution an emulsifier is added. The emulsifier causes the oil to disperse through the water in the form of small droplets. Soap is commonly used in various proportions as the emulsifying agent. Soluble oils provide an inexpensive lubricant for light cutting operations. Common applications of soluble oil includes grinding, milling, drilling and lathe turning. Each constituent of the mixture performs a specific function. Water, probably the most effective cooling agent, provides a medium to dissipate heat generated during cutting. The oil in the mixture lubricates the workpiece and cutting tool. Additional compounds are sometimes added to prevent rust or soften the water. It should be noted that when mixing the oil and water together, soft water is preferred and the oil is always added to the water. Various proportions of oil and water are used for various operations. A mixture of $10:1$ (10 parts water to 1 part oil) may be satisfactory for milling (higher cutting pressures) while a mixture of $100:1$ is satisfactory in grinding operations.

Chemical Additives

Chemicals are added to cutting oils to enhance certain desirable properties in the fluid. Chlorine and sulfur are two such chemical additives. Either chlorine, sulfur or a combination of the two are usually added to a mineral, a mineral-lard, or soluble oil. A chemical reaction occurs because of the pressure and heat of cutting which causes the chlorine or sulfur to form a sulfide or chloride film over the workpiece and the tool. The film reduces friction between the tool and workpiece, thus reducing the temperature in the cutting area. In addition to the reduction of friction and heat, the chemical film reduces the possibility of a built-up edge forming and improve surface finish. Primarily, the major difference between sulfur and chlorine is chlorine's ability to chemically react at a lower temperature.

Synthetic Cutting Fluids

Synthetic cutting fluids consist of some chemical mixed with water. Various chemicals are used depending upon the desired properties of the cutting fluid. Water provides the cooling properties needed in a cutting fluid. The chemical additives aid in reduction of friction, prevention of a built-up edge, and prevention of rust. Synthetic fluids are gaining in popularity as cutting coolants.

Solid Lubricants

Solid lubricants provide a convenient method of applying lubrication to the cutting tool. Solid lubricants are available in many forms and chemical compositions. Waxes and soaps are the most commonly used solid lubricants. The operator merely rubs the lubricant over the cutting tool, a convenient method. However, the lubricant wears off rapidly during the cutting operation and needs to be reapplied. Another solid lubricant that has received wide acceptance on some cutting tools is Teflon. *Teflon,* a trade name for fluorocarbon thermosetting plastics, is coated over the cutting tool then cured. Teflon provides a means of self-lubrication, while reducing friction. Teflon coatings are used for cutting tools on wood or nonmetallic materials. Teflon is not used on tools to cut metal because it would be rapidly worn away.

Kerosene

Kerosene or a mixture of kerosene and mineral oil, lard oil, or soluble oil provides a means of lubrication and cooling the cutting area. Kerosene or kerosene-based lubricants are used for cutting aluminum, aluminum alloys, and brass. Various proportions of kerosene and oil are used depending on the particular application. As the cutting action becomes heavier or severe, the proportion of cutting oils is increased.

Dry or Compressed Air

Compressed air is used in place of a cutting fluid for certain materials. As the air passes the workpiece and cutting tool, it has the tendency to lower the temperature of the work and tool through convection, and to remove the chips. Cast iron does not need lubrication because of the graphite contained in the metal, therefore it is machined dry or with compressed air. Brass and bronze are sometimes machined with compressed air. Compressed air is sometimes used when cutting wood and plastic to remove the dust, cool the tool, and the workpiece.

Application of Cutting Fluids

Cutting fluids are applied to the workpiece and tool in a variety of methods. The fluids can be applied by brush, or from an oil can; however, the economics of these methods do not lend themselves to production or automatic operations. Cutting fluids should be applied directly at the cutting action and at a rapid rate. The cutting fluid should be kept below 100°F [38°C] if at all possible to aid in cooling the cutting area. Common lubrication systems include flooding the work area, pressure jet, and mist lubrication systems.

Flooding. *Flooding* the cutting area with large volumes of coolant aims at reducing the heat being generated in the area. Lubricant is pumped from a reservoir to the cutting area and allowed to flood the work area. As the lubricant passes the cutting area, it absorbs some of the heat from the cutting action. In addition to removing the heat, it tends to lubricate the workpiece and the tool while washing away the chips. The lubricant is filtered and reused to conserve the cost.

Pressure Jet. *Pressure-jet* methods employ a pump and nozzle to direct a high speed jet of lubricant to the cutting action. The velocity of the lubricant leaving the nozzle should be high enough to enable the fluid to reach the cutting point. Because of the tool pressure on the workpiece, a high velocity for the fluid is required to enter this area. Usually, the fluid vaporizes when reaching the cutting area. As it vaporizes, it tends to remove heat generated during cutting.

Mist Lubrication Systems. *Mist lubrication systems* utilize compressed air and cutting fluids. The fluid is vaporized because of extreme pressure from the compressed air. The vaporized fluid and air are directed at the cutting action by nozzles. As the vapor passes the cutting tool and workpiece, heat is absorbed. Finite particles of the vapor also are allowed to penetrate into the cutting area providing needed lubrication.

REVIEW QUESTIONS

12-1. Explain the concept of machinability.

12-2. Identify and be able to explain what and how various factors influence machinability.

12-3. Explain the concept of orthogonal cutting.

12-4. Distinguish between the various types of chips and state their advantages and limitations.

12-5. List various types of cutting-tool materials and compare their characteristics.

12-6. Distinguish between relief angles and rake angles.

12-7. Differentiate between back and side rake.

12-8. What are the major types of lubrication? State their advantages and limitations.

12-9. Explain how lubrication is applied to the workpiece.

CHAPTER 13

PRINCIPLES OF MATERIAL CUTTING

Material cutting or removal is referred to as *machining* or as a *machining process* in manufacturing. "Machining" is a process where the size, shape, and/or finish of a material is changed to yield a consumer or industrial product. Material removal is accomplished through hand and machine operations utilizing a few basic cutting motions. Machining processes utilize basic cutting motions to produce flat, cylindrical, angular, and irregular surfaces.

CUTTING ACTION

Cutting action is the relative motion of a cutting tool in relation to the work material. Inherent in the cutting action is the motion of the tool into and through the work material. Basically there are two types of cutting action employed in machining processes. The types of cutting action are a *slicing motion* and *scraping motion*. Slicing motion utilizes a back and forth tool motion with low pressure being applied to the tool. Slicing motion is associated with sawing or cutting with a knife. Scraping action requires tool or workpiece motion and a relatively high pressure on the cutting tool. The cutting tool is forced into the workpiece under pressure scraping the workpiece. The scraping action and tool pressure remove the desired material in a form of a chip. Orthogonal cutting as described in Chap. 12 utilizes a scraping-cutting motion.

OTHER FACTORS IN MATERIAL REMOVAL

In addition to the cutting action, other factors enter into material removal and machining. Factors having an important role in material removal are cutting speed, feed, and depth of cut.

The *cutting speed* (CS) is the rate of the cutting motion expressed in feet per minute. Cutting speed refers to the relative speed between the workpiece and the cutting tool. Depending on the machine process being used the cutting speed can refer to the reciprocating, straight line continuous, or rotary motion of the tool and/or workpiece.

Depth of cut is the distance that the cutting tool projects into the workpiece. The depth of cut is expressed in inches or millimeters. The depth of cut represents one linear dimension in the machining operation.

Feed is the relative movement between the tool and the workpiece which allows new material to be exposed to the tool's cutting edge. Feed is therefore the amount of movement that new material advances into the cutting area per cycle of machine operation. Feed is expressed in a number of ways depending on the machine operation. On the shaper the feed is expressed as a unit per stroke of the tool while on the lathe it is expressed as a unit per revolution. The common unit of measurement in current use in the United States is the inch or thousandth of an inch. However, with the advent of the metric system of measurement, the unit may be referred to as the millimeter.

MOTIONS APPLIED TO MACHINING

The principles of cutting motion, feed, and depth of cut are applicable for virtually all conventional machining processes (Fig. 13-1). While the principles apply, their specific applications differ for various machining processes.

Regardless of the machining process, the tool and workpiece make contact while one or both are in motion yielding the desired surface. Common surfaces produced through machining operations include: flat, spherical, cylindrical, and angular, or a combination of these surfaces. When manufacturing products, machine tools are selected which yield the de-

Fig. 13-1 Common characteristics and interrelationships of selected machine tools.

Schematics of Selected Machine Tools	Common Name of Machine Tool	Available Motions at:	
		Workpiece	Cutting Tool
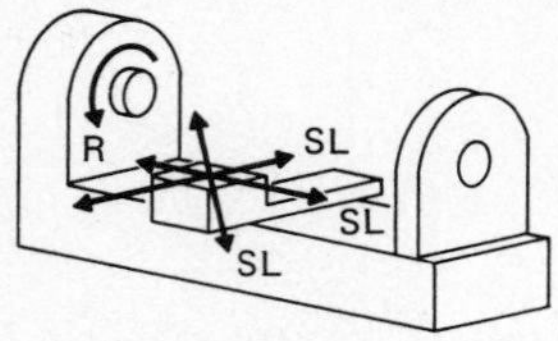	Lathes	Rotation (R)	Three straight line motions (SL)
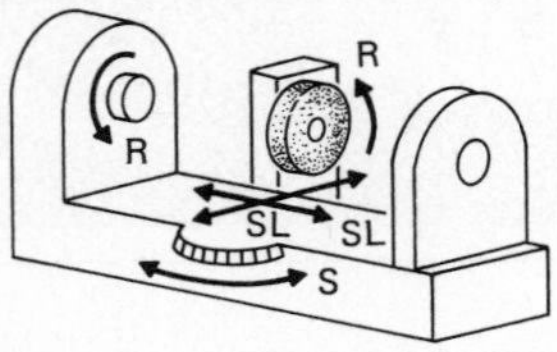	Grinders (cylindrical)	Rotation and one straight line	Rotation and one straight line
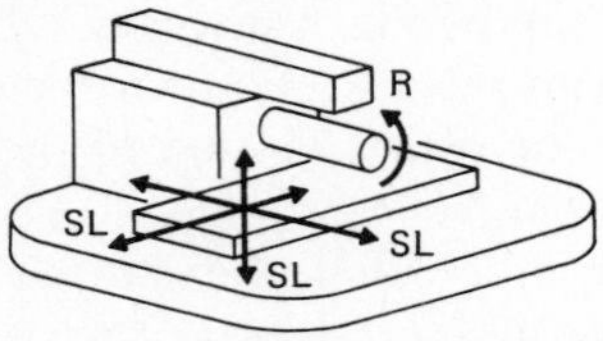	Milling machines (horizontal)	Three straight lines and rotation	Rotation only or rotation and one straight line
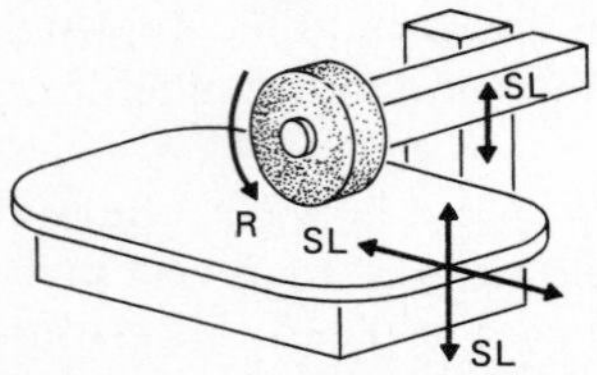	Grinders (surface)	Rotation or one or two straight line or rotation and straight line	Rotation or rotation and straight line
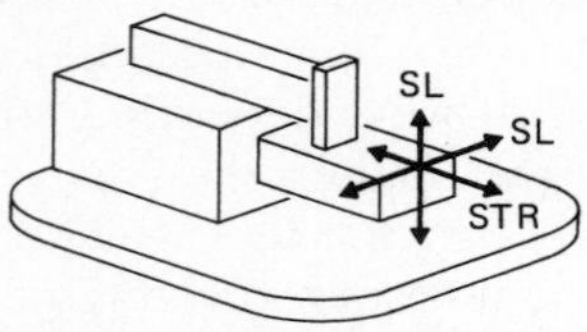	Shapers	Straight line or rotation	Straight line (stroke)
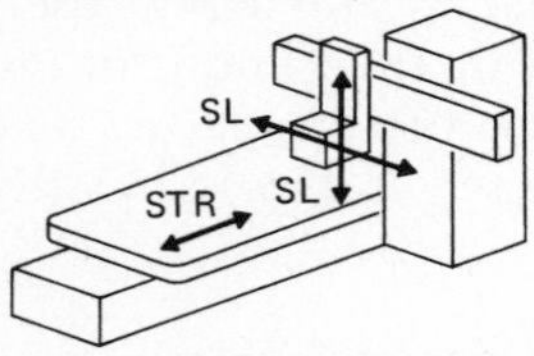	Planers	Straight line	Two straight lines
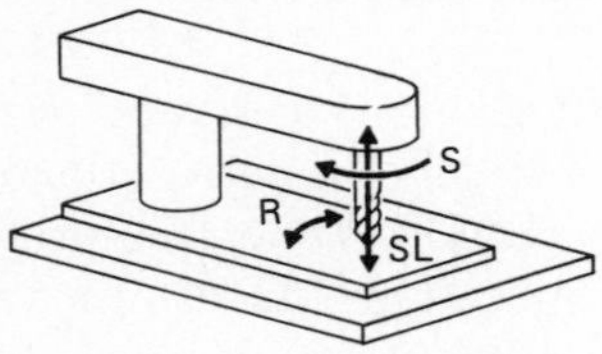	Drilling machines	Stationary or straight line	Rotation or rotation and straight line
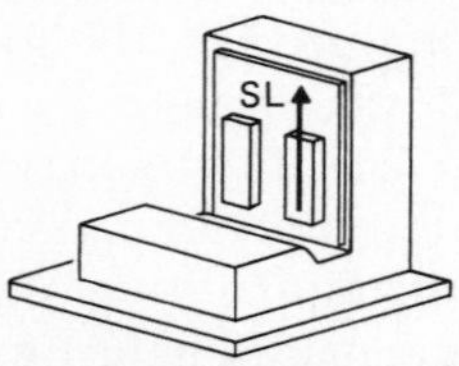	Broaches	Stationary or straight line	Straight line or stationary

Applied At:	Feeding Motion		Cutting Speed Obtained at:	Type of Cutting Action	Type of Cutting Tools Used	Relative Motion Between Work-Piece and Cutting Tool
	Type of Motion	When Applied				
Cutting tool	Straight line	During cutting action	Workpiece	Continuous	Single point	Rotation and one straight line
Cutting tool or work-piece	Rotation and straight line	During cutting action	Cutting tool (abrasive wheel)	Intermittent	Abrasive wheel	Two rotary and one straight line
Workpiece	Straight line or rotation	During cutting action	Cutting tool	Intermittent	Multiple point	Rotary and one straight line or two rotary
Workpiece and cutting tool	Straight line and rotation	During cutting action	Cutting tool (abrasive wheel)	Intermittent	Abrasive wheels	One or two rotary and one straight line
Workpiece	Straight line or rotation	Intermittently while not cutting	Cutting tool	Intermittent	Single point	Straight line
Cutting tool	Straight line	Intermittently while not cutting	Workpiece	Intermittent	Single point	Straight line
Cutting tool or work-piece	Straight line	During cutting action	Cutting tool	Continuous	Drills	Rotary and straight line
Cutting tool	Straight line	Intermittently while cutting	Workpiece or cutting tool	Intermittent	Multiple point (broach)	Straight line

sired surface within specified tolerances while being economically efficient. Machine tools are designed to perform specific operations, although auxiliary operations may be performed on many machines. Machine tools utilize a number of different cutting and feed motions to produce the desired surfaces necessary in manufacturing. A straight-line cutting motion and straight-line feed is usually employed in shaping and planing operation. A straight-line feed and rotating workpiece is employed on the lathe. Drilling operations employ a rotating tool, and straight line feed on a stationary workpiece. Each machine tool employs various motions to obtain the desired results. It is evident that a number of possibilities exist in the cutting motion and feeding motion on machine tools. The remainder of this chapter will discuss the relationship between tool and workpiece motions. In addition the typical machine applications and calculation of the cutting speed are presented (Fig. 13-2).

MACHINE TOOLS

Machine tools can be classified by (1) relative cutting motion, (2) relative feed motion, (3) type of surface produced, and (4) type of cutting tool used (Fig. 13-1). The type of cutting tool employed will form the basis for our discussion. Common types of cutting tools are single point, multiple point, drills, and abrasives.

Single-Point Tool Applications

Single-point cutting tools are used in shaper, planer, and lathe operations. The geometry of the single-point tool was discussed at length in Chap. 12. Probably the most common machine utilizing the single-point cutting tool is the lathe.

Lathes. The lathe consists of a drive mechanism, work-supporting device, and provisions to hold and feed the cutting tool. A drive mechanism located in the headstock of the lathe provides a means of rotating the workpiece. The drive mechanism also provides power through a gearing unit to feed the cutting tool. Lathes are designed to provide traverse, longitudinal, or angular straight-line feed movements. *Longitudinal* feeding causes the cutting tool to move parallel to the axis of the rotating workpiece. Since the workpiece is supported and rotated in a horizontal axis, the longitudinal feed is either from left to right or right to left. *Traverse feed* also called *cross feed* moves the cutting tool perpendicular to the workpiece. The traverse feed moves toward and/or away from the workpiece. An *angular feed* moves the cutting tool at virtually any angle between the longitudinal and traverse axis. Most lathes have a power-driven longitudinal and traverse feed with an operator (hand) fed angular feed. Figure 13-1 illustrates the principles of workpiece rotation and feed motions. The lathe is a versatile machine capable of producing cylindrical shaped products. The lathe is also used to produce internal cylindrical shapes through a *boring* operation. Internal and external conical (tapered) shapes can also be produced on the lathe. *Facing,* an operation performed on the lathe, is used to true the ends of the workpiece.

The preceding operations on the lathe are performed with single-point cutting tools. In addition to these operations the lathe can also be used for drilling, reaming, counter boring, counter sinking, threading (internal and external), and special formed shapes by using specially formed single-point tools or multiple-point cutting tools.

Cutting Tools. The lathe utilizes a number of single-point tool variations designed for special applications. Single-point tools are either left- or right-handed. Left-handed tools have the cutting edge on the right side of the tool and fed to the right. Right-handed tools have the cutting edge on the left side of the tool and are fed to the left. Lathe tools are further classified as roughing, turning, facing, boring, and threading. Roughing tools are designed for taking a heavy feed and depth of cut. Finishing tools are used for lighter cuts to produce a better quality surface finish. Round nose cutting tools are used for either left or right hand feed operations providing a relatively good surface finish. A *square nose tool,* commonly called a *cut-off tool,* is employed for cut-off operations.

Cutting Speeds and Feeds. In order to obtain the desired shape, size, and surface finish while increasing tool life and efficiency the proper cutting speed and feed are important considerations. Cutting speeds for the lathe are based on the materials hardness, cutting-tool material, feed, and depth of cut. Cutting speed is the rate, in surface feet per minute (sfpm), at which the cutting tool removes material from the workpiece. Formula 13-1 is used to calculate the cutting speed.

$$CS = \frac{\text{RPM} \times D \times \pi}{12} \qquad \text{(Formula 13-1)}$$

where

CS = cutting speed (sfpm)
RPM = revolutions per minute
D = diameter of workpiece (in inches)
π = 3.1416 (constant)

Fig. 13-2 Common characteristics and interrelationships of selected lathe operations.

Schematics of Selected Lathe Operations	Common Name of Operation	Class of Operation	Workpiece Motions	Cutting Tool		Machined Surfaces	
				Motions	Type	Shape	Type
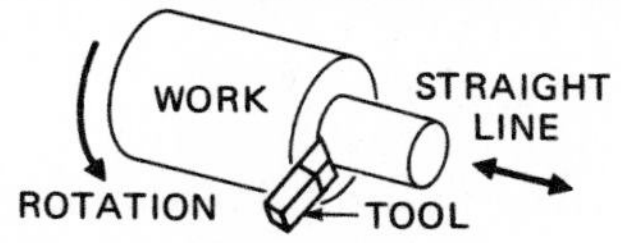	Straight turning	Traverse turning	Rotation	Straight line feeding parallel to centerline of workpiece rotation	Single point	Cylindrical (external)	Generated
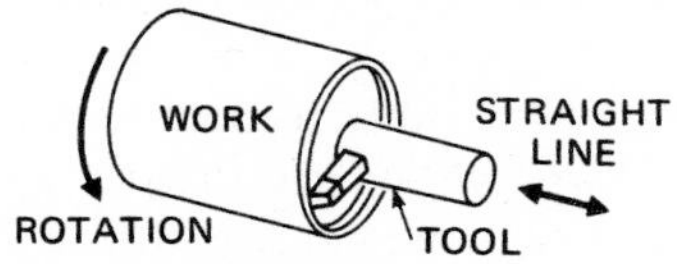	Boring	Traverse turning	Rotation	Straight line feeding parallel to centerline of workpiece rotation	Single point	Cylindrical (internal)	Generated
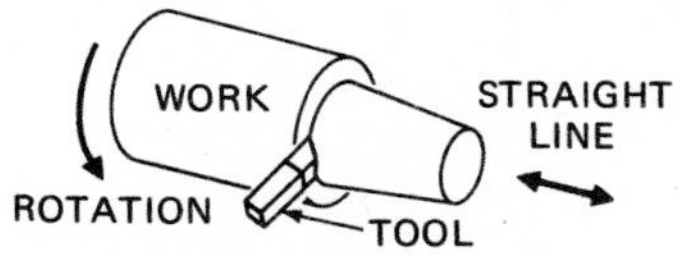	Taper turning	Traverse turning	Rotation	Straight line feeding at an angle to centerline of workpiece rotation	Single point	Conical (external)	Generated
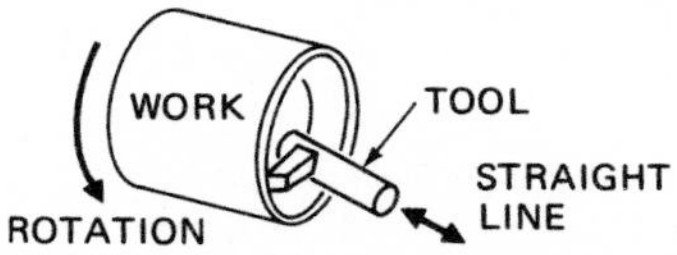	Taper boring	Traverse turning	Rotation	Straight line feeding at an angle to centerline of workpiece rotation	Single point	Conical (internal)	Generated
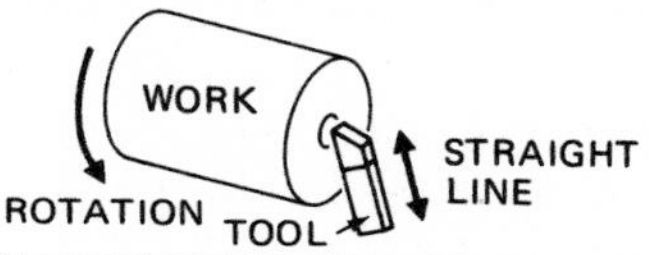	Facing	Traverse turning	Rotation	Straight line feeding perpendicular to centerline of workpiece rotation	Single point	Flat	Generated
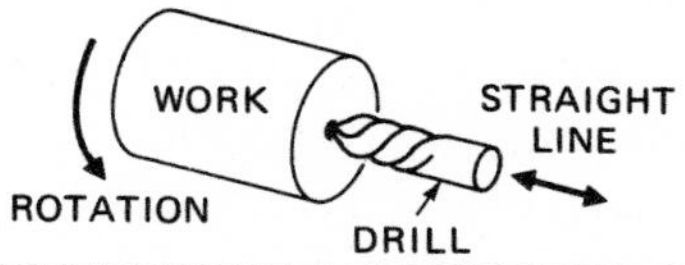	Drilling	Plunge turning	Rotation	Straight line feeding parallel to and concentric with the centerline of workpiece rotation	Drill	Straight hole	Generated
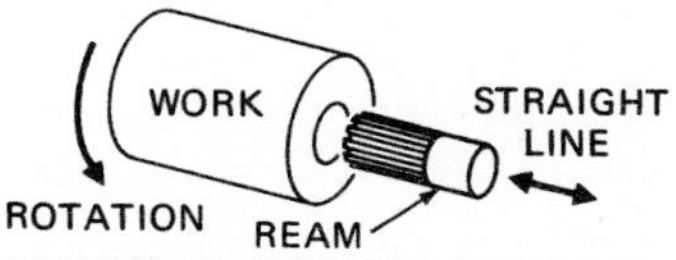	Reaming	Traverse turning	Rotation	Straight line feeding parallel to and concentric with the centerline of workpiece rotation	Reamer	Straight hole	Generated
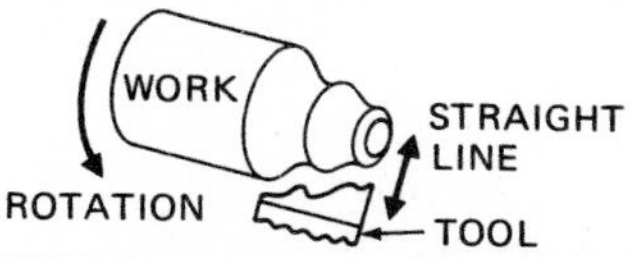	Forming (form turning)	Plunge turning	Rotation	Straight line feeding perpendicular to centerline of workpiece rotation	Single point	Selective	Formed
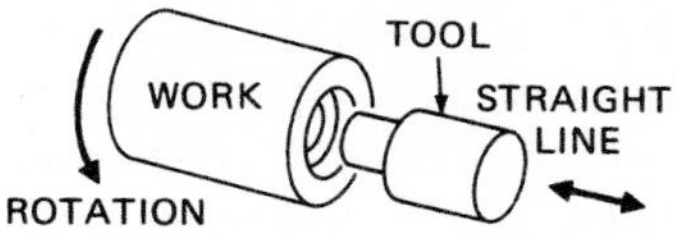	Counter-boring	Traverse and plunge turning	Rotation	Straight line feeding parallel to and concentric with the centerline of workpiece rotation	Counter-bore	Combination cylindrical hole and flat	Combination
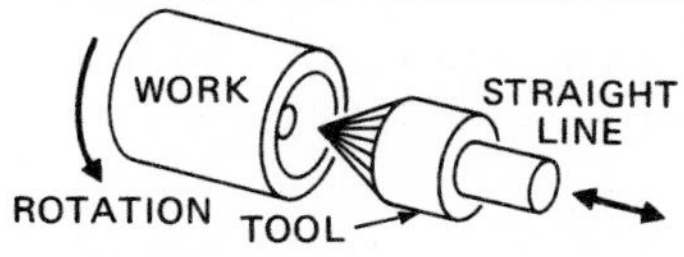	Counter-sinking	Plunge turning	Rotation	Straight line feeding parallel to and concentric with the centerline of workpiece rotation	Counter-sink	Conical (internal)	Formed
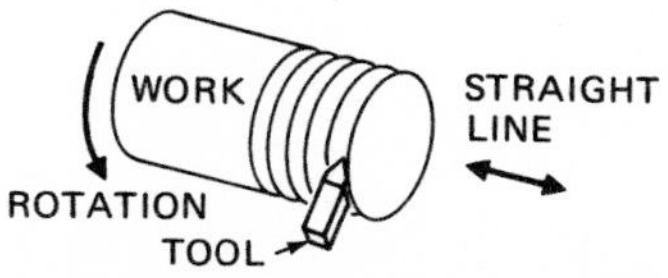	Thread turning	Traverse turning	Rotation	Straight line feeding parallel to centerline of workpiece rotation	Single point threading	Thread	Combination

To determine the RPM, when the desired cutting speed is known, use Formula 13-2.

$$\text{RPM} = \frac{CS \times 12}{D \times \pi} \qquad \text{(Formula 13-2)}$$

The "feed" refers to the distance that the cutting tool is fed through the workpiece. Feed on the lathe is expressed in inches/revolution (IPR). If the feed rate is 0.125 in, then the workpiece has to rotate eight times to machine a distance of 1 in. The time required to machine a desired distance is calculated using Formula 13-2.

$$T = \frac{L}{\text{RPM} \times F} \qquad \text{(Formula 13-2)}$$

where

T = time in minute
L = length of the cut in inches
RPM = revolution per minute (workpiece)
F = feed rate in thousandths

Depth of cut and the feed determines the size of the resultant chip. When determining the depth of cut the actual material removed is double the feed setting. For example, if the depth of cut is set for $1/16$ in, the actual material removed by the lathe is $1/8$ in because material is removed from both sides of the workpiece. Depth of cut will vary depending on the material's hardness, feed and type of cutting tool. Finally the size of the lathe and the available power will influence the maximum depth of cut possible.

Shapers. *Shapers* are used primarily to produce a flat surface utilizing a single-point cutting tool. The workpiece is mounted on a table which provides for the feed motion of the workpiece. A straight-line feed motion is used on the shaper. Horizontal and vertical feed motions are available on the shaper. Horizontal straight-line motion occurs perpendicular to the motion of the cutting tool. Vertical feed motion (usually hand operated) moves the workpiece up or down. The cutting tool moves in a straight-line motion across the workpiece. Cutting motion of the tool is provided by a mechanism which changes rotary (motor) motion to reciprocating (tool) motion. Shapers have an intermittent cutting action since cutting occurs only during the forward motion of the stroke. During the return no cutting occurs but the workpiece is moved in small increments perpendicular to the cutting stroke (Fig. 13-1). While the shaper is constructed primarily for cutting flat surfaces, irregular surfaces are possible. When producing curved or irregular surfaces, horizontal- and vertical-feed motion must be correlated. Shapers can also be

Fig. 13-3 Typical types of work produced on the shaper.

used to machine grooves, dovetails, angular surfaces and T-slots in the workpiece. Special irregular-shaped surfaces can be produced on a shaper by using special formed cutting tools. When formed cutters are used the machine capabilities are greatly reduced because of the feed motion employed. Internal shapes are possible on the shaper when the cutting tool is mounted on an external bar. Figure 13-3 illustrates some of the applications of the shaper. Cutting tools used in shaping are made from high-speed steel, cast alloys, or cemented carbides. Whatever material is selected it must be highly impact resistant. By nature of the shaper operation the cutting tool is exposed to a great deal of impact force at the start of the cutting stroke. *Cutting speeds are feeds.* Calculating the cutting speed for shapers is different than those made for turning, drilling, and milling. A number of factors must be considered when calculating shaper cutting speeds. Factors affecting the cutting speed are:

1. Work material: harder materials require slower cutting speeds than soft materials.
2. Cutting tools: the shape of the cutting tool can restrict the cutting speed. Special formed cutting tools require slower speeds.
3. Cutting-tool materials: high-speed cutting tools are capable of higher speeds.
4. Rate of material removal: greater feeds and an increased depth of cut reduces the cutting speed.

Speed of the shaper refers to the speed of the ram (drive mechanism) indicated in stroke per minute. The cutting action does not occur during the return stroke therefore the tool does not encounter resistance. During the forward or cutting stroke, resistance is encountered and therefore the cutting speed is reduced. A ratio of about 1.5 : 1 exists for the cutting time to the return time. Cutting speed for shaping is calculated by the following Formula (3-3):

$$CS = \frac{N \times L}{3/5 \times 12} \quad \text{or} \quad \frac{N \times L}{7.2} \qquad \text{(Formula 13-3)}$$

where

CS = cutting speed (ft/min)
N = number of strokes per minute
L = length of stroke in inches

Table 13-1 **Cutting Speeds for Shaping in Feet per Minute (FPM)**

Material	Carbon-Steel Tool	High-Speed-Steel Tool
Cast iron	20–30	65
Aluminum	50–60	100+
Brass	50–100	150+
Medium carbon steel	30–40	75–100

$3/5$ = actual proportion of stroke when cutting occurs

12 = conversion to inches

When the cutting speed is known for a material, the number of strokes per minute at which the shaper needs to run is calculated by using Formula 13-4.

$$N = \frac{CS \times 7.2}{L} \qquad \text{(Formula 13-4)}$$

Planers. *Planers* perform the same type of operations as the shapers. The planer is a heavy-duty machine used primarily for large-scale work. A major difference between the shaper and the planer is in the location and movement of the cutting tool and workpiece. Cutting tools are supported by a tool holder located on an over arm or vertical column. Feed motion is applied to the cutting tool rather than the workpiece as on the shaper. Horizontal and vertical straight-line feed motions are applied to the cutting tool. The workpiece is mounted on a large movable table. Cutting motion is supplied by the reciprocating table. As with the shaper the planer utilizes a straight line cutting motion perpendicular to the cutting tool. Both cutting action and feed are intermittent since feeding takes place on the noncutting return stroke (Fig. 13-1). Planers can be either mechanical or hydraulic drive and are available in various sizes and for specific purposes. The maximum size of work is determined by the capacity of the planer relative to width, height, and length of the machine. Single-point cutting tools similar to the shaper are used on the planer. Since the planer is designed for large scale work the cutting tool needs to be larger. Special shapes are available to produce grooves and other special cuts. The cutting tool must have high-impact resistance since the cutting action is intermittent.

Planers as with shapers employ a reciprocating cutting motion therefore the cutting strokes and return stroke occur at different rates. It is considerably more difficult to calculate the speed and feeds of the planer. Tables are available in machining handbooks with recommended speeds and feeds for various depths of cuts. Generally the heaviest possible feed

and fastest speed is advisable to reduce machining time.

Multiple-Point Tool Applications

Multiple-point cutting tools contain a number of single-point cutting edges. Common multiple-point cutting tools are saws, broaches, and milling cutters.

Saws. Saws are produced in a number of varieties with these basic cutting motions. Cutting actions employed in sawing are (1) reciprocating, (2) continuous straight line, and (3) continuous rotary (Fig. 13-4). *Reciprocating* cutting motion utilizes a slicing motion to cut the workpiece. The feed motion is applied to the blade perpendicular to the workpiece. Hack, coping, back, jig, saber, and hand held saws utilize a reciprocating cutting motion. *Continuous* straight line motion is accomplished by using a continuous flexible oval shaped blade. At the cutting area the saw is straight and under tension moving in one direction. Therefore, the cutting motion is continuous and in a straight line. Feed motion is applied to the workpiece in either a straight line or irregular motion. The band saw and chain saw are examples of the continuous straight line cutting motion. The chain saw used primarily for cutting trees has the feed motion applied at the cutting tool. A continuous-rotary-motion cutting-action saw uses a nonflexible circular-shaped blade. A series of single-point cutting edges are located along the blades circumference. Feed motion is applied in a straight-line motion at the workpiece. Primarily saws are used to cut material to length or width in order to prepare the workpiece for further processing. A secondary purpose of the saw is cut contours or profiles in the workpiece. Saws are constructed of various grades of steel depending upon their use. Carbon-steel blades prove to be satisfactory for soft materials while high-speed alloy blades are more suited for harder materials. Replaceable carbide tips are also available for circular saws to provide a very hard cutting surface.

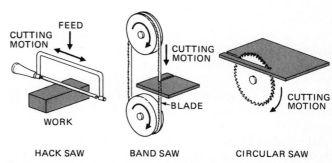

Fig. 13-4 Cutting actions employed in sawing.

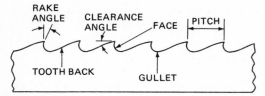

Fig. 13-5 Saw-blade nomenclature.

Saw-Tooth Nomenclature. Saw blades consist of specific parts similar to the single-point cutting tool (Fig. 13-5). The cutting edge of the blade (front edge) contains the blade's teeth. The *face* of the tooth is the forward edge which forms the chip. A curved area between the teeth at the base of the face is called the *gullet.* Primary purpose of the gullet is to remove the generated chip from the work material. *Side-clearance angle* is determined by the distance from the outer edge of the tooth to the blade-support backing. This clearance angle depends on the tooth offset. *Rake angle* refers to the angle made by the tooth measured perpendicular from the cutting edge and the back edge of the tooth. *Pitch* is the number of teeth per inch measured from the top of one tooth to the respective point on another tooth. *Gauge* refers to the thickness of the blades backing.

Tooth set, a common element among saws, refers to the offset of the cutting teeth on each side of the blade. Offset provides the necessary clearance between the saw's body and the workpiece. As the saw passes through the work material the tooth offset produces an enlarged path called the *kerf.* The kerf is the width of the removed material. Straight, wave, and raker are three types of tooth sets. *Straight-set* teeth consist of alternating left and right teeth. *Raker-set* teeth have an offset left, offset right, and straight configuration. *Wavy-set* saws have several teeth offset left, then several teeth offset to the right, and so forth.

Tooth form refers to the type of cutting edge on the saw blade. Common tooth forms are the buttress, knife, friction, scallop, claw, and spiral.

The *buttress*-tooth form consists of widely spaced teeth with large gullets (openings between the teeth). Large gullets enable the blade to remove the generated chips with a minimum amount of friction. Buttress-tooth-form blades are particularly effective when cutting wood, plastic, and other soft material.

Knife-edge-tooth forms have a straight edge with either a single- or double-bevel edge. Knife-edge-blade tooth forms are primarily used on band saw blades. Knife-edge blades utilize a slicing action leaving a small kerf and a minimum amount of chips. Knife-edge blades are useful in cutting fibrous material such as paper, textiles, cork, rubber, and bakery goods.

Friction-type blades are used to cut ferrous metals of irregular shape. Friction between the blade and metal workpiece cause the workpiece to heat up and melt the material. When friction blades are used, speeds of up to 15,000 ft/per minute are required. Friction blades are similar to abrasive disks having no teeth.

Scallop-edge blades or *wavy*-edge blades are similar to the knife edge except the edge is wavy and not straight. These blades are available with either a single or double bevel on the cutting edge.

Claw-tooth forms consist of a blade with positive rake angles on the cutting teeth. Positive rake provides a bitting action into the material therefore requiring little pressure on the blade. Claw-tooth blades are used on soft material such as nonferrous metals, woods, and plastics.

Spiral blades permit sawing in any direction. The blade consists of a raised double-helix cutting edge around the body of the blade. Primarily the preceding blades and tooth forms are used on band-saw blades. Hack-saw blades are produced in hand- and power-machine varieties. Hand hack-saw blades come in lengths from 10 to 12 in, with 14 to 32 teeth per in. Hand hack-saw blades are available with raker- and wavy-set teeth. Power hack-saw blades range from 12- to 30-in lengths with 3 to 14 teeth per in. Various materials require blades with a specific number of teeth per inch to be effective. The following list provides a basis for tooth selection.

> 3,4 teeth/in —cutting heavy solid bars, for example
> 6 teeth/in —machine steel, soft materials
> 10 teeth/in —high-speed steel, pipes, cast iron and so on
> 14 teeth/in —tubing, light structual shapes, for example

Hole saws (Fig. 13-6) are designed for cutting holes $8/16$ through 6 in diameters in material up to $1\frac{1}{8}$-in thick. The saw is used for a multitude of materials including steel, nonferrous metals, wood, and plastics. Hole saws are used on a drill press. Circular saw blades are available in various sizes and tooth configurations (Fig. 13-7). Circular saws are classified as: (1) solid, (2) segmented, (3) insert, (4) carbide tipped, and (5) abrasive. Solid circular saws are made of one-piece high-speed steel. Teeth are cut and ground on the periphery of the saw blank. Segmented circular saws have segments containing teeth which are attached to the body of the saw. An advantage of the segmented saw is that high-speed-steel segments can material being cut, and length of the broach. The first set of cutting teeth are called the *roughing* or *cutting teeth*. Roughing teeth are used to remove the most material. Following the roughing teeth are the *semifinishing teeth*. Within the semifinishing set of teeth the variation in tooth height is reduced. A minimum amount of work material is removed by the semifinishing teeth. The third set of teeth are the *finishing teeth*. Finishing teeth are all of the same height and used primarily to impart a high-quality surface finish. Depth of cut on a broach is regulated by the variations in cutting-teeth height. Each tooth removes a specific amount of material ranging from 0.002 to 0.0015 in per tooth. To calculate the length of the broach the total amount of material to be removed is divided by the depth of cut per tooth yielding the number of teeth required for the broach (Formula 13-5). Multiply this result by the pitch of the broach to determine its length (Formula 13-6).

$$T = \frac{Ct}{Cp} \quad \text{or} \quad L = \frac{Ct}{Cp} \times P \quad \text{(Formula 13-5)}$$

$$L = T \times P \quad \text{(Formula 13-6)}$$

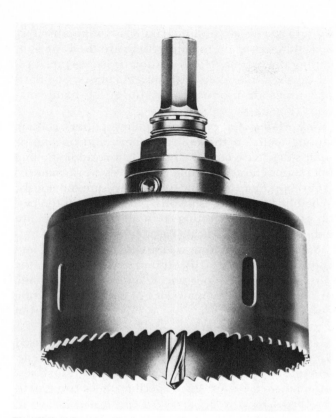

Fig. 13-6 Hole saw. (*Source: Simmonds Saw and Steel Co.*)

be attached to a tougher saw body. Carbide-tipped saws have carbide teeth brazed to the saw blank. Carbide tips provide excellent wearability when cutting soft materials. Carbide-tipped saws should not be used to cut harder materials because of their low shock resistance. Insert saw blades have provision for replaceable inserted teeth. Abrasive saws primarily used for cut-off work contain abrasive grains rather than saw teeth.

Broaches. Broaches are multiple-point cutting tools used to machine internal and external surfaces in metal and/or plastic. Broaching produces an accurate shape of high-surface quality. Either the workpiece or broach moves in a straight-line motion while the other remains stationary. Broaching is an intermittent-type cutting action with only one direction of travel used. Feed motion is applied at the broach through its unique tooth design. A broach consists of a series of cutting teeth along its length running perpendicular to the center line. Each successive series of teeth increase in height. As the broach passes the workpiece, each series of teeth removes additional amounts of work material. Series of teeth vary in number depending on the application, size, shape,

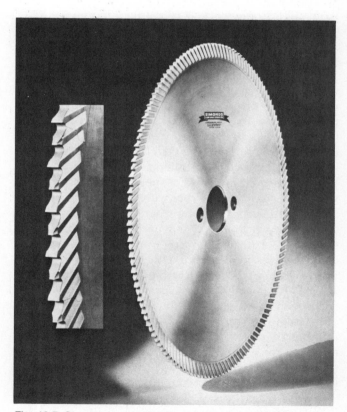

Fig. 13-7 Circular saw blade. (*Source: Simmonds Saw and Steel Co.*)

where

L = length of broach (in inches)
Ct = total material to be removed (in inches)
Cp = material removed per tooth (in inches)
P = pitch
T = number of teeth

Milling Cutters. Milling cutters provide a means of generating a flat or specially formed surface by progressively removing material from the workpiece. Intermittent cutting action is applied at the rotating milling cutter. Feed motion is usually applied at the workpiece in three straight-line motions. On special applications the workpiece may be rotated during the feed cycle. Milling cutters are used on either a horizontal or vertical milling machine. Horizontal machines have three straight-line motions and potential for rotary-feed motion. The milling cutter is mounted on an arbor which is in a horizontal position. Cutting teeth on the cutter are parallel to the axis of the arbor. Vertical milling machines also have three straight-line and rotary-feed motions available. The milling cutter is mounted perpendicular to the workpiece with cutting teeth both parallel and perpendicular to the work. Vertical-feed motion is available to the cutter on the vertical milling machine (Fig. 13-1). Milling cutters come in a variety of types and sizes for both the horizontal and vertical machines. Milling cutters are classified in two major groups: peripheral and face. *Peripheral cutters* have their cutting teeth located on the periphery of the cutter running parallel to the tool's axis. *Face milling cutters* have cutting teeth on both the periphery and face (end) of the tool. Conventional (up) and climb (down) are two methods employed in milling operations. During horizontal milling using a peripheral cutter, the workpiece may be fed either with or against the cutter rotation. In vertical milling when using a face-milling cutter, the workpiece is fed partly with and against the cutter's rotation. During *conventional milling* the cutter rotates in opposition to the direction of the feed. In other words the cutting teeth are moving upwards into the workpiece. *Climb milling* has the cutter rotating in the direction of the feed. The cutting teeth are moving down into the workpiece. There are a number of different milling cutters available, each designed for a specific purpose. Major types of milling cutters include plain, side, slitting saws, angular, form, and end and special cutters.

Plain Milling Cutters. Plain milling cutters are cylindrically shaped cutters having teeth on their periphery. Plain milling cutters are used to produce flat surfaces on the workpiece. Cutting teeth are either straight or helical, running parallel to the cutter's axis. Straight-teeth plain milling cutters are used for light cutting operations. Heavy cutting or excess-metal removal should be done with helical cutters. Standard plain mills are produced in right- or left-hand rotations (Fig. 13-8).

Side Milling Cutters. Side milling cutters contain cutting teeth on the mill periphery and on one or both sides. Side cutters are used in straddle-milling operations. Straddle milling removes work material from opposite sides of the cutter simultaneously. When heavy cutting is involved, the teeth on the cutter are staggered. Cutting teeth on the periphery are either straight or helical (Fig. 13-9).

Slitting Saws. Slitting saws are used for slotting and cut-off work (Fig. 13-10). Slitting saws either have the cutting teeth on the periphery, similar to a plain mill, or on the periphery and sides of the cutter. Slitting saws range in size from $\frac{1}{32}$- to $1\frac{1}{4}$-in thick and from 2 to 8 in in diameter.

Angular Milling Cutters. Angular cutters are used for milling angles, serrations, and grooves. The cutters consist of teeth on two or three adjacent surfaces. Angular surfaces are set at specific angles from 40 to 90° (Fig. 13-11).

Form Milling Cutters. Form milling cutters are designed to mill specially shaped surfaces. Formed cutters such as convex, concave, and irregular shapes are produced in various sizes (Fig. 13-12). In addition to these shapes and combinations, form cutters are designed to cut T-slots, threads, splines, woodruff keys, dove tails, and gears.

End Milling Cutters. End milling cutters have their teeth located on the periphery and on their ends. End mills are primarily used in vertical milling operations. Two varieties of end mills are the solid and shell types. Solid end mills have their teeth and shank as an integral unit. Shell end mills utilize a solid body with the teeth mounted on a shell. The body and shell are held together with an arbor. End mills have straight or tapered shanks, which are held in a chucking device. A

Fig. 13-8 Plain milling cutter. (*Source: Standard Tool Co.*)

Fig. 13-9 Staggered-tooth side-milling cutter. (*Source: Standard Tool Co.*)

Fig. 13-10 Metal slitting saw. (*Source: Standard Tool Co.*)

Fig. 13-11 Angular milling cutter. (*Source: Standard Tool Co.*)

number of end mills are produced including: two-, three-, four-fluted single or double end, ball end, carbide tipped, and special forms.

The milling cutter rotates at a constant speed (RPM) based upon the material's cutting speed and the type of cutting-tool material. An indication of the

(*a*)

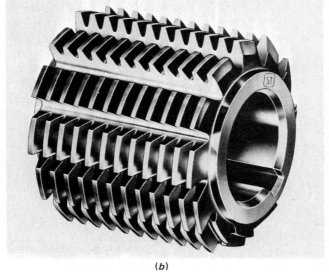

(*b*)

(*c*)

Fig. 13-12 Form cutters. (*a*) Convex cutter, (*b*) hob, and (*c*) gear cutter. (*Source: Standard Tool Co.*)

cutting speed can be obtained by using the Brinell hardness number. It should be noted that this number is only an indication since true cutting speed is dependent on the cutting-tool material. In order to determine the speed of the cutter's rotation, use Formula 13-7:

$$RPM = \frac{CS \times 12}{D} \qquad \text{(Formula 13-7)}$$

where

$CS =$
$CS =$ (cutting speed (surface feet per minute)
$D =$ cutter diameter (in inches)

Feed is motion supplied by straight-line action of the machine's table. Feed rate (F) of the operation is calculated by multiplying the feed rate per tooth (f), speed of the cutter (RPM), and number of cutting teeth (N). Formula 13-8 is used to calculate the feed rate (F).

$$F = f, N, RPM \qquad \text{(Formula 13-8)}$$

Feed rate per tooth is predetermined based upon the material being cut. The material of the cutter, and the type of cutter. Actual feed rates per tooth are calculated or found in machinist tables. Generally the feed rate per tooth ranges from 0.008 to 0.015 in for plain milling cutters. Stock-removal rate, at which workpiece material is removed, is calculated by Formula 13-9.

$$Rr = F\, d\, w \qquad \text{(Formula 13-9)}$$

where

$Rr =$ stock removal rate, cubic inches per minute
$F =$ feed rate, inches per minute
$d =$ depth of cut, inches
$w =$ width of cut, inches

Depth of cut for milling operations is dependent on the type of operation. A depth of cut of over an $1/8$ in is possible for roughing operations. Roughing operations are designed primarily to remove material at a fast rate. Finishing operations are used to impart good surface quality and dimensional accuracy. Depth of cut should be kept to a minimum (cut should not be deeper than $1/16$ in) with a slow feed; a higher cutter speed should be used (RPM) for finishing operations.

Drills

Drills are a type of multiple-point cutting tool used to produce internal cylindrical shapes and surfaces in the workpiece. Reamers, contersinks, counter bars, spot faces, and so on, are closely related to drills in that they use the same type of feed and cutting action as the drill. The workpiece is either held stationary or rotated during the cutting action. Cutting action is applied at the rotating cutting tool. Feeding action is applied to the cutting tool in a straight-line motion (Fig. 13-1). A *twist* drill is the most common variety of drill available. Drills have three principal parts: the body, shank, and point (Fig. 13-13). The body of the drill is the section from the point to the shank. Two or more helical grooves called flutes are located within the body of the drill. Flutes aid in curling and removal of the chip from the workpiece. In addition the flutes provide a means for lubrication to penetrate to the cutting area. Intersection of the flutes at the end of the drill (point) forms the cutting edge. Material between the flutes is called the *land*. At the edge of the land is a raised section called the *margin*. A clearance is formed by the raised margin to prevent the drill from binding in the hold produced. The *web* is the area in the center of the body between the flutes which provides rigidity to the drill. The point is formed by the intersection of the flutes and is ground to various angles, depending on the drill's proposed use. A sharp edge at the end of the drill point is the *chisel edge angle*. Cutting edges of the drill are located on the lead edge of the flutes. Cutting lips must be ground at the same angle to the drill a true size and shape hole. *Shank* of the drill is the position held in the chuck or jaws of the drive mechanism. Three standard shaped shanks are the straight, square, and taper. A square shank is primarily used in a bit hand-held brace. Straight shanks are commonly used in power-drill presses. A taper shank drill is held by means of a sleeve on the drill press.

Drill sizes are specified by either *number, letters,* or *fractional* sizes. Number drills are equivalent to wire-gauge sizes where the smallest drill is a number 80 (0.0135 in) to the largest number (0.2280 in). Following the number set is the letter drills. Letters from A to Z are used and range from 0.2340 to 0.4130 in. Fractional-sized drills range from $1/64$ to $3\,1/2$ in in diameter. In addition to the conventional designations, the drills are available in metric sizes. Drills are produced with special angle points to aid in cutting a variety of materials. Lip clearance provides the relief needed for the drill to penetrate the workpiece without interference.

When producing the clearance, an angle is ground from the lip to the heel of the point. Drills for general purpose use have clearance angles from 8 to 20°, depending on the drill size. Smaller drills need a larger clearance angle. Angles range from 7 to 20° for harder materials and from 10 to 25° for soft materials. The point angle which is measured from the cutting edge through the axis of the drill vary depending on the particular drill application. General purpose drilling should have an angle of about 118°. The point angle should be increased from harder materials and decreased for softer materials. Special types of drills are produced for various applications. Special drills include carbide tipped, three fluted, gun, straight fluted, subland, left hand, center drill, and many others. Carbide-tipped drills, as other carbide tools, have the tips brazed to the drill's body. Three-fluted drills sometimes called *core drills* are used to enlarge holes. Three flutes are provided to provide rapid chip removal. Drills are also available with a hole through the center of the body to supply lubricant to the cutting area. Other drills such as the center drill have a pilot-hole provision prior to the larger body of the drill. Step drills contain two or more diameters with a single set of flutes to drill multiple diameter holes. Subland drills are capable of drilling a multidiameter hole except that each diameter contains separate flutes. Figure 13-14 *a–i* shows some of the many types of drills available for special applications. Related drilling operations include counter boring, counter sinking, and spot facing. *Counter boring* provides an enlarged cylindrical hole having a flat bottom bearing surface. This enlarged section is located at the top of a predrilled hole providing a sunken surface for a nut or bolt. *Countersinking* produces a beveled section at the end of a drilling hole

providing a seat for a flat head screw or rivet. *Spot facing* similar to counter boring provides a smooth bearing surface of minimum depth around a predrilled hole.

Reamers are used to improve the internal surface and tolerance of a previously drilled hole. Exact size holes are possible when reamers are employed. When a reamer is to be used, the drilled hole should be slightly undersized. Tables in machinist handbooks provide recommended size differentials for drilling and reaming. In general holes smaller than one inch should be drilled undersized from 0.010 to 0.015 in and holes larger than one inch should be drilled 0.015 to 0.045 in undersized. Principal parts of the reamer are shown in Fig. 13-15. Reamers are constructed with either straight or special flutes. Chucking reamers are for general purpose applications having either straight or tapered shanks. Reamers are also designed for hand or machine use. The flutes contain the cutting edge and needed relief and clearance angles. Spiral fluted reamers find wider application in reaming harder materials. The spiral flutes enable the reamer to progress through the material reducing the force of the cutting action. Reamers are also classified according to the type of cut desired. *End cutting* reamers are used for reaming blind holes or where parallelism in the hole needs to be corrected. Tapered reamers are used to finish holes of various tapers. Roughing and finishing reamers are available in standard tapers such as Morse, Brown and Sharp, and so on. *Adjustable reamers* have blades or cutting edges that can be adjusted to obtain a greater range of sizes in one tool. *Expansion reamers* have the same capability; however, the range of sizes is limited to a few thousandths of an inch. *Shell reamers* provide a means where the cutting edges are located

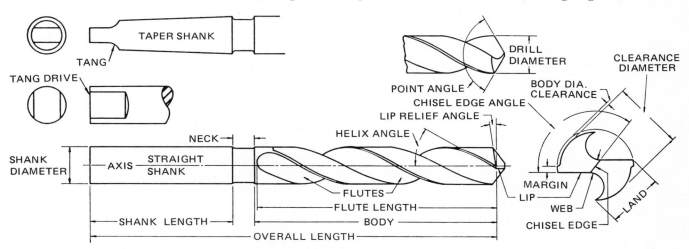

Fig. 13-13 Nomenclature of drill parts. (*Courtesy: Standard Tool Co.*)

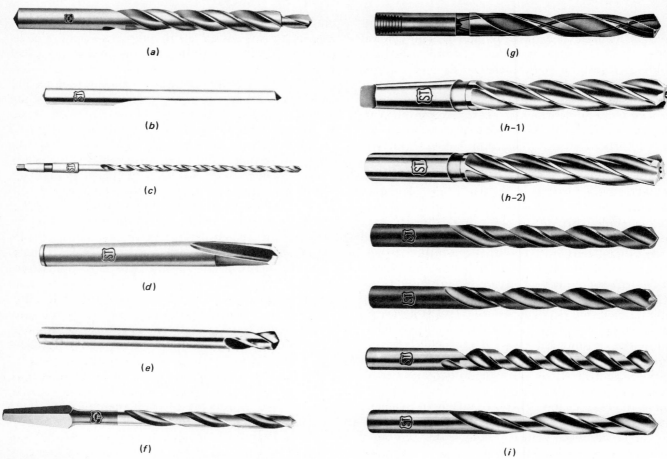

Fig. 13-14 Conventional and special purpose drills. (*Courtesy: Standard Tool Co.*):
(*a*) Double-margin step drill for drilling accurate holes.
(*b*) Half-round drill used with screw machines.
(*c*) Deep-hole (crankshaft type) drill for drilling holes deeper than 5 or 6 times the diameter.
(*d*) Rain drill used for hard, tough, high-strength material.
(*e*) Body drill used primarily to drill holes in sheet metal.
(*f*) Bit shank drill.
(*g*) Straight shank oil hole drill.
(*h*) Straight and taper-shank core drills with four flutes.
(*i*) Jobbers straight-shank drills: (1) regular, (2) heavy duty, (3) high helix, and (4) low helix.

on a shell which attaches to the reamers body. Shells are used primarily to reduce the cost of large reamers since only the shell has to be discarded when damaged.

Abrasives

Abrasives are multiple-point cutting edges which are bonded together to form a functional and highly efficient means of material removal. Abrasives in the form of a grinding wheel enables manufacturers to produce products of close tolerance and superior surface finish. Abrasives are employed in grinding and finishing operations on virtually any industrial ma-

terial. Grinding operations use an abrasive wheel as the cutting tool. Abrasives' particle (natural or synthetic) are bonded to form various sizes and shapes of wheels. Each abrasive grain on the surface of the wheel is a cutter. As the wheel revolves, thousands of minute chips are removed from the work material. Cutting action is applied to the abrasive wheel in a rotary motion. Some grinding machines also have provisions for a straight line and rotary motion at the abrasive wheel. Feed action is applied at either the wheel or the workpiece, depending on the machine used. Feeding motion is applied in either a rotary or a straight-line motion, applied during the cutting action. *Surface grinding*, producing a flat or formed surface, is performed in one of four different methods. The abrasive wheel is mounted either horizontally or vertically. The horizontal surface grinder has rotary cutting motion with both a horizontal and vertical feed provision. Work material is mounted on a table which has either straight-line horizontal motion or rotary horizontal motion. Vertical-type surface grinders utilize a horizontal rotary motion and vertical feed motion for the abrasive wheel.

Cylindrical grinding is used to grind the outside diameter of a cylindrical workpiece while the wheel and workpiece revolve. The workpiece is mounted between centers while rotating in an opposite direction of the abrasive wheel. Feed motion is applied to the grinding wheel in two straight line motions. Feed is either parallel to the axis of the workpiece or perpendicular to the workpiece providing feed along and into the work.

Internal grinding produces internal straight tapered or formed surfaces. The grinding wheel and workpiece revolve in opposition while a feed motion moves the wheel into the workpiece. Feed motion is also applied perpendicular to the workpiece center providing in feed.

Centerless grinding is similar to cylindrical except the workpiece is not mounted between centers. In centerless grinding the workpiece is supported by the workrest blade between a grinding wheel and regulating wheel. Cutting action is provided by the rotating grinding wheel while rotary and straight-line feed is provided by the regulating wheel (Fig. 13-16).

Abrasive belts are primarily used in finishing wood, plastic, and softer materials. Abrasive particles are applied to a flexible continuous belt of various lengths. The belt is driven and guided by two rotating cylinders which provide a continuous straight-line cutting motion. Feed motion is applied to the rotating belt in a straight-line rotary or irregular motion. Feed motion is also applied to the belt perpendicular to the workpiece.

Types of Abrasives. Abrasives are classified as either natural or synthetic. Natural abrasives were the first type used and include: flint, garnet, crocus, emery, corundum, and diamonds. Flint and garnet were widely used in the woodworking industry but have been gradually replaced by synthetic abrasives. Artificial abrasives include silicon carbide, aluminum oxide, boron carbide, and synthetic diamonds. Industrial applications use synthetic abrasives since they are harder than natural abrasives, except diamonds. Natural diamonds are expensive and have therefore been replaced by synthetic diamonds. Silicon carbide was the first synthetic abrasive produced by fusing silicon

sand and quartz in an electric furnace. Silicon carbide, a hard but brittle abrasive, is used in grinding wheels. The silicon carbide grinding wheels are used primarily for materials of low tensile strength such as cast iron, bronze, brass, glass, ceramics, plastics, and similar materials. Aluminum oxide is a mixture of bauxite (aluminum ore) and coke heated to extremely high temperatures in an electric furnace. Aluminum oxide is the most widely used abrasive in industry. While it is not as hard as silicon carbide, it is tougher and exhibits greater shock resistance. Aluminum oxide grinding wheels are used for grinding steel, alloy steel, space-age metals, and other materials of high tensile strength.

Boron carbide is the hardest synthetic abrasive except for the diamond. Boron carbide is used to grind very hard materials or as a powder for lapping and polishing techniques.

Diamonds, whether natural or synthetic, are the hardest known abrasives. Diamond chips are usually bonded to a metal wheel or dish and used as an abrasive for special applications. Diamond-wheel dressers are used to true other softer grinding wheels.

Bonding materials abrasive particles are formed into the desired shape and held together by a bonding material. Four basic types of bonds used are vitrified, resinoid, rubber, and shellac.

Vitrified clay bonding is the most commonly used for general-purpose grinding. Special clay and abrasive particles are bonded together forming a wheel with good strength, uniform structure, and porosity. A vitrified bond provides for a rapid rate of material removal.

Resinoid bonding provides a wheel of high strength designed for high-speed operation. The resinoid-bonded wheels remove material at high rate and are particularly suited for grinding rough spots in the workpiece.

Rubber bonds are used for cut-off wheels of high strength and toughness. The rubber-bonded wheel is also used in centerless grinding operations where a high-quality finish is required.

Shellac bonds are made from natural resins and designed for producing a good quality finish. Shellac-bonded wheels are being replaced by resinoid wheels but are still used for special applications.

Grades. Grinding-wheel grades give an indication of the bonding strength. Harder work materials require a softer grade wheel and vice versa. The proper grade wheel releases the abrasive particles as they become dull or used. Grade is important since an improper grade wheel can burn the workpiece or wear down too fast to be economical. Wheels are graded

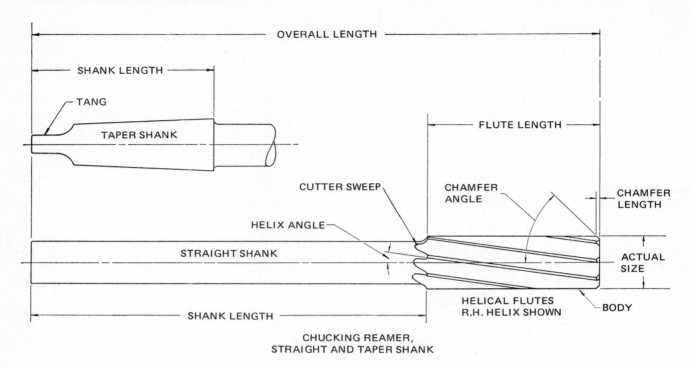

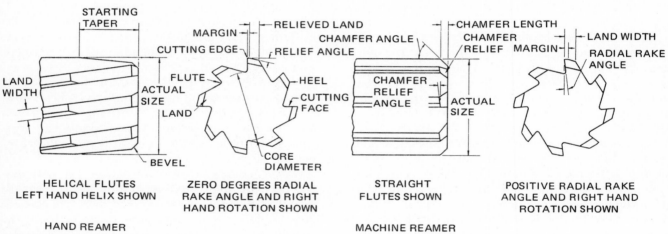

Fig. 13-15 Basic reamer nomenclature. (*Courtesy: Standard Tool Co.*)

from A to Z, with A being the softest and Z the hardest.

Grain Size. The grain size of the abrasive particle refers to the mesh size of the grain. Grain sizes are based on the number of openings per linear inch of separating screen. Designations for grain size range from 10 (coarse) to 600 (very fine).

Structure. The structure of the wheel refers to the density of the abrasive particles in relation to each other. Space must be allowed between the abrasive particles to provide chip removal and coating to the wheel. The wheel's structure is identified by a number

from 1 to 16. A number 1 structure refers to a dense structure, while 16 identifies an open-structure wheel.

Wheel Shapes and Sizes. Grinding wheels are produced in a number of different sizes and shapes (Fig. 13-16). The grinding edge may be flat or of various surface contours. Each wheel has specific applications to aid the manufacturing-producing finished products.

Wheel Identification. Grinding wheels are identified by a standard marking system. However the user

Fig. 13-16 Various types of grinding wheels and coated abrasives. (*Courtesy: The Carborundum Company.*)

should obtain manufacturers data before specifying a particular wheel. A standard marking may be :

A 60 - M8 -V32

where

> A = type of abrasive (A = aluminum oxide, C = silicon carbide, D = diamond)
> 60 = grain size (coarse = 10, very fine = 600)
> M = grade (A = soft, Z = hard)
> 8 = structure (0 = dense, 16 = open)
> V = bond (V = vitrified, B = resinoid, R = rubber, E = shellac)
> 32 = bond mixture (varies among manufacturers)

Miscellaneous Abrasives. In addition to grinding wheels, abrasives are available on cloth, paper, disk, belts, drums, and so on. Primarily these abrasives find specific application in the woods, plastics, and ceramics industries. Coatings are used instead of bonds and identification symbols differ. Additional information should be requested from specific manufacturers.

REVIEW QUESTIONS

13-1. Define cutting action.

13-2. Explain the relationship between cutting speed, depth of cut, and feed.

13-3. What machines utilize single-point cutting tools?

13-4. Differentiate between longitudinal, traverse, cross, and angular feed.

13-5. Calculate the RPM for the lathe when machining a material with a cutting speed of 80 and a 1½-in diameter.

13-6. Explain the relationship between cutting action and feed for the shaper, planer, and lathe.

13-7. Distinguish between reciprocating, continuous straight-line, and continuous rotary-cutting motion applied to saws.

13-8. Explain the difference between the various types of tooth forms relative to a saw blade.

13-9. Distinguish between straight-, wavy-, and raker-tooth set.

13-10. Explain the difference between conventional and climb milling.

13-11. Compare the design and operation principle between a drill and a reamer.

13-12. Compare cylindrical, internal, and centerless grinding relative to the type of feed and support system employed.

CHAPTER 14

MACHINE–TOOL TECHNOLOGY

In Chap. 12 the theories of machinability, chip generation, cutting-tool geometry, cutting-tool material, and lubrication were discussed relative to material removal. The cutting action, speed, feed, and principles related to the application of cutting tools and machining operations were presented in Chap. 13. This chapter will deal with the machine-tool operations and technology.

A *machine tool* is a power-driven machine that provides the needed motion to shape a workpiece by removing material. Machine tools are grouped into eight major categories. The categories of machine tools are (1) lathes, (2) shapers, (3) planers, (4) broaches, (5) mills, (6) drills, (7) saws, and (8) grinders.

Machine tools regardless of their purpose and design have common characteristics. All machine tools employ some form of power, usually an electric motor, to provide the necessary driving force needed in machining. Each machine tool usually has a separate power source. In early machine tools, power was delivered to a series of machines through an overhead drive system. Each machine in the series was connected by a pulley to a single drive shaft. In modern manufacturing a single-drive unit is economically unfeasible because if the power source malfunctions, the operation of a number of machines is jeopardized. Individual machine power sources add flexibility in operation and location of the machine.

Machine tools utilize the power to transmit motion to the workpiece and/or the cutting tool. In order to regulate the speed and the direction of this motion, controls are used on all machine tools.

Machine tools also require some means of transmitting the motion from the power source to the moving elements of the machine. In order to provide power transmission gears, pulleys, cams, and conversion devices (rotary to reciprocating) are employed. Hydraulic and pneumatic systems are often employed in feed and drive units of the machines. A common example is using a hydraulic drive system and pneumatic working holding devices on the shaper. Chucks, collets, vises, tool holders, dogs, clamps, mandrels, and arbors are only a few of the tools and work-holding devices used on most machine tools.

LATHES

The lathe is one of the oldest machine tools. Designed in 1797 by Henry Maudsley, it is probably the most versatile machine tool developed. Primarily the lathe was designed to perform turning, facing, and boring operations on cylindrical workpieces. However, operations such as drilling, reaming, tapping, knurling, grinding, milling, threading, and tapering are possible on the lathe when various cutting tools and attachments are used.

Lathes can be classified according to their drive mechanism (direct or indirect), feed mechanism (hand, power, or automatic), or production capabilities (non, semi, and production). Relative to production capabilities, lathes are classified as:

Nonproduction lathes
  Speed
  Engine
  Tool room

Semiproduction lathes
  Automatic tracer
  Turret
    Vertical
    Horizontal

High-production lathes
  Automatic turret
  Automatic screw machines

Lathe Structure

The lathe (Fig. 14-1), regardless of classification, has five basic parts or units. Major parts of the lathe are the headstock, bed, tailstock, carriage, and feed unit. The *headstock* unit contains the gears, pulleys, or a combination of both which drives the workpiece and the feed units. Located on the left side of the lathe, the headstock also contains the motor, spindle, speed selector, feed-unit selector, and feed-direction selector. The headstock provides a means of support and rotation to the workpiece by attaching a work-holding device to its spindle. The *bed* provides support for the other units of the lathe. V-shaped *ways* are located on the top of the bed providing alignment of the headstock, bed, and tailstock. The ways serve as an accurate guide for the lathe's carriage and tailstock.

The *tailstock,* usually located to the extreme far right side of the lathe, is capable of being moved and secured at various positions along the bed. Primarily,

the tailstock provides support to the outer edge of the workpiece. A handwheel allows for the extension of the tailstock spindle. The tailstock also contains adjusting screws which move the upper section laterally, thus altering alignment with the headstock. Lateral adjustment of the tailstock provides a means of cutting a taper on the workpiece.

The *carriage* consists of the apron (front side), saddle (top portion), feed mechanism, threading mechanism, compound rest, and provision to hold the cutting tool. Movement of the carriage in a longitudinal or traverse (cross-feed) direction is accomplished by hand or power feed.

Drive power is transmitted from the headstock by means of the lead screw and feed rod. *Power feed* is accomplished by engaging the clutch for the selected feed. The compound rest is capable of being rotated to various angles and secured for special applications. Cutting tools are mounted in a tool holder and secured to the top portion of the compound. *Hand feed*

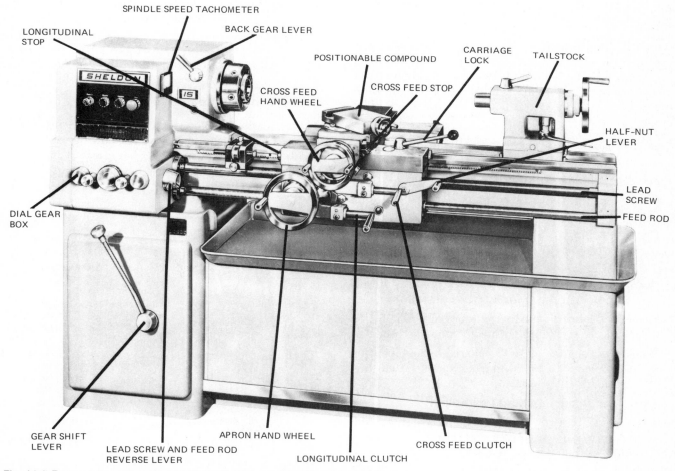

Fig. 14-1 Engine lathe with parts identified. (*Source: Sheldon Machine Tool Co.*)

is provided to the compound through use of a compound handwheel. The half-nut lever located on the front of the apron is used to engage the feed mechanism when cutting screw threads. A feed mechanism located in the headstock (gear box) allows the operator to select the desired feeds and threads specifications. Through a series of gears, feed per revolution of the spindle is set for both longitudinal and traverse feeding. Located near the feed mechanism is a lever which is used to reverse the direction of the lead screw and feed rod, thus reversing the direction of the longitudinal and traverse feed.

Lathe Capacity

The lathe's capacity is designated by the size of the maximum diameter of work that can be rotated (swing) and the maximum length of the workpiece that can be turned between centers. It should be noted that the maximum diameter of work that can be machined on the lathe is less than the swing diameter. The workpiece must be able to clear the carriage assembly, therefore reducing the maximum workpiece diameter by a few inches.

Auxiliary Equipment. In order to perform the various operations on the lathe a number of attachments are required. Work-holding devices, supporting devices, and tool holders are required to perform these operations. A few of the common attachments include:

Chuck—holds workpiece at the headstock, transmits motion.

Faceplate—holds workpiece at the headstock.

Centers—supports the workpiece at the head and tailstock.

Dog—attaches to faceplate and workpiece, transmits motion to work when mounted between centers.

Collet—hollow, compressible, work-holding device mounted in the spindle, transmits motion.

Center rest—supports the extended end of the workpiece when a tailstock cannot be used.

Follower rest—attaches to the carriage, enables long workpieces to be supported near the cutting action.

Turret—multiple indexing tool holder.

Knurling tool—two wheels which when forced against the revolving workpiece forms a knurled pattern in the part.

Boring bar—extended tool holder enabling internal boring of the workpiece.

Cut-off tool—single-point cutting tool used in cut-off operations.

Jacobs chuck—chuck used to hold drills, reamers, and so forth.

Mandrel—a bar used to hold the workpiece when machining a surface concentric to an inside diameter.

Nonproduction Lathes

Engine Lathe. The *engine lathe* is the most common type of lathe. An engine lathe contains basic components previously described and is capable of the operations already discussed (Fig. 14-1).

Speed Lathes. *Speed lathes* are used primarily for metal spinning, wood turning, or polishing operations. A speed lathe, as the name infers, is capable of achieving high speeds; it is designed for light cutting operations with a manually fed cutting tool.

Toolroom Lathes. *Toolroom lathes* are designed and built with greater accuracy for precision non-production-type work. This type of lathe is primarily used in making tools and dies or precision parts. In appearance the engine lathe and toolroom lathe are similar.

Semiproduction Lathes

Tracer Lathes. Automatic *tracer lathes* provide a means of controlling the lathe's cutting-tool movements to accurately duplicate the desired part. The tracer lathe is basically an engine lathe having a tracer attachment (Fig. 14-2). A template of the desired pro-

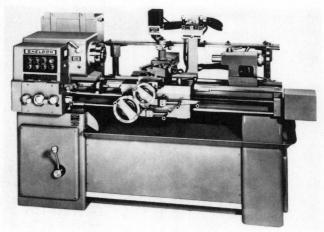

Fig. 14-2 Dual purpose lathe. It serves as precision engine lathe and as two-axis 180° continuous-tracer lathe. (*Courtesy: Sheldon Machine Co., Inc.*)

file is mounted on the tracer attachment. The tracer attachment enables the machine to accurately produce a number of parts exactly alike. An operator is needed only to change the workpiece and cutting tool and to make any minor adjustments. Hydraulic units from the tracer attachment are utilized to control the longitudinal and traverse feed motion of the lathe. Tracer lathes are available with multiple spindles providing for higher production.

Turret Lathes. *Turret lathes* have a multiple-tool indexing unit in place of the tailstock. Turret lathes are capable of controlling multiple cutting tools which can be selected and positioned in a minimum amount of time. In addition, the cross slide of the turret lathe is also designed for production having a four-position turret on the cross slide and a tool holder on the back side of the compound. A major advantage of the turret lathe is the reduction in the time for tool setup. Once the tools are placed in the turrets, they only have to be removed for sharpening. Turret lathes are classified as horizontal and vertical machines.

Horizontal Turret Lathes. *Horizontal turret lathes* are classified as either the ram or saddle types. These machines are further classified as either a bar or chucking variety. The *ram*-type horizontal turret lathe has a multiple-tool turret mounted on a slide or ram which fits on the saddle. The saddle is mounted on the ways. On this type of machine the turret and the ram move back and forth over the stationary saddle (Fig. 14-3). The *saddle*-type turret lathe has the turret mounted directly on the lathe's saddle. The saddle and turret move back and forth along the lathe's ways (Fig. 14-4). Because of its heavy construction the saddle-type

Fig. 14-4 Horizontal saddle-type-turret lathe. (*Courtesy: Gisholt Machine Co.*)

turret lathe is particularly well suited for heavier chucking materials or products which require long turning or boring operations.

Vertical Turret Lathes. The *vertical turret lathe* provides a design capable of handling large workpieces which would be difficult if not impossible to handle on a horizontal machine. A vertical turret lathe has a horizontal work-holding device which provides rotation to the workpiece. The turret (six sided) is supported over the workpiece by means of an overarm. A square (comparable to the cross-slide turret) is mounted on a vertical column alongside the workpiece. The vertical turret lathes have two basic designs: the single-station unit and the multistation unit. The *single-station* vertical turrets utilize one work-holding device. *Multiple-station* units have multiple spindles which index after each operation. Therefore, the use of the multistation vertical turret lathe increases the productive capabilities of the machine. With the completion of each indexing cycle, a part is completed. Figures 14-5 and 14-6 show single-station and multistation vertical turret lathes, respectively.

Automatic turret lathes are capable of performing all the necessary operations automatically. The tool is indexed either by mechanical means or numerical control. Figure 14-7 illustrates an automatic turret lathe that is controlled by a punch card (mounted on the right side of the control panel). The series of switches and buttons are for manual operation and setup of this type of machine.

Production Lathes

Automatic Chucking Lathes. The *automatic chucking lathe* (Fig. 14-8) is similar to a standard ram- or saddle-type lathe with the exception that the turret is mounted vertically. This machine does not have a tailstock like the engine lathe's. Feed motion is supplied

Fig. 14-3 Typical horizontal ram-type lathe. (*Courtesy: Sheldon Machine Co., Inc.*)

Fig. 14-5 Typical vertical-turret lathe. (*Courtesy: The Bullard Co.*)

to the turret unit which is capable of moving in and out depending on the selected operation. Automatic chucking lathes utilize a series of trip pins and blocks for controlling machine operations. In addition to the vertical chucking lathe, a horizontal lathe is also available.

Automatic Screw Machines. *Automatic screw machines* are designed for completely automated operations, including the feeding of work material into the holding device. Screw machines are controlled by a series of cams which regulate the machine's cycle. Screw machines are of the single- and multispindle varieties. Single-spindle automatics are similar to a cam-operated turret lathe, except for the location of the turret. The turret on the automatic screw machine rotates in a vertical axis rather than a horizontal axis as in the standard turret lathe (Fig. 14-9).

Swiss automatic screw machines differ from other types of automatic lathes in that the headstock provides the feed motion to the workpiece. The Swiss machine also uses a cam-operated tool-feed mechanism. The cams move the vertically supported cutting tool in and out, while the workpiece feeds past the tools. The specific shape of the cam determines the contour of the workpiece.

Multispindle automatic machines have from four to eight spindles which are indexed to various positions.

Fig. 14-6 Vertical multispindle chucking (lathe) machine. (*Courtesy: The Bullard Co.*)

Fig. 14-7 Control unit for numerically operated lathe using a punch card. (*Source: Jones and Lamson.*)

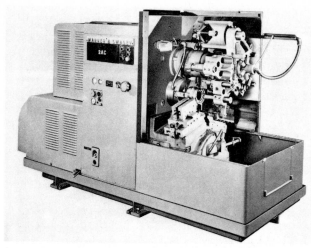

Fig. 14-8 Single-spindle chucking machine. (*Courtesy: The Warner & Swasey Co.*)

The spindles are carried and rotated by means of a spindle carrier (Fig. 14-10). As the spindles index, they expose the workpieces to different operations. At the end of one complete revolution of the spindle carrier, the workpiece is completed. Therefore, on an eight-spindle machine the workpieces index eight times to complete the machine's cycle. Each time the carrier indexes, a part is completed and discharged from the machine's spindle. Figure 14-11 shows a multispindle automatic machine during a typical operation. Various parts produced on the screw machine are illustrated in Fig. 14-12.

SHAPERS

Shapers are machine tools used primarily in the production of flat and angular surfaces. In addition, the shaper is used to machine irregular shapes and contours which are difficult to produce on other machine tools. Internal as well as external surfaces and shapes can be produced on the shaper. Common shapes produced on the shaper are flat, angular, grooves, dovetails, T-slots, keyways, slots, serrations, and contours. Single-point cutting tools similar to the type used on the lathe are used in machining most surfaces on the shaper. Contour surfaces can be produced with either a single-point tool or a formed cutting tool. When using a single-point cutting tool for contour surfaces, the depth of cut must be constantly regulated by the machine operator to achieve the proper contour. Profiling attachments can be used on some shapers to regulate the feed motion and provide duplication of other parts. Numerical control units also enable the production of irregular surfaces with constant regulation of the depth of cut.

Shapers are classified by the plane in which the cutting action occurs, either horizontal or vertical. In addition the horizontal-type shapers are further classified as push or pull cut. A *push-cut* shaper cuts while the ram is pushing the tool across the work, and a *pull-cut* machine removes material while the tool is pulled toward the machine. Vertical shapers use a pushing-type cutting action and are sometimes referred to as *slotters* or *keyseaters*.

Shaper Components

The horizontal shaper, probably the most common variety, consists of a ram, tool-holding assembly, table, and column. The ram pulls the tool assembly while moving back and forth along precision ways mounted on top of the machine's column. The length of the cutting stroke can be adjusted up to the maximum allowable for a particular model machine. The position of the ram over the workpiece is also adjustable. Either mechanical or hydraulic means are used to drive the ram. Hydraulically driven rams provide for a constant feed motion and rapid tool movement, which enables increased production.

The tool-holding assembly consists of a tool head and clapper box. A tool head has the capability of

Fig. 14-9 Single-spindle automatic-screw machine. (*Source: Brown & Sharpe Manufacturing Co.*)

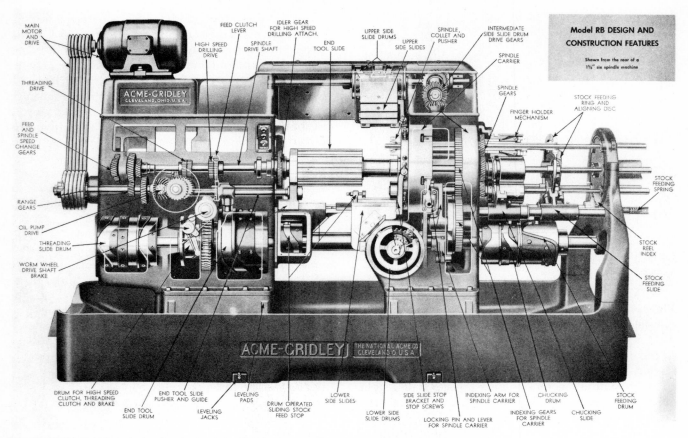

Fig. 14-10 Sectional view showing internal details of a six-spindle screw machine with principal parts named. (*Courtesy: The National Acme Company.*)

being rotated to various angles for cutting angular surfaces. A screw mechanism enables accurate feeding of the tool into the workpiece. The clapper box, mounted on the tool-head assembly, holds and supports the cutting tool and tool holder. In order to provide clearance for the cutting tool on the return stroke, the clapper box swings upward on a pivot-type mechanism.

Workpieces are mounted on the machine's table with clamps or vises. The table slides along cross rails mounted on the machine's column. Feed movement of the workpiece is provided by a cross-feed screw which passes through the table. Feed motion of the workpiece takes place prior to the cutting stroke. The depth of cut is regulated by the movement of the table or the tool head in a vertical direction.

The column forms the frame or support unit of the shaper. Enclosed in the column is the mechanical drive system which provides power for the ram and automatic feed systems. Located on the machine's base, the column supports the ram and the worktable.

The size of the shaper is designated by the maximum length of the stroke in inches. The length of the cross rails usually corresponds to the stroke length. Shapers range from 6 in for a light-duty machine to 36 in for a heavy-duty machine.

Types of Shapers

Horizontal Push-Cut Shapers. The previous discussion on the operation of the shaper describes the push-cut-type shaper and its construction. The cutting action occurs during the forward motion of the ram. During the forward motion the cutting tool is being pushed through the workpiece.

Horizontal Pull-Cut Shapers. Pull-cut shapers are larger than push-cut shapers and, therefore, the workpiece capacity is increased. The length of the stroke may be as long as 70 in. In order to provide support to the ram, an extended overarm is used. Cutting action on the pull-cut shaper occurs as the ram is drawn back toward the machine's column. A pull-cut shaper operates with less vibration because the ram is supported and the cutting force is toward the column of the machine.

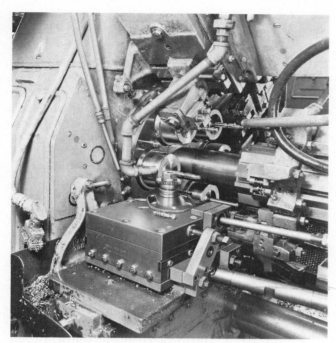

Fig. 14-11 Close-up of working area of a six-spindle screw machine with worm-generating attachment and parts produced shown. (*Courtesy: The National Acme Company.*)

Vertical Shapers. Vertical shapers utilize a ram which moves in a vertical plane. As in the horizontal shaper, the ram moves in a reciprocating motion. Cutting action occurs while the ram moves toward the workpiece. The vertical shaper's table is capable of moving in two straight-line motions and one rotary motion. Vertical shapers consist of the work table, ram, column, base, and tool-holding devices. Power for the ram movement is supplied by a crank, gear, or hydraulic system. Crank-drive units convert rotary motion to reciprocating motion by using a cam and pivoting arm. Heavy cuts are possible on the vertical

shaper because the cutting force is against the work, which is mounted on a heavy-duty table. Internal and external shapes and surfaces can be produced easily on the vertical-type shaper. Curved surfaces arcs and irregular contours are machined with ease because of the rotary motion of the table.

Special types of vertical shapers are the keyseaters, slotters, and gear shapers. The *keyseater* is a vertical shaper used almost exclusively for machining keyways. Figure 14-13 shows a *slotter* and Fig. 14-14 shows a *gear shaper*. Gear shapers use a specially formed cutting tool and can be used for both internal and external gears.

PLANERS

Planers are similar to shapers because both machines are primarily used to produce flat and angular surfaces. However, planers are capable of accommodating much larger workpieces than the shaper. In planer operations the workpiece is mounted on the table which reciprocates in a horizontal plane providing a straight-line cutting and feed action. Single-point cutting tools are mounted on an overhead cross rail and along the vertically supported columns. The cutting tools are fed into or away from the workpiece on either the horizontal or vertical plane, thus being capable of four straight-line feed motions.

Cutting speeds are slow on the planer because of the workpiece size and type of cutting tool being used. In order to increase the production of the planer, multiple tooling stations are employed. Two tooling stations are located on the overhead cross rails, with usually one tooling station on the vertical supports. Another method of increasing production on the planer is to mount a number of workpieces on the table at the same time. This method is only feasible when the workpieces require the same cut and

Fig. 14-12 Parts produced on an automatic screw machine. (*Source: Anaconda-American Brass Co.*)

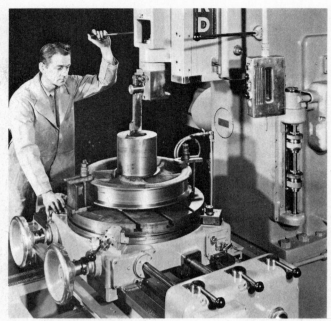

Fig. 14-13 Slotter (vertical shaper). (*Source: Rockford Machine Tool Co.*)

are relatively small in size. The planer size is designated by the maximum workpiece capacity of the machine. The height, width, and length of the workpiece that can be accommodated on the planer's worktable varies with the type of planer.

Types of Planers

Planers are classified as open-side, double housing, pit, and edge-type machines. Classification of the planer is based upon the machine's structure.

Open-Side Planers. *Open-side planers* have one vertical support column to which the cross rail is attached. An advantage of the open-side planer is that the workpiece can extend over the edge of the machine's table. Accessibility to the workpiece is increased on the open-side planer. A disadvantage to the open-side planer is that the cutting tool and cross rails are not adequately supported for extremely heavy-duty work. Figure 14-15 shows an open-side planer.

Double-Housing Planers. *Double-housing planers* have the cross rails supported on two vertical columns. Therefore the stability of the cutting tool is increased, allowing for heavier cuts to be taken. Four tooling stations can be used on the double-housing planer, two on the cross rails and one each on the vertical support columns. The size of the workpiece is,

however, restricted by the additional column and cannot hang over the edge of the table (Fig. 14-16).

Pit Planers. *Pit planers* are heavy-duty machines capable of accommodating large workpieces. The pit planer differs from the standard planner in that the table remains stationary while the column and cross rails reciprocate past the workpiece. Usually the bed of the machine is recessed below the floor, which facilitates mounting the workpiece.

Edge Planer. The *edge planer,* also referred to as a *plate planer* is used primarily for planing the edges of a large plate or slab. As with the pit-type planer, the work remains stationary while the cutting tool reciprocates.

BROACHES

Broaching is a process where a long multiple-point cutting tool is forced through a hole or past the surface of the workpiece. The broach contains a series of consecutive cutting teeth with each row of teeth being progressively higher. The varying height of the cutting teeth provides for material removal and the proper depth of cut. Broaching operations are used to produce flat or irregular internal and external surfaces. The contour of the broach's cutting edge determines the shape of the surface being produced. The surface produced on the workpiece is always the inverse of the broach's profile. Broaching is a continuous operation utilizing straight-line cutting motions applied to either the broach or the workpiece.

Broaches are designed to be either pulled or pushed past the workpiece. Broaching machines consist of a work-holding device, support column, and a tool or workpiece feeding unit. Workpieces are held in fixtures or mounted on the machine's table. The feed unit consists of a tool holder and some means, either mechanical or hydraulic power, to pull or push the broach. When a broach is pulled, some means to hold the broach is necessary in order to combat the pulling force required during the cutting action.

Broaching Machines

Broaching machines are classified as either vertical or horizontal. Horizontal machines are further classified as pull or continuous-cut machines. Power is supplied to the cutting unit either mechanically or hydraulically.

Vertical Broaching Machines. Vertical-pull machines are either the pull-up or the pull-down varie-

Fig. 14-14 Typical vertical-design gear shaper. Inset shows the work zone. (*Courtesy: Barber-Colman Company.*)

ties. The vertical broaching machine resembles a vertical shaper; however, a number of broaches can be used simultaneously (Fig. 14-17). *Vertical pull-down broaching machines* have a mechanism which lowers the pilot (small end of the broach) into a hole in the workpiece. Automatically operated broach pullers then

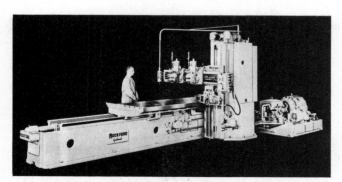

Fig. 14-15 Open-sided planer. (*Source: Rockford Machine Tool Co.*)

connect to the broach and pull it down through the workpiece.

Vertical Pull-Up Broaching Machines. *Vertical pull-up broaching machines* have the workpiece placed below the work table. The broach is fed into the workpiece from the bottom and connected to the broach-pulling mechanism and drawn through the workpiece.

Vertical Surface-Broaching Machines. *Vertical surface-broaching machines* are used to push the broach down toward the workpiece. The broach is mounted in a vertical slide unit mounted on the machine's column (Fig. 14-18). This type of machine is capable of heavy-duty operations. The table of the surface broaching machine is capable of lateral and rotary motion.

Horizontal Broaching Machines. *Horizontal broaching machines* are capable of performing internal and

Fig. 14-16 Double-housing planer. (*Source: Rockford Machine Tool Co.*)

Fig. 14-17 Vertical-broaching machine. (*Source: Footburt Division, Reynolds Metals Co.*)

Fig. 14-18 Vertical-surface-broaching machine. Inset shows close-up of working area and part being broached. (*Courtesy: The Lapointe Machine Tool Co.*)

surface operations on large workpieces (Fig. 14-19). Horizontal machines are used primarily for such internal operations as keyways, slots, and other irregular shapes. The capacity of the horizontal broaching machine is greater than the vertical machine. Since the horizontal machine is capable of larger workpieces, the broach can be larger and, therefore, usually only one pass of the broach is required.

Continuous Surface-Broaching Machines. *Continuous surface-broaching machines* are capable of increased production. In this type of machine the workpieces are loaded on a chain-type drive mechanism. The workpieces are pulled past the broaches on a continuous chain-type feed unit. Workpieces are loaded at one end of the machine and unloaded at the other end after passing beneath the broach.

MILLING MACHINE

Milling machines are probably the most versatile machine tools used in modern manufacturing with the exception of the lathe. Primarily designed to produce flat and angular surfaces, the milling machine is also used to machine irregular shapes, surfaces, grooves, and slots. The milling machine can also be used for drilling, boring, reaming, and gear-cutting operations.

Types of Milling Machines

A number of different types of milling machines are manufactured in order to serve the multitude of needs and the diverse applications. Milling machines are classified according to their structure and include column-knee, fixed bed, planer, and special machines.

Column-Knee. This type of milling machine consists of two main components—the column and the knee. The drive mechanism for the milling cutter is located in the column section of the machine. Two types of column drives are employed: vertical and horizontal. *Vertical mills* have the spindle and tool rotating around a vertical axis, while the *horizontal mill* has the axis of rotation in the horizontal position. On both the vertical and horizontal machines the knee section of the machine provides the means of holding and

Fig. 14-19 Horizontal-broaching machine. Inset shows close-up of working area and part being broached. (*Courtesy: The Lapointe Machine Tool Co.*)

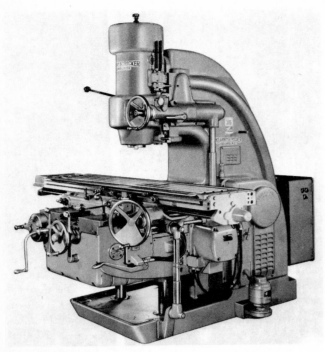

Fig. 14-20 Heavy-duty vertical knee-type milling machine. (*Courtesy: Kearney & Trecker Corp.*)

Fig. 14-21 A helical milling cutter on a horizontal universal knee-and-column milling machine. (*Courtesy: Brown & Sharpe Mfg. Co.*)

feeding the workpiece. Workpieces can be clamped to the table, held in a vise, between centers in an indexing-unit rotary-type vise, and specially designed fixtures. In all cases the work-holding device is secured to the machine's table. The milling machine's table is mounted on a saddle which in turn is mounted on the machine's knee. Movement of the workpiece is accomplished by either hand or power feeding the table (longitudinal), saddle (transverse), and knee (vertical). When only the three table movements are possible, the machine is of the plain column and knee variety (Fig. 14-20).

Universal column and knee milling machines increase the versatility of the standard column and knee machine. In addition to the three straight-line feed motions of the plain mill, the universal milling machine has a swivel-type table. The table can be rotated in the horizontal plane which enables the milling of a helix (Fig. 14-21).

Fixed Bed. The fixed-bed-type milling machine is designed for heavy-duty cutting operations and not flexability. The table located on the bed of the machine is limited to only straight-line feed motions. Vertical and traverse placement is accomplished by the movement of the spindle(s). This movement is accomplished during the machine setup and not during the milling operations, except on specially designed

machines. A fixed-bed machine capable of having vertical movement of the spindle during cutting is called a *rise-and-fall milling machine.* Figure 14-22 shows a fixed-bed-type milling machine.

Three types of fixed-bed machines are the *simplex, duplex,* and *triplex.* The simplex fixed-bed machine has one spindle mounted either vertically or horizontally. Duplex and triplex milling machines have two or three spindles respectively. Multiple spindles provide a means of machining more than one surface at a time, thereby increasing the productive capacity.

Planer Type Milling Machines. The planer type milling machine is similar to the planer except for the

Fig. 14-22 Heavy-duty multispindle numerically controlled profile-milling machine. (*Courtesy: Lucas Machine Division, The New Britain Machine Co.*)

cutting tool and feed. Rotary milling cutters are used rather than single-point cutting tools. The rotary cutters are capable of greater material removal than a single-point tool, therefore increasing the machine's capability. Spindles are located on the cross rail and vertical supports, as in the standard planer. Because milling operations require slower table speeds (reduced feed), the planer milling machine has a variable speed table. Two types of planner mills are the double housing (two vertical supports) and the open-sided (one vertical support) milling machine. Figure 14-23 shows a double-housing-planer milling machine.

Special Milling Machines. Milling machines designed for special applications are the rotary, duplicating, and gear-producing milling machines. The rotary milling machine provides for continuous milling operations. A continuously rotating table provides for a number of workpieces to be mounted on it in various holding devices. As the table revolves, the workpieces pass below the milling cutter. A number of milling cutters can be used to provide the roughing and finishing operations. The workpieces are loaded and unloaded continuously, either manually or automatically.

Duplicating mills (Fig. 14-24) enable a number of duplicate parts to be produced by using a tracer attach-

Fig. 14-24 Vertical-duplicating milling machine. (*Courtesy: Van Norman Machine Company.*)

ment or numerically controlled systems. A pattern made from metal, plastic, or suitable material is placed below the tracer attachment. The tracer passes over the pattern to control the movement of the milling cutters. Tracer units are designed for two-dimensional (engraver) or three-dimensional duplication.

A special type of milling machine commonly referred to as a *hobbing* machine is used primarily in the production of gears. Figure 14-25 shows a gear hobbing machine.

Fig. 14-23 Double-housing adjustable-rail planer-type milling machine. (*Courtesy: Rockford Machine Tool Co.*)

Fig. 14-25 Typical multicycle hobbing machine. (*Source: Barber-Colman Co.*)

DRILLING AND RELATED TOOLS

Drilling machines are used to perform one of the most common operations, that of drilling holes in virtually any material and workpiece. Operations such as boring reaming, tapping, counterboring, countersinking, and spotfacing, which usually follow a drilling operation, are also performed on a drill press. While some of these operations can be performed on other machines (milling machines and lathes), the drilling machine is the most commonly employed. Drilling machines are very simple in construction and operation. In addition to drilling machines, boring machines provide for accurate location and hole size in the workpiece. Since boring machines are similar to drilling machines, they are discussed briefly in this section. Drilling machines are available in many versions and can be classified as portable, upright, radial, multispindle, gang, automatic, and special machines.

Types of Drilling Machines

Portable Drills. Portable drills are designed for light-duty operations where stationary machine tools are not available or easily accessible. Portable drills are commonly powered by electric motors; however, pneumatic drills are available.

Upright Drilling Machines. Upright drilling machines consist of a base, column, head assembly, and work table. The work table is attached to a vertical support column which is secured to the machine's base. The table is adjustable relative to height and limited rotational movement. Workpieces are secured to the table by a number of holding devices, including vises, clamping devices, and special jigs and fixtures. Tool rotation is achieved by a pulley system located in the machine's head assembly. A chuck is attached to a rotating spindle which is located in a nonrotating quill assembly. Feed is controlled by an operator, either manually or automatically. Two types of upright drilling machines are the *bench* and *floor* models. The floor model stands higher than the bench model and is designed for heavier workpieces. The floor-model upright has a greater speed range than the bench model. Speed selection is achieved by a step-cone pulley system which transmits power from the motor to the spindle (Fig. 14-26).

Upright drilling machines can be further classified as sensitive or plain varieties. The *sensitive* drills are hand fed by the operator who regulates the feed on the basis of the feel of the cutting action. *Plain*, or standard, drilling machines utilize a mechanical feed

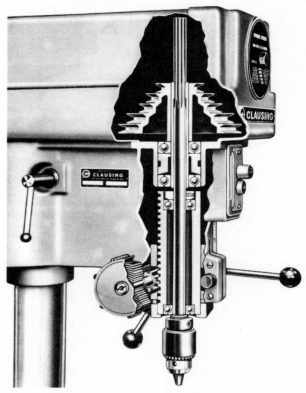

Fig. 14-26 Sectional view showing details of spindle construction of a typical 15-in drilling machine. (*Courtesy: Clausing Division of Atlas Press Company.*)

mechanism which feeds the drill at a constant preset rate.

Radial Drilling Machines. The radial drill is designed for drilling in large workpieces. The workpiece is located and secured to the base of the machine rather than being mounted on a table. The support column which can be rotated is considerably heavier than the column of the upright drilling machines. The radial arm is supported by the column and is capable of being raised or lowered. The spindle assembly is located in the overarm and can be moved along the arm assembly and adjusted vertically, horizontally, and laterally. Size of the radial drill press is designated by the maximum distance in feet from the spindle axis to the support column. Sizes of radial drills range from 3 to 12 ft.

Multiple-Spindle Drilling Machines. When a large number of holes are to be drilled in the workpiece at fixed locations, a multiple-spindle drill press is used (Fig. 14-27). The various drills are located and set in position in the spindle assembly and are driven and fed simultaneously into the workpiece. Drills are con-

Fig. 14-27 Multiple-spindle drilling machine. (*Source: Ettco Tool and Machine Co., Inc.*)

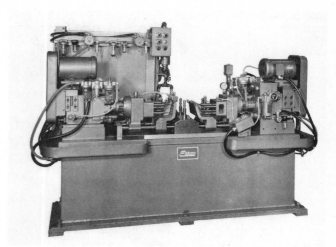

Fig. 14-28 Typical two-way horizontal multiple-spindle drilling machine. (*Courtesy: Ettco Tool and Machine Co., Inc.*)

nected to the drive spindle by a connector (similar to a universal joint) or through a series of gears.

Gang Drilling Machines. Gang drills consist of a single table along which are mounted a number of independent upright drilling heads. This type of machine is used when a number of operations are to be performed on a workpiece. The workpiece is simply slid along the table for the various operations, thus reducing tool change time or transfer time to another machine.

Special Drilling and Boring Machines. Drilling machines are designed to perform special operations

such as deep-hole drilling. When an excessively deep hole is required in the workpiece a *deep-hole drilling machine* is used. Horizontal drilling machines, either single- or multiple-spindle models, are very common in industrial manufacturing. Figure 14-28 illustrates the two-sided *multiple-spindle horizontal drilling machine*. Other special machines enable a number of other operations to be performed such as reaming, drilling, boring, milling, and tapping automatically (Fig. 14-29). *Jig boring machines* are specially designed machines similar to drill presses but are far more accurate in workpiece positioning and boring operations (Fig. 14-30).

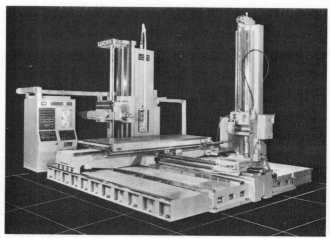

Fig. 14-29 Horizontal floor-type boring, drilling, and milling machine. (*Courtesy: Lucas Machine Division, The New Britain Machine Co.*)

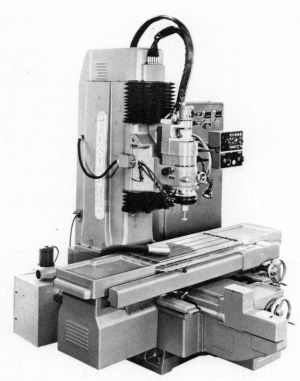

Fig. 14-30 Jig boring machine. (*Source: Fosdick Machine Tool Co.*)

SAWS

Sawing operations are performed primarily for cutting material to length for use in other operations. The saw's versatility enables the machine to be used in cutting irregular shapes and contours. Saws are multiple-point cutting tools that are available as hack saws, band saws, and circular saws.

Types of Saws

Sawing machines are classified as reciprocating, band, and circular saw. The hack saw is a reciprocating saw, commonly used to cut material to length. Band saws consist of vertical, horizontal, combination, and friciton saws. Circular saws include the radial arm, cold saw, and abrasive cut-off saw.

Reciprocating Saws.

The manual and power hack saws utilize a reciprocating cutting action applied to the hack-saw blade. The blade is mounted and reciprocates in a horizontal plane. The cutting motion is perpendicular to the plane of the workpiece, which is mounted in a vise. Feed motion is provided by the vertical movement of the hack-saw blade by either mechanical or hydraulic means.

Band Saws.

The band-saw machine uses a flexible continuous band with cutting teeth along one edge. Common types of band saws are the cut-off and contour models.

Cut-off band saws are classified as either vertical or horizontal saws. Vertical cut-off band saws have the blade moving in a vertical plane. The blade and its support unit can be tilted to angles up to 45°, providing for a bevel-type cut. Horizontal cut-off band saws are used for heavy duty cut-off work. On a horizontal band saw, the blade moves in a horizontal plane, perpendicular to the plane of the workpiece. Horizontal band saws have the flexibility of a hack saw, while providing the increased surface quality of the band saw (Fig. 14-31).

Contour and cut-off band saws are used for cutting irregular shapes and standard type cut-off work (Fig. 14-32). The table of the contour-type band saw can be tilted to various angles for cutting bevels and angles in the workpiece. Contour band saws usually have a flash welder and grinding unit attached as standard equipment to repair broken blades.

Friction-type saw blades can also be used on the contour band saw. In order to use the friction-type blade, the machine must be capable of operating at above normal speeds. Speeds over 12,000 surface feet per minute (sfpm) are required for friction sawing.

Circular Saw Machines.

The circular saw variety of machines includes cold sawing, abrasive disk sawing,

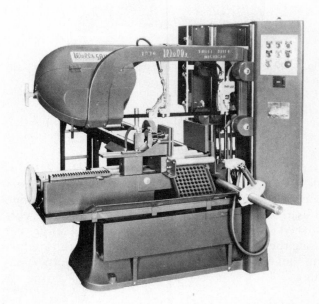

Fig. 14-31 Horizontal band-sawing machine with roto-veyor and automatic bar feed. (*Courtesy: Wells Mfg. Corp.*)

Fig. 14-32 Contour-cutting-band saw. (*Source: DoAll Co.*)

table saws, and radial arm saws. Cold sawing is commonly used in automated cut-off work. The type of saw blade used depends on the material being cut and speed of the machine. Cold sawing machines have a blade rotating in a vertical plane with the feed motion being applied to the saw in a straight-line horizontal plane. Friction saw blades can be used on high-operating circular saw machines specially designed for their use. As with the band saw, the friction circular saw provides a rapid means of cutting ferrous metals and some thermosetting plastics. Abrasive disk sawing uses a resinoid or rubber-bonded abrasive disk rotating at high speeds. The abrasive-disk sawing method performs rapid and precision cutting of metal and ceramics.

Table saws and *radial arm saws* are usually employed for cutting wood and plastic. The table saw has a rotating saw blade extending above the surface of the work table. Material to be cut is fed, either manually or automatically, past the blade in a straight-line motion. Various types of blades and attachments allow many types of operations to be performed on the table saw. The radial arm saw has a circular blade and motor assembly riding along an overarm assembly. The blade is fed into a stationary workpiece to perform the desired operation.

GRINDING AND RELATED MACHINE TOOLS

Grinding machines utilize abrasive grains, bonded into various shapes and sizes of wheels and belts to be used as the cutting agent. Grinding operations are used to impart a high-quality surface finish on the workpiece. In addition, the dimensional accuracy of the workpiece is improved since tolerances of 0.00001 in [0.00025 mm] are possible in grinding operations. Both internal and external surfaces can be ground by using the variety of grinding machines available. Related operations—which use abrasives in various forms such as paste, powder, and grains—include lapping, honing, and drum finishing.

Types of Grinding Machines

Grinding machines are classified according to the type of surface produced. Common surfaces and classifications of grinding machines are surface, cylindrical, and special machines.

Surface Grinders. *Surface grinders* are designed primarily to produce flat surfaces on the workpiece. Special and irregular surfaces may be achieved by using formed grinding wheels. Designation of the surface grinder is based upon the axis in which the grinding wheel and spindle rotate (horizontal-vertical). Surface grinders are further classified relative to the movement of the work table. Both the horizontal and vertical surface grinders have rotary or reciprocating tables. In either case the table movement is in the horizontal plane. Therefore the four main types of surface grinders are (1) horizontal-reciprocating, (2) horizontal-rotary, (3) vertical-reciprocating, and (4) vertical-rotary.

The surface grinder must have provisions for securely holding the workpiece on the machine table. If the work-holding device is adequate, excessive vibration may result yielding a poor surface finish on the workpiece. A common means of holding ferrous materials on the worktable is through the use of a magnetic holding device. Vises, clamps, and special fixtures in addition to vacuum holding tables are also used to secure other types of materials to the table.

Horizontal surface grinders (Fig. 14-33) utilize a standard shape or formed wheel which rotates and is fed into the workpiece. Regardless of the type of surface grinder that is used, the machine size is designated by the largest surface that can be ground on the machine.

Cylindrical Grinders. *Cylindrical grinders* are designed for grinding external and internal cylindrical surfaces. In addition to straight internal and external surfaces, the cylindrical-type grinder is used to grind tapered and irregular surfaces. Cylindrical grinders are classified as center type, chucking type, and centerless.

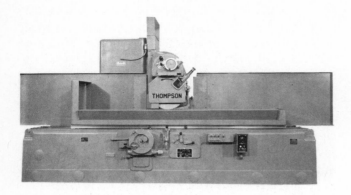

Fig. 14-33 Horizontal surface grinder. (*Source: Thompson.*)

Center-type grinders resemble an engine lathe in appearance and operation. A cylindrically shaped workpiece is mounted between the machine's headstock, tailstock, and centers. Rotation of the workpiece is provided by a faceplate and lathe-dog arrangement. The grinding wheel rotates in a vertical plane while being fed into the workpiece. Grinding occurs as the revolving wheel is moved parallel to the rotating workpiece. To provide an accurate surface, the grinding wheel reciprocates back and forth across the mounted workpiece. Rotation of the grinding wheel and workpiece is in opposite directions and at different speeds (workpiece revolves slower).

Tapered external surfaces are produced on a *universal* center-type grinder. This type of center grinder has provisions for moving the headstock and tailstock, providing the needed offset to grind a taper. Irregular shapes and threads are ground in the materials surface by using formed grinding wheels. When using specially shaped wheels, the reciprocating motion applied to the grinding wheel is not utilized. Only a straight line in feed is used when grinding the special shapes, sometimes called *plunge-grinding*. Thread grinding may use longitudinal feed motion when grinding long threads. To assure that the grinding wheel maintains the proper contour and shape, a specially designed wheel dresser (Fig. 14-34) or hardened form wheel is fed into the grinding wheel. This method assures that the wheel maintains the desired shape of the irregular surface.

Chuck-type grinding machines are used to grind internal and external surfaces in relatively small workpieces. The workpiece is held in a chuck or collet mounted in the headstock spindle. A grinding wheel is mounted on the longitudinal feed unit for external grinding, or on the tailstock for internal grinding.

Centerless grinding is used for both internal and external grinding of cylindrical parts where reference to a hole or a slot is not of prime importance. In centerless grinding the workpiece is not mounted between centers or in a holding device. Rather the workpiece is placed between two rotating wheels while being supported by a work rest. Two wheels are used in centerless external grinding, a grinding wheel and a regulating wheel. The regulating wheel is smaller than the grinding wheel and is mounted at a slight angle relative to the axis of the grinding wheel. A slow rotating motion applied to the regulating wheel provides control for the rotation and longitudinal feed motion of the workpiece. The angle at which the regulating wheel is mounted determines the longitudinal feed of the workpiece. Internal centerless grinding employs three control wheels and one grinding wheel. The workpiece is rotated and supported by the three regulating wheels. A rotating grinding wheel is fed longitudinally through the workpiece. Figure 14-35 illustrates the centerless grinding machine.

Special Grinding Machines. A number of grinding machines are used for specific application. A brief description of these machines is warranted. *Roll grinders* are capable of grinding extremely large workpieces

Fig. 14-34 Typical setup of Crushtrue grinding process. (*Courtesy: Automation and Measurement Division, Bendix Corporation.*)

Fig. 14-35 Close-up showing centerless grinding operation. (*Courtesy: The Carborundum Company.*)

Fig. 14-37 Typical vertical-abrasive-belt grinder with semiautomatic table. (*Courtesy: Hammond Machinery Builders, Inc.*)

such as rolling mills and calendering rolls (Fig. 14-36).

Multiple wheels and *multiple spindle grinders* are used for production-type work where a number of formed shapes or steps in the workpiece are to be ground. *Disk grinders* utilize an abrasive disk where the workpiece is held and fed manually past the abrasive disk. Disk grinders also referred to as *disk sanders* in the wood industry and are used basically for non-precision-type operations.

Fig. 14-36 Roll grinding machine. (*Source: The Cincinnati Milling Machine Co.*)

Belt grinders and *belt sanders* (Fig. 14-37) use an abrasive-coated continuous belt. This type of machine may be either horizontal or vertical and have various designs for the path that the belt follows. Belt abrasive machines are not classified as precision machines since they are hand fed.

Tool grinders (Fig. 14-38) are used for sharpening and grinding various cutting tools. Drills, milling cutters, broaches, lathe tools, and other specially designed tools are sharpened on the tool grinder. A number of work-holding devices such as vises, clamps, fixtures, and center-type holding devices are often used on the tool grinder. Indexing attachments provide a means of indexing the workpiece in order to grind helix flutes in drills and milling cutters.

Related Abrasive Machines

In addition to the grinding machines, abrasives are used in lapping, honing, drum finishing, and blasting operations.

Lapping. *Lapping* is performed by hand or machine to impart a superior-quality finish on the ground workpieces. In hand lapping, abrasive compounds are placed on a cast-iron lapping plate where the workpieces are rubbed in an irregular motion. Machine lapping uses a moving lapping plate and limited workpiece movement. The lapping compound is placed on the lapping plate which moves in a rotary and oscillatory motion. A hold-down plate is placed over the workpiece during the operation to ensure even pressure on the workpiece.

Honing. *Honing* is an abrasive finishing operation used on internal surfaces. Abrasive sticks (Fig. 14-39) are mounted on a free-floating holding device. The unit is placed in the hole and allowed to expand to the workpiece limits. The unit is then rotated while being moved in and out of the hole.

Drum Machines. *Drum machines* are used for fine finishing irregular-shaped workpieces. Workpieces are placed in a rotating drum with the abrasive particles. As the drum rotates, the workpiece and abrasives pass each other during the tumbling action, providing for material removal.

Blasting. *Blasting* is another means of finishing a workpiece through the use of abrasive particles.

Fig. 14-39 Microhoning tool for internal cylindrical honing. (*Courtesy: Micromatic Hone Corporation.*)

"Blasting," commonly called or referred to as "sand blasting," utilizes compressed air to force the abrasive particles through a nozzle at high speeds and force. As the abrasive particles hit the workpiece, small pieces of the work material are removed. Blasting is used for nonprecision cleaning and finishing the work material. Surface finish is dependent upon the particle size of the abrasive.

Fig. 14-38 Tool grinder. (*Source: Sheffield-Bendix.*)

REVIEW QUESTIONS

14-1. Define a machine tool. Identify the various categories of machine tools.

14-2. Identify common characteristics of machine tools.

14-3. Explain the operation of the following: lathe, shaper, planer.

14-4. Distinguish between semiproduction and high-production lathes.

14-5. Contrast the type of work produced on the shaper and planer.

14-6. List the types of planers. Compare each type relative to the machine's structure.

14-7. Differentiate between vertical, horizontal, push-cut, and pull-cut broaching machines.

14-8. Contrast the structure of the column-knee, universal, and fixed-bed-type milling machines.

14-9. Explain the similarities and differences between a drilling and broaching machine.

14-10. Differentiate between reciprocating, band, and circular saws.

14-11. Explain the operating principle of internal and external centerless grinding.

CHAPTER 15

SPECIAL MATERIAL– REMOVAL PROCESSES

MECHANICAL PROCESSES

The general trend in manufacturing has been in developing "chipless" processes to cut and form materials which are difficult to produce with conventional processes. Several new processes have been developed and are used industrially with machines commercially available. Other processes are still in the developmental stage. The processes presently used industrially are abrasive-jet machining (AJM), ultrasonic machining (USM), and fluid-jet machining (FJM). All these processes employ mechanical energy as the primary cutting energy. These processes are often referred to as "chipless" material-removal processes because there is no contact between the "cutting tool" and the work, and no chips are formed as with the conventional machining processes discussed earlier.

Abrasive-Jet Machining

Abrasive-jet machining (AJM) is similar in many ways to abrasive blasting used for cleaning purposes. However, it differs from blasting in terms of the abrasive used and the controls imposed on the abrasive action. The material-removal action (cutting) is based on the erosive or abrasive action of the abrasive medium. The process is relatively simple. Abrasive grains of predetermined shape, size, and type are forced with a high-velocity air stream (compressed air) through a nozzle against the work surface. As the abrasive grains come into contact with the work, they "chip" and sweep away very small particles of the work material (Fig. 15-1). This action creates the abrasive-jet machining which is the basic cutting or material-removal process. The only contact of the work and the "cutting tool" (abrasive) is between individual abrasive grains and work surface. Therefore, as the material is being removed, no chips are formed but instead very tiny powderlike particles which are blown away by the jet stream.

The composition, grain size, shape, strength, and mass-flow rate of the abrasive are critical variables of the process. The two most commonly used abrasives are aluminum oxide and silicon carbide. Some other types of abrasives are used for cleaning and polishing purposes. Among these are dolomite and sodium carbonate. The grain size of abrasives varies from 15 to 40 μ (microns), which is actually powder form (a micron is one millionth of a meter). The abrasive grains must be sharp rather than dull. The mass-flow rate is related to the air pressure and the material-removal rate. Figure 15-2 shows the relationship of the abrasive mass-flow rate, abrasive grain size, abrasive composition, and material-removal rate. The cutting speed of the abrasive may be varied by adjusting the nozzle-tip distance (NTD), abrasive flow, and air pressure.

The composition, size, geometry, and distance of the nozzle tip are also critical factors. Nozzles are made mostly from tungsten carbide and sapphire. The orifice shape of the nozzle is made either round or rectangular with sizes from 0.007 to 0.032 in [0.177 to 0.182 mm] in diameter for round shape and from 0.003 × 0.020 to 0.020 × 0.026 in [0.076 × 0.508 to 0.508 × 0.660 mm] for rectangular shape. Nozzles are mounted at right angles or straight heads (Fig. 15-3). Average nozzle life varies from 12 to 30 hours of operation for tungsten carbide nozzles and up to 300 hours for sapphire nozzles. The nozzle-tip distance from the work affects the material-removal rate and the diameter of cut. The emerging stream from the tip of the nozzle to about 0.060 in is cylindrical in shape, but the shape changes to a cone-shaped spray for longer distances. Nozzle-tip distances from 0.010 to 0.500 in are typical. Fig-

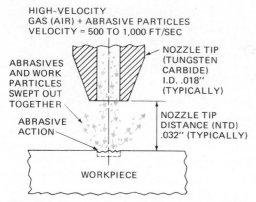

HIGH-VELOCITY
GAS (AIR) + ABRASIVE PARTICLES
VELOCITY = 500 TO 1,000 FT/SEC

NOZZLE TIP
(TUNGSTEN
CARBIDE)
I.D. .018″
(TYPICALLY)

ABRASIVES
AND WORK
PARTICLES
SWEPT OUT
TOGETHER

ABRASIVE
ACTION

NOZZLE TIP
DISTANCE (NTD)
.032″ (TYPICALLY)

WORKPIECE

Fig. 15-1 The abrasive-jet cutting process. (*ASTM.*)

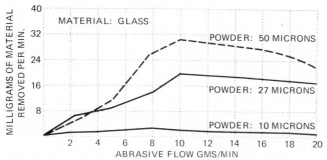

Fig. 15-2 Relationship of abrasive mass flow rate, grain size, composition and removal rate in glass. (*S. S. White Co.*)

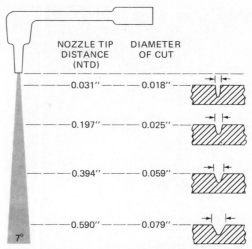

Fig. 15-4 Relationship of diameter of cut and nozzle-tip distance from work surface. (*S. S. White Co.*)

ure 15-4 shows the relationship between the diameter of cut and nozzle-tip distance.

Abrasive-jet material removal is accomplished by the abrasive-jet system shown in Fig. 15-5. The system is a bench-type machine consisting of the abrasive unit, dust collector, exhaust chamber, air compressor, and air filter. AJM has been successfully used for cutting and shaping hard materials such as glass, quartz, sapphire, and mica (Fig. 15-6*a*). It can be used for drilling and cutting thin sections of materials difficult to cut with conventional processes (Fig. 15-6*b*). AJM is used for the abrading and frosting of glass, etching, polishing, and cleaning of metallic and nonmetallic materials.

The basic limitations of the process are (1) low material-removal rate, which makes the process justifiable only for materials difficult or impossible to machine with other processes, (2) embedding of abrasive

ROUND ORIFICE (IN.)	MATERIAL	RECTANGULAR ORIFICE (IN.)	MATERIAL
.007 DIAMETER	CARBIDE	.003 X .020	CARBIDE
.008 DIAMETER	SAPPHIRE	.006 X .020	CARBIDE
.011 DIAMETER	CARBIDE	.006 X .060	CARBIDE
.018 DIAMETER	CARBIDE	.006 X .075	CARBIDE
.018 DIAMETER	SAPPHIRE	.006 X .100	CARBIDE
.026 DIAMETER	CARBIDE	.007 X .125	CARBIDE
.026 DIAMETER	SAPPHIRE	.010 X .030	CARBIDE
.032 DIAMETER	CARBIDE	.026 X .026	CARBIDE

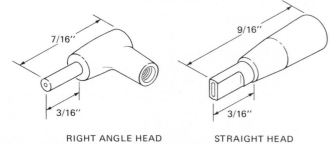

RIGHT ANGLE HEAD   STRAIGHT HEAD

Fig. 15-3 Types of nozzles, sizes, and material used in abrasive-jet machining. (*S. S. White Co.*)

Fig. 15-5 Typical air-brasive unit of an abrasive-jet machining system. (*S. S. White Co.*)

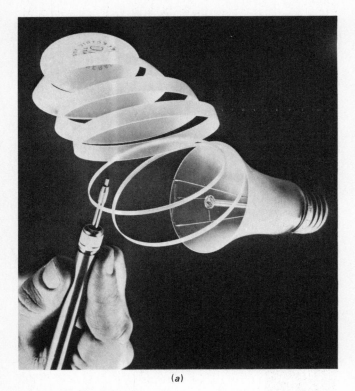

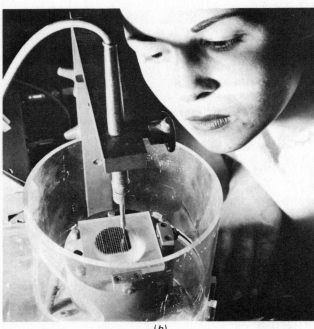

Fig. 15-6 Two typical applications of AJM using a straight-head nozzle. (*A*) Cutting glass; (*B*) machining a grid in brittle thin material. (*S. S. White Co.*)

grains in the work which requires careful cleaning after AJM has been completed, and (3) severe taper cutting in relatively thick material, or deep holes.

Ultrasonic Machining

As with AJM, ultrasonic machining (USM) has been developed and used for machining hard and brittle material which is difficult to machine with conventional processes. USM (also known as *impact grinding*) is based on the impact of abrasive grains driven by an oscillating cutting tool against the work surface (Fig. 15-7). The tool oscillates ultrasonically (at high frequencies of up to 20,000 cycles per second or Hertz, Hz) and forces the abrasive grains to impact against the work surface, causing a "chipping" or "grinding" which is responsible for the material-removal action of the process. The fast oscillating tool produces cavitation of the abrasive-liquid carrier necessary for material removal. This turbulent action of the slurry acts as a pump for the abrasive grains and the minutely cut particles in the cutting area. There is no physical contact between the tool and the work. However, the shape of the tool determines the precise impact pattern of the abrasive grains, which, in turn, determines the shape of the machined part. Therefore, the shape of the machined part is dependent on the shape of the tool.

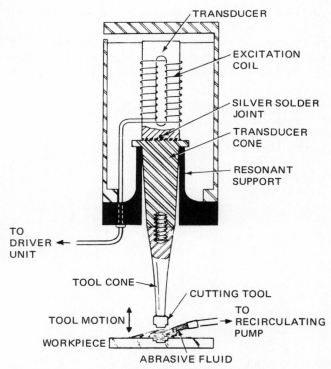

Fig. 15-7 Schematic showing the principal elements of an ultrasonic machining system. (*Raytheon*).

Fig. 15-8 Two cutting-tool profiles used to ultrasonically machine boron and tungsten composites with great economy. (*Bendix Corp.*)

The material-removal rate of USM depends on the amplitude of tool oscillations, impact force of the abrasive grains, and grain size. The material-removal rate is proportional to the square of the amplitude of the oscillations. The frequency of oscillations also affects the material-removal rate, especially in brittle materials where higher frequencies produce higher material-removal rates. The preferred tool materials are soft steel, annealed alloy steel, and stainless steel. The shape of the tool must be the mate of the surface to be machined. There are limitations on the profile of the tool (Fig. 15-8), but tool size is governed by the degree of accuracy and type of finish required. The tool is mounted to the tool holder by either brazing or soft soldering.

Among the most commonly used abrasives are *boron carbide, silicon carbide,* and *aluminum oxide,* with boron being the fastest cutting abrasive for the process. The grain size of the abrasive influences the material-removal rate and the surface finish (roughness), and it should be selected with care. The abrasive grains are mixed with water to form the *slurry.* Among the functions performed by the slurry are circulating the abrasive grains between the tool and work, removing worn grains, and cooling the work.

USM machines hard, brittle, or fragile material (Fig. 15-9) either difficult or impossible to machine with other processes. Among the materials successfully machined by USM are boron composites, molybdenum steels, ceramics, glass, tungsten carbide, quartz, synthetic ruby, and mica. Among the most common operations performed by USM are drilling

holes of any shape, blind, through tapered or multiple drilling; broaching and shaving; slicing of single or multiple parallel cuts; and dicing and coining dies of elaborate designs. All these operations are performed on the ultrasonic machining system illustrated in Fig. 15-10.

Some of the advantages claimed for USM are that the process (1) permits freedom of design for parts, (2) permits repair or alteration of hardened steel dies without annealing, (3) reduces machining time, (4) provides uniformity in duplicated parts, (5) is relatively safe to operate, and (6) does not affect the properties of the machined part. Among the limitations of the process are (1) relatively low material-removal rate, (2) hole depth limited by efficient circulation of the slurry, and (3) tool wear and fracture.

Liquid-Jet Material-Removal Process

Although still experimental, the liquid-jet material-removal process has potential for development into a powerful cutting process. Experiments have been performed on such materials as aluminum, acrylic, plastic, and rocks. Softer materials such as wood can be cut relatively easily.

Liquid-jet material removal employs a high-energy-density jet with velocities of up to 4500 feet per second (ft/s), which are capable of generating energy

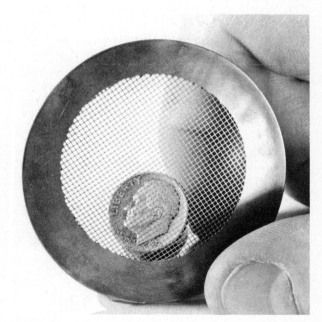

Fig. 15-9 Molybdenum disc 0.010 in thick and 1.5 in in diameter machined ultrasonically to produce 2.025 square holes each 0.052 in square and a grid of 0.0015 to 0.002 in thickness. (*Bendix Corp.*)

Fig. 15-10 Complete ultrasonic machining system (machine tool) used to machine the shapes shown in Figs. 14-8 and 14-9. (*Bendix Corp.*)

densities at the level of 10^6 hp/in². With these levels of energy almost all soft and medium-hard materials can be machined relatively easily.

Unlike the high-energy "thermal" processes of laser beam, ion beam, and plasma, which present the problem of temperature at the cutting zone, in liquid-jet machining there is no heat-affected zone. The material is cut by a cold liquid jet. Liquids that have been used with the process are kerosene, pentane, ethylene glycol in water, and plexol. Water alone cannot be used because it freezes at the extreme pressures (100,000–200,000 psi) [7000–14,000 kg/cm²] usually required by this process to cut medium-hard materials.

The liquid-jet material-removal system is relatively simple. It consists of the pump, reservoir, valve, pressure transducer, and the cutting nozzle. Figure 15-11 shows a schematic of the liquid-jet material-removal process.

ELECTROCHEMICAL MATERIAL-REMOVAL PROCESSES

The electrochemical metal-removal (ECMR) area is one of the most potentially useful areas of the new material-removal processes. All processes in the area of electrochemical metal removal are based on

Farady's law of electrochemistry which states that if two metals are placed in a conductive electrolyte and connected to a direct-current source, metal from the positive pole (anode) will be deplated and deposited on the negative pole (cathode). The processes based on electrochemistry have been advancing rapidly in the past few years. Among the factors contributing to the rapid development of ECMR are (1) the need for machining new materials, especially metals, which are difficult to machine with other processes, (2) the need to increase economy of operation, and (3) the need for part design flexibility. The most commonly used processes in ECMR are *electrochemical machining* (ECM), *electrochemical grinding* (ECG), and *electrohoning*.

Electrochemical Machining

Electrochemical machining (ECM) is based on the electrochemical action of Farady's electrochemical cell which consists of the anode, the cathode, and the electrolyte. In addition to the basic cell, the system contains some mechanical components to circulate the electrolyte and position the electrode called the *cutting tool* (Fig. 15-12). To form the electrochemical cell in ECM, the preshaped "cutting tool" (electrode) with high-surface finish is connected to the positive pole of the electrical source and forms the cathode of the system, while the work is connected to the negative pole and forms the anode (Fig. 15-12). A direct

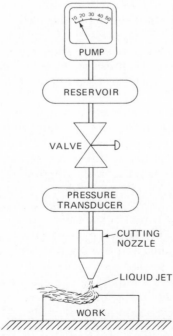

Fig. 15-11 Schematic showing the liquid-jet material-removal process.

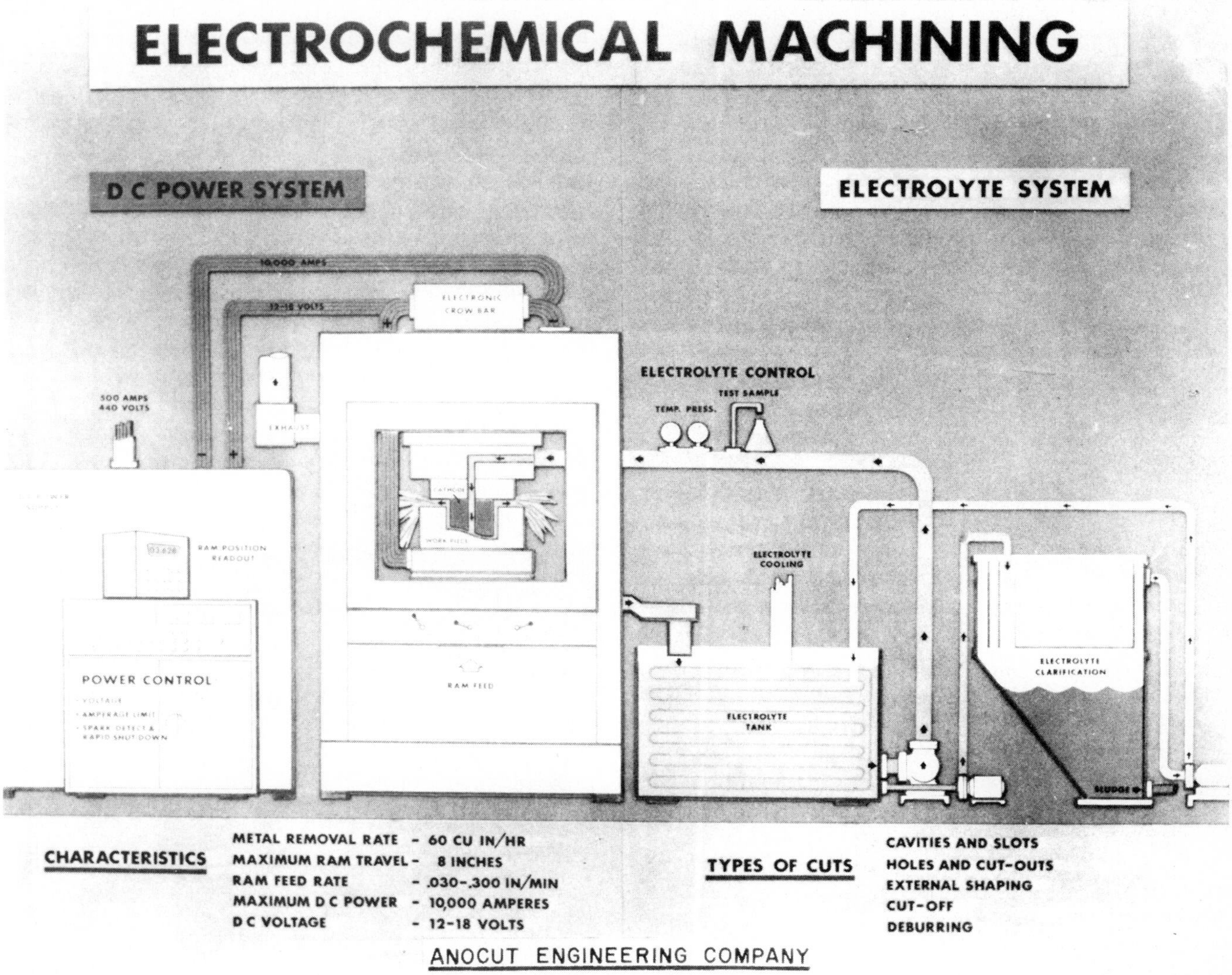
ELECTROCHEMICAL MACHINING
D C POWER SYSTEM
ELECTROLYTE SYSTEM
10,000 AMPS
12-18 VOLTS
ELECTRONIC CROW BAR
500 AMPS
440 VOLTS
ELECTROLYTE CONTROL
TEST SAMPLE
TEMP. PRESS.
ELECTROLYTE COOLING
ELECTROLYTE CLARIFICATION
ELECTROLYTE TANK
POWER CONTROL
CHARACTERISTICS
METAL REMOVAL RATE - 60 CU IN/HR
MAXIMUM RAM TRAVEL - 8 INCHES
RAM FEED RATE - .030-.300 IN/MIN
MAXIMUM D C POWER - 10,000 AMPERES
D C VOLTAGE - 12-18 VOLTS
TYPES OF CUTS
CAVITIES AND SLOTS
HOLES AND CUT-OUTS
EXTERNAL SHAPING
CUT-OFF
DEBURRING
ANOCUT ENGINEERING COMPANY

Fig. 15-12 Schematic showing a typical electrochemical machine tool system. (*Anocut Engineering Co.*)

current of constant intensity and low ripple is used as the power source. As the current flows through the constantly circulating electrolyte, it causes certain ions in the electrolyte to remove metal ions from the work on a deplating effect, the reverse of plating. The metal ions unite with the ions in the electrolyte, go into solution, and are flushed away from the gap formed between the tool and the work. This electrochemical action is the basic cutting force of the process. Because of the constantly circulating electrolyte, no metal is deposited on the tool's surface. The tool never touches the work, and therefore no tool damage from heat or sparking or tool wear occurs.

The gap between the tool and work surface is an important variable affecting the metal-removal rate. It has been found that in most ECM operations, the size of the gap is directly proportional to the voltage and inversely proportional to the feed rate and electrolyte resistivity. The tool-feed rate is controlled automatically by the moving ram and slides. Metal-removal rate and tool-feed rate for various metals can be calculated with satisfactory accuracy.

The electrolyte is another critical variable of ECM because it affects the metal-removal rate. For an effective and efficient operation, the electrolyte should possess certain characteristics. Among them are (1) good electrical conductivity to facilitate current flow, (2) economical in order to reduce operating cost, (3) readily available, (4) nontoxic for safe use, and (5) noncorrosive in order to avoid corrosion of work and tool surface. The most commonly used electrolytes are *sodium chloride in water, sodium nitrate, potassium*

Fig. 15-13 Large pinion dies of heat-treated die steel machined with ECM with surface finish less than 10 microinches (μin) in roughness and die cavities within 0.005 in precision. (*Anocut Engineering Co.*)

chloride, sodium hydroxide, sulfuric acid, and *sodium chlorate.* Sodium chloride in water and sodium nitrate seem to be the two most widely used electrolytes.

For an effective and efficient operation, the current flow should be maintained at a constant density at the tool. Although any conductive material can be used to make the tool, brass and copper are the two most commonly used. Since the tool actually reproduces itself in the work, tool design is an important factor. The tool must have a high quality surface finish to be able to produce good finish on the work surface (Fig. 15-13). Machine tools for ECM vary in design and size, depending on the type of work processed and tolerances desired. A typical machine tool consists of a table to mount the work, a ram with plates to mount the tool and the various controls to position the tool, the electrolyte pumping system, and the power-supply system (Fig. 15-14). ECM has been used to successfully perform such operations as turning, die sinking, profiling and contouring, multiple-hole drilling, and broaching. Figure 15-15 shows typical parts machined with ECM operations.

Among the advantages claimed for ECM are virtually no tool wear, machining done basically below 200°F [111°C], and complex shapes with single-axis movement concentrated on relatively small areas of the work. Some of the limitations of the process are high initial cost, high cost for tool production, sludge problems, not very accurate process (0.005 to 0.002), expensive fixtures for work holding and corrosion, and safety protection problems.

Electrochemical Grinding

Electrochemical grinding (ECG) is a variation of the electrochemical material-removal process. In theory it may be regarded as a combination of high-speed electroetching and concurrent grinding. ECG involves a grinding wheel connected to the negative pole (cathode) of the direct current source, the work connected to the positive pole (anode), and the constantly circulating electrolyte to form the basic electrochemical cell (Fig. 15-16). The grinding wheel is made of steel and it is covered with insulating abrasive particles. The insulating abrasive particles form the necessary "gap" between the tool (wheel) and work. The electrolyte is forced through this "gap" and creates the electrochemical action which is the primary cutting action. The secondary cutting action is created by the abrasive particles of the wheel. Thus material is removed by both electrochemical and

Fig. 15-14 Double-A-frame ECM-production-machine tool. (*Anocut Engineering Co.*)

Fig. 15-15 Typical parts machined by ECM and tools used in machining the parts. (*Anocut Engineering Co.*)

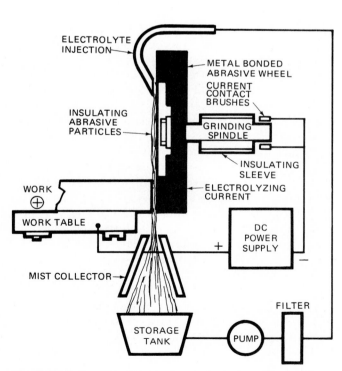

Fig. 15-16 Schematic of electrochemical grinding. (*Cincinnati Milling Machine Co.*)

abrasive actions although the metal removed by abrasion is relatively slight. Therefore, electrochemical grinding is a way of removing material by using the principle of conventional grinding reinforced by electrochemical action. ECG is applicable to any conventional grinding operations such as face grinding, surface grinding, internal and external grinding, and to a lesser extent, form grinding. ECG employs machine tools basically similar to those used in conventional grinding (Fig. 15-17).

ECG is used mostly for grinding fragile parts which are difficult to grind with conventional grinding, heat sensitive parts which may be affected by the heat generated with conventional grinding, carbide cutting tools, component sizing, cutting laminations and thin tubing, and grinding slots and grooves in collets and threaded parts.

Among the advantages claimed for ECG are the following: heat-sensitive parts can be ground completely cold; work is free from burrs; cutter life can be predicted; fragile and thin parts, such as honeycomb and thin-wall tubing, can be ground without distortion; and low grinding-wheel cost and nonproductive time reduces the overall operating cost of the process. Some of the limitations include: relatively wide contact area is required to draw amperage, only certain types of forms can be ground, grinding small interval diameter holes is impractical, and wheels need occasional deplating and conditioning.

Fig. 15-17 Electrochemical-grinding-machine tool with automatic indexing rotary fixtures and special workholders set up for grinding small steel-circuit parts. Inset shows a carbide boring tool ground with ECG in about 1½ minutes. (*Hammond Machine Builders.*)

Electrochemical Honing

Electrochemical honing (ECH) combines the metal-removal capabilities of ECM with the accuracy and surface-finish characteristics of electropolishing and conventional honing. The process operates on the same basic principles of ECG and ECM. The machine tools used are similar to those of conventional honing. ECH has been successfully used for internal honing with limited application for external and flat-surface honing.

CHEMICAL MATERIAL-REMOVAL PROCESSES

The art of chemical material removal has been known and practiced for thousands of years for a variety of applications. However, chemical material removal as an industrial process is a relatively new process. Chemical material removal is practiced today in three variations: *chemical machining* or *chemical milling, chemical blanking,* and *chemical engraving.* All these processes employ basically the same "cutting" action which is accomplished either by an acid or alkaline chemical solution called the *etchant.* The basic process consists of the tank, the chemical solution, and the

work (Fig. 15-18). The "cutting" or material removal action in this process is strictly chemical action. The part to be processed is immersed into the tank containing the acid or alkaline chemical solution, and the chemical etches or dissolves the exposed surface of the material. Unlike the other material-removal processes, chemical machining employs no shaped tool of any kind. Where pockets or reliefs are to be made, the part is coated with an appropriate chemical resistant material before immersion into the chemical solution. Chemical material-removal processes are mostly used for metals. However, other materials such as glass and silicon have been successfully processed with this method.

The two most important variables of chemical material removal are the chemical solution (etchant) and the etchant-resistant material called the *maskant* or *resist.* The main purpose of the resist is to protect parts of the surface from the chemical attack of the etchant. Thus the etchant will attack the surface selectively, forming the desirable shape of the part. Therefore, the resist must be highly resistant to the chemical attack of the etchant. The three most commonly used types of resists are *maskants, photo resists,* and *silk-screen resists.*

Maskants are made of synthetic material such as vinyl, neoprene, and butyl. They are applied on the

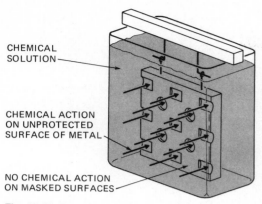

CHEMICAL
SOLUTION

CHEMICAL ACTION
ON UNPROTECTED
SURFACE OF METAL

NO CHEMICAL ACTION
ON MASKED SURFACES

Fig. 15-18 The basic chemical material-removal unit. (*Anocut Engineering Co.*)

work surface by flow, dip, spray, or roller coating. The thickness of the film formed on the work surface varies from 0.001 to 0.005 in [0.025 to 0.127 mm], depending on the type of maskant used, the material etched, and the type of etchant used. However, a gallon of this material may cover from 30 to 40 ft² [2.7 to 3.6 m²] of work surface. This type of resist is used for etching parts which are relatively large, have many irregularities, require deep etching, and require more than one etching step in the material-removal areas. After the etching has been completed, the maskants are usually peeled off or dissolved from the unetched areas of the work surface.

Photo resists are photographic resists which produce resistant images on the work surface by means of photographic techniques. They can be applied on the work surface by dipping, spraying, flowing, roller coating, or laminating. They are used for etching parts which are very thin and require very close dimensional tolerances and high rates of production.

Screen resists are materials which are applied on the work surface through the silk screening process used in the printing industry. This is the fastest method to apply resists and produces surfaces which are more resistant than photo resists.

In selecting and applying the resists, a number of factors should be considered. The number of parts required to be produced is an important factor. If relatively few parts are required, the maskant may be the ideal resist. Chemical resistance is another critical factor. The ability of the resist to withstand the attack of the etchant may vary from a few minutes to a few hours. Size, shape, and accuracy of the parts produced is another factor for consideration. Some resists are peeled off while others are dissolved in order to remove them.

Many types of etchants are used and are commercially available at different concentration levels. Al-

though the material to be etched is the governing factor in the selection of the type of etchant, many other factors need to be considered. Among these are depth of cut, which may be as much as 0.500 in [12.7 mm], the type of resist available, and the surface finish required. The speed of material removal should also be considered.

Chemical Machining or Milling

Chemical machining is the most widely used variation of the chemical material-removal process. Chemical machining is variously called *chemical milling, photoforming, chemical cutting, photo etching,* and *photo mechanical duplicating.* Chemical machining is the process for removal of materials by chemical action of an etchant. The process is suited primarily for production of flat, relatively thin parts of almost any configuration. The amount of material removed or depth of cut is controlled by the immersion time of the part in the etchant. The unetched areas of the part are controlled by covering them with a chemically resistant material. The basic steps of the process are cleaning the material, applying the resist, scribing, etching the part, and removing the resist. Figure 15-19 shows a typical flowchart of chemical machining using photo resists.

Cleaning of the work surface is important to assure reliable adhesion of the resist. All oxides, oil, or grease are removed from the surface. Cleaning can be accomplished by any cleaning processes described in Chap. 21 of this book. However, solvent and vapor degreasing processes are preferred. Sometimes it may be necessary to use more severe cleaning media such as flash etching or alkaline cleaning. After cleaning, the parts must be thoroughly rinsed to remove all solvents or etchants that may be trapped on the surface.

Applying the resist is accomplished by either photographic, silk screen, or masking methods by means of dipping, spraying, or roller coating. It may be necessary to apply more than one coat, depending on the material etched and the etchant used. After the resist is applied, it is thoroughly dried or cured.

If the masking method is used to apply the resist, the next step is to scribe the image of the part to be etched. Images of the part are placed on the work surface by means of templates to guide the scribing of the resist with an appropriate knife or scraper. If photo resist has been used, the images have been photographically exposed on the work surface and therefore no scribing is necessary.

After the work surface has been prepared with the resist, the part is immersed into the etchant for a pre-

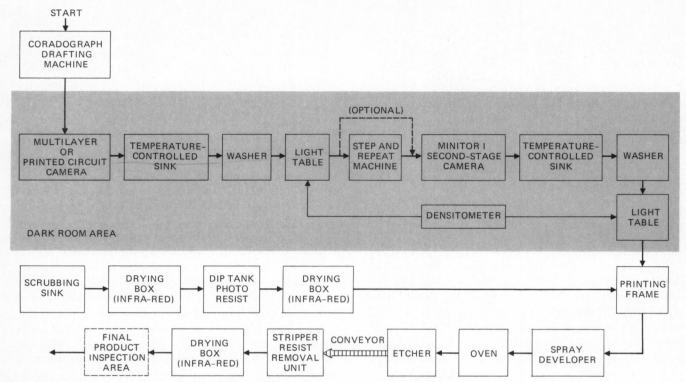

Fig. 15-19 Typical flowchart of chemical-machining process using photoresist. (*Chemcut, Division of Centre Circuits, Inc.*)

determined time. For uniform etching, the parts are moved in the etchant or the etchant is lightly agitated. In many applications the etchant may be sprayed onto the parts instead of immersing the parts into the etchant. Figure 15-20 shows a conveyorized etching machine. After the appropriate etching has been achieved, the parts are thoroughly rinsed to remove all traces of the etchant from the work surface.

The final step in the process is to remove the resist. This can be accomplished by either hand stripping or by immersion into a suitable dissolving solution. The parts are then cleaned and further processed as necessary.

Chemical machining is a relatively precise process, with average dimensional tolerance from 0.001 to 0.005 in [0.025 to 0.127 mm]. Depth of etching may vary from 0.001 to 0.500 in [0.025 to 0.127 mm]. The process provides flexibility of design, and it is not limited to metallic material, though metals form the major part of this type of machining. The process is free of heat and therefore parts chemically machined are free of stresses that may develop due to heat. Chemical machining is not limited to any size of parts. Parts of small and large sizes can be processed in the same way. Figure 15-21 shows typical parts produced by chemical machining. The process is more suitable

for machining such parts as nameplates, instrument panels, printed circuits, instrument boards, aircraft-wing skins, and so on. Some of the limitations of chemical machining are that machining of corners is difficult to control and inside corners are machined

Fig. 15-20 Typical conveyorized chemical machining unit. (*Chemcut, Division of Centre Circuits, Inc.*)

Fig. 15-21 Typical parts produced by chemical machining. (*Chemcut, Division of Centre Circuits, Inc.*)

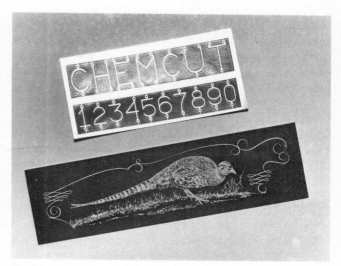

Fig. 15-22 Typical parts produced by chemical-engraving process. (*Chemcut, Division of Centre Circuits, Inc.*)

with a spherical shape while outside corners remain sharp. Surface finish in some materials is difficult to control, and holes and deep and narrow cuts are difficult to machine or control.

Chemical Engraving

Chemical engraving is another variation of the basic chemical material-removal process. It differs from chemical machining only in the depth of cut. In chemical engraving the primary purpose is to engrave an image on the work surface, such as a nameplate, shown in Fig. 15-22. Chemical engraving usually machines the part to a certain depth, but seldom are very deep cuts made. Engraving can be accomplished either with recessed or raised images. In recessed images the image is machined, while in raised images only the background of the image is machined.

Chemical Blanking

Chemical blanking is the third variation of the basic chemical material-removal process. Unlike chemical engraving, where parts are engraved into the work surface, in chemical blanking the parts are machined through (blanked). Chemical blanking is based on the same material and procedures employed by chemical machining and engraving. The only difference is that the parts are machined (blanked) through the base material. Figure 15-23 shows typical parts produced

by chemical blanking; both sides of the stock must be protected for preferential removal of material.

ELECTRODISCHARGE METAL–REMOVAL PROCESS

The electrodischarge metal-removal process (EDM) is the most widely used of the new and special processes. The phenomenon of EDM has been carefully studied and possibly is better understood than any of the other processes. EDM is based on the electrical-discharge principle. When a pulsating dc source (current) is connected to two conductors (electrode and work) and its potential (voltage) is impressed for a sufficient period (on-time) across a certain gap formed between the electrode and the work, an avalanche of electrons in the form of a spark crosses the gap, vaporizing or melting a small amount of metal

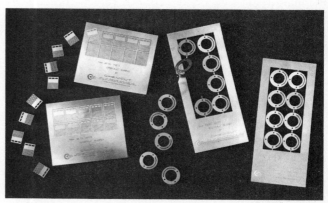

Fig. 15-23 Typical parts produced by chemical-blanking process. (*Chemcut, Division of Centre Circuits, Inc.*)

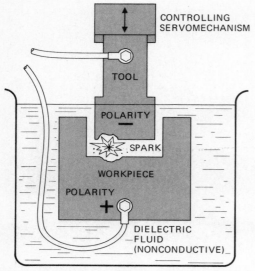

Fig. 15-24 Schematic showing a typical EDM machining system. (*Anocut Engineering Co.*)

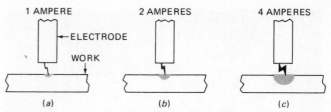

Fig. 15-25 Relationship of metal-removal rate and amount of electric energy used.

on both the electrode (tool) and the work. If this electrical action is controlled in a dielectric fluid, the pulse energy can be concentrated on a small area and will effectively remove metal (Fig. 15-24) by creating a crater in the work. Thus EDM is a complex thermoelectric process in which the metal-removal rate depends on the characteristics of the pulse energy, the gap, the characteristics of the electrode/work combination, and the characteristics of the dielectric fluid. EDM is used to machine only metals and alloys and cannot be used with nonmetallic materials.

The dc pulsation source is used to charge the capacitors which are responsible for discharging the avalanche of electrons to form the sparks. Individual sparks occur at frequencies of up to 10,000 Hz (hertz or cycles per second) and current intensities of over 10^6 amperes per square inch (A/in^2). The volume of metal removal depends on the pulse energy as shown in Fig. 15-25. The metal-removal rate is directly related to the electric energy used. The surface finish is related to the discharge frequency as shown in Fig.

15-26. There is also a relationship between the discharge frequency (sparks per second) and the surface finish in EDM.

The dielectric fluid is an important variable. The main function of the dielectric is to concentrate the pulse energy of each spark on the closest points between the electrode (tool) and the work. However, the dielectric has three other functions: (1) it flushes or removes the chips, (2) it cools the work, and (3) it insulates the electrode and the work. Thus the dielectric affects the electrode wear, metal-removal rate, and the heat-affected zone of the work. Dielectric fluids used today are mineral and silicone oils, kerosene, ethylene glycol, polar fluids, glycerol, water, and sodium silicate solutions.

Unlike chemical machining, where no heat is produced, EDM is a thermal process and, therefore, the machined surface is affected by the heat; the surface material is actually melted. The heat effect on the machined surface is more pronounced at high metal-removal rates, and it can be reduced by low removal rates and by using electrodes which produce stable machining conditions. The machined surface resulting from EDM consists of two layers at the heat affected zone: the outer layer, which is relatively hard, and the inner layer, which is less affected by heat and is not as hard.

There are several types of power-supply circuits used in EDM machine tools. Each of the circuits has advantages and limitations, depending on factors considered. Among the most commonly used types of

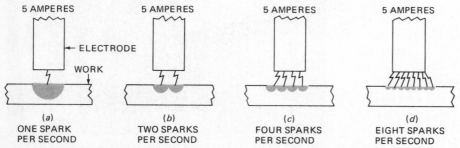

Fig. 15-26 Relationship of discharge frequency (sparks per second) and surface finish.

Fig. 15-27 A large-size-production EDM-machine tool with a traveling bolster which allows easy loading and unloading of parts with an overhead crane. (*The Cincinnati Milling Machine Co.*)

Fig. 15-28 EDM-machine tool supplied with a solid-state power supply. Inset shows a typical setup of the electrode and the work mounted in position for EDM. (*The Cincinnati Milling Machine Co.*)

circuits used are the basic, resistance-capacitor, rotary impulse generator, controlled pulse-vacuum tube, and controlled pulse-transistor.

EDM machine tools are available in a variety of sizes and designs, depending on need. In selecting an EDM machine tool several factors must be considered. Among those are the production rate and dimensional tolerances required, the size of the work and the size of the electrode required. Figure 15-27 shows a large-production-type machine tool and Fig. 15-28 shows a precision toolroom machine. Most machine tools consist of a table to mount the work, a tank to hold the dielectric fluid, a ram with a plate to mount the electrode (tool), a power supply system to provide the required pulsating energy, a dielectric fluid system, and various controls to position and feed the electrode.

The electrode material is a critical variable in EDM, and therefore its selection must be carefully considered. Graphite has been one of the most commonly used electrode materials. The reason for using graphite is that it does not melt as metallic electrodes do,

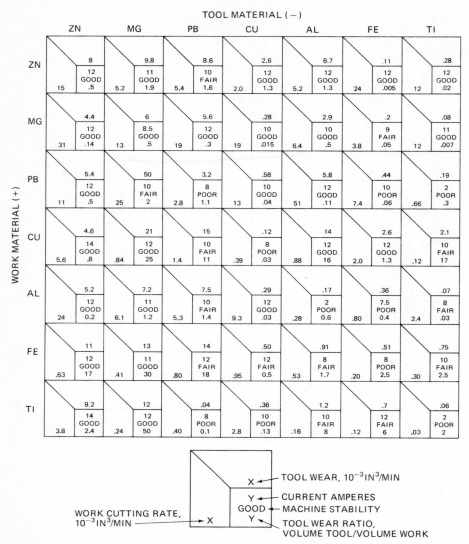

Fig. 15-29 Selection chart for electrodischarge machining of various electrode-work material combinations. (*The Cincinnati Milling Machine Co.*)

Chart: TOOL MATERIAL (−) across top, WORK MATERIAL (+) down side.
Cell legend: X = Tool wear, 10^{-3} in³/min (top); Y = Current amperes; GOOD = Machine stability; Y = Tool wear ratio, volume tool/volume work (bottom); X (left) = Work cutting rate, 10^{-3} in³/min.

Each cell listed as: [work cutting rate] ; [tool wear] / [amperes] [stability] / [tool wear ratio]

WORK \ TOOL	ZN	MG	PB	CU	AL	FE	TI
ZN	15 ; 8 / 12 GOOD .5	5.2 ; 9.8 / 11 GOOD 1.9	5.4 ; 8.6 / 10 FAIR 1.6	2.0 ; 2.6 / 12 GOOD 1.3	5.2 ; 6.7 / 12 GOOD 1.3	24 ; .11 / 12 GOOD .005	12 ; .28 / 12 GOOD .02
MG	31 ; 4.4 / 12 GOOD .14	13 ; 6 / 8.5 GOOD .5	19 ; 5.6 / 12 GOOD .3	6.4 ; .28 / 10 GOOD .015	3.8 ; 2.9 / 10 GOOD .5	12 ; .2 / 9 FAIR .05	12 ; .08 / 11 GOOD .007
PB	11 ; 5.4 / 12 GOOD .5	25 ; 50 / 10 FAIR 2	2.8 ; 3.2 / 8 POOR 1.1	13 ; .58 / 10 GOOD .04	51 ; 5.8 / 12 GOOD .11	7.4 ; .44 / 10 POOR .06	.66 ; .19 / 2 POOR .3
CU	5.6 ; 4.6 / 14 GOOD .8	.84 ; 21 / 12 GOOD 25	1.4 ; 15 / 10 FAIR 11	.39 ; .12 / 8 POOR .03	.88 ; 14 / 12 GOOD 16	2.0 ; 2.6 / 12 GOOD 1.3	.12 ; 2.1 / 10 FAIR 17
AL	24 ; 5.2 / 12 GOOD 0.2	6.1 ; 7.2 / 11 GOOD 1.2	5.3 ; 7.5 / 10 FAIR 1.4	9.3 ; .29 / 12 GOOD .03	.28 ; .17 / 2 POOR 0.6	.80 ; .36 / 7.5 POOR 0.4	2.4 ; .07 / 8 FAIR .03
FE	.63 ; 11 / 12 GOOD 17	.41 ; 13 / 11 GOOD 30	.80 ; 14 / 12 FAIR 18	.95 ; .50 / 12 FAIR 0.5	.53 ; .91 / 8 FAIR 1.7	.20 ; .51 / 8 POOR 2.5	.30 ; .75 / 10 FAIR 2.5
TI	3.8 ; 9.2 / 14 GOOD 2.4	.24 ; 12 / 12 GOOD 50	.40 ; .04 / 8 POOR 0.1	2.8 ; .36 / 10 POOR .13	.16 ; 1.2 / 10 FAIR 8	.12 ; .7 / 12 FAIR 6	.03 ; .06 / 2 POOR 2

but it goes directly into the vapor phase. Other materials such as copper, aluminum, manganese, and zinc are being used. Figure 15-29 shows the operating and performance characteristics of selected electrode-work material combinations for optimum machining results. Electrodes are produced by using any of the conventional manufacturing processes. Since the electrode is impressed into the work, it is made of opposite configuration than the required machined surface. Figure 15-30 shows a typical EDM electrode and the part (die) produced with that electrode.

EDM is used to produce both tools and parts, with the production of tools outweighing the production of parts, particularly precision molds for the plastics industry (Fig. 15-31). Some of the applications of EDM are piercing and machining odd shapes and

Fig. 15-30 An intricate part (die) machined with EDM and the carbon electrode used in this operation. (*The Cincinnati Milling Machine Co.*)

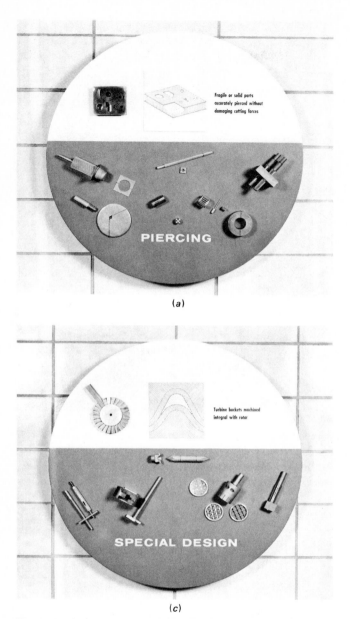

(a)

(c)

Fig. 15-31 Some applications of EDM showing parts produced and the electrodes used to produce them. (*The Cincinnati Milling Machine Co.*)

parts of special designs (Fig. 15-32). EDM is suited for production of parts with odd shapes which would be either impossible or economically prohibitive to produce by other processes. It is also suited to produce fragile parts, made from such materials as honeycomb laminates, which cannot be produced by conventional processes. However, EDM cannot be used to produce parts or tools from nonmetallic materials such as plastics and ceramics.

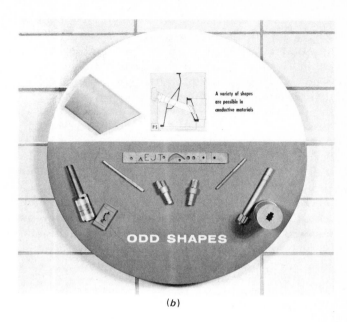

(b)

MICROMACHINING AND RELATED PROCESSES

Several unique material-removal processes have been developed in the last few years. Some of them have been used on a limited scale industrially while others are still in an experimental stage. As a group, these processes have certain unique characteristics: (1) all are primarily used in situations where no other process can do the job, (2) all are of the high-thermal-en-

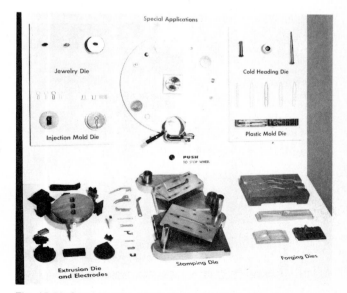

Fig. 15-32 Some typical applications of EDM in the die-making industry. All these dies can be made using the machine tool illustrated in Fig. 14-27. (*The Cincinnati Milling Machine Co.*)

ergy type, (3) the material-removal rates are relatively low, (4) their principles of operation can be explained and understood only at a subatomic level since they are based on the electron-movement theory, (5) there is no contact between the work and the "tool," (6) all these processes have some effect on the properties of the material due to heat, and (7) in addition to material removal, most of them can be used for other processes as well. Among those processes which have been used either experimentally or to a limited industrial scale and which appear to have potential for the future are (1) the *electron beam material removal* or *electron beam machining* (EBM), (2) laser beam material removal or *laser beam machining* (LBM), (3) ion beam material removal or *ion beam machining* (IBM), and (4) plasma arc material removal or *plasma arc machining* (PAM).

Electron Beam Material Removal

Electron beam material removal (EBM) is based on the principle of electron movement. Electrons generated in a suitable electron gun are accelerated by a high potential (voltage) to form a relatively narrow, round beam which is focused and directed by an electromagnetic field to bombard the work surface with high speeds (about one-half the speed of light). As the

Fig. 15-34 Setup of a nonvacuum electronbeam unit. (*Westinghouse Electric Corp.*)

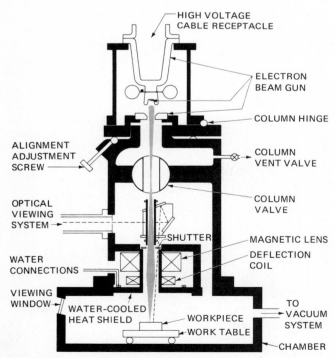

Fig. 15-33 Schematic diagram of electron-gun column with principal parts identified for electronbeam machining applications. (*Hamilton Standard.*)

high-speed stream of electrons which have mass (electron beam) bombards the work surface, its kinetic energy is transformed into thermal energy hot enough to melt or vaporize the material at the point of contact. The electron-generating gun (Fig. 15-33) consists of the cathode, the electron-emitting hot tungsten filament, and the anode. The electrons emitted by the tungsten filament are negatively charged and pass through the anode, which is positively charged. This phenomenon accelerates the electrons, providing them with the high kinetic energy to be transformed into thermal energy.

Since this emission of the electrons results in X-ray radiation, the unit must be shielded with appropriate X-ray insulating material (lead) to absorb the radiation. In addition, if handling of the work is required during processing, suitable handling equipment must be used. There are no personnel in the room during machine operation. The X-ray radiation and the special handling equipment required make electron beam material removal a very unique process (Fig. 15-34).

This process has been successfully used for drilling holes of 0.005 to 0.001 in [0.025 mm] in diameter in material of 0.010- to 0.125-in [0.254 to 3.175 mm]

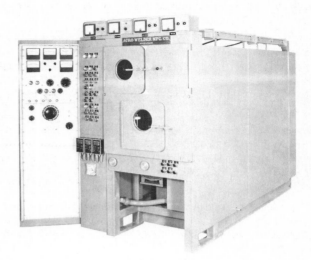

Fig. 15-37 Typical hard-vacuum electronbeam unit. (*Acro Welder Mfg. Co.*)

Fig. 15-35 Close-up of the beam of an electronbeam unit in performing a machining operation. (*Sciaky Bros.*)

Fig. 15-36 Soft-vacuum electronbeam unit. (*Sciaky Bros.*)

thick, for cutting narrow slots in relatively thin material (up to 0.005-in [0.127 mm] thick), and for milling small profile-shaped holes (Fig. 15-35). In milling and slotting operations the work remains stationary and the electron beam is programmed to move and cut the required profile. The beam can successfully cut steels, ceramics, and glass. However, with the high-thermal-energy levels provided by this process, any known material can be instantaneously vaporized or melted. Some advantages and limitations of the electron beam material removal process as outlined by the Society of Manufacturing Engineers (SME) are given in Table 15-1 (see page 250). Three types of electron beam equipment are commercially available: (1) non-vacuum equipment (Fig. 15-34), (2) soft-vacuum equipment (Fig. 15-36), and (3) hard-vacuum equipment, a nearly "perfect" vacuum (Fig. 15-37).

Laser Beam Material Removal

The high-energy output and reliability of some types of *lasers* (light amplification by the stimulated emission of radiation) resulted in the use of this powerful and flexible tool in many different industrial applications, including material removal (Fig. 15-38). The laser is basically a light source which: (1) has a high degree of directivity or collimation, (2) is highly monochromatic, having a single wavelength, and (3) has the ability to generate extremely high-peak powers. The light beam emitted by the laser is easily collimated (sharply focused) because its rays are nearly parallel to each other and diverge only slightly as they travel. Because of the small ray divergence the laser beam can be focused to an extremely high-inten-

Table 15-1 Advantages and Limitations of the Electron Beam Material-Removal Process

Advantages	*Limitations*
1. Most precise cutting tool available.	1. Only relatively small cuts are economically feasible since the material-removal rate is approximately 0.1 mg/sec or approximately 0.0001 in 3/min.
2. Cuts holes of very small size (down to 0.002 in in diameter).	
3. Cuts any known material, metal or nonmetal, that will exist in high vacuum.	2. Holes produced have a slight crater where beam enters work and also has small taper (4° included angle). Hole geometry in depth direction varies with material thickness because the beam fans out above and below its focal point. This beam divergence tends to produce an hourglass shape, especially in small, deep holes.
4. Excellent for micromachining.	
5. No cutting-tool pressure or wear.	
6. Cuts holes with high depth to diameter ratios (200:1 ratio).	
7. 0.001-in wide slots can be machined.	
8. Because of small beam diameter (0.005 in), extremely close tolerances can be held (+0.00005 to 0.0002 in). Positioning can be held to +0.0005 in or better with handling devices	3. High equipment cost.
	4. Highly skilled operator required.
	5. Usually applicable to only thin parts, 0.010 to 0.250 in range.
9. Very adaptable to automatic machining.	
10. Drills holes and end-mill slots or orifices that cannot be machined by any other process.	
11. Distortion-free machining of thin foils and hollow wall parts.	
12. Precise control of energy input over wide range.	
13. Extremely fast cutting speeds per hole, averaging 1 sec/hole. Cutting speeds, depending on material composition and thickness, range from less than 2 to more than 24 in/min. Largest part of cycle time is setup and chamber-pump downtime.	
14. No physical or metallurgical damage results. Heat-affected zone practically non-existent.	

SOURCE: SME.

sity light capable of producing temperatures to melt or evaporate any known material (even a diamond). Therefore, the output of the laser is highly directional because it is monochromatic and coherent as well. Because of this high degree of concentration, all laser beam energy can be collected with simple optics and focused on the work surface, creating the laser beam material removal process or laser beam machining (LBM).

The process is based on the emission of photons from the atom as electrons change energy levels (Fig. 15-39). An atom's orbital electrons jump to higher energy levels (orbits closer to the nucleus) by absorbing discrete or fixed quantities of stimulating energy. The extra energy may be transmitted to the atom by heating, by electrical discharge, or by radiation from other atoms which have previously acquired extra energy. The extra energy in the laser is transmitted to the atoms by radiation. When the atom's orbiting electrons change from a lower to a higher energy level, the atom is "excited" and radiates a quantum of energy. If, while the atom is in the excited state, an-

other quantum of energy is absorbed by the electron, two quanta of energy are radiated and the electron drops to a lower energy level. The energy radiated this way has the same wavelength as the stimulating energy. As a result the stimulating energy is captured, amplified, and intensified into a high-power beam. This is the principle on which the laser beam material removal process is based.

There are two basic types of laser systems used in material removal operations: the optically pumped solid-state laser or *ruby laser* and the continuous $CO_2 - N_2$ molecular laser or *gas laser*. The gas-laser system is a relatively new system and one with great promise for future material removal processes. The ruby laser is one of the oldest and most widely used. The basic system of the ruby laser for material removal is illustrated in Fig. 15-40. The lasing in this system begins when an atom is excited by the flash lamp and emits a photon (quantum of energy) parallel to the axis of the ruby. The emitted photon stimulates another atom in its path to contribute a second photon, in step, and in the same direction. This pro-

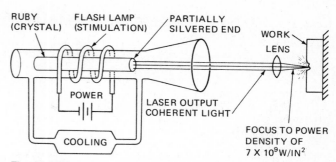

Fig. 15-40 Schematic showing a typical ruby-laser system used for material-removal processes.

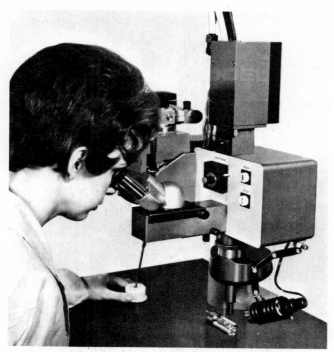

Fig. 15-38 Complete laser-beam unit. Inset shows close-up of unit in operation. (*Electronic Division, Union Carbide Corp.*)

cess continues as the photons are reflected back and forth between the ends of the ruby. The beam builds up until amplification is great enough to pass through the partially silvered mirror at the right end of the ruby, resulting in a narrow, paralleled, concentrated, coherent beam of light ready to be focused by the lens on the work surface.

Among the operations successfully performed with the laser beam material removal process are drilling holes, machine slots, and various contours. Materials successfully cut with laser beams are metallic and nonmetallic; however, it is possible to "cut" any material, vaporizing or melting it with the laser beam.

The laser beam process holds promise for the future unmatched by the other micromachining processes.

Among the advantages and limitations of the laser beam material removal process as outlined by the Society of Manufacturing Engineers (SME) are those given in Table 15-2.

Ion Beam Material Removal

Ion beam material removal or ion beam machining (IBM) is a relatively complex phenomenon and has not been entirely understood. Its further study, understanding, and industrial applications are matters of future concern. Although the process has been successfully demonstrated in the laboratory, no equipment for it has been manufactured and used industrially. The extremely low metal-removal rates limit the process to only micromachining applications which cannot be performed by other processes.

Plasma Arc Machining

Plasma arc material removal or plasma arc machining (PAM) is not considered a micromachining process; neither should it be considered entirely a new process. The plasma arc principle has been known and

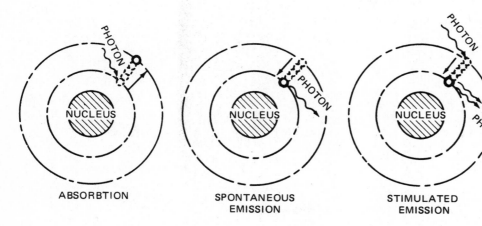

Fig. 15-39 Stimulating energy due to the movement (change) of electrons from one energy level to another. (*SME.*)

Table 15-2 **Advantages and Limitations of Laser Beam Material-Removal Process**

Advantages	*Limitations*
1. No direct contact between tool (laser) and workpiece.	1. Low overall efficiency.
2. Machining and welding through optically transparent materials.	2. Pulsed mode operation (solid state).
3. Welding and machining of areas not readily accessible.	3. Practically limited to thin-sheet plate or wire fabrication.
4. Melting or vaporizing of any known material.	4. Holes machined not always round or straight.
5. Easy welding of dissimilar materials.	5. Control of hole size and weld size difficult.
6. Refractory metals easy to work with.	6. Slow repetition rate.
7. Machining of brittle, nonmetallic, hard materials.	7. Durability and reliability limited.
8. Welding and machining in any desired atmospheric environment.	8. Short life of flash lamp.
9. Small heat-affected zones, and negligible thermal damage or effect on adjacent regions.	9. Necessity for careful control of pulse length and power intensity to obtain a desired effect.
10. Machining extremely small holes and precision welding of small sizes.	10. Effective safety procedures required.
11. Easy control of beam configuration and size of exposed area.	11. High cost.

SOURCE: SME.

used in limited industrial applications for years. Plasma oxy-fuel cutting of hard-to-cut metals is one example of the early industrial applications. However, the plasma arc metal removal is relatively new, although it is based on the oxy-fuel-cutting torch principle. *Plasma* has been defined by SME as a gas

heated to a high temperature to become partially ionized.

The plasma arc metal-removal process is relatively simple. A volume of gas is bombarded by electrons produced by an electric arc. As the electrons collide with the gas molecules, they produce dislocation and partial ionization of the gas molecules. This is accomplished in a torch as shown in Fig. 15-41. An electric arc is produced in the torch opening located behind the nozzle and is filled with compressed gas. The gas is heated by the electric arc and becomes plasma gas before it is forced through the nozzle. The heated gas with the electric arc as a plasma jet is forced through the nozzle duct of the torch on the work surface. The heat generated on the work surface by the plasma jet,

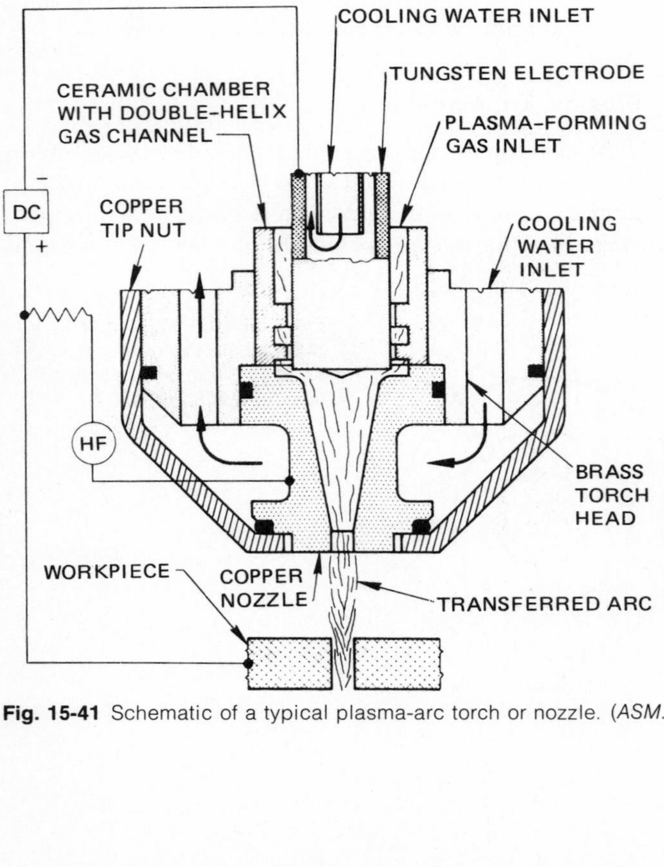

Fig. 15-41 Schematic of a typical plasma-arc torch or nozzle. (*ASM.*)

Fig. 15-42 Plasma-arc endless conveyor chain-type automatic-production unit for hard surfacing engine valve cross-heads at rate of eleven per minute. (*Acro Welder Mfg. Co.*)

due to plasma gas and electron bombardment from the electric arc, is sufficient to melt the work material, thus creating the plasma metal-removal process.

The plasma arc is considered a high-rate metal-removal process. However, it heats the work to the point of altering its metallurgical characteristics and it produces extremely rough surfaces. Among the gases successfully used in plasma arc machining are hydrogen (H_2), nitrogen (N_2), and oxygen (O_2). Plasma arc has been used with such machining operations as turning, milling, planing, and cutting (see Fig. 15-42 on page 252). However, in all those operations, plasma arc has been used as a roughing rather than finishing process. The process must be further developed and understood before it can be used for finish machining applications.

REVIEW QUESTIONS

15-1. Compare the new and special material-removal processes with the conventional machining processes in terms of cost, extent of application, and quality of work produced.

15-2. Explain why abrasive-jet machining cannot be used successfully with relatively soft materials.

15-3. Discuss the advantages and limitations of the ultrasonic material-removal process.

15-4. Discuss the factors contributing to the rapid development and expansion of the electrochemical material-removal processes.

15-5. List and explain the characteristics of the electrolyte used in ECM operations.

15-6. State and explain the factors to be considered in selecting a resist for chemical material-removal processes.

15-7. Discuss and compare chemical milling, chemical engraving, and chemical blanking.

15-8. Explain the reasons why EDM is used only for machining metallic materials.

15-9. Discuss the advantages and limitations of EDM.

15-10. State and explain the characteristics of micromachining processes.

15-11. Discuss the advantages and limitations of the electron beam material-removal process.

15-12. Explain the laser beam principle.

15-13. What are some of the advantages and limitations of the laser beam material-removal process?

15-14. Why is plasma-arc considered a roughing rather than a finishing process?

15-15. Explain the principle of the ion beam material-removal process.

CHAPTER 16

THE TECHNOLOGY OF JOINING PROCESSES

Almost every manufactured product in today's society is made up of more than one piece. A modern automobile is composed of several thousand pieces which are fastened together to make a single product. The method used to fasten or join the many parts of an automobile must, in some instances, hold the piece very rigidly, while in other instances, the piece must be held in a specific position but be free to move as needed. Many fastening methods must allow for disassembly of the product for servicing or replacement of parts and must often be capable of adjusting the position of a piece to compensate for normal wear within the product.

The many diverse fastening requirements are met by a multitude of permanent and nonpermanent joining or fastening techniques and devices. The product engineer or designer must be familiar with the characteristics, advantages, and limitations of all the fastening techniques and devices in order to select and specify the one most appropriate for a given application.

In general, joining and fastening methods can be classified into broad categories of either *mechanical fastening* or *bonding* techniques. Mechanical fastening techniques may be either permanent or nonpermanent types while bonding techniques are considered to be of the permanent type. Permanent joining or fastening methods do not allow for adjusting, removing, or separating the parts once they are fastened together, such as the frame of an automobile.

Bonding processes consist of two categories, *cohesion* and *adhesion*. A common example of cohesion is *welding*. The cohesion processes feature the fusing of two or more pieces into a continuous, or monolithic, single piece. Cohesion is widely used in joining industrial products, particularly metals, although plastics are also extensively welded. Advantages of welding include speed, efficiency, and the adaptability to automated assembly processes.

The most frequently automated welding processes for metal products are the variations of resistance welding such as spot welding and seam welding. The ultrasonic process is the fastest welding process for plastics. The process causes a metal stylus to vibrate rapidly (20,000 Hz) and to create heat which causes the plastics to soften and fuse. Taillight reflectors, fishing corks, and toys are typical applications.

The welding of plastics is done by both heat and pressure processes. The pressure systems employed are much like spot or seam weldings except that electrical current is not used to generate heat within the material. However, plastic is also welded by the applications of solvents much like gluing. In solvent bonding, solvent is applied to the plastics to literally dissolve the plastics. Then when the two or more pieces of the softened plastics are pressed together, complete fusion occurs.

The following two chapters will deal with the cohesion processes in some detail; the emphasis of this chapter will be upon mechanical fasteners and adhesive bonding.

Adhesion differs from cohesion because a substance that is entirely different from the joined pieces is used to bond, or glue, the pieces together. Adhesion includes the use of glues and adhesives and soldering and brazing. Brazing and soldering (which are actually the same thing but at different temperatures) depend upon the adhesion of a material such as brass or copper to other metals such as steel by capillary action, just as with glue and other adhesives.

Mechanical fastening is also composed of two broad groups, joints and fasteners. Joints feature the mechanical configuration of the pieces so that the parts are supported or retained by physical interference. Included are *seaming* and *joinery*.

The mechanical fasteners group includes all types of screws, pins, nails, staples, rivets, spring clips, and other such mechanical devices which join two or more parts by the addition of a separate piece which serves as the binding element.

MECHANICAL FASTENING

The broad group of mechanical fastening includes all methods of using mechanical devices to hold two or more parts together. As a broad group, the advantages are the adaptability and versatility of mechanical fasteners in terms of size, shape, strength, cost, and the ease of assembly using both common and specialized tools. An additional factor can include the ease of service or repair by utilizing certain types of fasteners.

Threaded Fasteners

The category of threaded fasteners consists of any mechanical device utilizing the principle of the inclined circular plane to apply pressure. This group includes many different varieties of fasteners such as screws for wood, plastics, and sheet metal, as well as bolts, machine screws, and nuts.

Bolts and Machine Screws. Bolts and machine screws are very similar in shape and use. They consist of a threaded shaft, usually straight and without a point, and a head. The threads are regular and designed to match a threaded hole in either the body of the piece or a special piece called a *nut*. Figure 16-1 top shows some of the most common types of bolts and machine screws while Fig. 16-1 bottom shows the variety and range of size and shapes that may be used.

Bolts and screws are most commonly made of steel, although almost any material can be used when desired. Both rough and finished threads are now made by rolling or forging. These types of threads make up perhaps the bulk of the common steel "bolts" now used. Machined threads are cut from bar stock on specialized screw-cutting machines.

Sizes are determined by the diameter and the length. Flat head-screw length includes the head, but the length of all other shapes includes only the length of the body (see Fig. 16-2). Generally machine screws are smaller than $1/4$ in in diameter and are identified by a gauge number rather than actual diameter. Bolts, on the other hand, are generally larger than $1/4$ in [0.64 cm] in diameter and are identified by the actual diameter in fractions, decimals, or metric measures. Another general distinction between bolts and screws is whether or not the piece can be reached from only one side or from both sides with wrenches or screw drivers. Screws can be reached from only one side, while bolts can be reached from two. However, there are several exceptions to both definitions.

(a)

(b)

Fig. 16-1 Threaded fasteners differ in size and shape.

Bolt and Screw Variations. There are several variations that are distinctive. These include carriage bolts, lag screws, stud bolts, thumb screws, stove bolts, cap screws, and tap bolts. The standard machine bolt has a thin hexagonal head. The thread is only formed on about 40 percent of the body.

A *carriage bolt,* as shown in Fig. 16-3, has a rounded head that cannot be tightened normally. You will note that below the head is a square section. Carriage bolts are made for fastening wood to metal. A hole is drilled in both pieces and the bolt driven through from the wood side so that the round head is flush. However, the square underhead is wedged into the round hole in the wood and holds the bolt so that the nut can be tightened from the other side.

The *lag screw* is yet another variation. As in Fig.

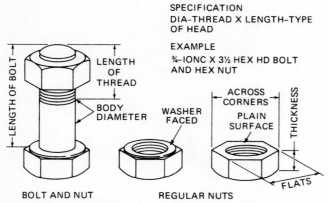

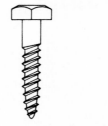

Fig. 16-2 Specifications of a bolt and nut assembly.

Fig. 16-4 A lag screw.

16-4, the thread is tapered and shaped like a wood screw. It is used for fastening heavy pieces to wooden surfaces, such as the base of a machine to a wooden floor. Lag screws can also be used to fasten heavy objects to concrete by using lead inserts that expand tightly against the sides of a drilled hole as the screw is tightened. These are called *lag shields.*

Stud bolts, as shown in Fig. 16-5, have no heads and are threaded on both ends. One end has more threads than the other. This type bolt is used in machine assembly where small variances in distance might make the use of machine bolts impractical. As shown, it is also easier to align and assemble a top piece where the studs are projecting than to try to align holes and bolts which are not visible.

Tap bolts resemble machine bolts except that the en-

tire length of the body is threaded. *Stove bolts* are small bolts, usually $^3/_{16}$ or $^1/_4$ in [0.48 or 0.64 cm] in diameter, with round or flat heads. They are made with a coarse thread. *Cap screws* are identical to machine bolts except that they have shaped heads (such as the fillister or socket) and were originally designed for use where the head was to be recessed.

Thumb screws are small bolts with flared, wing-shaped heads that are used where frequent adjustments are needed. Generally the forces involved are small, allowing the screws to be set with the hands. Figure 16-6 shows two types of thumb screws.

Nuts. In using machine threads, a matching thread must be used. There are various ways of doing this. First, the hole into which the bolt is inserted has been drilled and tapped (internally threaded). In the second, two pieces may be held together tightly by inserting a bolt completely through holes in both pieces and applying the force by using a threaded piece on the end of the bolt. This threaded piece is called a *nut.* There are several different variations in the use of

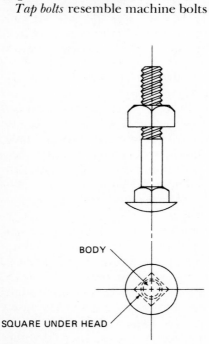

Fig. 16-3 A carriage bolt.

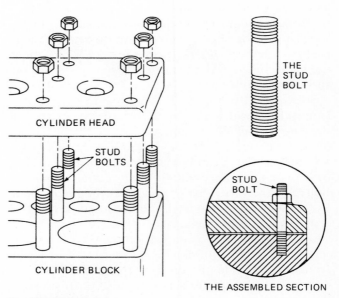

Fig. 16-5 Stud bolts.

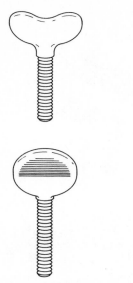

Fig. 16-6 Thumb screws.

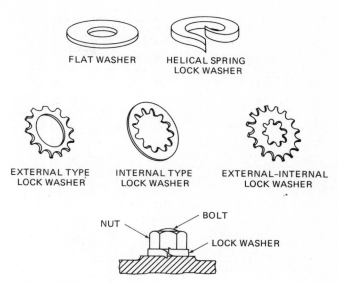

Fig. 16-8 Types of washers.

nuts which will be explained shortly. The third way is the use of a nut variation where small sheet-metal stampings are used. The edges of the hole of the sheet-metal stampings are offset slightly to fit into the helix of the screw thread. These pieces are often referred to by workers as *burrs* and are classed as single-thread fasteners. Several different variations are commonly employed in assembly operations.

Common Nut Types. A large variety of nuts are available for use. The most common are shown in Fig. 16-7 and include the square machine nut, the hexagonal nut, a thinner nut known as a jam nut, a "wing" nut that can be tightened with fingers, and a special nut called a *castellated nut* that has slots cut in it for the insertion of retaining pins.

When nuts and bolts are subjected to vibration which could cause them to loosen, a washer may be used to keep the nut in place. Types of washers are

illustrated in Fig. 16-8. In addition to reducing the loosening of a fastener from vibration, washers help distribute the pressures evenly and help provide a better thrusting surface.

Plain washers are flat and should slip easily over the threads. They are available in a wide range of sizes and several thicknesses and widths for special applications.

Lock washers are used to further reduce the chance of loosening due to shock, vibration, or movement from temperature changes. Lock washers may be of the helical spring type or the toothed shape. In either case, the sharp edges cut into the surface to provide a nonslip action and also act as springs to keep pressure between the fastener and the piece.

There are numerous other types of washers designed for special uses.

Floating Nuts. In rapid, mass-assembly operations using threaded fasteners, a great deal of time can be spent in aligning the fastener with the threaded hole. To minimize time losses, special nut variations are used. Because these variations are made so that they can move, laterally, radially, or both, they are called *floating* nuts. When the threaded fastener is positioned, the nut can "float" to align itself easily. This allows greater locational-fit tolerances and reduces "thread matching" time.

Several types of special nuts are made, as shown in Fig. 16-9. The nut shown in the figure has three advantages in assembly work: (1) it floats and can ease alignment, (2) it is pressed into place in a drilled hole, eliminating expensive machining, welding, or double-side wrench assembly, and (3) it features precision cut-threads that provide maximal strengths. Further,

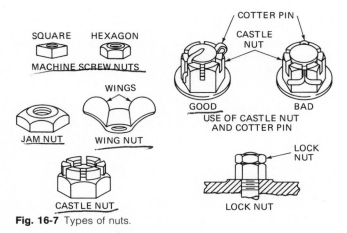

Fig. 16-7 Types of nuts.

Fig. 16-9 Floating nuts minimize alignment in assembly.

such systems are suited for repeated fastening and unfastening.

Clip Fasteners. Clip fasteners, as shown in Fig. 16-10a usually feature floating action, are very economical as they are made from stamped sheet metal rather than machine threaded pieces, and are easily and quickly installed. They can be installed by hand as in Fig. 16-10b or by machine as in Fig. 16-10c.

Captive Nuts. Captive nuts are those that are fastened permanently to a section of the assembly. They reduce assembly time because no special hole threading is required and strong steel threads can be inserted into softer aluminum, brass, or plastic materials. Captive nuts can increase the "pull-out" and torque strength many times more than a tapped hole in a soft material. The nut, as shown in Fig. 16-11, is pressed into a previously drilled hole, and cold flow of the softer material fills the coaxial groove which provides "pull-out" strength. Torque resistance is provided by the hexagonal shape of the face. Captive nuts can also be obtained with floating capacity, as previously illustrated in Figure 16-9.

Quick Releases. Quick releases are threaded systems featuring long screw-pitch systems so that a quarter or half turn will disengage the thread. The quick-release feature is ideal for the assembly of inspection panels, service outlets, and other areas where frequent opening and closing are expected. Modifications can be made to the heads of the screws so that common tools will not open the unit. This allows greater security and reduces tampering or mischief.

The ISO system specifies all the diameters in metric units for uniform use in all nations. The old pipe-thread system, however, has been adopted by the ISO.

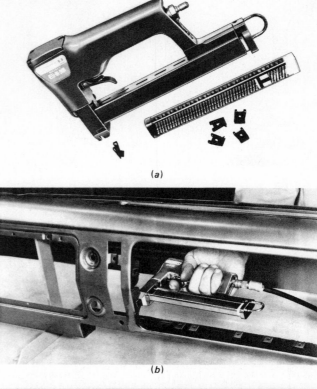

Fig. 16-10 (a) Gun and magazine loaded with clip-type fasteners. The actuated gun automatically applies prepackaged fasteners to panels as fast as operator can locate and trigger the gun. (b) Worker applies fasteners to an automobile dashboard. Fasteners can be applied as close as 1-in centers. (c) Developed for high-speed assembly lines which require a high number of fasteners per assembly, the air-actuated continuous-feed delivery system receives its supply of U-type fasteners from a vibratory bowl through a flexible plastic delivery tube. The bowl can be refilled without interrupting assembly operations. (*Courtesy: Eaton "Tinnerman" Fasteners.*)

Fig. 16-11 Captive nuts increase thread strength for thin or soft materials. (*Courtesy: Precision Metal Products Co.*)

The ramifications of such adaptations for international trade are obvious. The table in Fig. 16-12 compares the ISO and unified thread systems.

Sheet-Metal Self-Tapping Screws. Thread-forming sheet-metal screws are driven into drilled or punched holes in thin or soft metal. The holes are made slightly larger than the minor diameter of the screw so that the external screw threads displace metal to form their own shallow mating threads into one or more of the parts being assembled. Figure 16-13 shows the more common thread styles. Type A screws have sharp points and type B and C screws have finer threads. Using sharp-pointed screws makes alignment easier during assembly.

There are also thread-cutting types of screws in this general class. They differ in that the screws actually cut threads by chip removal rather than displacement. Some relief is needed for chip forming, so these threads have formed slots in the thread body to provide relief and to expose a cutting edge. The different types relate to point shape and the manner in which the relief and cutting edge is formed.

Metallic-Drive Screws. Metallic-drive screws, such as in Fig. 16-14, are made of hard steel, form their own threads by displacement, and are driven in much the same manner as sheet-metal screws. However,

Fig. 16-12 Screw-thread comparison.

Unified Screw Thread	Decimal Equivalent (DIN)	Unified Converted mm	Nominal Diameters ISO	Unified Screw Thread	Decimal Equivalent (DIN)	Unified Converted mm	Nominal Diameter ISO
—	0.0394		1.	$7/16$	0.4375	11.113	
—	0.043	1.1		—	0.4724		12.
—	0.047		1.2	$1/2$	0.5	12.7	
—	0.055		1.4	—	0.5512		14.
#0	0.060	1.524		$9/16$	0.5625	14.288	
—	0.063		1.6	$5/8$	0.625	15.875	
—	0.0709	1.8		—	0.6299		16.
#1	0.073	1.854		$11/16$	0.6875	17.463	
—	0.078		2.	—	0.7087		18.
#2	0.086	2.184		$3/4$	0.75	19.05	
—	0.0866	2.2		—	0.7874		20.
—	0.984		2.5	$13/16$	0.8125	20.638	
#3	0.099	2.515		—	0.8662		22.
#4	0.112	2.845		$7/8$	0.875	22.225	
—	0.1181		3.	$15/16$	0.9375	23.813	
#5	0.125	3.175		—	0.9449		24.
—	0.1378		3.5	1.	1.0	25.4	
#6	0.138	3.505		$11/16$	1.0625	26.988	
—	0.1575	4.		—	1.063		27.
#8	0.164	4.166		$11/8$	1.125	28.575	
—	0.177	4.5		—	1.181		30.
#10	0.190	4.826		$13/16$	1.1875	30.163	
—	0.1969	5.		$11/4$	1.25	31.75	
#12	0.216	5.486		—	1.2992		33.
—	0.2362		6	$15/16$	1.3125	33.338	
$1/4$	0.250	6.35		$13/8$	1.375	34.925	
—	0.2756		7	—	1.4173		36.
$5/16$	0.3125	7.938		$17/16$	1.4375	36.513	
—	0.315		8.	$11/2$	1.5	38.1	
$3/8$	0.375	9.525		—	1.5354		39.
—	0.3937		10.				

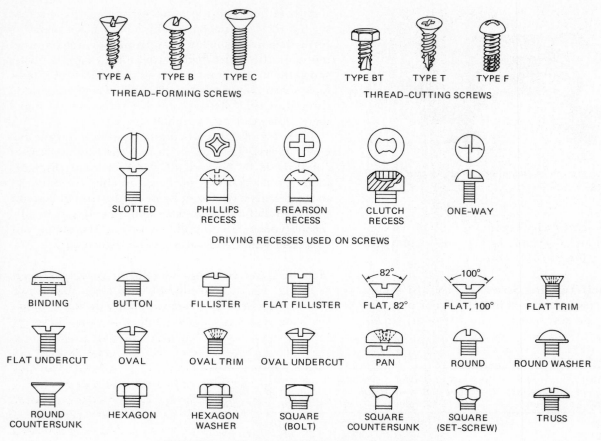

TYPE A TYPE B TYPE C
THREAD-FORMING SCREWS

TYPE BT TYPE T TYPE F
THREAD-CUTTING SCREWS

SLOTTED PHILLIPS RECESS FREARSON RECESS CLUTCH RECESS ONE-WAY
DRIVING RECESSES USED ON SCREWS

BINDING BUTTON FILLISTER FLAT FILLISTER FLAT, 82° FLAT, 100° FLAT TRIM

FLAT UNDERCUT OVAL OVAL TRIM OVAL UNDERCUT PAN ROUND ROUND WASHER

ROUND COUNTERSUNK HEXAGON HEXAGON WASHER SQUARE (BOLT) SQUARE COUNTERSUNK SQUARE (SET-SCREW) TRUSS
KINDS OF HEAD STYLES USED ON THREADED FASTENERS

Fig. 16-13 Sheet-metal screw factors.

they have fewer threads and are designed to be used on thick metal castings and plastics. They are widely used to attach labels, to attach identification and information plates to machines, in plastic toy assembly, and in attaching decorations. They are for permanent assembly and are not intended for disassembly.

Wood Screws. Wood screws have a shape similar to type A or B sheet-metal screws. Figure 16-15 shows the three major head shapes for wood screws. The sizes are described by gauge size in which a smaller number indicates a screw with a smaller diameter. For example, a number 10 wood screw is much larger in

diameter than a number 4 wood screw. The gauge number of the screw refers to the major diameter of the screw and generally conforms to the American standards for wire gauges. Wood screws are extensively used in woodworking industries. However, they are usually unsuited for metals or plastics because they are not hardened to form their own internal threads in these materials during driving.

Staples

Staples are also extensively used in industrial assembly operations. They can be applied in wood, paper, metal, plastics, cloth, and a variety of other materials. Staples are frequently applied in metal-working operations, such as the assembly of home appliances in-

Fig. 16-14 Metallic-drive screws.

OVAL HEAD ROUND HEAD FLAT HEAD

Fig. 16-15 Typical wood screws.

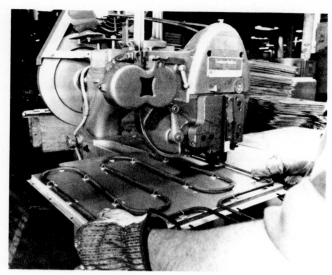

Fig. 16-16 Staples are used to "stitch" the heating element to an oven panel. (*Courtesy: Interlake, Inc.*)

cluding such appliances as ovens. Staples are driven with special machines rather than by hand in such cases. Metal stapling is also called metal *stitching.* Figure 16-16 shows a typical stitching operation.

Rivets

Rivets are metal pins much like bolts except that they have neither threads nor heads shaped for driving. Rivets are generally made of soft, malleable metals so that they may be easily worked. They are inserted into drilled or punched holes; the body is then expanded to fit tightly against the side of the hole, and the protruding end is expanded out to form a head or small flange. The result is a tight pin with heads at both ends.

Although welding has supplanted riveting in many cases, rivets are still used extensively in aircraft, hard-to-weld metals, and many other metal-fastening operations. Rivets also are excellent for fastening materials such as plastic, cloth, and leather to wood or metal parts. Special rivets and riveting machines maintain riveting as a viable, fast, strong, and economical fastening method for a widespread variety of applications.

When riveting most types of metal shapes, common rivets, as in Fig. 16-17, are appropriate. For some structural uses, rivets should be installed hot. Because metal shrinks when it cools, the hot rivets will hold tighter when they cool. Rivets used on bridges, tall buildings, and similar structures are set while hot. Common rivets are described by body diameter and by length.

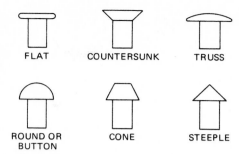

Fig. 16-17 Shapes of common rivets.

Rivets used for thin materials should have wider heads than the common rivets. The tinners' rivet, shown in Fig. 16-18, is used for such applications. Tinners' rivets are sized by the weight of 1000 rivets, in pounds.

The spacing and length of rivets is important in designing applications for most riveting situations. Generally, rivets should be no closer together than 3 times the rivet diameter, nor any farther apart than 24 times the diameter. The rivet should protrude approximately $1\frac{1}{2}$ times its own diameter in order to provide enough material to form the head. The hole itself should normally be located a minimum of three diameters from the edge or corner of the material being joined. In assembling with rivets, generally, a hole slightly larger than the rivet diameter is drilled or punched into the material. The reason for a slightly larger hole is that a $\frac{1}{4}$-in rivet would barely fit into a $\frac{1}{4}$-in hole. However, when allowances are made for mass-produced items in which the nominal size may be either slightly larger or slightly less in terms of either the hole or the rivet, the matching of larger rivets with smaller holes would result in approximately 50 percent of the rivets unable to fit into the hole. Therefore, a slightly larger hole will offset these problems.

Tubular and special rivets, as shown in Fig. 16-19, are used as eyelets, have special appliance uses, and are used for many types of tools and hardware items. Tool handles, knife handles, spring attachments,

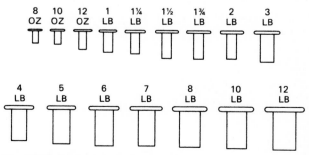

Fig. 16-18 Tinners' rivets (actual size).

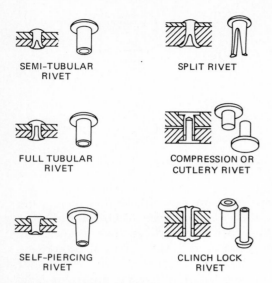

Fig. 16-19 Special and tubular rivets.

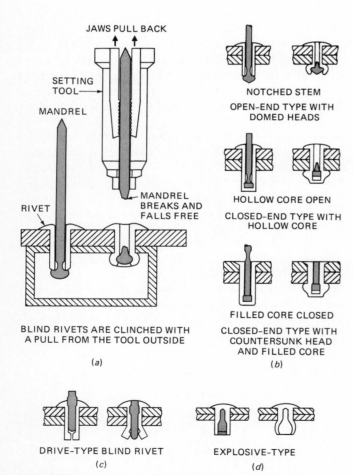

BLIND RIVETS ARE CLINCHED WITH
A PULL FROM THE TOOL OUTSIDE

(a)

DRIVE-TYPE BLIND RIVET

(c)

EXPLOSIVE-TYPE

(d)

Fig. 16-20 (a) The "mandrel" type blind rivet. Blind rivets are clinched with a pull from the tools outside. (b) Shapes of mandrel-type blind rivets. (c) Blind rivet. (d) Explosive type.

notebooks, and identification tags on machines and cabinets are typical examples of "special" rivets.

Most rivets are inserted from one side and must be clinched or set from the back side, thus requiring access to both sides. However, a special class of rivets, known as *blind* rivets, is used so that all operations are done from one side. The implications for aircraft assembly alone are obvious. Blind rivets, as in Fig. 16-20a, b, c, and d, require special tools for installation.

Rivets may also be combined with other shapes, such as threaded nuts, eyes, hooks, and so forth.

Retainers

Almost any device designed primarily to hold something in position can be classed as a retainer. A retainer could align a pulley on a rotating shaft, keep a shaft from turning inside a gear or pulley, keep a lever from sliding out of place, or even keep a group of wires together while they are being installed in an automobile. Let us look at some devices in this fascinating group: pins, keys, retaining rings, push nuts, snap bushings, staking, and wire and tubing retainers.

Pins. Most pins consist of a shaft inserted through a shaft. The pin may also go through a gear as in Fig. 16-21 to hold the gear in place. Such pins may be either straight or tapered. A straight pin is often made from a rolled sheet so that it expands to grip the sides of the hole more firmly. Sometimes pins are designed to shear under stress so that the forces involved will not break the more expensive parts such as the gear or shaft. Once the load is reduced, the pin can be replaced and the operation can continue. Such pins are called *shear pins.*

Pins can also be used to keep castellated nuts from loosening, parts from slipping off the ends of shafts, and so on. These pins, called *cotter pins,* are shown in

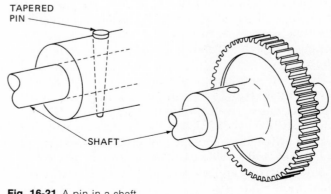

Fig. 16-21 A pin in a shaft.

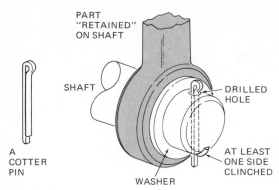

Fig. 16-22 Cotter-pin use.

Fig. 16-22, and are made from soft steel. Once inserted into a drilled hole, they are clinched by bending at least one "leg."

Keys. Keys, as shown in Fig. 16-23, are used to keep gears, pulleys, and other rotating parts in proper positions on shafts. The key fits in a *keyway* on the shaft

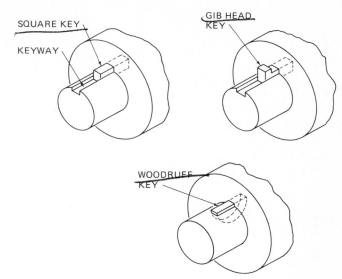

Fig. 16-23 Types of keys.

	BASIC **N5000** For housings and bores			BOWED **5101** For shafts and pins			REINFORCED **5115** For shafts and pins			TRIANGULAR NUT **5300** For threaded parts	
INTERNAL	Size Range	.250—10.0 in.	EXTERNAL	Size Range	.188—1.750 in.	EXTERNAL	Size Range	.094—1.0 in.	EXTERNAL	Size Range	6-32 and 8-32 10-24 and 10-32
		6.4—254.0 mm.			4.8—44.4 mm.			●			1/4-20 and 1/4-28
	BOWED **N5001** For housings and bores			BEVELED **5102** For shafts and pins			BOWED E-RING **5131** For shafts and pins			KLIPRING **5304** T-5304 For shafts and pins	
INTERNAL	Size Range	.250—1.750 in.	EXTERNAL	Size Range	1.0—10.0 in.	EXTERNAL	Size Range	.110—1.375 in.	EXTERNAL	Size Range	.156—1.000 in.
		6.4—44.4 mm.			25.4—254.0 mm.			2.8—34.9 mm.			4.0—25.4 mm.
	BEVELED **N5002** For housings and bores			CRESCENT® **5103** For shafts and pins			E-RING **5133** For shafts and pins			TRIANGULAR **5305** For shafts and pins	
INTERNAL	Size Range	1.0—10.0 in.	EXTERNAL	Size Range	.125—2.0 in.	EXTERNAL	Size Range	.040—1.375 in.	EXTERNAL	Size Range	.062—.438 in.
		25.4—254.0 mm.			3.2—50.8 mm.			1.0—34.9 mm.			
	CIRCULAR **5005** For housings and bores			CIRCULAR **5105** For shafts and pins			PRONG-LOCK® **5139** For shafts and pins			GRIPRING® **5555** For shafts and pins	
INTERNAL	Size Range	.312—2.0 in.	EXTERNAL	Size Range	.094—1.0 in.	EXTERNAL	Size Range	.092—.438 in.	EXTERNAL	Size Range	.079—.750 in.
		●			●			●			2.0—19.0 mm.
	INVERTED **5008** For housings and bores			INTERLOCKING **5107** For shafts and pins			REINFORCED E-RING **5144** For shafts and pins			HIGH-STRENGTH **5560** For shafts and pins	
INTERNAL	Size Range	.750—4.0 in.	EXTERNAL	Size Range	.469—3.375 in.	EXTERNAL	Size Range	.094—.562 in.	EXTERNAL	Size Range	.101—.328 in.
		19.0—101.6 mm.			11.9—85.7 mm.			2.4—14.3 mm.			●
	BASIC **5100** For shafts and pins			INVERTED **5108** For shafts and pins			HEAVY-DUTY **5160** For shafts and pins			PERMANENT SHOULDER **5590** For shafts and pins	
EXTERNAL	Size Range	.125—10.0 in.	EXTERNAL	Size Range	.500—4.0 in.	EXTERNAL	Size Range	.394—2.0 in.	EXTERNAL	Size Range	.250—.750
		3.2—254.0 mm.			12.7—101.6 mm.			10.0—50.8 mm.			6.4—19.0 mm.

Fig. 16-24 Types of retaining rings. (*Waldes-Kohinoor, Inc.*)

and into a slot in the part. The key helps transmit the power so that the part and shaft do not rotate independently. The square key (also called the *feather*) is the most common type.

Retaining Rings. Retaining rings are relatively recent in industrial applications. Originally, retaining rings were simply springlike circular fasteners inserted into a machined groove or slot. They were characterized by low strength and inexpensive applications. However, careful engineering and the use of alloys and tempered steels has developed a new look for the retaining ring. Now, such simple devices as those shown in Fig. 16-24 can provide strength greater than that of the regular machined material. One of the main advantages of using retaining rings, as discussed in the previous chapter, is that it reduces the complexity of both the forming and assembly operations. This can provide savings in material, labor, and time, and in ease of service as indicated in Fig. 16-25a and b.

Retaining rings can be installed manually, as shown in Fig.16-26, or may be installed with special tools as in Fig. 16-27, or in production situations as shown in Fig. 16-28 and 16-29.

Retaining rings may be installed internally within a drilled or bored hole, or in the more common external applications. Retaining rings have shown considerable versatility and adaptability in replacing other types of fasteners.

Push Nuts. Push nuts are the same as self-locking, external retaining rings and feature similar aspects in application and in use. Push nuts were originally a product of comparatively low strength and limited applications but have been carefully engineered to provide great strength and considerable breadth of applications. The principle of operation is simply that the hole in the push nut is actually smaller than the shaft to which it is attached. However, the interior portions of the hole in the push nut are bent slightly. This allows the push nut to be jammed forcefully onto the end of the shaft. However, if force is exerted in the opposite direction, the angle of the small pieces around the shaft is such that leverage is exerted to make these elements grip more tightly. The push nut cannot be used where access is desired, as the nut must be destroyed in order to be removed. Push nuts may be simple, flat sheet-metal stampings or may be capped pieces for use on the ends of rotating shafts that are exposed.

Snap Bushings. Snap bushings serve several functions in industrial applications. They can be used as

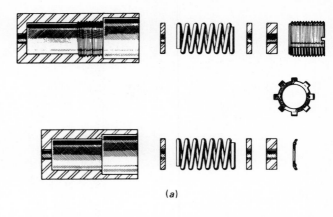

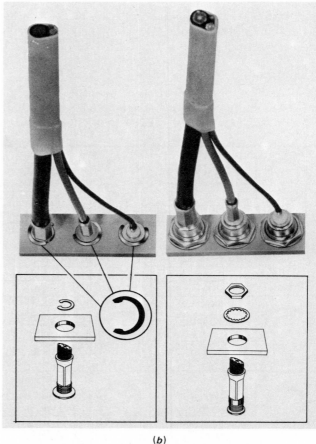

Fig. 16-25 (a) Threaded internal fasteners are costly because of expensive internal threading operation. Simplify by substituting a self-locking retaining ring. (b) Use of retaining rings can reduce assembly complexity. This change resulted in a 25 percent savings. (© 1971 and reprinted with permission of Waldes-Kohinoor, Inc.)

true bushings, that is, as bearing surfaces for rotating shafts, or they can be used for alignment and centering purposes for nonrotating shafts. The snap bushing involves the use of a hollow tube pushed through a formed hole in the workpiece and retained

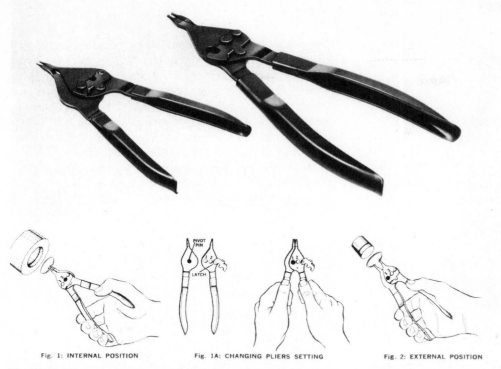

Fig. 1: INTERNAL POSITION   Fig. 1A: CHANGING PLIERS SETTING   Fig. 2: EXTERNAL POSITION

Fig. 16-26 Manual installation of either interior or exterior retaining rings. (© 1973 and reprinted with permission of Waldes-Kohinoor, Inc.)

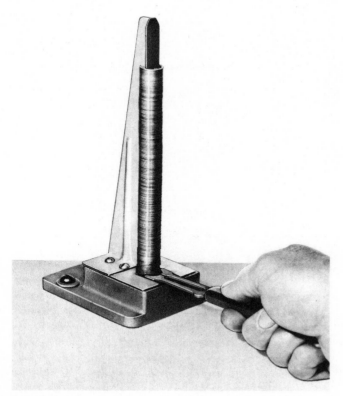

Fig. 16-27 A simple device to speed manual installation of retaining rings can result in unexpected savings. (© 1962 and reprinted with permission of Waldes-Kohinoor, Inc.)

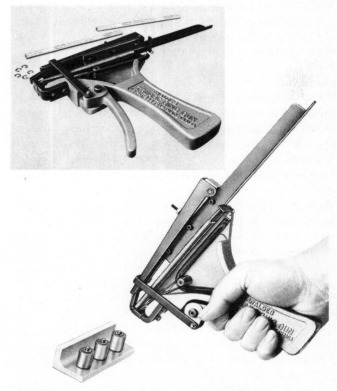

Fig. 16-28 Easy loading semiautomatic "guns" for fast, flexible installation. (© 1966 and reprinted with permission of Waldes-Kohinoor, Inc.)

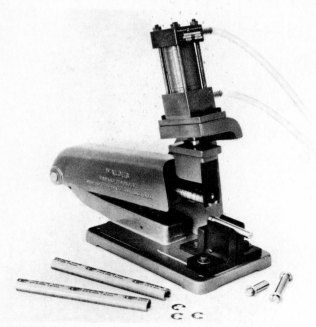

Fig. 16-29 A fully automatic retaining-ring installation system. (© *1974 and reprinted with permission of Waldes-Kohinoor, Inc.*)

by the use of retaining rings, special threaded caps, and other devices. Figure 16-30 shows a typical snap bushing. Snap bushings can also be used to cover sharp edges of holes in thin metal casings. Wires can then be inserted without danger of cutting the insulation.

Staking. Staking is a process where two or more parts are permanently fastened. One or more prongs in one piece are passed through a matching hole in the second piece, and the prong is clinched by a staking punch. The prong may be bent as a cotter pin or expanded like a rivet. The process is fast and cheap.

Wire and Tubing Retainers. In recent years a group of retainers has been developed to hold bundles of wires and tubes in position. The devices, including straps, clamps, and retainers can serve to bind the bundle for ease in assembly, handling, and installation and can serve to position the bundle within the assembly. Figure 16-31 shows an adjustable *retaining strap*. The ratchet-type adjustment allows one size strap to be used in all situations. The use of only one size reduces inventory problems and assembly confusion as there are no sizes to become mixed and the device can only be assembled in one way.

The combination clamp and retaining strap is shown in Fig. 16-32. The pin can be used to hold the

bundle together and position the bundle on the wall or panel of a larger assembly.

Rigid clamps are used in a variety of situations where weight, impact, or vibration conditions are more severe. They can be adapted for wire, tubing, heavy pipes, and multiple group assembly.

All the recent wire and tubing retainer devices have shown great usefulness in automotive, space, and aircraft industries. They are used to group and secure oil lines, fuel lines, hydraulic lines, electrical wires, control cables, and other similar items.

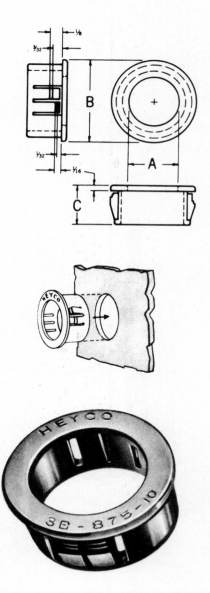

Fig. 16-30 Snap bushings provide bearing surfaces for thin sections.

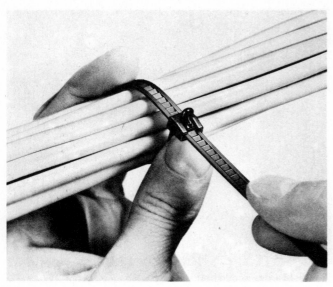

Fig. 16-31 A versatile retaining strap bundles rods, tubes, or wires neatly. (*Eaton.*)

Fig. 16-32 A combination clamp and retaining strap. (*Eaton.*)

Clips and Closures

Clips. Clips are used to position and hold two or more parts together without the use of other fasteners. They rely on spring-developed friction for maintaining their position. Clips, of the types shown in Fig. 16-33 are used extensively in low-stress assemblies such as electronic circuit boards, automotive trim and dash mounting, and appliances. Clips feature fast, simple installation and good durability when correctly applied.

Closures. Closures differ from most fasteners in that they are made to offer a quick-opening feature. Closures are used to hold cabinet doors shut, fasten flaps, hold removable items in place, and have other simple uses. Closures include catches, hook and loop systems, and other previously discussed items such as quick-release screw systems.

Catches. Catches are most frequently used in industrial assemblies as door and panel closures. There are three main types of catches used, although there are numerous variations of each.

Spring catches offer a positive-locking feature and cannot be disengaged except by physical manipulation.

Friction catches utilize a spring action, as shown in Fig. 16-34, but can be disengaged by a strong push or pull. Spring-tensed rollers, plungers, and ball stops are all used to reduce wear and stress fatigue of parts.

Magnetic catches have nearly the same characteristics as friction catches (see Fig. 16-35). However, unlike friction catches, they have no moving parts that can wear out.

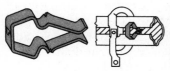

DART CLIP CUTS ASSEMBLY COSTS 80%. THIS TINNERMAN DART CLIP REPLACED A CONVENTIONAL THREADED HOLE AND SCREW ASSEMBLY FORMERLY USED AS A GLIDE-STOP BY A MANUFACTURER OF DRAPERY HARDWARE. IT SLIPS INTO PLACE EASILY, QUARTER TURNS TO HOLD SECURELY, AND CAN BE EASILY REMOVED WHEN NECESSARY. IT REDUCES PARTS HANDLING AND SIMPLIFIES INSTALLATION BY THE CUSTOMER.

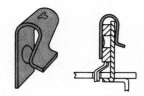

TINNERMAN FASTENER REPLACES FOUR PARTS. . .FOR A RECORD PLAYER MANUFACTURER WHO NEEDED AN INEXPENSIVE YET EFFICIENT METHOD OF MOUNTING A PLASTIC TRANSISTOR AND PROVIDING AN EFFECTIVE HEAT SINK. THIS STANDARD TINNERMAN FASTENER WAS SELECTED TO HOLD THE TRANSISTOR FIRMLY AGAINST A VERTICAL TAB. IT REPLACED A SCREW, TORQUE WASHER, LOCK WASHER, AND NUT; MADE A SIMPLER ASSEMBLY, AND REDUCED PARTS COSTS.

Fig. 16-33 Clips can speed and simplify assembly processes. (*Courtesy: Eaton Tinnerman Fasteners.*)

MECHANICAL CATCH   SPRING CUSHIONED
DOUBLE-ROLLER CATCH

Fig. 16-34 Friction catches.

Hook and Loop Systems. Hook and loop fasteners are used in both flexible and rigid systems, but they offer greater advantages in closing flexible units. Figure 16-36 illustrates a hook and loop system. They are made of plastic, originally for use in space flight, and offer great versatility. They are noncorroding, allow for infinite adjustability, and are light in weight. They may be attached by mechanical means, including sewing and stapling, or by use of adhesives. They can be bonded to almost any substance in almost any shape without machining or extensive surface preparation.

Joints and Seams

Joints and seams are used extensively in the assembly of sheet metal and wood products. They are called *joints* when used in heavier metal applications and in wood products. They are called *seams* when utilized in sheet-metal products.

Joints and seams are shapes formed to help the assembly offset the forces it will withstand. Sheet-metal seams are often used without additional fastening or adhesion processes. Most joints in wood and metal fabrication also require the use of additional fasteners or adhesives.

Figure 16-37 shows several of the more common seams used in sheet-metal products. Typical applications include making "tin" cans, buckets, air condi-

Fig. 16-36 Hook and loop reclosable fasteners. Several versions of these come with adhesive backings. They include pressure-sensitive, heat-activated, and solvent-activated adhesives. Applications range from garment closures to holding headrest in aircraft. This photo merely displays the product with close-up photography. (*Courtesy: 3M.*)

tioning and heating ducts, metal office furniture, and so forth. Joints are also formed by swaging and HERF processes. These are discussed in detail in the chapters dealing with those processes.

Examples of wood joints are shown in Fig. 16-38. These are extensively used in furniture manufacturing. In most cases, wood joints are held by adhesives or mechanical plates, brackets, or braces screwed or stapled in place.

ADHESIVE BONDING

Adhesive bonding, as previously explained, uses an adhesive material, or glue, to join other pieces together. The adhesive material does not mix with the other materials but flows into the tiny irregularities on the surface and clings tightly.

When dealing with adhesives, many terms are used interchangeably, such as glue, cement, and adhesive. Adhesion includes the use of glues and certain operations often considered in other classes. These are soldering operations and include the common "soft" sol-

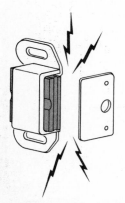

Fig. 16-35 A magnetic catch.

STANDING   GROOVED   FOLDED   DOUBLE SEAM

Fig. 16-37 Common sheet-metal seams.

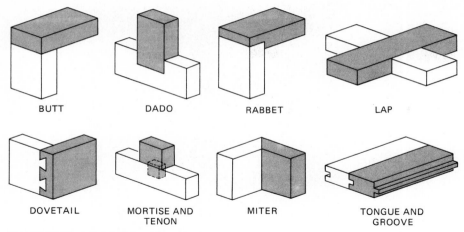

Fig. 16-38 Basic joints in woodworking.

der made from lead and tin. However, adhesion also includes the "hard" solders, such as silver solder and brazing. All soldering and brazing processes use heat to spread an adhesive material (the solder) over a surface.

Soldering

Soldering, as previously mentioned, includes two subclasses, "hard" and "soft" solders. The differences include the relative strength and hardness concerned, as well as the working temperatures. Soft solders are not as hard nor as strong and are applied at much lower temperatures.

Soft Solder. Soft solder is made from lead and tin. Its most common industrial application today is in making electrical connections. There are several different compositions of solders which vary the proportions of lead and tin. The purpose of alloying the two materials is to produce the *eutectic* composition. This is a term which means that the melting point of the resulting alloy is less than that of either of the original materials and occurs at a single temperature rather than at a slush range. Generally, the greatest effect in lowering the temperature occurs when the metals are mixed equally. Lead melts at 621°F [327°C] and tin at 429°F [220°C]. Solder (⁵⁰/₅₀) melts at 400°F [204°C].

When solder is used, two rules must be followed. The materials at the joint must be as hot as the temperature which is used to melt the solder, and all elements must be kept spotlessly "clean." Frequently, a cleansing agent known as a *flux* is used to help keep the materials free ("clean") of oxides and other matter. When joining common metals such as copper, zinc, steel, and so forth, the flux is made from various types of acids. However, when electrical work is used,

acids cannot be used as fluxes. This is due to the chemical and corrosive action of the flux on the conductive elements. The resulting oxides are nonconductive and can cause the soldered electrical connection to stop conducting electricity. Therefore, a flux made from rosin is used when soldering electrical work.

Solder can be applied in several ways. It can be applied by a tool known as a *copper* which heats the work and melts the solder. The soldering "iron" and the soldering "gun" are variations of this. In assembling printed circuit boards in electronics industries, the solder is applied by floating the connected but unsoldered assembly upon a pool of molten solder. The printed circuit board will not sink because it is much lighter than the solder and the solder only adheres to the exposed copper circuits "printed" on the bottom. This technique is quicker and uses less solder than using any type of "iron."

Hard Solder. Hard solders include brazing, silver solder, and other similar solders. Brazing requires the use of special brass or bronze alloys to join iron and steel components and is normally thought of as a welding operation because it is often done with welding equipment. It is fast and has great strength. It is discussed in some detail in the chapters on welding processes.

Other hard soldering operations are generally done to assemble various parts together permanently. Items include jewelry, gun sights, engraved stamps, and their presses, and so on. Generally, the workpieces are assembled with a thin layer of solder between the pieces to be joined. Clamps or wires are used to hold the pieces securely in place, and heat is applied. The heat melts the solder between the pieces and adheres to each, creating a tight joint. This operation is usually called *sweating.* Advantages of such operations include the strength, hardness, and durability of the joint. The thinner the joint, the stronger the assembly will be.

Adhesives

The original adhesives were generally called *glues.* These were derived from animal sources and were used in a hot, melted form. They were used extensively on wood, leather, and paper. They generally required a rough, porous structure for adhesion to occur. The need for a rough, porous surface eliminated adhesives as a satisfactory joining material for any smoothly surfaced material. As a class, animal and animal-derivative glues were not waterproof and generally had poor stress resistance.

In more recent years, however, modern chemistry has created a wide range of adhesive substances suitable for industrial applications. The use of adhesives for joining both porous and nonporous materials is widespread. Metals, glass, and ceramics are commonly joined using modern adhesives which can also be both waterproof and shock resistant. Modern adhesives are widely used in situations where stresses occur, such as in aircraft industries. However, joints must still be carefully designed to minimize the stress because the adhesive generally has less strength or stress resistance than the material being joined.

The use of adhesives opens a promising approach to conserve materials, reduce manufacturing costs, and yet provide good strength with little or no structure modification for purposes of joining. When materials are normally welded, riveted, or bolted, special flanges or overlapping seams are required. Also, after joining the parts, the heads and beads of these joining processes protrude from the surfaces. With the use of adhesives, however, most of these protuberances are eliminated. Adhesive bonding can be used to increase strength and stiffness; reduce the number of parts; lower first cost and maintenance cost; lower assembly times; and reduce susceptibility to corrosion, vibration, and noise which, in turn, reduces inspection of mechanical fasteners that may loosen in use. Using sandwich configurations to laminate wide sheets together with adhesives increases both shear and compressive buckling strengths in modern materials. Adhesives also allow thin sheets to be formed into complex shapes without the need of cutting or forming operations by pressing the stacked pieces into shape before the adhesive hardens. The use of adhesives allows different materials such as aluminum (which is easily welded) and titanium (which is not easily welded) to be joined.

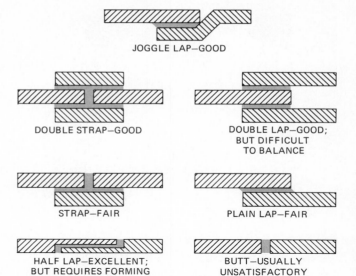

Fig. 16-39 Types of adhesive-surface joints.

Adhesive joints can be classified according to the configuration of the glue or bonded joint. The surface joint is used where two sheets are joined to each other. Figure 16-39 shows some types of commonly used adhesive-surface joints. Core-to-face bonding occurs when a filler interior is joined to a face material. An example is the use of hollow core doors as indicated in Fig. 16-40. In both cases consideration must be given to the types of stress encountered in the joint. Figure 16-41 illustrates typical types of stresses imposed on adhesive joints. Naturally, the choice of adhesives depends in some degree upon the type of material to be joined, the type of joint, and the type of stress to which the joint will be subjected. Generally, in selecting a joint, good mechanical joining is desired. This means simply that the two pieces should fit together well without the adhesive. If the joint can be subjected to moderate stresses without the adhesive, this also increases the resulting strength.

Advantages and Limitations of Adhesives. The use of adhesives provides a number of advantages in terms of manufacturing. As previously mentioned, it

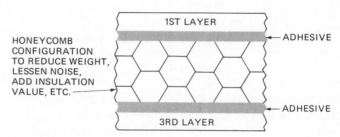

Fig. 16-40 Adhesive-core joint.

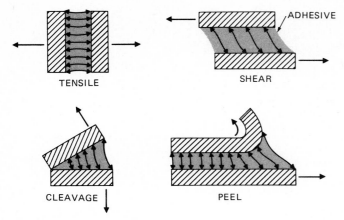

Fig. 16-41 Types of stresses on adhesive joints.

increases the adaptability of machines and can reduce tooling cost and increase the desirable characteristics of the materials. Almost any material or any combination material can be joined using adhesives. Many adhesives do not require any setting time, the use of complicated catalytic agents, or the use of heat to cure and harden the joint. However, all these factors singly or jointly may be required depending upon the type of adhesive used. Adhesives also allow the joining of thin and delicate materials, including such things as metallic foil, plastics, and cloth, to a variety of heavier sections, including sections of the same material or of different material. Continuous adhesive joints distribute the stress much better than single-point mechanical fasteners and better fatigue resistance is generally obtained because of this dispersion of stress. Smooth contours and shapes are common and no added oper-

ation, such as the holes for rivets, is required. The use of adhesives can also provide both thermal and electrical insulation as well as acting as a damper for vibrations. The use of adhesives for electrical insulation also provides protection against galvanic action where two metals of dissimilar nature are joined. Further, the labor cost is frequently reduced because no intricate machining or measuring is required, thus lessening the skill required by the labor force in the operation.

Limitations are largely low strength or poor adhesion. As a class, adhesives retain relatively low physical strength, but the distribution of the adhesive over a wide surface generally offsets this. For adhesives that bond strongly to nonporous structures, great expense is also common. Some of the more expensive adhesives can cost hundreds of dollars per liter. Obviously such materials are not used haphazardly. Some adhesives also deteriorate when exposed to air, light, heat, vibration, or other situations. When the use of adhesives is being considered, it is best to consult a representative from a company that manufactures such adhesives. The number of different formulations is considerable. One major adhesive supplier, for example, manufactures more than three thousand different tyes, all of which have different applications and limitations. Adhesives may also require special storage and safety conditions which must be considered.

Types of Adhesives. There are several ways of classifying adhesives. The charts in Tables 16-1, 16-2,

Table 16-1 **Chemical Classes of Adhesives**

	Natural	*Thermoplastic*	*Thermosetting*	*Elastomeric*
Common names	Casein (milk, etc.)	Polyvinyls	Phenolic	Rubber
	Hide (ureic)	Acrylics	Urea	Neoprene
	Starch	Cellulose nitrates	Alkyd	Silicone
	Rosin		Epoxies	Butadiene
	Asphalt		Resorcinol	groups
			Melamine	
Most common form	Liquid	Liquid	Liquid	Liquid
	Powder			
Major use	Furniture	Furniture	Furniture	Clothing
	Household	Decorations	Plastics	Tires
	Paper products	Household	Novelties	Sealants
Materials joined	Wood	Wood	Phenolics for	Wood
	Paper	Cork	metal and	Rubber
	Textiles	Paper	glass	Fabric
			Alkyds—metal	Foil
			Epoxies—plastic	Plastics
			Urea and mela-	
			mine for woods	
			Others—varied	
Strength	Usually low	Good	Good	Poor

Table 16-2 **Comparisons of Adhesives**

Adhesive	Major Advantages	Major Limitations
Animal and fish glue	Strong in wood and paper	Not moisture resistant
Starch glue	Good on paper	Not strong or moisture resistant
Acrylic group	Clearness, flexibility	Not strong or heat proof
Polyvinyl group (acetate and butyral)	Tough, strong in wood, adaptable to liquid or hot-melt use	Low heat and moisture resistance
Silicones	High peel strength, resistance to all environments	Expensive
Cellulose group	Low cost, quick air drying	Low strength, low heat resistance
Epoxy—nylon	High tensile and peel strength, tough	Poor low-temperature resistance and moisture resistance
Epoxy—phenolic	High tensile and peel strength, tough, good environmental resistance	Handling and mixing
Urea-formaldehyde	Low cost, moisture resistant in wood, good strength for wood	Poor flexibility, drying time required
Elastomers	Flexible, good shear and impact strength	Low tensile strength; mixing and handling

and 16-3 show three systems. Generally, adhesives are categorized as thermoplastic or thermosetting. Thermoplastic adhesives are those adhesives which become plastic, or soft, when subjected to heat. Thermosetting plastics are those which dry or harden completely and do not soften when exposed to heat or radiation. Both types may be applied as liquids or as solids and must be given time to cure, set, or dry. Both generate better strength and resistive characteristics when pressure is applied and retained during their curing, or setting, process (the thinner the joint the stronger the bond.) Thermoplastic materials are frequently used as air-drying emulsions or solutions that achieve their strength when a solvent they contain evaporates. Some adhesives are also used as a fusible solid that will liquefy when heated, flow in the

joint, and solidify when cooled. *Pressure-sensitive films* are also made using thermoplastic adhesives. Pressure-sensitive films are tapes or sheets which are simply cloth, plastic, or paper backings with adhesives on one side or both sides.

The thermosetting adhesives can be hardened by heat or by catalytic agents that cause chemical hardening. Generally, thermosetting adhesives are stronger and more durable than the thermoplastic types. Both can be made to be waterproof and water resistive.

Included in the thermoplastic adhesives are the animal and vegetable materials of organic nature. These include animal glue made from hides and hooves, vegetable glues, animal-derivative glues made from things such as milk and egg whites, and vegetable-derivative glues made from such things as the latex or sap from rubber trees. The "rubber" cements, both natural and synthetic, are also classed as *elastomerics*. The thermosetting group is largely composed of synthetic, or manufactured, materials. However, synthetic thermoplastic types are also widely utilized. The synthetic glues include sodium silicate (waterglass), the ureic-formaldehyde grouping, resorcinol, the alkyds, vinyls, acrylics (which are generally the clearest of the adhesives), cellulose adhesives (such as model airplane cement), melamine, polyesters, polyurethane, epoxies, and silicone derivatives. Each exhibits different characteristics, advantages, and limitations, as shown in Tables 16-1 through 16-3.

Table 16-3 **Forms of Adhesives**

Form	Advantages	Other
Liquid	Easy to apply; flow and vicosity controllable	Most common form, best for hand application
Powder	Mix as needed, longer shelf life	Curing initiated by mixing or heating
Films (tape and sheet)	Uniform thickness, quick, no waste	Limited to flat areas, low strength
Paste	Good shear and creep strengths, less "drying" time needed	Can be applied by machines in production

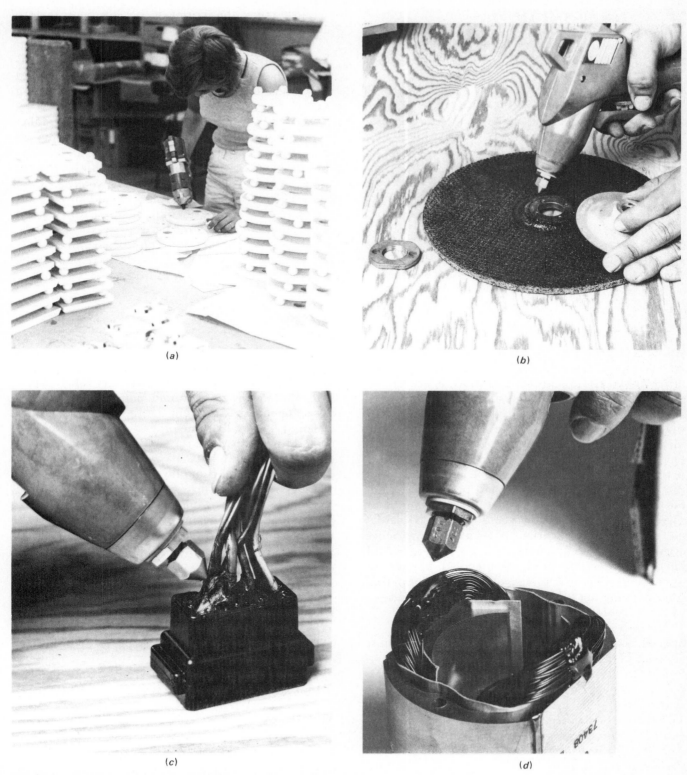

Fig. 16-42 Hot-metal adhesives are used widely for a variety of applications. (*Courtesy: Nordson Corp.*) (*a*) Hot-metal adhesives "glue" feet onto wooden toys; (*b*) hot-melt adhesives are used to fasten a metal disc to a fiber wheel; (*c*) hot-melt adhesives used as "potting" for electrical components; (*d*) hot-melt adhesives can be used to fasten and position electrical wiring.

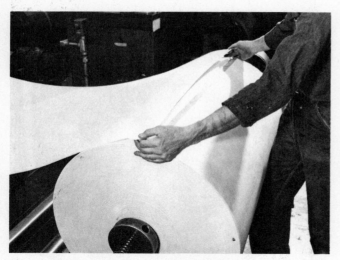

Fig. 16-43 Adhesive films are used in industry. A double-coated tape is used here to splice webs of paper in paper and printing plants. There are sophisticated systems to accomplish "flying splices." (*3M.*)

Applying Adhesives. Generally, the adhesive film is applied after the parts are well formed. The film generally should be kept as thin as possible. However, because adhesives vary so widely in characteristics, uses, and advantages, there is logically a wide variation in the methods for applying them. Common methods include spray, brush, and roller coatings. But, where heat and pressure are to be used for curing, adhesives in the form of films or sheets may be inserted between the parts to be joined, heated, and pressure applied. Such a method is frequently used for the bonding of plywood, laminated plastics, and laminated plastic-metal combinations. Various types of hot applicators are used, as shown in Figure 16-42*A–D*, where heated adhesives are applied and, when cooled, become set or hardened. Some bonding is also done by preparing or moistening the surfaces treated with dry adhesives with a solvent (Fig. 16-43).

Special Considerations. The user of adhesives must also consider rate of use and the *shelf life* of the material. This term means that the material will decompose and lose its desirable characteristics over a period of time. The useful life of the adhesive varies with the particular type used. This may vary from a relatively short period of a few weeks to several years. Another factor in the curing and the shelf life of these materials is in the "pot life" of these materials. This is particularly true for those materials that react to catalytic agents or to hardening when exposed to air, radiation, heat, and so on. Generally, the faster the curing time, the shorter the pot life. Some materials, such as the resins used in fiberglass work, set up in a matter of minutes while others may take months. It becomes important to purchase adhesives in amounts that can be used up before deterioration begins and to mix quantities that can be used before hardening occurs.

One additional factor in the application of adhesives concerns worker protection. When solvents are used, dangers arise from inflammability. In all cases solvents may cause skin damage, and the inhalation of fumes may also be dangerous. Some catalytic hardening agents can cause permanent and progressive eye damage as well. Generally, manufacturers point out precautions considered necessary in the use of the adhesive, and the specifications of manufacturer's representative should be consulted before putting the material into the production process.

SEALANTS

Sealants are similar to adhesives, and it is sometimes difficult to detemine whether a material is being used as a sealant or an adhesive. The primary function of a sealant, however, is not to hold materials together but to seal openings to prevent the passage of either gas or liquid. Generally some other system provides mechanical strength while the sealant material provides the tight seal. Sealants may have certain adhesive characteristics but generally do not harden or set. They generally remain soft and flexible to provide continuous sealing against air and vapors and to adjust to dimensional changes without losing seal integrity. Sealants may include a wide variety of chemicals, plastic, and rubber compounds to keep out moisture, air, and chemicals or to retain pressurized systems.

REVIEW QUESTIONS

16-1. How are fasteners classified?

16-2. How are adhesives classified?

16-3. Distinguish between fasteners and adhesives.

16-4. What are the advantages of mechanical fasteners?

16-5. What are the advantages of using adhesives?

16-6. What are the advantages and limitations of threaded fasteners?

16-7. List the types of mechanical fasteners.

16-8. Is a joint a "fastener"? Explain your answer.

16-9. How are sealants used?

16-10. What is a "blind" rivet?

16-11. How are machine screws, wood screws, and sheet-metal screws different?

16-12. How is brazing different from welding?

16-13. Why are "standard" fasteners desirable?

16-14. What is the purpose of a washer?

16-15. How can bolts be attached to thin pieces?

16-16. How can a thin piece support a rotating shaft?

16-17. What is a retainer?

CHAPTER 17

INTRODUCTION TO WELDING TECHNOLOGY

Welding is the process of permanently joining two or more pieces of material together by the application of heat, pressure, or both. The American Welding Society has defined welding as ". . . the process of joining two or more pieces of material, often metallic, by localized coalescence or union across an interface.*" In welding, common edges or surfaces are generally melted and fused together (fusion welding); however, several techniques are used to join materials by the application of heat and/or pressure without the pieces being melted (nonfusion or solid-state welding). Welding, properly done, using fusion or nonfusion processes will result in a joint as strong or stronger than the weakest member of the joint. Although welding was first used to join pieces of metal, any material which will melt and fuse with itself or another material can be welded.

The technology associated with welding and the sophistication of the equipment used for welding has improved tremendously during the past few decades. The Space Age provided a need for joining new materials which were not previously joined by welding. As the need for joining new materials grew, the technology and equipment for welding these materials was also developed and advanced. Also, processes and equipment for making these welds faster than ever before are constantly being introduced into the welding field.

Recorded history places the origin of welding around 1300 B.C. During this period of time the famous Damascus steel was made by welding layers of iron and steel together. This material was made into swords and daggers of unexcelled excellence. Developments in the art of welding were nil until the early 1800s when new sources of heat were developed. Edmund Davy discovered acetylene gas and Humphrey Davy produced an electric arc.

For over 2500 years welding had been done by hammering two pieces of red-hot metal together. But, with these two new sources of heat the way was opened for new improved methods of welding. However, it was late in the nineteenth century (1892) when a Russian by the name of V. G. Slavianoff told of a process for electrowelding which he developed four years earlier. Also, in 1892, an American from Detroit, Michigan, by the name of C. L. Coffin was granted a patent for a bare, metal-electrode arc-welding process. From this time until World War I, welding was used mostly as a repair technique, but with the demands of war for faster production techniques, welding was used as a production technique to speed up fabrication processes.

Many improvements in welding equipment and welding supplies were marketed during the early 1900s. In 1930, Hobart and Devers patented welding with an electric arc in an inert-gas shield. This process proved to be very successful for welding the hard-to-weld metals such as aluminum and magnesium which were being used by the aircraft industry. By the mid-1950s, welding in an inert-gas shield was widely used by industry, and a number of different welding machines were available for this type of welding. Other developments in welding have resulted in over 25 different welding processes. Chapter 18 will deal with these processes in detail.

METALLURGICAL CONSIDERATIONS IN WELDING

The production of the coalescence, called the weld, requires the application of heat, pressure, or both, and

---

* Arthur Phillips, ed., *Welding Handbook*, 6th ed., American Welding Society, New York, 1968, p. 32.

involves both physical and chemical reactions. In welding, these reactions must be controlled in order to produce satisfactory welds.

Physical Reactions

A number of welding processes require the material being welded to be heated until the joint is melted (Fig. 17-1). These processes are called *fusion welding processes* since the materials, when melted, fuse or mix together. As the joint cools, the material which was melted from each piece being joined solidifies or freezes and thus welds (joins) the pieces together. The physical changes in the material of melting and resolidifying are necessary in fusion welding processes; however, they are accompanied by other reactions most of which are undesirable and must be controlled when welding.

Expansion and Contraction

The exact dimensional size of a piece of material is dependent upon its temperature. As the temperature of a piece of material increases, its size also increases and as its temperature decreases, the size of the piece of material decreases. The amount of change per inch in the size of a piece of material for a change in temperature of 1° is called the *coefficient of thermal expansion*. The coefficients of thermal expansion for some of the materials commonly joined by welding are shown in Table 17-1. The coefficient of thermal expansion varies according to the type of material being heated; however, the expansion is generally more pronounced for metals and plastics than for other materials. Metals and plastics also constitute the two materials most commonly joined to themselves by welding, and metals are by far the most commonly welded materials. Therefore, the major portion of this chapter is devoted to welding as it relates to metals. The last section of this chapter is devoted to welding plastics.

The fact that the coefficient of thermal expansion varies for different metals creates some problems when dissimilar (different kinds of) metals are welded together. When dissimilar metals are joined and then heated, one piece of metal expands faster than the

(a)

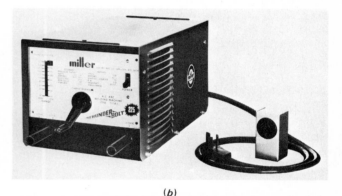

(b)

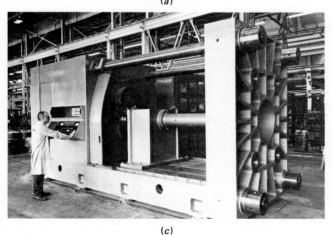

(c)

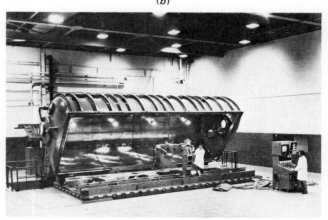

(d)

Fig. 17-1 Metal parts fastened together by a variety of welding processes: (a) Metal parts fastened together by welding to make a truck body; (b) Welding machine for arc welding; (c) Friction welder; and (d) A high-vacuum welding unit. (*Miller Electric Manufacturing Co.*)

Table 17-1 **Coefficients of Thermal Expansion for Commonly Welded Materials at 68°F (do not remain constant with temperature)**

Material	Per °F	Per °C
Aluminum	0.00001234 or (1.234 × 10⁻⁵)	0.000022 or (2.2 × 10⁻⁵)
Copper	0.00000887 or (8.87 × 10⁻⁶)	0.000019 or (1.9 × 10⁻⁵)
Cast iron	0.00000556 or (5.56 × 10⁻⁶)	0.000012 or (1.2 × 10⁻⁵)
Steel	0.00000636 or (6.36 × 10⁻⁶)	0.000013 or (1.3 × 10⁻⁵)

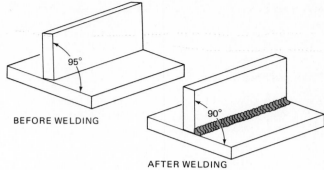

BEFORE WELDING

AFTER WELDING

Fig. 17-3 Allowing for the normal warpage of metals in a tee joint.

other and causes the pieces of metal to warp or bend (Fig. 17-2). This same warpage occurs when two metals cool after being joined hot, as is the case in fusion welding. One metal contracts faster than the other, causing internal stresses in the joint and warpage of the metal pieces. Similar warpage from the internal stresses in metal pieces occur when metal pieces joined by fusion welding differ considerably in their thicknesses. The larger, heavier piece of metal cools slowly and contracts after the thinner piece has cooled. This requires the metal pieces to be joined at an angle larger than the one desired, and then the normal warpage is relied on to bring the pieces into the desired angle when the joint is cool (Fig. 17-3).

The problem of warpage is often reduced or eliminated by allowing a gap in the joint to be fusion welded or by preheating (heating before welding) heavy sections and post heating (heating following welding) thin sections in the metals being welded. Also, warpage is usually eliminated when the joint is welded by one of several nonfusion welding techniques.

Metals are usually refined from ores which consist of the metallic element being refined, other metallic elements, and oxygen, all of which are chemically bound together. In order to get what we call a metal from the ore, it must be refined or reduced to the metallic state. Removal of the oxygen from a metal's ore

causes a metal to have an affinity for oxygen, and it will absorb oxygen from the air or water whenever possible. As a metal's temperature increases, its affinity for oxygen increases and its ability to absorb oxygen increases. However, oxygen absorbed into a weld introduces brittle oxides. Therefore, in order to weld a joint satisfactorily, the atmosphere and its oxygen must be kept away from the joint until the joint cools.

The process of keeping the atmosphere with its high-oxygen content away from a joint being welded is called *shielding*. Shielding may be done in several ways; however, the use of some type of gas shield is the most common. Inert gasses (argon and helium), CO_2 (carbon dioxide) gas, CO_2 gas mixtures, and fluxes which melt and form a gas or liquid shield are all used extensively in welding. Some welding is being done in a vacuum chamber where no air or gasses exist and, therefore, the weld area does not need to be shielded.

Non-fusion-welding processes are accomplished at lower temperatures than fusion-welding processes; therefore, oxidation of the weld metal is less of a problem. Some non-fusion-welding processes are completed at room temperature through applications of pressure. Thus, oxidation of the metal during these welding processes is of little concern. However, the joint must be clean and free of oxidation immediately before beginning the welding process. *Wire brushing* is the preferred method of cleaning the joint to be welded because chemical cleaning leaves a film or residue on the joint which is detrimental to a good weld. Also, the use of sand blasting or abrasive cleaning, unless followed by wire brushing, allows nonmetallic inclusions to cover the weld area.

The different shielding techniques used in welding and the different ways in which pressure or heat is applied to the joint accounts for all the many different types of welding. Each type of welding will be explained in Chap. 18.

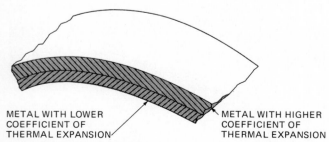

METAL WITH LOWER COEFFICIENT OF THERMAL EXPANSION

METAL WITH HIGHER COEFFICIENT OF THERMAL EXPANSION

Fig. 17-2 Warpage of dissimilar metals which have been joined together.

Joint Designs

A number of different joint designs are used in welding to reduce warpage and to assure 100 percent penetration of the base metal (Fig. 17-4). Five basic types of joints are used in welding: butt, corner, edge, lap, and tee (Fig. 17-5). These joints can be used in combination to form other joint designs for special purposes. For example, the lap and corner joints are combined to form a corner-lap joint (Fig. 17-6).

In addition, the five basic joints can be prepared and welded in various ways to produce a number of weld types. Some of the more common types of welds are shown in Figs. 17-7, 17-8, and 17-9. The grooved welds (square, single, double, single-bevel, double-bevel, single-U, double-U, single-J, and double-J) are designed to permit 100 percent penetration of the base metal when butt welding thick pieces together. Certain welding processes are capable of more penetration than others, and the type of weld is determined by both the thickness of the metal and the capability of the welding processes to be used. The size of the groove is also determined by the thickness of the base metal and the welding process to be used.

The *fillet weld* (Fig. 17-8) is used to join two surfaces at angles close to 90°. The weld metal itself forms a triangular bead fused into the workpieces across the length of the joint. Fillet welds generally require the addition of filler metal so that the bead can taper smoothly into the surfaces being joined. A pronounced angle between the weld metal and either of the surfaces being joined should be avoided.

The flare-vee, flange-edge, and bead welds are used to join pieces of material as shown in Fig. 17-9. The pieces are butted together without grooves and are joined with a single-weld bead. These welds can

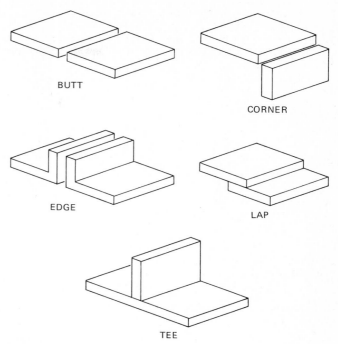

Fig. 17-5 Five basic joints used in welding.

be made with or without filler metal; however, filler metal is generally used in the bead and flare-vee welds. The weld metal may be weaved as shown in Fig. 17-10 to provide a larger weld bead for joints requiring extra filler metal.

When welding thick materials, the joint is grooved. The grooves are often large and require multiple welding passes in order to completely fill the joint with weld metal. Multiple-pass welds are made according to the diagram in Fig. 17-11. The root pass is made first to assure complete penetration of the base metals. Each weld must be clean and free of slag or foreign matter prior to each of the welding passes made. If the slag from a welding pass is not removed prior to a subsequent pass, slag is trapped in the weld metal, thereby reducing the strength of the weld. Multiple-pass welds, like single-pass welds, must be

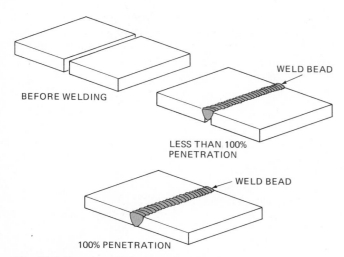

Fig. 17-4 Complete (100 percent) penetration.

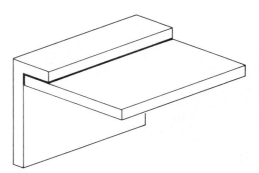

Fig. 17-6 Corner-lap joint.

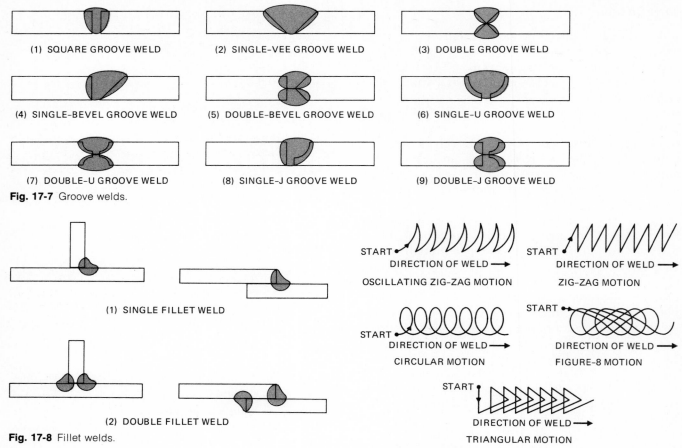

Fig. 17-7 Groove welds.

(1) SQUARE GROOVE WELD

(2) SINGLE-VEE GROOVE WELD

(3) DOUBLE GROOVE WELD

(4) SINGLE-BEVEL GROOVE WELD

(5) DOUBLE-BEVEL GROOVE WELD

(6) SINGLE-U GROOVE WELD

(7) DOUBLE-U GROOVE WELD

(8) SINGLE-J GROOVE WELD

(9) DOUBLE-J GROOVE WELD

(1) SINGLE FILLET WELD

(2) DOUBLE FILLET WELD

Fig. 17-8 Fillet welds.

START DIRECTION OF WELD →
OSCILLATING ZIG-ZAG MOTION

START DIRECTION OF WELD →
ZIG-ZAG MOTION

START DIRECTION OF WELD →
CIRCULAR MOTION

START DIRECTION OF WELD →
FIGURE-8 MOTION

START DIRECTION OF WELD →
TRIANGULAR MOTION

Fig. 17-10 Weaving motions for welding.

free of voids (hollow areas in the weld) and slag inclusions (pockets of slag in the weld) in order to achieve maximum strength and corrosion resistance.

Welds can be completed in four basic positions: (1) flat position, (2) vertical position, (3) horizontal position, and (4) overhead position (see Fig. 17-12). If at all possible, welds are made in the flat position because the welder has better control over the molten puddle. In the flat position, the pull of gravity increases weld penetration and helps keep the molten puddle in the joint. However, in the vertical, horizontal, and overhead welding positions, gravity pulls the molten puddle out of the joint, thus making it more difficult for the welder to manipulate the molten puddle and fill the joint properly.

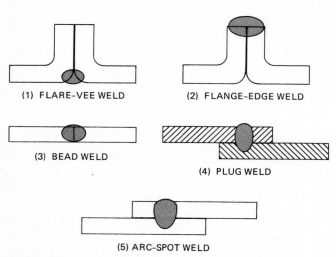

(1) FLARE-VEE WELD

(2) FLANGE-EDGE WELD

(3) BEAD WELD

(4) PLUG WELD

(5) ARC-SPOT WELD

Fig. 17-9 Special types of welds.

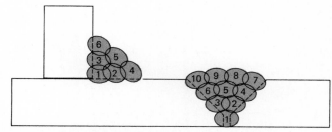

Fig. 17-11 Multiple-pass welds.

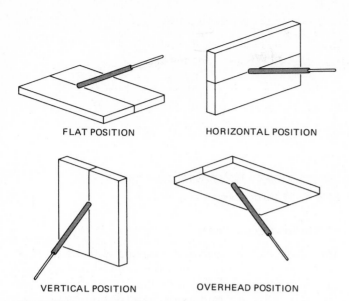

Fig. 17-12 Four basic welding positions.

INSPECTION AND TESTING OF WELDS

The quality of a weld is often difficult to determine by simply looking at it. In most instances, the quality of a weld is critical, may be the difference between success and failure, and is often a matter of life and death. Surface cracks and irregularities in a weld can be detected visually with the aid of chemicals or a magnetic field. However, visual techniques offer no insight into the weld metal beneath the surface. Hidden or subsurface defects in a weld are detected by destructive testing or nondestructive testing using sophisticated equipment. These tests are necessary in instances where welds must meet stringent structural requirements or certain building codes. This is why welders are certified as being able to produce quality welds.

Destructive Testing

In *destructive testing* methods, the welded piece is placed under a predetermined stress which bends, stretches, breaks, or destroys the piece being tested. Since the piece under testing is destroyed, destructive testing is usually done on a sampling basis. That is, every weld being made is not tested but rather one out of ten or one out of a hundred, or a weld is periodically selected and tested. If the part being welded is large or expensive, a small section of the weld, or a weld made under the same conditions and in the same material as the part being welded, is all that is tested or the weld itself is X-rayed.

In many cases, destructive tests are used on specimen pieces to determine if a welding process, proce-

dure, material, or welder is satisfactory for a certain application. Destructive tests include: fracture tests, tensile tests, bend tests, and metallographic tests.

Fracture Tests

A *fracture test* is widely used to indicate the general quality of a welded joint. The welded joint to be tested is clamped tightly in a vise; it is then broken or severely bent with a hammer or bending tool (Fig. 17-13). A representative weld or a short section of a weld is all that is tested by a fracture test. The weld which is destroyed in a fracture test is inspected visually for defects.

Tensile Test

The strength of a welded joint is determined by a *tensile test*. In a tensile test, the welded joint or a small section of it is broken by tensile forces (Fig. 17-14). The broken weld is examined visually for defects, and in

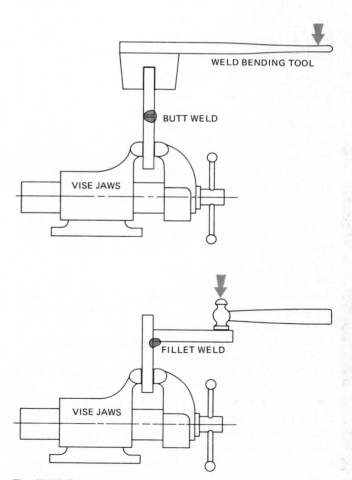

Fig. 17-13 Fracture test.

TENSILE (PULLING) FORCE

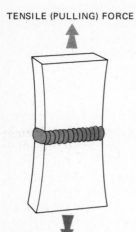

TENSILE (PULLING) FORCE

Fig. 17-14 Tensile test.

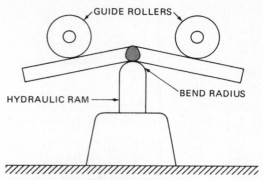

Fig. 17-15 Guided-bend testing fixture.

this respect the tensile test is similar to the fracture test. However, the tensile strength, yield strength, and percent of elongation for the welded joint are also determined during the tensile test. These values may then be compared with previously determined standards to indicate the general quality of the weld.

Bend Test

The physical condition and approximate strength of a weld can be tested quickly and easily by the *bend test.* A sample weld from a quantity of welded parts is usually taken. The weld is bent to a predetermined radius in a press and then inspected for cracks, poor penetration, and uneven stretching. Welds are often broken following the bend test to yield additional information about the weld.

Two types of bend tests are commonly used in testing welds: the free bend test and the guided bend test. The guided bend test is made on standard-size weld specimens. The welded specimen is bent to a $^3/_4$-in [1.91-cm] radius in a bending fixture by mechanical or hydraulic pressure. Guided bend tests must meet the standards established by the American Welding Society (AWS) for such tests. The results of one bend are then readily compared to the results of other bend tests. Figure 17-15 shows a typical guided-bend testing machine.

Free bend tests are often made in a vise with a bending tool (Fig. 17-13). Tests made in this way give an indication of the general quality of a weld and cannot be readily compared with the results of other bend tests. The free bend test does not require elaborate equipment and is easily conducted by the welder on the job. It is a very fast and popular method of testing welds.

Metallographic Tests

Metallographic tests are very exacting and require special equipment and trained personnel for their completion. Metallographic tests may be macroexaminations or microexaminations. Macroexaminations are performed with up to a 10X magnification. Figure 17-16 shows a microscope used in metallographic tests.

The weld to be tested by metallographic methods must be cut at right angles to the weld and have a sample removed. The edge of the sample is filed, sanded, or machined until it is perfectly smooth. Macroexamination of the sample will show the structure

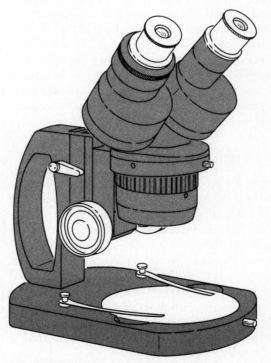

Fig. 17-16 Microscope for macroexamination of metal specimens. (*Olympus Corp.*)

of the weld area and any defects, such as slag inclusions, voids, or cracks, which it contains.

A weld to be microexamined is first prepared the same as it was for macroexamination but must undergo an additional fine polishing stage to produce a mirrorlike finish. This fine finish is usuallly accomplished using diamond abrasive particles. The sample is then etched with an acid solution. The grain structure of the weld is now clearly visible when the sample is viewed through the metallographic microscope. The heat-affected area in the parent materials, along with exact size and shape of the weld metal, can also be clearly seen. In microscopic examination, a large grain size indicates improper welding or treatment.

Methods exist whereby a very small specimen from a welded part can be removed for metallographic tests without destroying the piece. This makes metallographic tests useful as either a destructive or nondestructive testing method, since the welded piece can be put in service after the specimen has been removed or the test may be performed at the site of the actual weld.

NONDESTRUCTIVE TESTS

Nondestructive tests are widely used to determine the quality of a finished weld since the weld which is tested is the same one put in service. These tests are often referred to as inspection techniques because the weld is not cut, bent, broken, or otherwise destroyed. Nondestructive tests are capable of detecting cracks, porosity, inclusions, lack of fusion, or any general type of discontinuity. However, nondestructive tests cannot determine the actual strength of a welded joint.

A number of nondestructive tests are available for welds. The more common ones are visual inspection, magnetic-particle inspection, liquid-penetrant tests, ultrasonic inspection, and radiographic tests.

Visual Inspection

Visual inspection is the quickest, easiest, and least expensive of all nondestructive tests and is the most widely used. However, it yields very little information about the internal structure or strength of the weld. Errors in weld preparation, alignment, fit up, warpage, undercuts, overcuts, and penetration can be detected and corrected early, in many cases, before all the welding is completed.

Visual inspection is often aided by a magnifying lens, gauge, or other visual aid (Fig. 17-17). Leaks in welded pressure vessels are detected by pressurizing the part and submerging it in a tank of water or by

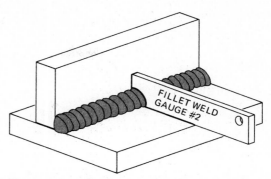

Fig. 17-17 Using a gauge to inspect a weld visually.

brushing soapy water on the welded joint. Only low pressures (20 lb [9.1 kg] or less) should be used for this test. Air pressure escaping through a leak in a welded container results in an easily detected string of bubbles with either the submerged or soapy-water technique. Containers which must be tested at pressures over 20 psi [1.4 kg/cm²] should be tested with water or oil under pressure since they do not compress and they present the hazards of an explosion from air compressed to high pressures.

Magnetic-Particle Inspection

Magnetic-particle inspection can detect surface discontinuities such as seams, porosity, slag inclusion, and lack of fusion. This process is applicable only to ferromagnetic materials, but the part may be of any size or shape. Ferromagnetic materials are those materials (primarily iron and steel) which are capable of attracting iron when subjected to magnetic lines of force.

In magnetic-particle inspection, magnetic particles similar to iron filings are placed on the surface to be tested, and a magnetic field is established in the metal to be tested. The magnetic particles are attracted to surface discontinuities by the magnetic field and cause these defects to be easily detected (Fig. 17-18).

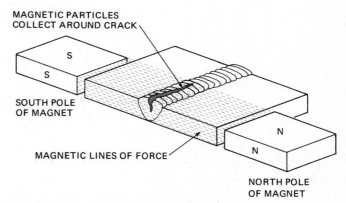

MAGNETIC PARTICLES
COLLECT AROUND CRACK

S
S

SOUTH POLE
OF MAGNET

N
N

MAGNETIC LINES OF FORCE

NORTH POLE
OF MAGNET

Fig. 17-18 Crack detected by magnetic-particle inspection.

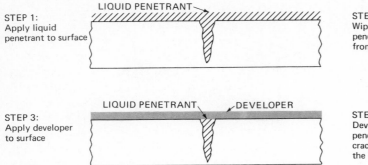

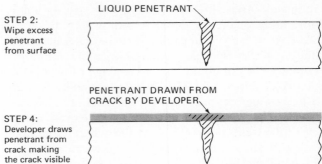

Fig. 17-19a Crack detected by liquid-penetrant test.

Fig. 17-19b Steps in liquid-penetrant inspection of a weld.

1. Thoroughly clean the surface with solvent cleaners. Remove all traces of dirt and films (one of the most important steps).
2. Apply the liquid penetrant to the surface to be inspected.
3. Allow the penetrant to remain on the surface long enough for it to seep into any defects.
4. Remove excess penetrant and clean the surface with appropriate cleaners.
5. Apply the developer to the surface.
6. Inspect the surface for discoloration of the developer since it will draw the penetrant from any cracks or defects in the surface. If a fluorescent penetrant is used, the surface must be inspected under a block light.

The magnetic field required by magnetic-particle inspection can be established in the piece being tested by induction from an electric current or by direct contact with a permanent or electromagnetic magnet. The magnetic field should be at a right angle to the defect for the best detection. Since most magnetic materials retain some magnetism after being tested by this process, a workpiece should be demagnetized following the test.

Liquid-Penetrant Tests

Liquid-penetrant tests are capable of detecting very small surface discontinuities such as cracks, porosity, or incomplete fusion. The surface to be tested must be free of dirt, paint, and oil films. A thorough cleaning of the weld surface before testing is the most critical and important aspect of this test. After the surface has been cleaned, an oil base, water-soluble penetrant is applied to the weld area. Following a soaking time, the penetrant is washed off with a recommended solvent. The penetrant is drawn into any surface opening in the weld area and will not be removed from these spots by the solvent washing. Therefore, when the weld has dried, a powdered, absorbent material, called the *developer,* draws the penetrant from the surface openings and makes these openings visible.

The penetrant contains either a dye or a fluorescent substance that is readily absorbed by the developer. Dyes stain the developer thus making a surface defect readily visible to the unaided eye. On the other hand, fluorescent penetrants require a black light to make the defects glow or fluoresce. Both types of penetrants are widely used in testing welds for surface discontinuities.

Liquid-penetrant tests are applicable to most materials which are weldable (Fig. 17-19a and b). They have been used successfully for ferrous metals, nonferrous metals, glass, and plastics. The testing procedure is basically the same regardless of the type of material or joint being tested. One of the most useful tests for liquid penetrants is a leak test. In a leak test, the penetrant is put on one side of the joint and the developer on the other side. Leaks are then readily detectable by traces of the penetrant showing in the developer. In the case of a pressurized container or system, a measured amount of penetrant is placed in the device, and it is pressurized in the normal manner or with air pressure. Leaks are then easily detected when the developer is applied to the outside of the device.

Ultrasonic Tests

Ultrasonic tests in welding are used primarily to test the soundness of welds. These tests are fairly new in the welding area and are not widely understood. However, ultrasonic tests are capable of detecting both surface and subsurface discontinuities without damage to the piece being tested. The equipment requirements are minimal, and ultrasonic testing equipment may be a self-contained portable unit or a larger stationary unit (Fig. 17-20).

In ultrasonic testing, a beam of high-frequency sound is transmitted through the weld area by a transducer. This beam passes through the welded material with little loss, unless a defect is encountered, and is reflected back to the transducer. The

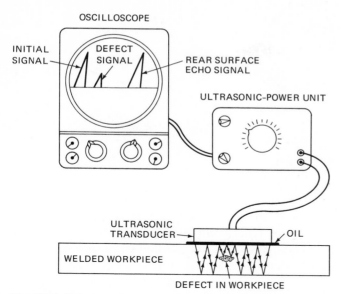

Fig. 17-20 Ultrasonic circuit for testing welds. The trace on an oscilloscope in an ultrasonic test.

transducer converts reflected waves into an electrical signal which is displayed on the cathode ray tube of an oscilloscope. A defect in the weld will show up as a small peak between the main signal and the echo signal from the rear surface of the workpiece (Fig. 17-20). The horizontal distance between the initial signal and the rear surface echo is a function of the time required for the sound wave to be reflected back to the transducer and can be used to determine the depth of a defect.

Evaluating the displayed results of an ultrasonic test requires an experienced and well-trained person. A person experienced in ultrasonic tests of welds can pinpoint defects in welds and also determine the size and shape of each defect.

Ultrasonic testing of welds is gaining popularity wherever welding is done because of the versatility and information provided by this process. Also, these tests are quick and do not damage the workpiece. Practically any material can be tested ultrasonically at a comparatively low cost. However, a high degree of skill is required to interpret the oscilloscope readings and convert them into indications of weld quality.

Radiographic Tests

Radiographic tests use X rays or gamma radiation to explore the depths of a welded workpiece and determine weld quality. These tests provide a permanent film record of the defects in a welded workpiece. Radiographic tests are expensive and require considerable time and equipment.

In radiographic tests, a piece of radiographic film is placed on one side of a welded workpiece. The other side of the workpiece is subjected to the energy of X rays or gamma rays (Fig. 17-21). The energy absorbed by the workpiece depends upon its density and thickness. Thicker or more dense workpieces absorb more energy than thinner or less dense workpieces; therefore, there is less energy to expose the radiographic film. When the exposed film is developed, the areas of low density will be dark and indicate defects in the weld area.

With either the X-ray machine or the radioactive source of gamma rays, the operator must be shielded from exposure to the rays with lead or concrete and steel. All these factors make radiographic tests the most expensive type of nondestructive tests. However, in spite of the cost factor, radiographic tests are the most widely used nondestructive testing methods and are used for practically all materials and joint types.

WELD DEFECTS

A part fabricated from two or more pieces joined by properly welded joints is as strong as the part made from a single piece of the same material. However, a number of defects appear in improperly welded joints which adversely affect the strength and quality of the joint, thereby making the joint weaker than it should be. The defects which occur with the most frequency in welded joints are described in the following paragraphs.

Cracks

Cracks are the most detrimental type of weld defect; therefore, they are a major concern in welding. Most weld specifications say that welds must be crack-free. A crack is a separation or rupture which occurs in or near the weld metal due to stresses which are created

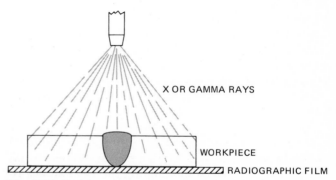

Fig. 17-21 Radiographic tests for weld defects.

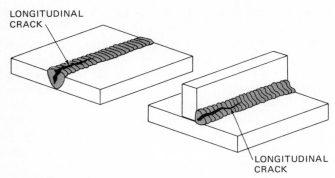

Fig. 17-22 Longitudinal cracks in welds.

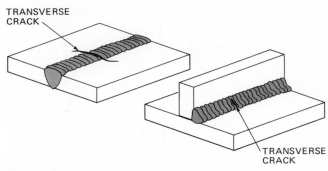

Fig. 17-23 Transverse cracks.

in the welded piece. Cracks are sometimes large and readily discernible; however, they are often very fine hairline cracks which are very difficult to detect. They are generally longitudinal (Fig. 17-22) or transverse (Fig. 17-23) and may vary considerably in their depth and length.

Cracks may be caused in a number of ways; however, they are usually a result of improper welding techniques for a particular joint and material. Overheating is the most common cause of cracks. Excess heat generated when making a weld creates a higher degree of expansion. The expansion, in turn, results in an excessive amount of stress in the welded joint. Excess heat also causes other elements to be attracted into the weld area which may cause stresses that result in a crack in the weld. The second most common cause of cracking is too rapid cooling or uneven cooling. The rate of cooling for a weld in a given type of metal is directly related to the amount of internal stress generated in a welded joint and is therefore related to the number and size of cracks which occur in the weld. The material being welded determines the correct temperature and cooling rate which should be used for a given type of weld in order to minimize the probability of a crack in the weld.

Porosity

Porosity is a result of a large number of small voids or gas pockets in the weld area which reduces the strength of a weld (Fig. 17-24). Most materials react chemically by absorbing other elements or generating gas when heated. Upon cooling the gas or elements thus attracted form porosity in a weld. Porosity may occur in varying amounts and may be evenly dispersed throughout a weld or concentrated in a small area.

Porosity is often caused by improper cleaning of the weld area. The dirt or other elements create a gas when burned, thus contaminating the weld and causing porosity. Porosity may also be caused by inade-

quate shielding of the weld area. Shielding keeps other elements from being absorbed into the weld, thereby reducing the possibility of porosity being created in the weld.

Cold Shut

A *cold shut* is an area where incomplete melting and a lack of fusion occurs in a weld. Generally the weld metal laps over the base metal, but they are not completely fused together. A cold shut often occurs where a weld is restarted. If the weld metal is not completely remelted into a molten puddle before restarting a weld, a cold shut is likely to occur. This is the most usual cause of cold shuts in welding.

Inclusions

Inclusions are another type of defect which severely weaken a weld. Inclusions occur when a solid material unlike that being welded is inadvertantly included in the weld metal. Dust, ceramic particles, or slag are the inclusions most frequently found in welds. Inclusions may be dispersed throughout the weld area; however, they most often occur as pockets (concentrated areas) or layers within the weld area (Fig. 17-25).

Lack of Fusion

The full strength of a welded joint is not achieved unless 100 percent fusion exists throughout the joint. Poor or low penetration results in a *lack of fusion*. Joints are sometimes welded on both sides of the workpiece to assure complete fusion and adequate

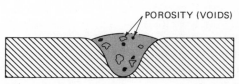

Fig. 17-24 Weld porosity.

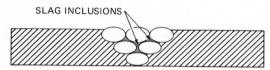

Fig. 17-25 Slag inclusions.

strength. The depth of penetration is affected by the joint preparation, especially joints in thick materials. An inadequately prepared joint is the most usual cause for lack of fusion in welding.

Undercut

A weld is *undercut* when the portion of the joint which has been welded is not completely filled in by the weld metal (bead). Figure 17-26 illustrates an undercut weld. Since undercutting reduces the cross-sectional area of the workpiece, the strength of an undercut joint is considerably less than the strength of a properly welded joint.

Undercutting is generally caused by poor control of the welding torch or electrode by the welder. When the welder holds the electrode or torch at an improper angle or uses the wrong travel speed, the resulting weld is generally undercut. Also, undercutting often results when the welding heat is too high for the existing welding conditions.

Welds defective due to undercutting are easily detected by visual inspection techniques. However, the effect of undercutting on the strength of the joint is not so easily determined. The loss in strength of a joint due to undercutting is often compounded by internal stresses in the joint and can only be determined by destructive-testing procedures.

The common welding defects, their causes, and corrective action to be taken are shown in Fig. 17-27. Many of these defects which sometimes occur in welds are difficult to detect. Therefore, inspecting welds and testing welds have become critical aspects in the field of welding. The number or size of allowable defects is often written into the specifications for welding structures or is contained in a building code. On other jobs, the minimum strength which must be

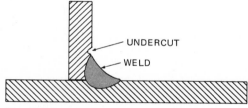

Fig. 17-26 Undercut weld.

Fig. 17-27 Causes and corrections for common weld defects.

Defect	Probable Cause	Corrective Action
Cracks	Poor welding action	Use appropriate action
	Wrong type filler metal	Use correct filler for metal being welded
	Poor joint preparation	Use correct joint spacing Vee joint properly
	Overheating weld	Decrease welding heat
	Stressed joint	Use multiple passes Pre and Post heat weld area
Porosity	Poor joint preparation	Clean weld area thoroughly
	Dirty filler metal	Clean the filler metal
	Insufficient shielding	Use correct electrode for position and metal or increase shield gas flow
Lack of fusion	Welding heat too low	Increase welding amperage or voltage
	Improper joint preparation	Vee the joint properly
	Arc too long or flame too far away from joint	Reduce arc gap or bring flame closer to joint
Cold shut	Improper restarting of weld	Ensure complete melting of previous weld where new one is to start
	Arc gap too long	Reduce arc gap
	Improper addition of filler metal	Add filler metal only in a molten puddle
Inclusions	Inadequate removal of slag from weld area	Remove all foreign matter and slag from weld area before welding
	Poor joint preparation	Clean and vee weld area
Undercuts	Incorrect welding procedure	Hold electrode or flame at proper angle
	Excessive welding heat	Reduce welding amperage or voltage or decrease tip size
	Improper welding speed	Use recommended welding speed

attained by the weld may be specified. Regardless of whether or not a weld must meet certain written standards, a welder is expected to produce quality welds of high strength and wants the weld made to measure up to these expectations. The welder may have to be certified to show he or she can hold to these standards. The inspection and testing procedures described in the above paragraphs are the way a welder or welding inspector determines if a weld reaches acceptable standards.

WELDING PLASTICS

All plastics are divided into one of two groups: thermosetting plastics and thermoplastics. Thermosetting plastics set up or cure when heated and once they have cured they will not remelt, soften, or fuse to another material. Thus thermosetting plastics are not weldable. On the other hand, thermoplastics soften when heated and will fuse with another thermoplastic if subjected to light pressure while both pieces are soft. To determine if a piece of plastic is a thermoplastic or a thermosetting plastic, heat a small rod or nail until it is blue, around 500°F (260°C), and touch it against the piece of plastic. If the nail starts to melt into the plastic, it is a thermoplastic; however, if the heated rod doesn't affect the plastic, it is a thermosetting plastic. Thermosetting plastics cannot be welded by the procedures discussed in the following paragraphs and must be glued, bonded with adhesives, or fastened with mechanical fasteners. In addition to these methods of joining plastics, most thermoplastics may be welded.

Not all thermoplastics can be welded since the plastics to be welded must be heated and a fairly low heat destroys some thermoplastics. Also, some plastics oxidize and will not bond. The plastic manufacturer's recommendations should always be followed when welding plastics. However, the general procedure is to heat the joint until it is soft and tacky. Light pressure is then applied to the joint which causes the pieces of plastic to fuse together. Filler rod may or may not be added to the joint depending upon whether or not the joint needs to be filled or strengthened.

In welding plastics, there are three basic ways in which the joint is heated. The joint may be heated by friction, by direct contact with a heated iron, or by hot gas or air. These three ways of heating the joint constitute the three common methods for welding plastics. Regardless of the welding procedure used, pressure must be applied to the joint to ensure fusion of the pieces being joined. The amount of heat and pressure required for satisfactory welds is determined by the type of plastic and type of joint. The manufacturer's recommendations for each type of plastic should be used as the reference for any welding application. However, in general, the heat required for welding plastics is relatively low, in the approximate range of 400 to 600°F (200 to 325°C). Table 17-2 gives the approximate welding temperatures of some commonly welded plastics. The pressure required for welding plastics is also fairly low, 10 to 150 psi (7031 to 105,465 kgs per m²).

Friction Welding

Generally *friction welding* is used for welding joints where one member of the joint is circular. It is also possible to make circular welds on noncircular pieces provided one member of the joint can be spun or rotated. In friction welding, one member of the joint (usually the circular one) is rotated and pressed against the other member of the joint. Friction between the two pieces heats the joint to the proper welding temperature, and pressure is held on them until they are fused together. It is essential that the rotating member be stopped from rotating as the pressure required for fusion is applied. Friction welding, when properly done, results in an excellent joint because it is possible to have 100 percent fusion across the weld interface, and the pressure flushes out the joint of undesirable contaminants.

Friction welding is a popular manufacturing process for plastic products because of its several advantages and relatively few disadvantages. Among the advantages associated with friction is its speed and economy. Only a few seconds are required to complete a joint. No fuel, filler rod, or elaborate equipment is required for friction welds. However, one of

Table 17-2 **Approximate Welding Temperatures for Common Thermoplastics**

Type of Plastic	°F	°C
Polyethlene (low density)	500	260
Polyethlene (high density)	510	266
P V C	475–550	246–283
Polypropylene	550	288
Polyvinylidene fluoride	600	316
Acrylic	600–650	316–343
Chlorinated polyester	600–650	316–343

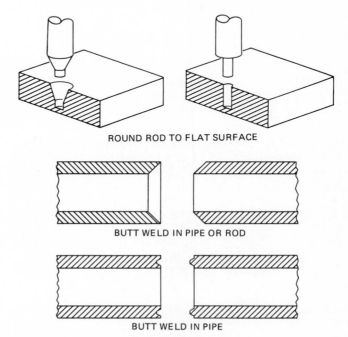

ROUND ROD TO FLAT SURFACE

BUTT WELD IN PIPE OR ROD

BUTT WELD IN PIPE

Fig. 17-28 Special joint designs for friction welding.

the two members being joined must be spun, thus limiting the size of the article which is practical to weld by friction. Another disadvantage is the small ring of flash (excess plastic squeezed out of the joint) which occurs as the joint is completed. Also, although a circular butt joint is the basic joint design for friction welding, many variations have been developed which require special and critical preparation prior to welding (see Fig. 17-28).

Hot-tool welding

In *hot-tool welding,* a heat source such as a soldering iron, strip heater, hot plate, and so on, is brought into contact with, or to within 1/8 of an inch [0.032 cm] of, the joint being welded. When the joint reaches the correct temperature, pressure is applied to the pieces of plastic being welded until they fuse together. The pressure should be retained on the joint until it has cooled. Both the time and pressure required for hot-tool welding are relatively low. Figure 17-29 shows a typical hot-tool welding application.

The time between heating the joint and applying pressure to the joint is the most crucial aspect of hot-tool welding. This time period must be short, about 1 or 2 seconds for most joints. If an excessive amount of time elapses between heating the joint and applying pressure to it, poor fusion will result and the joint will be weak.

Although the temperature required for welding plastics is considered to be low, 400 to 700°F degrees (200 to 375°C), it must be hot enough to cause the plastic to melt and become sticky. This causes problems when the heating tools make contact with the joint because they tend to stick together. A number of lubricants and parting agents have been used in an effort to keep the plastic being welded from sticking to the heating tool; however, they proved to be detri-

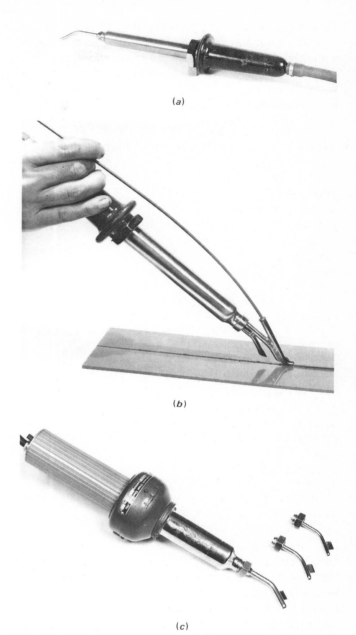

(a)

(b)

(c)

Fig. 17-30 Hot-gas welding torches used for welding plastics. (a) Hot-gas torch. Uses separate unit to heat and blow gas through the torch. (*Laramy Products.*) (b) Hot-gas torch with filtered-rod guide tube. Gas is heated and blown through the torch by a separate unit. (*Laramy Products.*) (c) Self-contained hot air torch, heater, and blower. (*Leister Corp.*)

mental to good weld strength and their use was abandoned.

The hot tool welding process works well for thin-sheet stock and for butt welds on plastic pipe. The process is readily adaptable to high-volume production and automation since the entire process can be completed in a few seconds. Also, the welded piece

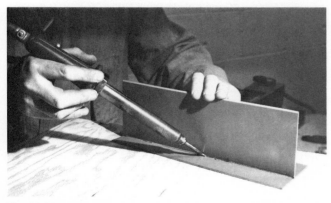

Fig. 17-29 Hot tool welding a T joint. (*Laramy Products.*)

can be put into service after cooling for only 10 to 15 s. The required heating and cooling time is determined by the type of plastic, the type of joint, and the thickness of the plastic being welded.

Hot-Gas Welding

A special welding torch is used for *hot-gas welding* plastics (Fig. 17-30*a*, *b*, and *c*). This torch heats a gas or air as it is blown through it. The tip of the torch allows the welder to direct the stream of hot air onto the joint being welded. Most torches use an electric heating element to heat the air or gas as it is blown through the torch; however, torches which use a gas flame for heating the air are also available. The torch must be capable of heating the air to temperatures up to about 800°F (425°C). Also, provisions must be made to adjust and accurately control the temperature of the stream of hot air provided from the torch.

In joining plastic pieces by hot-gas welding, the joint is first prepared by sanding, filing, cutting, or sawing. Figure 17-31 shows some common joints for plastics. Joints in thick plastics are beveled to form a vee joint while thin plastics may be positioned with a small gap between the square edges of the pieces to be welded. When the joint has been properly prepared, the pieces are tacked in position at strategic spots (usually around the corners or near the ends). Tacking is accomplished by fusing the members of the joint together at various spots along the joint. These spots then hold the pieces in proper alignment as the weld is made and allow the joint to be moved as needed by the welder (Fig. 17-32*a* and *b*).

The next step is to weld the joint. This is done by

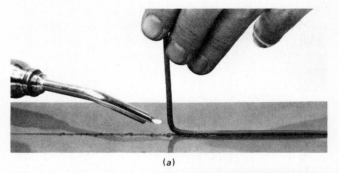

(a)

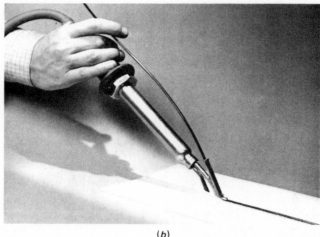

(b)

Fig. 17-32 Adding filler rod in plastic welding. (*a*) Butt weld without guide tube; (*b*) butt weld with guide tube.

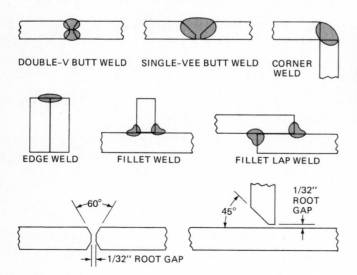

DOUBLE-V BUTT WELD   SINGLE-VEE BUTT WELD   CORNER WELD

EDGE WELD   FILLET WELD   FILLET LAP WELD

60°
←1/32″ ROOT GAP

45°
1/32″ ROOT GAP

BOTH MEMBERS VEE GROOVED
Fig. 17-31 Joints for welds in plastic.

Fig. 17-33 Welding plastic pipe using a hot-gas torch with filler rod guide. (*Laramy Products.*)

Fig. 17-34 Probable cause and corrective action for the common defects found in plastic welds.

Defect	Probable Cause	Corrective Action
Poor Fusion	Too much heat	Reduce torch temperature
	Travel speed too fast	Slow travel speed
	Filler rod too large	Use correct size filler rod
	Nonuniform pressure on filler rod	Hold filler rod with light and steady pressure
	Poor joint preparation	Vee joint at proper angle Use correct gap in joint
Charing	Too much heat	Reduce welding temperature
	Travel speed too slow	
	Jerky torch movement	Hold torch further from work
		Alternate heating filler rod and workpiece
		Speed up welding speed
		Heat joint and filler rod the same length of time (one-half second each)
Cracking	Wrong welding temperature	Use recommended welding temperature
	Wrong type filler area	Use correct filler area for plastic being welded
	Chemical reaction or oxidation	Use recommended welding procedures
	Stressed joint	Allow for expansion of joint when clamping or welding a joint in plastic
Warping	Overheating the joint	Control the heat and use as little heat as possible
	Welding temperature variation	Keep welding temperature uniform for entire joint
	Poor joint preparation	Follow recommended procedures for preparation of joints
	Poor clamping or tacking of joint prior to weld	Make strong tack welds or clamp securely at critical points before welding

heating the joint with the hot-gas torch until the surface of the joint becomes tacky. Care must be taken not to overheat the joint since this will char the plastic and cause a poor joint. The heating is started at one end of the joint, and when the joint reaches the proper temperature, a plastic filler rod is held at equal angles to the pieces being joined at 90° to the joint lengthwise (see Fig. 17-33). The heat is directed alternately to the joint and filler rod at approximately one-half second intervals. Constant pressure is main-

tained on the filler rod as it and the joint are heated. When the proper temperature for welding is reached, the pressure on the filler rod causes it to flow into the joint and fuse the members of the joint together. Too little pressure on the filler rod will result in poor fusion, and too much pressure on the filler rod will cause poor fusion or bulging of the joint, depending upon the temperature of the joint.

The variables of temperature, welding speed, angle of filler rod, and pressure on the filler rod must be controlled properly for satisfactory weld to be made in plastics. Failure of the welder to control these variables will result in defective welds. Defects of warpage, incomplete fusion, cracking, and so on are similar to those found in metals and are detected in a similar manner by destructive and nondestructive testing. These testing techniques were discussed in the preceding section of this chapter. Figure 17-34 presents the possible causes and the corrective action required for some of the defects commonly found in welded plastics.

REVIEW QUESTIONS

17-1. Describe the process of welding as it applies to metals and plastics.

17-2. What happens to a metallic structure of a piece of metal undergoing welding?

17-3. Why do metals expand when they undergo welding, and what does that mean in terms of design?

17-4. What type of inspection methods would you use to check the soundness of welds of a building?

17-5. Examine a welding joint and identify some of the defects by visual examination.

17-6. Identify the most suitable welding process for welding aluminum, steel, and cast iron.

17-7. What role does the coefficient of thermal expansion play in welding of metals?

17-8. What would you do to prevent warpage of an assembly consisting of dissimilar metals welded together?

CHAPTER 18

WELDING PROCESSES AND EQUIPMENT

The development of welding as a method of joining metal parts together for both fabrication and repair work has been a steady parade of new welding machines, more and better supplies and improved welding processes (Fig. 18-1a–c). Many of the welding processes which have been developed are very specialized and usually require specialized equipment. However, each new process which is developed overcomes a limitation, weakness, or problem area associated with the existing process.

The major differences in the welding processes and equipment which have been developed relate to:

1. The use and sources of heat for welding.
2. The use and sources of pressure required for welding.
3. The way the weld area is shielded from atmospheric contamination.
4. The welding situation for which the technique is suited.

SOLID STATE WELDING

In solid state welding (pressure welding), joining is accomplished through the application of heat, pressure, or both; however, the workpiece is not melted. A number of processes have been developed for joining various materials to meet different job specifications. As shown in Fig. 18-2, forge welding, friction welding, cold welding, diffusion welding, ultrasonic welding, and explosive welding are the basic types of solid state welding.

Forge Welding

Forge welding is believed to be the oldest form of welding. In forge welding, the workpieces are heated in a forge. Early forges were coal fired, while modern forges burn gas or heat by induction. The pieces to be welded must be heated uniformly across the joint interface until they are hot enough so their shape can be readily changed. At this stage in the forge welding process the workpieces are removed from the heat and are overlapped for welding. Pressure is then applied until a union of the workpieces is achieved.

The pressure for forge welding can be applied by hammering on an anvil (hammer welding), by squeezing between dies (die welding), or by rolling between rolls (roll welding). In hammer welding, the hammer blows may be supplied by hand or from modern automatic or semiautomatic hammers. Although hammer welding was widely used to join metal parts by the village blacksmith, modern fabrication and repair shops have largely replaced it with other welding processes.

Die welding has limited but important applications in modern industry. It provides a means of forming and joining workpieces at the same time.

Roll welding is used extensively in the cladding of steel sheets. Cleaned sheets of steel and cladding material are sandwiched together and uniformly heated to around 2000°F [1032°C]. Pressure is applied by rolling until the cladding material and base metal are joined.

Heating is critical in all forge-welding processes. If the surfaces to be joined are underheated, the surfaces will not bond together. Overheating or burning the surfaces to be welded will result in a rough, brittle joint that has little strength. Not only is the temperature critical in forge welding but the temperature of the surfaces to be joined must be uniform in order to achieve uniform bonding of the surfaces.

The joint to be forge welded must also be clean and free of oxides or scale. When a metal is heated, it reacts with the oxygen in the air to form oxides on the metal's surface. These oxides inhibit the bonding of the workpieces. Therefore a flux, usually borax, is

(a)

(b)

Fig. 18-1 (a) Complete MIG welding unit with constant potential power source. (*Miller Electric.*) (b) An inertia welding machine. (*Caterpillar Tractor Co.*)

used to remove surface oxides and prevent further oxidation.

Both ferrous and nonferrous metals have been successfully forge welded. Typical joint designs are shown in Fig. 18-3. Joint designs must allow intimate contact of the surfaces to be welded and often have a slightly convex surface.

Friction Welding

Friction welding, also called *inertia welding* (Fig. 18-4), is a relatively simple process. Basically it consists of rotating, at high speed, one of the two pieces to be welded while the other piece is stationary. Frictional forces cause the contacting surfaces to heat up. When the proper temperature for welding is reached, the rotating piece is stopped and the two pieces to be welded are pressed tightly together. Generally the pieces are heated up to the welding temperature, which is well below their melting temperature, in a few seconds. Thus a joint can be made very rapidly by friction welding.

In friction welding, the parts are joined across the entire interface (Fig. 18-5). This produces a very strong joint and is used to butt weld rods and cylindrically shaped pieces. It was first used commercially to weld tool joints to drill pipe for use in drilling oil

wells. Current uses for friction welding include the joining of steering shafts to gear assemblies, valve stems to valve heads, gears to hubs, yokes to drive shafts, drills to shanks, axle shafts to hubs or gears, shanks to reamers, and many other uses.

Friction welding has been successfully used to weld a number of materials. Although it is most often considered in relation to joining various types of steel, it can be used to join nonferrous metals or dissimilar metals such as copper to aluminum, brass to steel, titanium to aluminum, stainless steel to zirconium, and others. Friction welding is also used extensively to join most thermoplastics.

Cold Welding

The joining of two metals by the application of pressure alone is known as *cold welding*. Cold-pressure welding and solid-state bonding are names also given to this process. In cold welding, the metals to be joined are overlapped or butted together. High pressure is then exerted on the joint until the pieces bond together. The pressure required for cold welding depends on the thickness of metal being welded, area of the weld itself, and the type of metal being welded. Pressures of 400,000 psi [28,000 kg/cm²] are not uncommon for aluminum, and about three times this

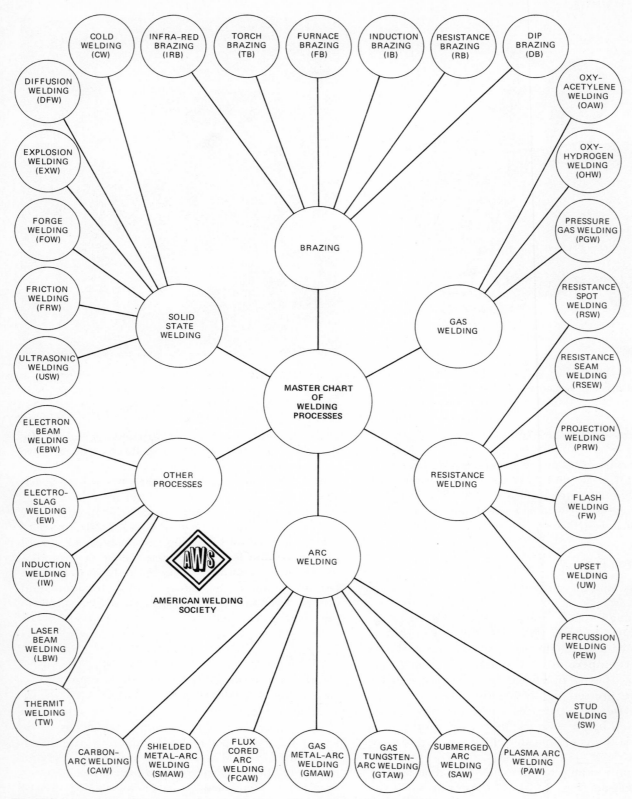

Fig. 18-2 A complete welding process chart. (*Courtesy of the American Welding Society.*)

294

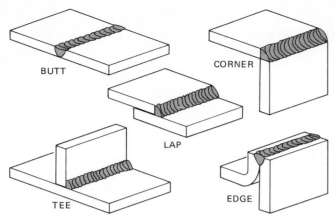

Fig. 18-3 Basic types of joints.

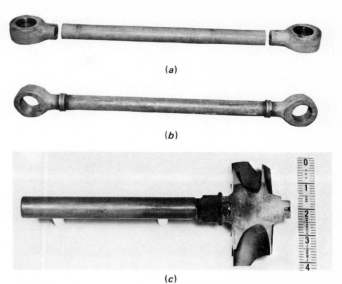

Fig. 18-5 A typical inertia-welded part. (*a*) Before welding, (*b*) after welding. (*c*) cut-away of a shaft welded to a turbine wheel by inertia welding. (*Courtesy Caterpillar Tractor Co.*)

amount is required to cold weld copper. Any metal that is plastic enough to flow and not become brittle under pressure can be cold welded. Aluminum, copper, almost pure lead, zinc, nickel, silver, platinum, gold, cadmium, and most of their alloys can be cold welded to themselves or to each other.

Metals to be cold welded must be clean and should be wire-brushed just before welding. Mechanical cleaning has proved to be superior to chemical cleaning for pieces to be cold welded, because most chemicals leave a residue on the surface.

Cold welding has found its widest application in joining wires and in joining wire to thin sheets of metal. Also, heat-sensitive containers and aluminum food containers are joined by cold welding. Parts which are easily damaged by heat, such as semiconductors used in the manufacture of transistors, can be cold welded without damage from the heat, flame, and sparks normally associated with welding.

Diffusion Welding

Diffusion welding or *diffusion bonding,* as it is sometimes called, is based on the metallurgical phenomenon of the diffusion or <u>movement of atoms from one piece</u>

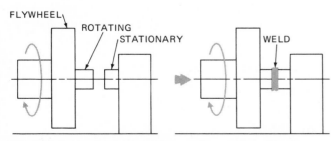

Fig. 18-4 Inertia (friction) welding unit in operation. (*Courtesy Caterpillar Tractor Co.*)

<u>of metal to another.</u> This atom movement is possible only if the metal pieces are extremely clean and fit together perfectly. The diffusion of atoms occurs more rapidly at higher temperatures; therefore, diffusion welding is generally done at elevated temperatures, yet below the melting point of the metals being joined. Pressure is used in diffusion welding to bring the metal surfaces into very close contact. The surfaces to be welded must be within an atom's width of each other.

Many dissimilar metals which were previously thought to be unweldable are now being joined by diffusion welding. However, all metals cannot be joined by diffusion welding since their atomic movement is slow, thus limiting the diffusion that can occur.

Joints made by diffusion welding may be as strong as the base metals, or they may be weak and brittle. The problem is in determining which type of weld was achieved since nondestructive tests rely on voids and cracks which are nonexistent in diffusion-welded joints. Destructive testing of randomly selected samples seems to be the best method of determining the quality of diffusion-welded joints.

Ultrasonic Welding

Ultrasonic welding employs a high-frequency vibrating tool to impart vibratory (ultrasonic) energy into the weld area (Fig. 18-6). This energy causes a minor <u>decrease in the thickness of the metals in the weld area,</u> and this action in turn bonds the pieces of metal to-

(a)

(b)

Fig. 18-6 Complete ultrasonic welding unit. Shows close-up of working zone. (*Courtesy Aeroprojects, Inc.*)

gether. Fluxes, filler metal, or heat are not required in ultrasonic welding.

Practically all the nonferrous metals and iron are capable of being joined by ultrasonic welding. Figure 18-7 shows typical common applications of ultrasonic welding. Most of its present industrial applications involve foils and thin pieces of metal. Present equipment is only capable of joining metals of 0.1 in (0.25 cm) or less in thickness. If the metals being joined are of different thicknesses, only one of the two pieces need be 0.1 in (0.25 cm) or less in thickness. There is no apparent lower limit to the thickness

of metals which can be joined by ultrasonic welding. Wires less than 0.0005 in [0.013 mm] and foils less than 0.0002 in [0.005 mm] in thicknesses have been successfully welded by ultrasonic methods. Figure 18-8 shows typical parts welded ultrasonically.

Light cleaning of the surfaces to be welded ultrasonically is the only surface preparation required for ultrasonic welding. The vibratory energy used in the welding process breaks up surface oxides and films normally found on metal surfaces. Joints are of an overlapping design, and welding time is less than a second.

Explosive Welding

Many metals have been successfully welded using the energy from an explosion. *Explosive welding* is generally used to weld plates of metal together in the form of a sandwich or to clad one sheet of metal onto another sheet or plate. In joining two plates by explosive welding, the plates are placed at an angle to each other. Explosive material is placed around the outer surfaces of the two pieces. When this material is detonated, the explosive force slaps the two plates together. The heat and surface ripples created by the dissipation of the explosive force through the surfaces of two metal plates cause a weld to be formed at their interface.

In cladding metals by use of explosives, the sheet of metal to be clad is placed on an anvil or flat base. A sheet of cladding metal is then placed over it, and a sheet of explosive material is placed over the cladding metal. The metals are bonded together by detonating the explosive material progressively from one edge of the material to the other.

Explosive welding is very dangerous and must be done by specially trained workers. Also, many states and municipalities have laws governing the use of explosives which require special permits and safeguards for their uses.

All the preceding solid-state-welding methods are nonfusion (the base metals are not melted) processes. They are finding more and more applications in the aerospace and electronic component industries. Most of these processes have some distinct advantages over other welding processes due primarily to the lower temperatures involved in their execution. Some of their advantages include:

1. High efficiency.
2. Reduced oxidation.
3. Little shrinkage or heat cracks.
4. Precise control over the variables of pressure, time, and temperature.

(a)   (b)

Fig. 18-7 Applications of ultrasonic welding.

5. Joining across the entire interface.
6. Metals dissimilar readily joined.

The major disadvantages to solid-state-welding processes are that (1) they are limited to thin stock or special geometrics and (2) joints must be fit to close tolerances.

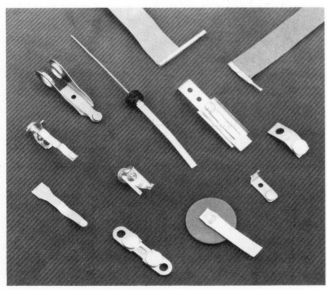

Fig. 18-8 Typical parts welded by ultrasonic welding unit. *(Courtesy Aeroprojects, Inc.)*

RESISTANCE WELDING

Resistance welding includes welding processes in which the metal is heated by passing an electric current through the joint. The resistance of the metal to the flow of an electric current causes the metal at the joint's interface to melt. The molten metal in the joint is then allowed to resolidify, thus joining the workpieces together. Joints are made very quickly by resistance welding and therefore it is used extensively in mass production. As seen in Fig. 18-2, spot welding, seam welding, projection welding, flash welding, upset welding, and percussion welding make up the resistance-welding processes.

Spot Welding

Spot welding is the most commonly used resistance welding method. It is used to produce overlapping joints in sheet metal. In spot welding, the joint to be welded is placed between the spot welder's electrodes, called *tips* (Fig. 18-9*A*, *B*, and *C*). One of the tips is movable and light pressure is applied to the joint by lowering it with mechanical, pneumatic, or hydraulic pressure. When pressure is against the joint, a low-voltage (2 or 3 volts, V), high-amperage (around 100,000 amperes, A) current is passed through the joint from one tip to the other. The resistance of the

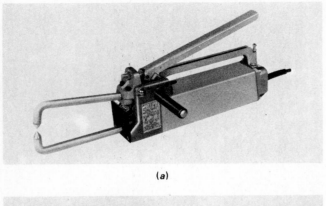

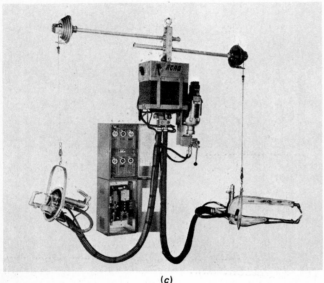

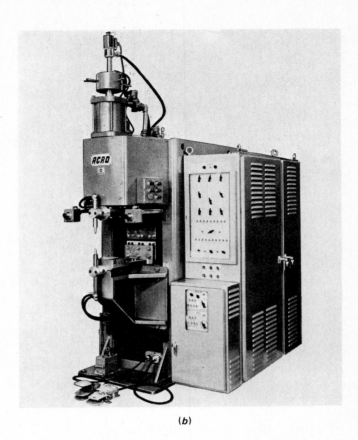

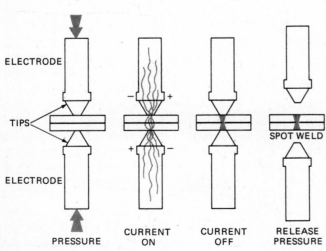

Fig. 18-9 (a) Portable spot-welding unit. (*Miller Electric.*) (b) Complete 150 KVA three-phase spot-welding unit. (*Courtesy Acro Welder Mfg. Co.*) (c) Complete dual air-hydraulic portable-production-gun spot-welding unit. (*Courtesy Acro Welder Mfg. Co.*)

Fig. 18-10 Stages in spot welding.

metal to the high electric current causes the joint to get hot and melt. When the metal at the joint melts, the flow of current is stopped, but the pressure is retained on the joint by the tips until the metal in the joint has solidified. This spot of resolidified metal in the joint is the spot weld, and it permanently joins the pieces of metal together. Figure 18-10 shows the spot-welding process.

Seam Welding

Seam welding is done on a machine similar to the one used in spot welding, except it has wheels instead of tips (Fig. 18-11). The workpieces are placed between the wheels which are held against the joint with pressure (Fig. 18-12). The joint is moved between the wheels at a constant speed while the welding current flowing through the joint from one wheel to the other

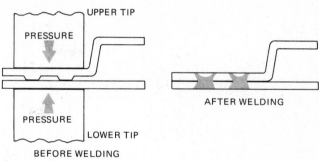

Fig. 18-13 Projection welding.

Fig. 18-11 Complete 150–250 KVA, 54-in throat, upper gear-driven circular seam-welding unit. (*Courtesy Acro Mfg. Co.*)

is switched off and on. The result is a continuous seam weld made by what would appear to be overlapping spot welds.

Seam-welded joints are gas and water tight and are used extensively in the manufacture of tanks, cans, and mechanical tubing.

Projection Welding

In *projection welding,* one of the members to be joined is designed to have small dents or projections like little feet on the joint (Fig. 18-13). A high-amperage electric current is passed between the tips causing the

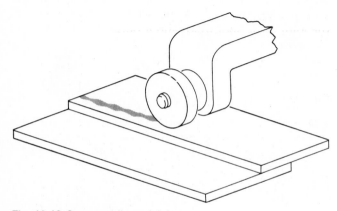

Fig. 18-12 Seam welding a joint.

projections to be melted and fused into the other piece of metal. The heat of the electric current is concentrated in the projection because it is the part of the joint which makes electrical contact between the two parts.

Several projection welds can be made at the same time. This makes projection welding highly desirable for large-volume production. Also, the electrode tips for projection welding last much longer than those used for spot welding. Some projection welding tips also align the pieces to be welded; therefore, in projection welding, welds will always be in the exact location desired. Projection welds are easily made on dissimilar metals and metals of different thicknesses. The projection is made on the metal of least resistance or, in the case of different thicknesses, on the thicker piece since these are the members which will normally require more heat to reach the correct welding temperature.

Flash Welding

Flash welding is also called *flash-butt welding* since a butt joint is the only type of weld made by this process. The process involves fastening two pieces of metal into the flash-welding machine (Fig. 18-14). The machine brings the pieces of metal together end-to-end, and an electric current is passed through them. As the temperature of the pieces being welded increases, their edges melt and the two pieces are separated by a narrow gap. The current arcs or flashes across this gap, and in so doing, it generates an intense heat between the two pieces. The fusion temperature required at the joint is reached very quickly. As the fusion temperature is reached, the two pieces remain in the welding machine under light pressure until the joint has resolidified.

Flash-butt welding results in complete fusion of the end surfaces of two pieces of metal. The metal near the joint will be upset (made shorter and larger) by the pressure applied to the joint (Fig. 18-15). This

Fig. 18-14 Automatic flash-welding machine. (*Courtesy Acro Welder Mfg. Co.*)

slight bulging of the joint requires a machining operation to restore the metal pieces to their original size. The metal thus is as strong as a single piece of metal of the same size.

Flash-butt welding is used extensively for joining band-saw blades, sheet-metal parts, wire, and rod into ring shapes.

Upset Welding

Upset welding is very similar to flash-butt welding and has been, for the most part, replaced by flash-butt welding. In upset welding heavy pressure is exerted on the joint before, during, and following the flow of current through the joint. The metal at the joint interface heats due to its resistance to the flow of an electric current, but no flash occurs. The power requirements and heat time for upset welding is much greater than is required for flash-butt welding. The higher cost of upset welding is the major factor influencing the decline in popularity of upset welding in favor of flash-butt welding.

Percussion Welding

Percussion welding employs an arc or flash produced by the discharging of an electrostatic capacitor through a joint to be welded. This rapid discharge of stored electrical energy produces sufficient heat for fusion of the interface between two metal pieces which have pressure percussively applied to them.

The percussion force must be at or immediately following the discharge of the capacitor through the joint. Percussive welding is used to join rod, pipe, or tubing to each other or to a flat surface. The process is applicable only to butt-type joints of two separate pieces of metal with surface areas of ½ in² [3.25 cm²] or less. Since the heat-affected area is extremely shallow, usually less than 0.01 in (0.025 cm) percussion welding can be used to join heat-treated steels or cold-worked metals without affecting the metals' mechanical properties.

Percussion welding can be used to weld dissimilar metals not considered weldable by most other welding processes. It is more expensive to percussive weld metals than to flash weld them; therefore, percussion welding is generally used only for joining metals which are not weldable by flash welding or in situations where flashes must be avoided.

All the resistance welding techniques are fast, permanent methods of fastening two metals together. They are used for both butt and overlapping joints. The training of workers for resistance welding is relatively simple and only requires a short period of on-the-job training. Since most resistance welders are of the automatic or semiautomatic type, the chance for human error is relatively low.

Resistance welding has a number of advantages and limitations when compared with other welding processes. The major advantage of resistance welding is that the welded joint does not increase the total weight of the finished object. For this reason, resistance welding is used extensively in the transportation and aerospace industries. Also, most resistance welds are made very rapidly with a low probability of human error. On the other hand, resistance welds are limited to relatively thin or small areas and can be done only on materials which are conductors of electricity.

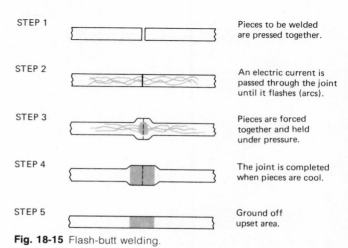

STEP 1		Pieces to be welded are pressed together.
STEP 2		An electric current is passed through the joint until it flashes (arcs).
STEP 3		Pieces are forced together and held under pressure.
STEP 4		The joint is completed when pieces are cool.
STEP 5		Ground off upset area.

Fig. 18-15 Flash-butt welding.

GAS WELDING

Gas welding includes all the welding processes where the source of heat is a gas flame or flames. The joint may be made with or without filler metal and with or without pressure. A fuel gas, such as acetylene gas, propane gas, natural gas, or mapp gas, is burned in conjunction with oxygen gas. The oxygen gas may be in the form of compressed air, but pure oxygen is more commonly used.

In gas welding, the fuel gas must be uniformly mixed with oxygen. This is accomplished in a mixing chamber which is made as a part of a blowpipe or torch (Fig. 18-16). The torch provides the welder with a means of moving, directing, or manipulating the flame. When the mixture of fuel gas and oxygen is ignited, the oxygen supports the combustion of the fuel gas and produces high-flame temperatures, causing the mixture to burn with a much hotter heat. Table 18-1 shows the flame temperatures for the gases commonly used in gas welding.

The characteristics of these fuel gases are shown in Table 18-2 (see page 302). Acetylene gas is used for gas welding considerably more than the other three gases because when combined with oxygen in the proper proportions, it produces a flame about 1000°F [537°C], hotter than either natural gas or propane. However, acetylene gas is very unstable at pressures over 15 psi [1.05 kg/cm²] and must be stored in specially constructed containers (Fig. 18-17).

The container for acetylene is a steel cylinder which has been filled with a porous substance that is saturated with acetone. Acetone has the ability to absorb

Table 18-1 Flame Temperatures for Common Fuel Gasses in Oxygen

5600°F or 3093°C—acetylene gas
5300°F or 2927°C—mapp gas
4600°F or 2538°C—natural gas
4580°F or 2527°C—propane gas

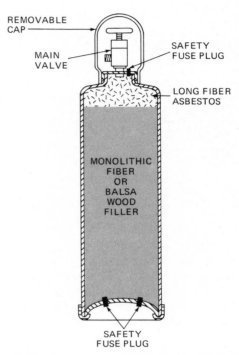

Fig. 18-17 Cross-sectional view of a typical acetylene cylinder. Some cylinders have a metal ring welded around their tops instead of the removable cap to protect the main valve.

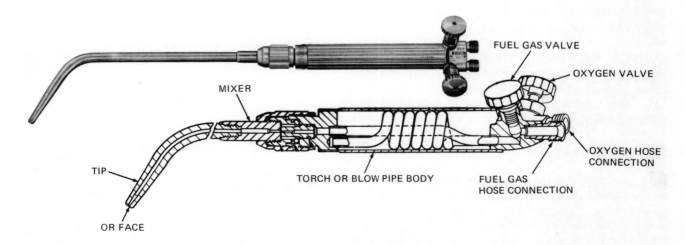

Fig. 18-16 Typical all-purpose welding torch. (*Courtesy: Linde Division, Union Carbide Corp.*)

Table 18-2 **Comparison of Welding Fuel Gasses**

	MAPP Gas	Acetylene	Natural Gas	Propane
Safety				
Shock sensitivity	Stable	Unstable	Stable	Stable
Explosive limits in oxygen, %	2.5–60	3.0–93	5.0–59	2.4–57
Explosive limits in air, %	3.4–10.8	2.5–80	5.3–14	2.3–9.5
Maximum allowable regulator pressure psi	Cylinder (225 psig at 130°F)	15	Line	Cylinder
Burning velocity in oxygen, ft/sec	15.4	22.7	15.2	12.2
Tendency to backfire	Slight	Considerable	Slight	Slight
Toxicity	Low	Low	Low	Low
Reactions with common materials	Avoid alloys with more than 67% copper	Avoid alloys with more than 67% copper	Few restrictions	Few restrictions
Physical properties				
Specific gravity of liquid ($^{60}/_{60}$°F)	0.576	—	—	0.507
Pounds per gallon liquid at 60°F	4.80	—	—	4.28
Cubic feet per pound of gas at 60°F	8.85	14.6	23.6	8.66
Specific gravity of gas (air = 1) at 60°F	1.48	0.906	0.62	1.52
Vapor pressure at 70°F, psig	94	—	—	120
Boiling range, °F, 760 mm Hg	−36 to −4	−84	−161	−50
Flame temperature in oxygen, °F	5301	5589	4600	4579
Latent heat of vaporization at 25°C, BTU/lb	227	—	—	184
Total heating value (after vaporization BTU/lb	21,100	21,500	23,900	21,800

SOURCE: Mapp Products, booklet #AD6-Mapp-1001B, p. 10.

acetylene gas and stabilize it under pressure. To avoid drawing the acetylene out of the cylinder, only one-seventh of the cylinder's volume of acetylene can be used per hour. In an attempt to overcome the problems associated with the storage and use of acetylene gas, a new gas called *mapp gas* was developed. Mapp gas is stable under pressure and can be stored safely at high pressures. Also, a typical full cylinder of acetylene gas weighs 240 lb [109 kg] and contains only 20 lb [9 kg] of acetylene, while a comparable full cylinder of mapp gas weighs 120 lb [55 kg] and contains 70 lb [32 kg] of mapp gas. On the other hand, mapp gas requires over twice the amount of oxygen required by acetylene gas to produce a neutral flame for welding or heating.

Acetylene Gas

Acetylene gas is a compound of carbon and hydrogen whose chemical symbol is C_2H_2. It is a colorless gas which is lighter than air. Acetylene is highly combustible and has a very distinctive odor. At high temperatures, over 1435°F [764°C] or at pressures above 15 psi [1.05 kg/cm²], acetylene gas becomes unstable and the result may be an explosion with or without the presence of oxygen. When mixed with oxygen, it burns with the hottest flame of all the industrially used gases. Acetylene is produced by a chemical reaction between calcium carbide and water. The chemi-

cal equation which explains the reaction is:

$$CaC_2 + 2H_2O \longrightarrow C_2H_2 + Ca(OH)_2$$

Calcium carbide plus water forms acetylene and calcium hydroxide

For welding purposes, acetylene is purchased in 10 to 300 cubic-foot (ft³) cylinders (Fig. 18-18), or it is generated by the user in a carbide-to-water acetylene generator. Although the generation of acetylene gas by the user must be at a low pressure and is not as convenient as the use of cylinders, it is less expensive for large volume users to generate their own acetylene. Also, acetylene cannot be withdrawn from its cylinder at a high rate or the acetone in which it is absorbed will be drawn out with it. Therefore, large-volume users would have to have several cylinders in use at one time or have one extremely large cylinder of acetylene.

Propane Gas

Propane gas is produced from gas mixtures obtained from oil and gas wells. The chemical symbol for propane is C_3H_8, and it is used for heating applications and as a preheating fuel for gas-cutting operations. Propane is seldom used as a fuel in the gas welding of steel because it is not as hot as acetylene and has less than one-half the usable heat of an acetylene flame.

Propane also requires about four times the oxygen required by acetylene to burn with a neutral flame. Propane gas is supplied in hollow steel cylinders containing 20 to 100 lb; large-volume users can have a tank which is filled from a bulk delivery truck.

Natural Gas

Natural gas is found deep in the earth and is brought to the surface from gas wells. The gas supplied by these wells is distributed in pipelines. Actually natural gas is made up of several gases in different proportions depending upon the geographic location of the well. Its major components are ethane (C_2H_6) and methane (CH_4) gases. Natural gas produces approximately the same flame temperature and usable heat as propane; however, it requires less than one-half the oxygen required by propane to produce a neutral flame. Natural gas and propane are used for the same purposes; however, not all gas consumers have access to a pipeline which carries natural gas. Therefore, these potential users of natural gas must use the less economical propane gas for their heating requirements.

Mapp Gas

Mapp gas is a liquified compound of acetylene, methylacetylene propadiene, or just mapp for short. Mapp gas is a fairly recent development in industrial fuel gases. It is safer than acetylene and heats faster than either propane or natural gas. Mapp gas is supplied in either steel cylinders or bulk tanks. Since mapp gas is stable under pressure, it is supplied in hollow steel cylinders which weigh about one-fourth the weight of an acetylene cylinder of the same size. Although mapp gas is less expensive than acetylene, it has less than one-half the usable heat of acetylene. Also, a neutral flame of mapp gas uses over twice the oxygen per cubic foot than is required by acetylene. Small mapp-gas outfits like the one shown in Fig. 18-18 are very popular and practical for the hobbyist and people who enjoy home crafts.

Oxygen

Regardless of the fuel gas used, *oxygen* is required for the proper and effective combustion necessary to produce the high-temperature flames required in gas welding. Oxygen causes a fuel gas to burn faster, which results in a higher-flame temperature. Oxygen gas is odorless, tasteless, and colorless. It is produced from the atmosphere by an air-liquification process.

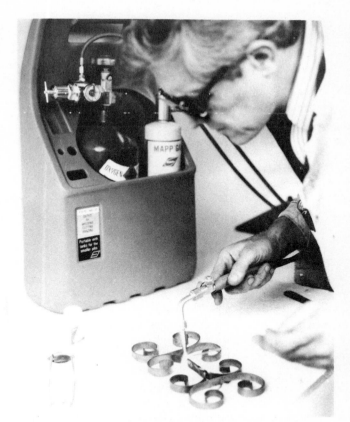

Fig. 18-18 Portable welding outfit with refillable 20-cu-ft oxygen cylinder and 1-lb (8.8 cu ft) disposable cylinder of mapp gas. (*Airco, Inc.*)

Oxygen gas is supplied at high pressures (approximately 2200 psi [154 kg/cm²]) in hollow, seamless, steel cylinders (Fig. 18-19). Figure 18-20 shows the great variety of cylinders available today. The safety valve built into the cylinder's main valve will pop open to relieve excess pressure if the cylinder pressure exceeds safe working limits. Oxygen is available in many different cylinder sizes; however, most welding shops use the 110 ft³ or 220 ft³ [3.3 or 6.6 m³] sizes.

Figure 18-21 shows the basic components necessary for gas welding. The relatively high pressures of the gases used in gas welding must be reduced to a usable pressure. These gases must also be supplied to the torch in a constant and uniform flow. Both of these functions are performed by gas regulators (Figs. 18-22 and 18-23).

GAS REGULATORS

Gas regulators are mechanical devices which regulate the flow of gases to the welding torch. There are two types of regulators, single stage and two stage. The

Gas-regulator connections and adjusting screws must be kept free of oil or grease. Oil or grease will serve as a fuel for high-pressure oxygen and can result in an explosion or fire.

GAS WELDING TORCH

The gas welding torch (Fig. 18-16) consists of fuel and oxygen hose connections, fuel and oxygen needle valves, a mixer, a tip, and a handle. The hose connections have left-hand threads for the fuel gas hose and right-hand threads for the oxygen hose. These connections fasten the gas hoses to the torch's needle valves. The needle valves independently control the flow of fuel and oxygen through the torch. The mixer thoroughly mixes the fuel and oxygen gases to ensure complete combustion and rapid burning of the fuel. The tip directs the flame toward the area to be welded. Tips are usually made of copper or a copper alloy. The opening or orifice in the end of the tip determines the volume of the flame. Larger tips have larger orifices and are capable of heating larger workpieces; however, the temperature of the flame is not

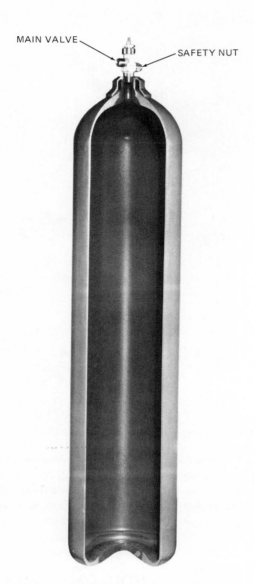

Fig. 18-19 Cross-sectional view of a typical oxygen cylinder. (*Courtesy: Pressed Steel Tank Co., Inc.*)

MAIN VALVE

SAFETY NUT

two-stage regulator provides more accurate gas regulation than the single-stage regulator. However, it is also more expensive than the single-stage regulator. Two-stage regulators are generally used to regulate the gases for small precise torches and for machine welding and cutting equipment. In order to ensure that oxygen regulators or hoses cannot be interchanged with the fuel regulator or hose, the inlet and outlet connections for fuel have left-hand threads, while the oxygen regulator and hose have right-hand threads (Fig. 18-24). The hoses are also color coded for easy identification. Red hoses are for fuel gas, and green or black hoses are for oxygen.

Fig. 18-20 Various kinds and sizes of compressed gas cylinders. Oxygen cylinders are seamless and have a removable cap to protect the main valve. Acetylene cylinders have seams and either a metal ring or removable cap to protect the main valve. (*Courtesy: Pressed Stell Tank Co., Inc.*)

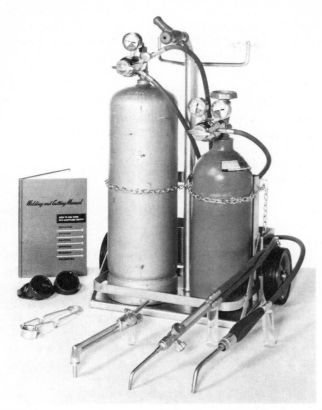

Fig. 18-21 Complete portable oxyacetylene heating, welding, and cutting unit for general maintenance work. (*Courtesy: Linde Division, Union Carbide Corp.*)

Fig. 18-23 An acetylene regulator with cylinder pressure gauge (0 to 400 psi) and working pressure gauge (0 to 15 psi).

affected by the size of the tip but the amount of energy at that temperature is high. See Table 18-3 for recommended tip sizes for oxyacetylene welding.

Gas welding torches are injector-type torches or equal-pressure-type torches. The injector torch is capable of operating with 1 psi [0.07 kg/cm^2] of acetylene pressure and 10 to 40 psi [0.7 to 2.8 kg/cm^2] of oxygen pressure. This is a necessary feature when an acetylene generator is used. The equal-pressure torch

Fig. 18-22 Oxygen regulator with cylinder pressure gauge (0 to 3000 psi) and working pressure gauge (0 to 100 psi).

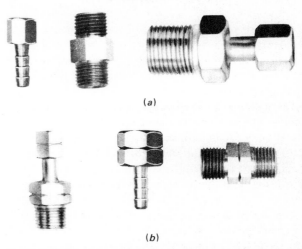

(a)

(b)

Fig. 18-24 Gas line connectors. (a) Groove indicates left-hand threads, used only for fuel gas; (b) connectors for oxygen hoses, right-hand threads.

Table 18-3 **Recommended Tip Sizes for Oxyacetylene Welding**

Metal Thickness	Tip Number	Orifice Diameter
Up to 1/32 in	000	0.021 in
1/64 to 3/64 in	00	0.028 in
1/32 to 5/64 in	0	0.035 in
3/64 to 3/32 in	1	0.040 in
1/16 to 1/8 in	2	0.0465 in
1/8 to 3/16 in	3	0.595 in
3/16 to 1/4 in	4	0.073 in
1/4 to 1/2 in	5	0.089 in
1/2 to 3/4 in	6	0.1065 in
3/4 to 1 1/4 in	7	0.1285 in
1 1/4 to 2 in	8	0.136 in
2 to 2 1/2 in	9	0.1405 in
2 1/2 to 3 in	10	0.144 in
3 to 3 1/2 in	11	0.147 in
3 1/2 to 4 in	12	0.1485 in

uses a supply pressure of 10 to 40 psi [0.7 to 2.8 kg/cm²] of pressure for both fuel and oxygen gases.

OXYACETYLENE WELDING

Oxyacetylene welding is the most commonly used type of gas welding. In this process the metal being welded is heated to the welding temperature by the combustion of oxygen and acetylene gas. This combustion produces a flame with a temperature of approximately 5600°F [3076°C] when equal parts of oxygen and acetylene are provided to the flame through the welding torch. This type of flame is called a *neutral flame*. The flame temperature can be increased to about 6300°F [3465°C] by increasing the amount of oxygen supplied to the flame. A flame with an excess of oxygen is called an *oxidizing flame*. A *reducing* or *carburizing flame* is produced when the amount of acetylene supplied to the flame exceeds the amount of oxygen supplied to the torch. The temperature of a carburizing flame is less than the neutral or oxidizing flame; however, the possibility of burning up the workpiece is much less with a slightly carburizing flame.

Oxyacetylene flames of various sizes or volumes are produced by different sized tips. A large flame is no hotter than a small flame; however, it is capable of heating a larger volume of metal in a given period of time. The oxyacetylene welding tip sizes and the thicknesses of metal recommended for each tip size are shown in Table 18-3. The hottest part of the oxyacetylene flame is the very tip of the small inner blue flame. The maximum amount of heat is transferred to the workpiece when the tip of the flame's inner cone just touches the surface of the workpiece.

In oxyacetylene welding the joint is melted by a flame produced by burning oxygen and acetylene gases. The welder directs the flame by means of a torch so that the inner cone of the flame almost touches the surface of the joint. When the joint melts, a small puddle of molten metal is formed. The flame is moved slowly along the joint producing small puddles of molten metal continuously for the length of the joint. Filler metal which has the same composition as the metals being joined may be added to the molten puddle as the puddle is moved across the joint. The use of a filler metal in welding helps to fill the joint so the finished weld is slightly thicker than the metal being joined. As the torch is moved slowly along the joint, the puddle of molten metal solidifies behind the torch and bonds together the pieces of metal being welded.

Oxyacetylene welding can be performed by the forehand technique or the backhand technique (Fig. 18-25). In the forehand technique, the torch points in the direction the weld is being made. This technique is used for metals less than 1/8-in [0.32-cm] thick since the flame preheats the metal ahead of the joint and does not melt the metal as deeply as the backhand technique. In the backhand technique, the flame points toward the metal already welded. This keeps the metal molten for a longer period of time and results in deeper penetration than is practical with the forehand technique. Backhand oxyacetylene welding is generally used for metals 1/8 in [0.32 cm] or more in thickness.

OXYHYDROGEN WELDING

Oxyhydrogen welding uses a flame produced by the combustion of oxygen and hydrogen gas. The oxyhy-

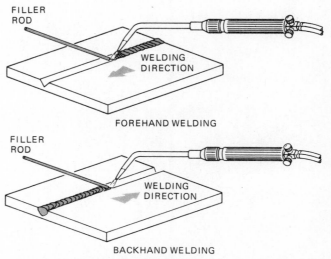

FILLER ROD

WELDING DIRECTION

FOREHAND WELDING

FILLER ROD

WELDING DIRECTION

BACKHAND WELDING

Fig. 18-25 Oxyacetylene welding techniques.

drogen flame is very difficult to see and gives no indication of the proportions of oxygen and hydrogen being burned. The flame is a fairly low temperature flame and is used primarily in the gas welding of aluminum, lead, and magnesium.

Oxyhydrogen welding is similar to oxyacetylene welding. The torch, hoses, mixers, and tips are used for either oxyacetylene or oxyhydrogen welding; however, a special regulator must be used for hydrogen gas. The flux, joint design, welding techniques, and the use of filler metal is also the same for oxyacetylene and oxyhydrogen welding.

Oxyacetylene welding can be used instead of oxyhydrogen welding for any application. The major advantage of oxyhydrogen welding is in its relatively low temperature flame which is easier to control when welding thin pieces of metal, especially thin pieces of metal which have a low melting temperature.

PRESSURE GAS WELDING

Metal pieces are joined by pressure gas welding by two basic methods, the closed-joint method and the open-joint method. In the closed-joint method, the pieces to be welded are brought together and held under medium pressure while the joint is heated around its outer surface. When the proper temperature for welding is reached, the metal on each side of the joint is upset (made shorter and larger) by the pressure used in the process. The two pieces of metal are bonded together over the entire interface of the joint; however, the bulge produced by the upsetting must be removed by machining to restore the welded pieces to their original thickness.

In the open-joint method of pressure gas welding, the surfaces to be joined are heated to a molten state. When the surfaces to be joined are melted, they are brought together under pressure. Upsetting of the metal near the joint occurs, and when the joint resolidifies, the metal pieces are joined together across the entire joint interface.

Although the central section of the joint interface is not melted in the closed-joint method, the entire area of the abutting surfaces are bonded together by either type of pressure gas welding. Tests have shown that pressure gas welds are of a very high quality and have strengths superior to those obtained by many other welding processes. The characteristic bulge produced at the joint by pressure gas welding is the major disadvantage associated with the process, but it is part of the reason for the superior weld strength.

The heat for pressure gas welding is produced by the combustion of a fuel gas and oxygen much the same as in other gas-welding processes. A special water-cooled oxyacetylene torch with multiple flames is generally used in the open joint method. Pipe, tubing, and bars are heated by a split torch in the form of two semicircles for pressure gas welding.

A wide range of low- and high-carbon steels, alloy steels, and several nonferrous metals are welded by pressure gas welding. *End-butt welds* in pipe, tubing, bars, structural shapes, rings, and chain links are the most common pressure gas welds.

In all gas welding processes, the preparation of the joint before welding is critical. The joint must be free of dirt, oil films, and oxides. A poorly cleaned weld will result in incomplete fusion of the joint, blowholes in the joint, weld porosity, or slag inclusions in the weld. Any of these imperfections result in a weakened weld which will fail sooner than a properly welded joint.

The joints for gas welding must be spaced or grooved to permit full penetration of the joint. Specifications for joint preparation are based on the workpiece's thickness and should be strictly adhered to.

ARC WELDING

The heat for arc welding is produced by an electric current jumping or arcing across a gap in an electric circuit. The arc created by an electric current is one of the hottest heat sources for welding, around 9000°F [4963°C]. The arc is also highly localized and can be concentrated in a small spot. The arc produces a small puddle of molten metal at the point where it strikes the base metal (workpiece) or joint. By moving the arc slowly along the joint, the joint is melted, the pieces being welded fuse together, and when the puddle resolidifies, the pieces are joined together.

A complete welding circuit for arc welding (Fig. 18-26) provides a path for an electric current to flow

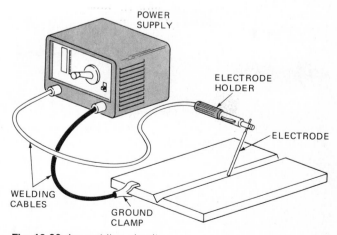

Fig. 18-26 Arc-welding circuit.

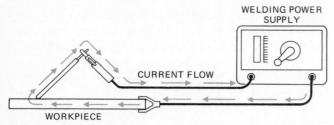

Fig. 18-27 Direct-current reverse polarity (dcrp) arc-welding circuit.

from the power supply through the electrode, to the joint through the base metal ground clamp, and back to the power supply. The arc is established between the electrode and the joint by breaking the circuit with a gap a fraction of an inch in length. If the gap is too long, no current will flow in the circuit and there will be no arc. On the other hand, if there is no gap, current will flow in the circuit but there will be no arc. The current must jump the gap to create an arc. If the gap is too short (electrode too close to the joint), the arc will not be able to develop maximum heat. The actual gap length is determined by the type and size of electrode being used and the welding situation.

A variation of the basic arc-welding circuit is used in some arc-welding processes. The arc is established between two electrodes in these processes. The heat generated by the arc melts a puddle in the joint. Filler metal may be added to the puddle if desired.

Electrodes for arc welding may be consumable or nonconsumable. Consumable electrodes are melted by the heat of the arc. The melted electrodes serve as filler metal for the weld and should be the same type of metal and composition as the metal being welded. On the other hand, nonconsumable electrodes are not designed to melt and form a part of the weld. A nonconsumable electrode is generally made of carbon or tungsten.

The welding current provided by the power supply or welder may be direct current (dc) or alternating current (ac). Direct current flows in only one direction in the welding circuit. Direct current straight polarity (dcsp) flows through the electrode to the base metal (Fig. 18-27). Reverse polarity results in deep penetration with a narrow weld. Direct current reverse polarity (dcrp) flows through the base metal to the electrode. The dcsp arrangement results in a wide weld with shallow penetration and is used mostly for thin metals and for some machine welding where fast welding speeds are required. Alternating current welding combines or averages both of the direct current welding polarities. When the welding power supply operates from a 60 Hz power line (most electrical power lines in the United States are 60 Hz while

many other countries supply 50-Hz electrical power), the polarity of the welding terminals alternate from positive to negative 120 times per second. Therefore, since electric current flows from the negative terminal toward the positive terminal, the alternating current in the welding circuit flows from the electrode to the base metal 120 times each second and from the base metal to each electrode 60 times each second. The resulting weld has medium width, medium penetration, and an average welding speed. Welding with alternating current also eliminates the problem of magnetic arc blow, which often accompanies dc welding.

Magnetic arc blow is a condition that occurs when a magnetic field is established between the electrode and the joint being welded. The magnetic field deflects the arc away from the joint, making it difficult or impossible to maintain a stable arc and weld the joint uniformly. The constant flow of dc welding current through the workpiece and arc will establish a magnetic field between the electrode and joint. On the other hand, alternating current is constantly changing directions and is not a totally steady current. Therefore, it will disrupt a magnetic field rather than establish one.

The welding machine or power supply which provides the welding current may be a motor-generator type, transformer type, or rectifier type. The motor-generator-type welding machine shown in Fig. 18-28 uses an electric motor to drive a dc generator. The motor-generator-type welder shown in Fig. 18-29 uses a gasoline engine to drive either an ac or dc generator. This type of welder is available in a number of sizes. The output of the ac generator can be converted to dc by a rectifier circuit, thus allowing a choice of output currents.

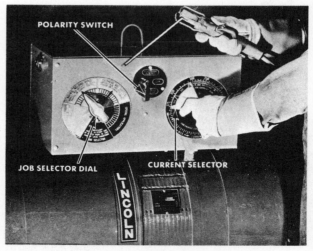

Fig. 18-28 Motor-generator arc welders.

Fig. 18-29 Typical gasoline-engine-driven welding unit. (*Courtesy: Miller Electric Manufacturing Co.*)

The transformer-type welding machine (Fig. 18-30) produces alternating current for welding. It is the smallest, lightest, and least expensive type of welding machine. This machine takes electric power directly from the line supply of electrical power companies and transforms it into a welding current. The output current of transformer-type welders is supplied in a number of ampere settings. This provides different heat ranges for making welds with different sizes of electrodes and on workpieces of varying thicknesses.

The rectifier-type welding machine provides ac or dc welding power. Generally a flip of a switch is all that is necessary to change from an ac output to a dc output. The rectifier-type welding machine is basically a transformer-type welder to which an electric-rectifier circuit has been added. The rectifier-type welder is more efficient than the motor-generator-type welder for providing dc welding power. Also, the transformer-type welder operates very quietly compared to the motor-generator-type welder. However, it must be connected to a power company's lines

Fig. 18-30 Transformer-type ac welding machine. (*Miller Electric.*)

Table 18-4 **Duty-Cycle Conversion Factors**

If the Present Duty Cycle is:	A 100% Duty Cycle is Obtained by Operating the Welder at:
60%	0.75 × rated amperage
50%	0.70 × rated amperage
40%	0.55 × rated amperage
30%	0.50 × rated amperage
20%	0.45 × rated amperage

while a generator powered by a gasoline engine can operate independently.

All three types of welding machines control either the current or the voltage and have a duty cycle. The constant-current welder is used for most manual welding operations. The constant-voltage-type welding machine is used for automatic and semiautomatic welding applications. The duty cycle is the ratio of actual welding time to 10 minutes. Therefore, a welding machine with a 60 percent duty cycle could be used at its maximum rated output for 6 minutes out of every 10 minutes. The duty cycle is concerned with actual arc or welding time, as opposed to the amount of time the welding machine is turned on. Usually the constant-voltage welding machines used for automatic or semiautomatic welding have 100 percent duty cycles. The duty cycle of a welder is increased if it is used at less than its rated capacity. Table 18-4 shows how the duty cycle for most industrially used machines can be increased to 100 percent.

Carbon Arc Welding

Carbon arc welding includes all the welding processes which employ a nonconsumable (does not melt and become a part of the weld metal) carbon electrode. Current from the welding machine arcs to the workpieces, or in the case of twin carbon arc welding, to the other electrode. The heat of the arc is 7000 to 9000°F [3853 to 4963°C], which quickly melts the joint in the workpieces and allows them to fuse together as the arc is moved the length of the joint. Filler metal may be added to the molten puddle if desired (Fig. 18-31).

The carbon electrode allows the welder to manipulate and direct the arc while serving as the arcing point for the electric circuit. Electrodes are available in sizes from 1/8 in to 1 in [0.32 to 2.54 cm] in diameter as determined by the amperage required for the thickness of the metal being welded. Air-cooled electrode holders are used to hold the single or twin carbon electrodes.

Carbon arc welding has been generally replaced with other arc welding processes. However, there are

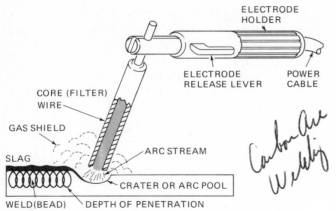

Fig. 18-31 Welding pipe using a coated electrode. (*Miller Electric.*)

some applications where a high rate of heat input is required and carbon arc welding can be used to advantage. Because of its high heat input, carbon arc welding speeds are high and the problems created by prolonged heating are much reduced. The twin carbon arc is used primarily for heating heavy metals where a gas torch is not available or is not adequate.

Shielded–Metal Arc Welding

Shielded–metal arc welding is the most common type of arc welding and uses a stick electrode made of metal wire. The metal wire is coated with a flux which melts and forms a shield to protect the weld area from atmospheric contamination. The metal electrode also melts, and it combines with the molten base metal in the weld area as filler metal. The heat required to melt the base metal, electrode, and electrode's flux coating is from an arc produced by an electric current jumping a gap between the electrode and base metal. The arc has a temperature between 7000 and 9000°F [3853 and 4963°C], which is sufficient heat to melt all the common metals. A representation of the shielded–metal arc welding process is shown in Fig. 18-32.

Shielded–metal arc welding requires six basic components: (1) the power supply, (2) power cables, (3) electrode holder, (4) ground clamp, (5) electrode, and (6) the metal to be welded. These components used in the proper way by a skillful operator or welder are capable of joining or repairing metal parts with strength equal to the strength of the pieces themselves. Many factors affect the weld produced by shielded–metal arc welding which require a knowledge of metallurgy or physics beyond the scope of this book. However, the basic welding concerns related to the six preceding components will be presented in the following paragraphs.

Power supplies for shielded–metal arc welding may be of the motor-generator, transformer, or rectifier type. These power supplies are the basic types for all arc welding and were explained in previous paragraphs of this chapter. The motor-generator-type welder is noisy and is mainly used in outdoor locations away from sources of commercial electrical power and for applications where adjustment of both the welding voltage and current is desired. The transformer-type welder is the least expensive and is used extensively by farms and small repair and job shops. The rectifier-type welder is used if dc or both ac and dc power is required and the noise level must be low.

Power supplies for shielded–metal arc welding applications provide welding current outputs in the range of 15 to 500 A and 14 to 40 V. They provide a constant current of a preselected value. The output voltage varies in relation to the arc length. When the welder is on but not being used for welding, an open-circuit voltage of 80 V is common. A voltage of this magnitude could result in an electrical shock; therefore caution is recommended on the part of the welder. When an arc is struck, the voltage drops and varies in relation to the arc length. A short arc results in a low arc voltage and shallow penetration of the weld.

Electrodes first used for metal arc welding were bare, without a flux coating. Bare electrodes were unsatisfactory and resulted in weak, rough welds and

Fig. 18-32 Shielded-metal arc welding with a coated (stick) electrode.

caused many people to believe that metal arc welding was not practical. Coating the bare electrodes with chemical fluxes overcame the objections to metal arc welding. Time and use has shown that metal arc welding with coated electrodes (called *shielded metal arc welding*) is not only a practical way of joining metal parts, but is fast, economical, and reliable as well.

The flux coatings used for electrodes vary considerably in chemical composition. However, they all basically perform the following functions:

1. Produce a gas shield to protect the molten metal and the arc from atmospheric contamination

2. Increase the flow of electric current across the arc gap and help stabilize the arc.

3. Chemically clean or refine the molten-weld metal and produce a slag coating to protect the metal from oxidation until cool.

4. Control weld characteristics such as bead shape, solidification time, and ease of striking and maintaining an arc.

Manufacturers of electrodes have developed their own chemical formulas for flux coating. Therefore, the welder has a wide range of electrodes from which to choose for a given application. An electrode classification system has been developed by the American Welding Society (AWS) in an effort to help welders identify various electrodes and provide some type of continuity to the selection of electrodes. Figure 18-33 provides an explanation of the AWS electrode classification system.

Electrode holders for shielded−metal arc welding hold the bare end of the electrode and make the necessary electrical connection from the power cable to the electrode. The holder also forms a handle for the welder to use in manipulating the electrode. Some holders have provisions for locating the electrode at various angles, while other holders require that the electrode be bent to the desired angle (Fig. 18-34).

A welding power cable and ground clamp connect the workpiece to the power supply. A second power cable connects the power supply to the electrode holder. The ground clamp may be as simple as a bolt and nut to hold the power cable to the workpiece. However, the ground clamp is usually a spring-loaded clamp which can be moved to another workpiece quickly and easily (Fig. 18-35). The power cable is a stranded copper wire insulated with a rubber coating. Higher welding currents and longer cables require larger diameter and more expensive cables than are required by shorter cables and lower welding currents. Table 18-5 gives some of the standard sizes and lengths for power cables carrying a given amperage.

Shielded−metal arc welding is usually done on ferrous (iron base) metals. Some nonferrous metals, such as aluminum, brass, or bronze, were mainly joined by shielded−metal arc welding; however,

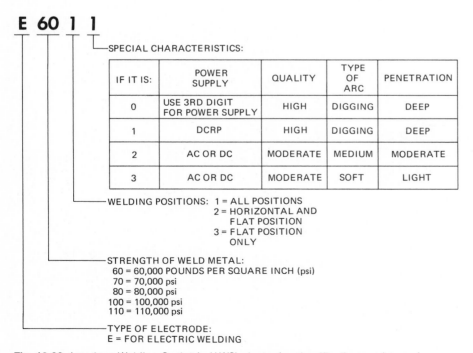

IF IT IS:	POWER SUPPLY	QUALITY	TYPE OF ARC	PENETRATION
0	USE 3RD DIGIT FOR POWER SUPPLY	HIGH	DIGGING	DEEP
1	DCRP	HIGH	DIGGING	DEEP
2	AC OR DC	MODERATE	MEDIUM	MODERATE
3	AC OR DC	MODERATE	SOFT	LIGHT

SPECIAL CHARACTERISTICS:

WELDING POSITIONS: 1 = ALL POSITIONS
2 = HORIZONTAL AND FLAT POSITION
3 = FLAT POSITION ONLY

STRENGTH OF WELD METAL:
60 = 60,000 POUNDS PER SQUARE INCH (psi)
70 = 70,000 psi
80 = 80,000 psi
100 = 100,000 psi
110 = 110,000 psi

TYPE OF ELECTRODE:
E = FOR ELECTRIC WELDING

Fig. 18-33 American Welding Society's (AWS) electrode classification number code.

Fig. 18-34 Electrode holder for arc welding. (*Jackson Products.*)

Table 18-5 **Recommended Welding Cable Sizes**

Amperage	Distance from Welding Machine, in Feet								
	50	75	100	125	150	175	200	225	250
100	4	4	2	2	1	1/0	1/0	2/0	2/0
150	4	2	1	1/0	2/0	3/0	3/0	4/0	4/0
200	2	1	1/0	2/0	3/0	4/0	4/0		
250	2	1/0	2/0	3/0	4/0				
300	1	2/0	3/0	4/0					
350	1/0	3/0	4/0						
400	1/0	3/0	4/0						
450	2/0	4/0							
500	2/0	4/0							
550	3/0								
600	4/0								

other methods have been developed which are far superior for welding these metals. Special precautions must be exercised when welding cast iron. These precautions include the use of special electrodes along with preheating, postheating, or peening of the weld area to reduce severe internal stresses in the base metal.

Flux–Cored Arc Welding

Flux–cored arc welding is a shielded–arc welding process which uses a consumable electrode wire (melts and forms part of the weld bead) which is filled with flux. Although the electrode wire is ⅛ in [0.32 cm] or less in diameter it is made in the form of a small tube which is filled with flux (Fig. 18-36). Flux-cored electrode wire is supplied in spools, and its composition should match the type of metal being welded.

The flux–cored arc welding process originally used an external shielding gas of carbon dioxide which was sprayed on the weld area during welding to shield out atmospheric contamination. The development of improved fluxes has largely replaced the externally supplied gas shield with an internally generated one. As the flux core of the electrode melts, it mixes with the molten puddle of the base metal and gives off a gas which drives the atmosphere away from the weld area. A more complete shield is provided when the atmosphere is forced away from the weld area from within the joint itself. Some of the flux does not vaporize but floats on the molten base metal to form a protective coating of slag over the weld when solidified.

Fig. 18-35 Typical ground clamps. (*Forney Industries.*)

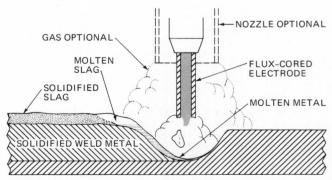

Fig. 18-36 Flux–cored arc welding.

Flux–cored arc welding is in essence an automated or semiautomated version of shielded–metal arc welding and is used to replace the shielded–metal arc welding process for high-production applications or where X-ray-quality welds are required. A wire feed and control unit is added to the basic arc welding machine for flux–cored arc welding. The power supply is of the constant-voltage-type machine in a 200 to 1000 percent duty-cycle model. The welding machine may supply ac or dc of either polarity; however, dcrp is the most common in the United States. Since the wire feed unit feeds from a large spool of electrode wire instead of the stick electrode, the stick electrode holder is replaced with a gun.

Gas–Metal Arc Welding

Gas–metal arc welding is also referred to as *MIG* (metal inert gas) *welding*. The process uses the heat of an electric arc from a continuously fed electrode wire and the base metal. The electrode is consumed in the process and serves as filler metal in the joint. Shielding for the arc and molten weld metal is an inert gas externally provided to the weld area. Inert gases such as argon, helium, and CO_2 (carbon dioxide) do not mix with other gases. When forced onto the weld area under pressure, any one of them will force the possible contaminants in the atmosphere away from the weld area.

The MIG welding equipment includes the basic 100 percent duty-cycle rectifier type power supply, electrode wire feed and control, welding gun, electrode wire, and the shielding gas (Fig. 18-37).

The power source or welding machine used for MIG welding is of the constant-voltage type. The actual output voltage is adjusted by the machine's rheostat voltage control. The output voltage varies very little from the selected value throughout the welding process. However, the value of the welding current is determined by the speed of the wire feed. At any set speed of wire feed, the welding machine supplies the current necessary to maintain a steady arc. The current is dc with reverse polarity and may vary from a maximum of 150 to 1000 A depending upon the capacity of the machine.

Two wire-feed systems are commonly used in MIG welding. Each system has an adjustable wire-feed rate. One system is attached to the welding machine and pushes the electrode through the cable and gun to the workpiece (Fig. 18-38, top). In the other system, the feed mechanism is an integral part of the gun (Fig. 18-38, bottom). The gun holds a small spool of electrode wire which is pulled by drive rollers near the nozzle portion of the gun and is fed through the nozzle to the workpiece. The wire-feed gun is used primarily to feed small-diameter aluminum wire which cannot be pushed very far without buckling.

The MIG welding gun has a copper tip which makes a sliding electrical contact between the welding machine's power cable and the electrode being fed to the workpiece. A metal nozzle protrudes from the gun and surrounds the tip, which causes the shielding

Fig. 18-37 MIG welding components.

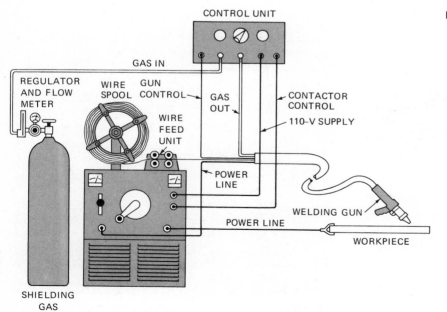

CONTROL UNIT
GAS IN
REGULATOR AND FLOW METER
WIRE SPOOL
GUN CONTROL
GAS OUT
CONTACTOR CONTROL
110-V SUPPLY
WIRE FEED UNIT
POWER LINE
WELDING GUN
POWER LINE
WORKPIECE
SHIELDING GAS

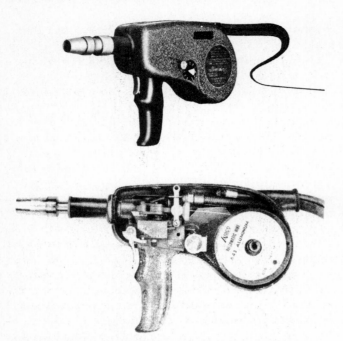

Fig. 18-38 *Top,* the MIG welding gun; *bottom,* the cutaway view shows the 4-in diameter spool of wire in the gun.

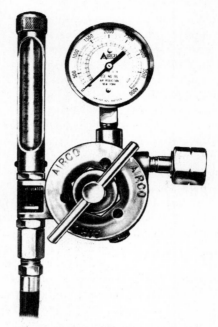

Fig. 18-39 A combination gas regulator and flow meter. (*Airco Welding Products.*)

gas to be directed in a concentrated area around the arc. The nozzles on low-amperage guns are air-cooled, while the nozzles of high-amperage guns are water-cooled. The gun's handle allows the welder to manipulate the electrode and welding arc, or in the case of automatic MIG welding, special clamps and brackets replace the gun's handle.

Electrode wire for MIG welding is a solid bare wire which serves two functions in the welding process. The wire is melted by the heat of the arc and is fused in with the base metal to fill the joint (filler metal). The wire also serves as one of the points in the welding circuit which is in contact with the arc (electrode). The electrode wire must be compatible with the metals being welded. When the metal pieces being joined are the same type of metal, the electrode wire should also be the same. Electrode wire is available in several diameters and spool sizes. However, the 1 lb, 4-in diameter and 15 lb, 12-in diameter spools are the most common.

The shielding gas for MIG welding is an inert gas which displaces the atmosphere surrounding the weld area, thereby avoiding contamination of the weld metal by oxygen and nitrogen from the atmosphere. The shielding gases commonly used for MIG welding are helium, argon, and carbon dioxide (CO_2). Argon and CO_2 or argon and helium are often mixed together to provide a shielding gas superior to either gas alone. Shielding gases are supplied under pressure in hollow steel cylinders. The gas cylinder's pres-

sure is reduced to a usable pressure, and the flow of gas is regulated by a combination regulator–flow meter in MIG welding (Fig. 18-39).

Gas–Tungsten Arc Welding

Gas–tungsten arc welding is also called *TIG* or *Heliarc welding.* It is a fusion-welding process where the weld is produced by heating the joint with an electrical arc

Fig. 18-40 Welding aluminum-tubing joints with TIG (heliarc) process. (*Courtesy Linde Division, Union Carbide Corp.*)

between the workpiece and a nonconsumable tungsten electrode (Fig. 18-40). The weld area is shielded from atmospheric contamination by an inert gas. Filler metal may or may not be added to the joint; however, if used, it is dipped into the molten puddle by hand. Filler metal is added to the joint in much the same way it is added in flame welding with the oxyacetylene torch. The components for TIG welding include the torch, filler rod, welding machine, welding cables plus hoses, and a regulated supply of inert gas. A typical TIG welding setup is shown in Fig. 18-41. Several items of optional equipment may be used with the basic TIG welding components. Some welding machines are equipped with gas and water valves, timers, and a remote amperage control. Also, a water-cooled torch and a recirculating water-cooling system are additional equipment often used in TIG welding.

The power supply for TIG welding can be a rectifier or motor-generator type (Fig. 18-42). TIG welding is done with ac or with dc with straight or reverse polarity. The type of current required for TIG welding is determined by the metal being welded. Table 18-6 gives the recommended type of current for some common metals. Usually the power supply is capable of superimposing a high-frequency (hf) current over

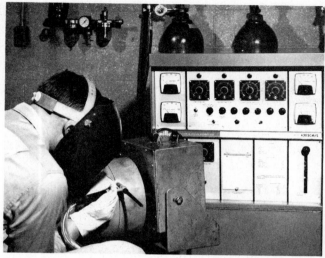

Fig. 18-42 TIG power supply 500 A ac/dc rectifier unit. (*Miller.*)

the welding current. A high frequency is helpful when starting the arc, and since an ac welding current is constantly stopping and starting, a high-frequency current is used constantly with ac TIG welding.

The TIG welding torch may be air-cooled or water-cooled. The air-cooled torch becomes large and bulky

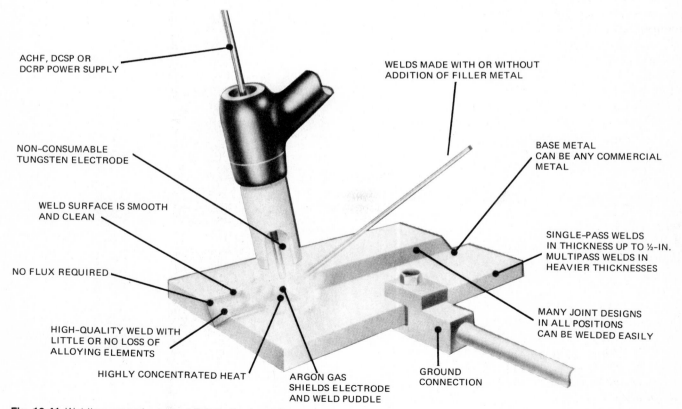

ACHF, DCSP OR
DCRP POWER SUPPLY

NON-CONSUMABLE
TUNGSTEN ELECTRODE

WELD SURFACE IS SMOOTH
AND CLEAN

NO FLUX REQUIRED

HIGH-QUALITY WELD WITH
LITTLE OR NO LOSS OF
ALLOYING ELEMENTS

HIGHLY CONCENTRATED HEAT

ARGON GAS
SHIELDS ELECTRODE
AND WELD PUDDLE

GROUND
CONNECTION

WELDS MADE WITH OR WITHOUT
ADDITION OF FILLER METAL

BASE METAL
CAN BE ANY COMMERCIAL
METAL

SINGLE-PASS WELDS
IN THICKNESS UP TO ½-IN.
MULTIPASS WELDS IN
HEAVIER THICKNESSES

MANY JOINT DESIGNS
IN ALL POSITIONS
CAN BE WELDED EASILY

Fig. 18-41 Welding zone of a typical TIG (heliarc) welding setup.

Table 18-6 **Current Recommendations for T16 Welding Various Metals**

Material	Alternating Current	Direct Current	
	With High-Frequency Stabilization	Straight Polarity	Reverse Polarity
Magnesium up to 1/8 in thick	1	N.R.	2
Magnesium above 3/16 in thick	1	N.R.	N.R.
Magnesium castings	1	N.R.	2
Aluminum up to 3/32 in thick	1	N.R.	2
Aluminum over 3/32 in thick	1	N.R.	N.R.
Aluminum castings	1	N.R.	N.R.
Stainless steel	2	1	N.R.
Brass alloys	2	1	N.R.
Silicon copper	N.R.	1	N.R.
Silver	2	1	N.R.
Hastelloy alloys	2	1	N.R.
Silver cladding	1	N.R.	N.R.
Hard-facing	1	1	N.R.
Cast iron	2	1	N.R.
Low-carbon steel, 0.015 to 0.030 in	2	1	N.R.
Low-carbon steel, 0.030 to 0.125 in	N.R.	1	N.R.
High-carbon steel, 0.015 to 0.030 in	2	1	N.R.
High-carbon steel, 0.030 in and up	2	1	N.R.
Deoxidized copper	N.R.	1	N.R.

KEY:
1—Excellent operation.
2—Good operation.
N.R.—Not recommended.

for high-amperage ratings; therefore, the water-cooled torch is the most popular type of torch. Two hoses are connected to the torch for the inlet and return of the water coolant. The inlet hose also contains the power cable from the welding machine. This cable can be of a smaller size if it is water-cooled.

The torch holds the tungsten electrode and connects the power cable and electrode together. A ceramic cup or nozzle fastened to the torch surrounds the exposed end of the electrode (Fig. 18-43). The cup directs the shielding gas around the electrode and over the weld area.

The tungsten electrode used in TIG welding is made of a special tungsten alloy with a melting point over 6000°F [3312°C]. The electrode is nonconsumable (does not melt) and should not touch the workpiece or molten puddle. Electrodes contaminated from touching the workpiece must be ground off until the contaminated area has been removed.

A filler metal may or may not be used in TIG welding. The joint design is the determining factor as to whether or not filler metal should be used. The composition of the filler metal should match that of the metal being welded.

The weld area is shielded by an inert gas: helium, argon, or a mixture of the two. The gas is stored under pressure in hollow steel cylinders. A regulator and flowmeter control the flow of gas onto the weld area. Gas flows in the range of 15 to 30 ft³/h (hour)

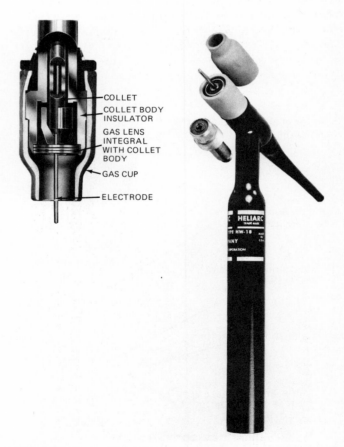

COLLET
COLLET BODY INSULATOR
GAS LENS INTEGRAL WITH COLLET BODY
GAS CUP
ELECTRODE

Fig. 18-43 Typical TIG (heliarc) torch. Inset shows internal details of the torch. (*Courtesy Linde Division, Union Carbide Corp.*)

[or 0.5 to 0.9 m³/h] are usually sufficient for TIG welding.

The TIG welding process was developed for welding the hard-to-weld metals used by the aviation and aerospace industries. However, it proved to be a practical way of joining all types of metals. Many dissimilar metals can be welded by the TIG process.

Submerged Arc Welding

The arc in *submerged arc welding* is shielded by a blanket of powdered flux. An electric arc between a bare metal electrode and the workpiece provides the heat for this welding process. The process got its name from the fact that the arc is completely submerged in a layer of powdered flux and is not visible as the weld is made. The wire electrode is melted by the heat of the arc and fuses in with the base metal to help fill the joint (Fig. 18-44).

Submerged–arc welding applications are fully automatic or semiautomatic since the actual welding head is moved over the workpiece on a carriage or the workpiece is fed on the carriage under the welding head (Fig. 18-45). The carriage may be on tracks, rails, the work itself, or it may be guided electronically. The workpiece and welding head is positioned so the welding is always done in the flat position, since the flux is gravity fed onto the joint just in front of the electrode (Fig. 18-46).

A constant-current power supply with ac or dc power is used for submerged arc welding. The current range required is determined by the size of the electrode being used and ranges from 150 A for a

Fig. 18-45 Complete submerged–arc welding unit. (*Courtesy Linde Division, Union Carbide Corp.*)

¹/₁₆-in-diameter electrode to 1400 A for a ¹/₄-in-diameter electrode. For thick joints where a high deposition rate is required, more than one electrode is used, and they may be powered by the same power supply or by a separate unit for each electrode. A single welding machine is used to power six electrodes for some hard-surfacing applications.

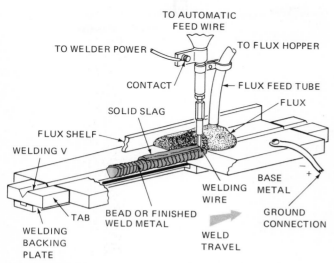

Fig. 18-44 Schematic diagram of submerged–arc welding process with principal parts identified. (*Courtesy Linde Division, Union Carbide Corp.*)

Fig. 18-46 Welding a pipe with submerged–arc welding unit. (*Courtesy Linde Division, Union Carbide Corp.*)

Originally submerged arc welding was designed to weld heavy sections of mild steel; however, the process is applicable to any type of metal which is over $\frac{1}{8}$-in thick. Fluxes and filler wire are presently available for almost any metal or metal alloy.

The weld produced by submerged arc welding is exceptionally sound, extremely smooth, and of high quality in all respects. However, the flux drastically affects the weld quality. If the flux is too deep, the arc will be confined and will not develop its full heat. The result is a rough, stringy weld. If the flux layer is too shallow, flashing will be visible and metal will be splattered around the weld. The weld will also be unnecessarily porous. The welding conditions of amperage, electrode diameter, speed, and type of metal will determine the proper flux depth.

Plasma Arc Welding

The *plasma arc welding* (PAW) process produces a weld by the heat from a constricted arc between an electrode and the workpiece or the electrode and the nozzle which constricts the arc (Fig. 18-47). The arc, electrode, and weld metal are shielded from the atmosphere by an externally supplied inert gas and ionized plasma gas coming from the constricting orifice. Pressure and filler metal may or may not be used in plasma arc welding.

Basically, the plasma arc process is similar to the TIG welding process. Both processes use a nonconsumable tungsten electrode and an inert-gas atmospheric shield. However, in plasma arc welding the arc is used to heat and ionize the plasma gas (helium or hydrogen) which is forced through the nozzle. The plasma gas forms a high temperature (6000 to 100,000°F), [3315 to 55537°C], high-velocity stream of ionized gas particles. PAW is capable of full penetration in metals above $\frac{3}{32}$-in [2.38 mm] in thickness.

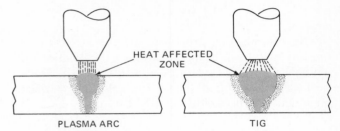

Fig. 18-48 Comparison of weld beads produced by plasma arc.

The resulting weld is a narrow, uniform bead with 100 percent penetration (Fig. 18-48).

A constant current or a motor-generator or rectifier dc power supply is used with PAW. A high frequency current is used to help start the arc. Two types of arcs are established in PAW as determined by the type of torch being used. The transferred arc torch constricts the arc but allows it to be established between the electrode and workpiece. The nontransferred arc torch establishes the arc within a special water-cooled copper nozzle and only the plasma stream (arc) contacts the workpiece. The transferred arc torch is used for most cutting and welding applications.

The plasma arc was first successfully used for welding in 1963. Therefore, PAW equipment, supplies, and applications are still being developed at a rapid pace. Its extremely high temperature of up to 100,000°F [55537°C] shows considerable promise for high-speed welding and cutting applications.

STUD WELDING

The process of stud welding is a specialized application of arc welding in which an arc is established between the workpiece and the end of a stud. When the workpiece and the stud are heated to the proper tem-

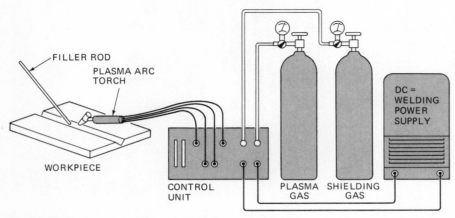

Fig. 18-47 Plasma arc welding components.

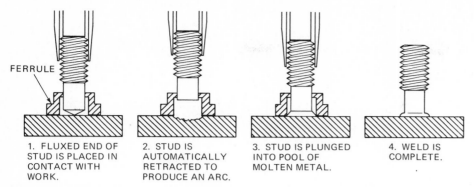

FERRULE

1. FLUXED END OF STUD IS PLACED IN CONTACT WITH WORK.

2. STUD IS AUTOMATICALLY RETRACTED TO PRODUCE AN ARC.

3. STUD IS PLUNGED INTO POOL OF MOLTEN METAL.

4. WELD IS COMPLETE.

Fig. 18-49 Basic steps of stud-welding process. (*Courtesy: Nelson Stud Welding Division, Gregory Industries, Inc.*)

perature they are brought together under pressure. When the joint cools, the stud is welded to the workpiece.

In stud welding, the stud to be welded is held in a portable or machine-mounted gun. The exposed end of the stud is passed through a ceramic ferrule which does the following: (1) shields the arc flash, (2) shields the weld area from atmospheric contamination, (3) contains the molten metal to the weld area, (4) protects the surrounding surface from sparks and spatters, and (5) concentrates the arc's heat. The stud gun is positioned to properly locate the stud, and the gun is triggered. The gun automatically lifts the stud as the current arcs between the end of the stud and the workpiece, then forces the stud against the workpiece when the joint reaches the proper welding temperature. After the joint has cooled, the gun and ceramic ferrule are removed to expose the stud, which is butt welded to the workpiece. Figure 18-49 shows the stages in stud welding.

Stud welding was originally developed for use in shipbuilding but has been expanded to the welding of many types of fasteners (Fig. 18-50) to metal pieces for use in the automotive, appliance manufacturing,

Fig. 18-50 Typical studs for stud welding. (*Courtesy: Nelson Stud Welding Division, Gregory Industries, Inc.*)

Fig. 18-51 Complete production stud-welding unit. (*Courtesy: Nelson Stud Welding Division, Gregory Industries, Inc.*)

construction, prefabrication, and aerospace industries. The basic equipment required for stud welding is shown in Fig. 18-51.

OTHER WELDING PROCESSES

A number of welding processes exist which are not directly related to the processes previously described. Most of these processes are specialized and have found only limited industrial applications. Some of the processes described in the following paragraphs are still in developmental stages and are considered

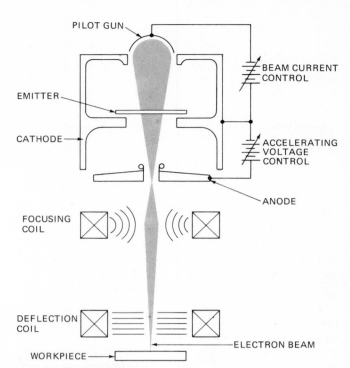

Fig. 18-52 Diagram of a electron-beam welding gun.

somewhat experimental. But it is possible that these very processes may be the major welding processes of the future.

Electron Beam Welding

In *electron beam welding*, a high energy stream of electrons (negatively charged particles or electricity) is given off by a hot-metal cathode. The electron stream is focused on the joint causing it to melt, and the members of the joint fuse together (Fig. 18-52).

Electron beam welding is capable of pinpoint accuracy (Fig. 18-53). However, rather extensive electrical and electronic equipment is required in the process.

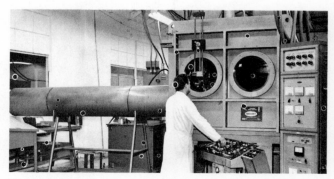

Fig. 18-53 An electron-beam welding machine. (*Sciaky Bros.*)

Also, a chamber for placing the joint in a vacuum is required, because any type of gas would affect the electron beam. Because of these equipment limitations, the uses of electron beam welding are rather limited.

Laser Beam Welding

Laser beam welding is similar to electron beam welding except the laser uses a beam of light instead of an electron beam (Fig. 18-54). The light energy used by a laser has been produced by several techniques; energy from a ruby optical laser has been successfully used to weld metals. A pulse of energy from the ruby laser is of short duration and is repeated after a cooling cycle much like the tick of a clock. A single pulse lasts for about $1/10{,}000$ of a second, and when focused on a joint in metal parts, it is capable of melting a very concentrated spot (0.0025 in diameter) [0.06 mm] of metal almost instantly.

Laser beam welding is done at normal atmospheric pressure, thus eliminating the time-consuming vacuuming step required in electron beam welding. Laser beam welding has found limited application in industry at the present time. It has been used to weld fine wires and weld connections on miniaturized electronic components. Also, laser beam welding has been the subject of a number of recent experimental projects for the aerospace and electronics industries.

The laser beam has also been used successfully to cut a wide variety of materials including woods, metals, plastics, and composites.

Electroslag Welding

Electroslag welding is a fusion welding process which uses the heat of molten flux to melt the base metals and filler metal. The flux is melted by its resistance to an electric current which must flow through it in order to flow from the electrode to the base metal. Shielding of the weld area is produced by the pool of molten flux covering the weld and the water-cooled copper shoes on each side of the joint (see Fig. 18-55 on page 322).

Electroslag welds are made on end from the bottom up. The sides of the joint are closed by copper shoes which cover the open sides of the joint. The shoes may be either fixed or sliding. Fixed shoes are as long as the joint is tall, while sliding shoes may be fairly short since they move with the weld. The wire electrode is fed into the joint and fuses with the base metals to fill the joint under a pool of molten flux. In electroslag processes, the electrode is fed through a guide which also melts to help fill the joint. This welding process is called *consumable guide electroslag welding.*

Heavy plates are quickly and easily joined edge to edge in a single pass by electroslag welding. The process is used successfully on all types of steel and nickel chromium alloys which are high in nickel. Joints from

Fig. 18-54 Laser-beam welding machine. (*Union Carbide Corp.*)

4 in to 10 ft [10.16 cm to 3.05 m] in length, and joints in metal up to 12 in [30.48 cm] thick are quickly and easily made by the electroslag process.

Induction Welding

Induction welding is similar to resistance welding except the electric current is induced into the joint without physical contact by tips or electrodes.

When an electrical conductor is placed in a changing magnetic field, a current is caused to flow (induced) into the conductor. Resistance to the flow of this induced current will cause the conductor to heat up. Induction heating of metals has been used for a number of years. Induction furnaces are commonly used in the metal-casting industry to melt down metals for casting.

In induction welding, a high current is simultaneously induced into selected areas of the pieces being joined. These areas quickly fuse together, or they are pressed together in order to produce a welded joint.

Induction welding has found broad applications in the manufacturing and fabrication industries. It is fast and has only localized areas which are heated; therefore, distortion is low. Most metals in thin sections can be welded by induction welding, which makes induction welding especially applicable to sheet metals.

Thermit Welding

Thermit welding uses the heat from a superheated molten metal and metal oxide to fuse metal parts together. Thermit welding may be accomplished with or without the application of pressure and filler metal.

Thermit mixtures are basically iron oxide and aluminum in a powdered form. When the thermit mixture is ignited, it burns and produces a temperature of around 5000°F [2760°C]. This heat is sufficient for fusion welding of all types of iron and steel.

In thermit welding, the parts to be welded are aligned with a gap between the parts. A mold is made

Fig. 18-55 Operator positions copper mold for consumable guide electroslag-welding application. (*Courtesy: Linde Division, Union Carbide Corp.*)

to surround the parts but not fill the joint. The mold is such that the thermit mixture when molten will flow into the joint. When the mold is properly prepared, the thermit mixture is ignited and cast into the mold between the parts to be joined. The thermit mixture has sufficient heat to fuse the edges of the pieces being welded. After the metal has cooled, the mold is opened and the welded piece is removed.

Thermit welding requires that enough metal be removed from the pieces to be joined so that there will be a gap in the joint. It is usually used for repair welding and is used extensively for repairing broken parts when their alignment is critical. The railroads and rapid transit systems use thermit welding to join the ends of rails to make continuous stretches of rails. Also, the construction industry thermit welds reinforcing bars into a single bar to meet building code requirements for continuous concrete reinforcement.

REVIEW QUESTIONS

18-1. Differentiate between pressure welding and ultrasonic welding.

18-2. In what applications is explosive welding usually employed and why?

18-3. How does gas welding differ from arc welding in terms of structural changes of the base metal and types of metals that can be welded?

18-4. How does propane gas differ from acetylene? Which one is more efficient in terms of temperature?

18-5. Why is the electrode in arc welding coated with flux? What would happen if noncoated electrode is used?

18-6. Outline the basic functions of a flux-coated electrode.

18-7. Differentiate between electron beam welding and laser beam welding.

18-8. Why is polarity important in welding different types of metals in arc welding?

18-9. Explain what we mean by a E6011 electrode.

18-10. What is the difference between a flux-covered electrode and flux-coated electrode?

CHAPTER 19

THE TECHNOLOGY OF ASSEMBLING MANUFACTURED PARTS

As explained in Chap. 1, there are two broad types of production. The first is called *custom,* or *job shop,* production and consists of the manufacture of a few products, one unit at a time. Job-shop production involves fewer automated processes and greater numbers of hand operations, fittings, and inspections. Job shop or custom work is often equated with quality work. This is sometimes, but not always, the case. The Rolls Royce automobile is typical of this type of manufacturing. However, custom-produced items include large industrial machines such as multiton presses, hydroelectric turbines, ships, and other large objects. Generally, custom production is small in volume because the demand for the product is not large. The limited demand makes the cost of an automated or mechanical assembly uneconomical.

LINE PRODUCTION

The other type of production is *line production,* often called *mass production.* This type of production is typified by the assembly of most American automobiles, household appliances, and other such items. This type of production makes it possible for the manufacturing organization to make a substantial total profit from a very small profit per item that is multiplied by an extremely large production rate.

Line production derives its name from the assembly line of parts moving past work stations. Henry Ford pioneered this concept and was able to reduce the cost of the automobile from a luxury item to an item within reach of almost every worker in the United States. By moving the parts to the worker, time for assembly was greatly reduced simply because no time was lost by the assemblers moving back and forth. Further time savings were made by studies that found ways of reducing reaching, lifting, sorting, and positioning actions by workers. The time taken to reach for a wrench can be very costly when repeated

thousands of times. Thus, moving the wrench closer can reduce costs and increase production rates. These types of studies literally gave birth to the field of modern industrial engineering.

Several factors are important to both profitable and successful operations. These areas include: interchangeability of similar parts, division of labor and processes into tiny elements that can be quickly performed at one place, movement of the materials to the work station, efficient use of time and the placement of the work station, minimizing the amount of set-up time through the use of jigs and fixtures, lowering material cost by purchasing in large quantities, reducing the amount of labor through efficiency studies to the lowest possible amount; reducing the cost of labor by reducing the need for highly skilled craftspeople in the assembly line, and the utilization of specialized tools to increase the rate of work performance.

Automated and semiautomated machines are also used to eliminate unproductive time and to increase the speed of assembly. However, the more specialized the equipment, the more critical certain essential factors become. These include the need for a mass market to consume vast quantities of the product, increasing cost of labor skills, and the need for redesign to minimize assembly processes.

Types of operations that are most adaptable to automated processes include gathering, measuring, filling, holding, bending, folding, spot and seam welding, cutting, stamping, turning, and packaging. However, almost any operation can be automated, provided the demand justifies the expense of the equipment.

All these factors must be incorporated into the production without sacrificing the quality of the performance of the intended part. Quality, however, does not necessarily mean the best possible performance; quality simply means the minimum amount of processed product that can successfully accomplish the purpose of the product.

ORGANIZATION OF THE ASSEMBLY PROCESS

The term *joining* usually relates to the attachment of one individual piece to another. The term *assembly* usually refers to the accumulated process of building a complex product from a number of pieces.

A single piece is called a *piece part* and can refer to anything from a tiny screw to the engine block of an automobile. Piece parts are often put together into integral units called *subassemblies*. For example, the engine block becomes a part in the whole engine, which in turn becomes a part of the automobile. Figure 19-1 shows how subassemblies, made on *feeder lines,* are combined into the final product.

A wide variety of methods are used in assembly processes. Assembly processes utilize fasteners and fastening methods to assemble various parts and subassemblies into a finished product. The particular assembly method utilized is generally designed for specific purposes.

Special considerations involved in the tooling processes for line production include the speed of assembly, machine cost, and the ability to incorporate design changes. Faster assembly speeds reduce the labor cost involved. However, speed of assembly is generally acquired through the use of more labor or through the adaptation of special equipment (Fig. 19-2). Special equipment, of course, adds cost to the tooling processes. Where this machine cost exceeds the typical labor cost, it is more economical to use labor for the assembly process. However, where the cost of

Fig. 19-2 Auto bodies are moved to automatic welding stations on a conveyor. (*Courtesy Chevrolet Division of General Motors.*)

machinery is less than the wages and other miscellaneous costs associated with labor, the automated processes are more economical and generally installed, provided machinery and equipment may be designed and purchased that will adequately perform the operations. Another feature in tooling processes is the adaptability or availability of standard tools. It is much more economical to use standard tools and to adapt them to situations than to have a manufacturer design and make a limited number of specialized tools.

Typical types of tools adapted for use on assembly lines include driving tools for bolts, nuts, and screws, as the "drill" shown adapted for use to drive screws in Fig. 19-3. Drivers, such as those in the illustration, may be manufactured with single heads or driving elements to drive a single nut, bolt, or screw. However, multiheaded units, may also be purchased. These special tools can drive several nuts or screws at the same time. Another type of system that is used in assembly to reduce time but which uses standard tools is the use of the low-pitch screws and threads. These low-pitch screws need only a part of a turn to engage the device. These screws make assembly very quick; however, low-pitch screws generally have less strength than standard screws. Low-pitch screws are used in soft materials such as plastics or where there is a low-tensile-strength requirement. These threads are very useful in assembling access panels—such as in electrical equipment—which must be removed frequently for modification, repairs, or adaptations.

PRODUCT REDESIGN

Products can also be redesigned to eliminate costly operations without reducing the quality of perform-

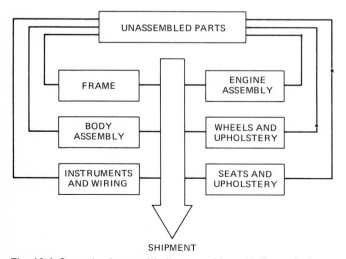

Fig. 19-1 Several subassembly lines combine with the main line.

Fig. 19-3 Using a standard tool in assembly work.

Fig. 19-4 Machined shoulders are replaced with savings in material, tools, and time. Grooving for ring can be done during a cut-off, or other machining, operation.

ance. Redesign can often reduce the cost of a product by tremendous amounts. Such savings are advantageous in two ways. First, the lower cost per item can allow the firm to lower its price to its customers and consumers. Lowering the price makes the manufacturing company more competitive allowing it to increase its sales and its production rates. The second factor is that the manufacturer can have more profit margin. By lowering the production cost without lowering the cost to the consumer, the profit margin can be greatly increased. Normally, for competitive purposes the manufacturer in such a situation will reduce the price to the consumer by a lesser amount than the actual cost reduction. This lets the manufacturer both lower the cost to the consumer and increase the profit margin. The reader must remember that within the capitalistic society, manufacturing organizations exist for profit making.

Figure 19-4 illustrates how an assembly can be redesigned. Notice that the redesign involved the reduction in machining and the waste of materials. A real case history is shown in Fig. 19-5 where a redesign resulted in a total savings of 30 percent. A summary of savings through redesign is illustrated in Fig. 19-6.

SELECTING THE METHOD OF ASSEMBLY

The selection of the assembly method and the particular type of fastening or joining methods involves a consideration of several factors. Of course, the labor cost, material cost, and the cost of the tooling processes are the primary factors. However, the number to be produced and the market potential of such a product must be considered. It may well be more economical to make a new product with available equipment and tools until the market potential is assessed. A limited number of parts would be produced ini-

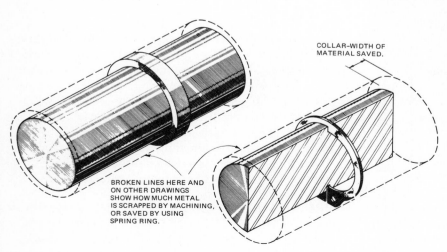

COLLAR-WIDTH OF MATERIAL SAVED.

BROKEN LINES HERE AND ON OTHER DRAWINGS SHOW HOW MUCH METAL IS SCRAPPED BY MACHINING, OR SAVED BY USING SPRING RING.

Fig. 19-5 The use of retaining rings on the front-wheel assembly of a wheel chair resulted in a 30 percent savings on parts and labor. (ⓒ *1972 and reprinted with permission of Waldes-Kohinoor, Inc.*)

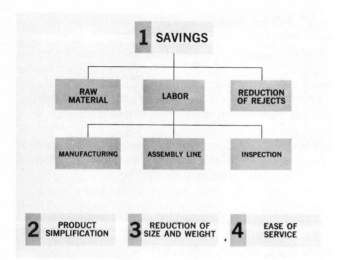

Fig. 19-6 Summary of savings with retaining rings. (© *1964 and reprinted with permission of Waldes-Kohinoor, Inc.*)

tially, and if full-scale production is deemed advisable and profitable, then more efficient tooling for greater production-rate systems could be developed.

Once the decision to produce in quantity has been made, the product can be redesigned and special machines purchased. However, the selection of the assembly process also involves the consideration of the auxiliary equipment and facilities. Auxiliary equipment and facilities include the storage of supplies and the related assembly tools or processes. For example, to glue wooden pieces of furniture together, the traditional processes involve applying the glue to the joints by hand, clamping the pieces firmly, and then moving the clamped pieces to a storage area for drying. Some glues require clamping for periods up to 24 hours so that to produce several hundred items per day, the manufacturer would require as auxiliary equipment several hundred sets of clamps and a great amount of storage space. The cost of the auxiliary tools, in this case clamps, and the storage space could be reduced by using microwave glue-drying machines, by using adhesives that dry or cure more rapidly, or by using screws or other types of mechanical fasteners rather than glue.

PHYSICAL CONDITIONS

Another consideration in assembling operations is that of "fit." The term *fit* or *fitting* means the preparation of two or more parts that touch. There are two terms in fitting to be remembered; the class of fit refers to how accurately, or smoothly, the parts touch or fit together. The type of fit refers to the physical relationship of the parts that fit together. These considerations affect two aspects of assembly. It is generally more expensive and difficult to make parts with greater accuracy, and it is more time consuming (and more costly) to put them together with accuracy.

Classes of Fits

The *class of fit* refers to the degree of accuracy. The class of fit which exists in an assembly operation is the result of the production of the mating pieces within the maximum and minimum size limitations specified on the working drawings. Some parts have to fit together tightly and accurately and are generally referred to as *Class I* fits. Lower class-fitting situations are those assemblies in which less accuracy is needed. The amount of variation in size or location that is acceptable is called the *tolerance*. The amount of tolerance depends upon the type of item being assembled. The classes of fit and size limitations vary widely from one situation to another. For example, a tolerance of 0.001 in [0.025 mm] is a relatively large tolerance on a cylinder that is only 0.125 in [3.175 mm] in diameter. However, a tolerance of 0.001 in [0.025 mm] is relatively small on a cylinder which has a diameter of 15.000 in [38.1 cm].

Types of Fits

The *type* of fit relates to the physical position of the parts rather than to the degree of accuracy as does the class of fit. There are three types of fits:

1. RC—running and sliding fits.
2. L—locational fits. There are three subtypes:
   LC—locational clearance.
   LT—locational transition fits.
   LN—locational interference fits.
3. FN—force fits.

There are two terms involved in the RC category. The *sliding fit* is a snug or close fit with little clearance between two parts so that they move or rub together with little or no lateral movement. Examples include the pistons in a cylinder of an automobile or the parts of a machine lathe sliding upon the bed. The *running fit* is used where one part, such as a shaft, is turning within a bearing. The bearing and shaft must fit together tightly enough to keep the shaft from wobbling or moving laterally but must not be so tight that the shaft will not turn.

Locational fits (L) are used to locate and position joining parts which must be fitted together rigidly and accurately. Locational clearance fits (LC) are used in the assembly of stationary parts, such as the assembly of the head of an automobile engine to the block. Locational interference fits (LN) are fit describing the accuracy of the location and the rigidity of the joining parts. The rigidity is the more important item involved because the amount of stress or force applied to one part must be transmitted to the mating part. A very simple example of such fits is the joining of the pedal onto the sprocket assembly of a bicycle.

Locational transition fits (LT) are ranked between clearance fits and interference fits. They are an intermediate class used where accuracy and location are important and where a small amount of transmission of force is involved, but where some degree of clearance is also desirable.

Force fits (FN), frequently called *shrink fits,* include several subclasses of fits involving the interference or transferal of force of the adjoining parts. Generally, in making force fits, the hole size is made to a standard or basic size while the shaft is machined slightly larger. The parts are assembled using some type of force such as a driving hammer, called a *drive fit;* pressing together with a large press such as an arbor press, called a *force fit;* or where the assembly process is done by heating or cooling to make one part shrink tightly onto another, called *shrink fits.* Generally, an item with a hole machined to a standard size such as a pulley or a gear can be easily fastened to a shaft by heating the gear or pulley. The heat causes the item to expand which causes the hole to become slightly larger. The object is then slipped over the shaft and allowed to cool and shrink on the shaft.

WITHSTANDING FORCES

Four types of forces must be considered in assembling operations. The reason is that some fastening processes can withstand certain types of forces better than others. Thus, each type of assembling process has certain advantages and limitations in relation to the various types of forces that will be employed on the product (see Fig. 19-7). Tension forces are forces that tend to pull things apart. Compression forces push things together, and shear forces act as a cutting process. The fourth force is a rotational force causing rotary motion or "spin."

These forces may act independently or in combination. For example, peeling may be a combination of rotary spin and tension. An axle of a machine must transmit rotational forces and withstand shear and compression forces generated by the weight of both the machine and its load.

MATERIAL CHARACTERISTICS

Selection of the assembly process must include a consideration of the characteristics of the materials being used for both the fasteners and the product. Some materials will withstand vibration, while others will not; some materials work when forces are exerted upon them, while others do not; some materials undergo plastic deformation when an excessive stress is exerted; some materials can be welded and others cannot. Thus, the selection of the assembly process must include the consideration relevant to the type and characteristics of the materials involved in the assembly. Other special characteristics include whether or not the materials will conduct either electricity or fluids; whether the materials are to be permanently attached or whether the elements can be removed for service or modification; and what limitations are placed upon weight, space, the complexity of design, the closeness of fit, and the tolerances involved. All these factors combine to form decision elements that must be included in selecting the assembly methods.

Another characteristic that must be included when working with metals is the characteristic of work hardening. This is a characteristic where repeated stressing or exerting of any type of force has a tendency to cause the material to harden. The corollary of hardening, brittleness, is the major factor involved. As the material operates in its normal situation, stresses of a normal operation have a tendency to harden the material and therefore make it more brittle. Another term, in these cases, is *fatigue.* While this frequently is not a major consideration, in situations where the operating characteristics of the assembly are expected to be performed at or near the peak failure limits in terms of the strength of the material, the factor of work hardening in either the assembly process or the normal operating conditions must be considered. For example, rivets, when set in the assembly

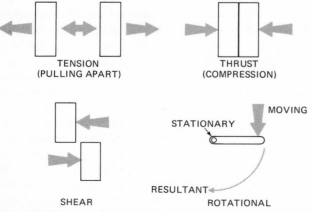

TENSION
(PULLING APART)

THRUST
(COMPRESSION)

SHEAR

STATIONARY

MOVING

RESULTANT

ROTATIONAL

Fig. 19-7 Types of forces.

process, compress the material around the rivet holes and can cause a slight hardening in these areas. Continued stress, then, around the rivets serves to increase the work hardening and brittleness characteristics causing, ultimately, failures in the materials due to the assembly process. Normally, rivets are quite strong and when not used at the peak tolerance limits do not exhibit these characteristics.

The assembly method selected must consider all of these characteristics. Assembly methods commonly used in industrial processes involve the use of joints, catches and clasps, mechanical fasteners with threads, nonthreaded mechanical fasteners such as rivets and other devices, adhesive bonding, and welding. Adhesive bonding is a misleading classification because it includes three areas that are not normally related. Adhesive bonding includes the most obvious grouping of the use of adhesives; however, adhesive bonding also includes the operations of brazing and soldering. Adhesives involve using a third substance to hold two separate materials together.

HUMAN FACTORS

There are also certain problems involving the people who work on production lines. In order to do their best work, humans generally want to feel that they are involved and that their efforts visibly contribute to the quality of the product. Where assembly methods reduce the feeling of involvement of the individual, enthusiasm and performance often deteriorate. This is evidenced by high absenteeism rates of the workers, poor workmanship, and increases in the number of faulty and reject parts. These in turn increase the cost of production which, in turn, calls for closer quality controls. This can cause workers to attempt to play games to circumvent the inspectors and can also result in direct or indirect sabotage. Sabotage, which is normally considered to be an act of an enemy, may actually be the result of a competitor hiring an individual to cause damage. However, indirect sabotage occurs when noncaring workers, in an attempt to be more human on the production line, do things that would not normally be done. For example, many people have heard the stories of the coke bottle inserted in the door panel of an automobile so that some mechanic would have a difficult time later on tracing the funny rattle in the door. This "joke" was the result of an attitude developed in a worker in a nonhumanized job who was making the subconscious attempt to humanize it. Attempts have been conducted recently to give production-line workers more of a view of the total process, more of an identity in the process, and more involvement. Such innovations have actually increased production rates and reduced rejection rates, although the assembly method might be a little less efficient in terms of mechanical processes. Engineers and technicians today must consider the reactions of the humans involved in the tooling sequences.

Training is another factor involving humans that must be considered in maintaining production rates. Workers must be trained efficiently. However, people employed on production lines frequently do not have strong academic backgrounds. Directions must be complete, and the process involved in the training must be such that the workers can easily attain the desired production rates. Some engineers refer to this as the "idiot factor." Humans characteristically make mistakes. Even the best workers cannot work with 100 percent accuracy 100 percent of the time. Therefore, the setups for the tooling operations must be done in such a manner that judgments and involvement are restricted so that workers either do not make mistakes or cannot make mistakes. For example, a worker drilling holes will be assigned to drill only one diameter of hole so that the worker cannot mistakenly drill the larger hole in place of the smaller hole.

JIGS AND FIXTURES

Jigs and fixtures are work-holding devices used in assembly of products. They offer two advantages. First, they reduce the time and labor costs of production by making setup quick and easy. Secondly, they reduce the chance of error from measurements or incorrect cuts.

For a machine to make a cut upon a workpiece, the workpiece must be firmly fixed in place. When done by hand as in custom work, this "setup" time is normally very involved and takes a long period. However, by using special holding devices (jigs or fixtures), this setup time can be reduced to a minimum. Even though making the work jig or fixture may take some time and investment, if several thousand items are to be produced, they allow the machine to work efficiently and not remain idle during the time that would be required for individual setup.

The two terms *jig* and *fixture* have different meanings. A "fixture" is defined as a device used in the assembly of an item and a device that holds the work while some operation is done to the item being assembled, as in Fig. 19-8. A "jig" differs from a fixture because a device of some type for guiding a cutting operation is built into the jig, as shown in Fig. 19-9. Thus, a fixture simply holds work in position, and a jig both holds work and guides a cutting operation. Actually, in real practice, there is little distinction

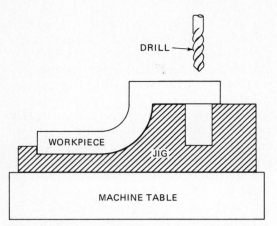

Fig. 19-8 A fixture holds the workpiece. The drill must be positioned properly.

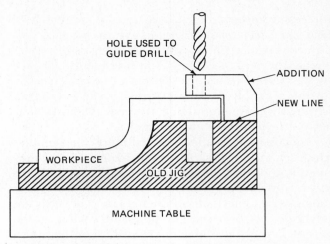

Fig. 19-9 A jig holds the workpiece and also guides the operation.

made between the two types of devices, and they are generally treated together as a class of work and time-saving devices that are part of every tooling operation.

Jigs and fixtures also allow greater production rates and lower the cost of assembly by lessening the number of hand operations that must be carefully measured, located, and accurately performed. Precision is substituted for human judgment which can be, and often is, in error.

REVIEW QUESTIONS

19-1. What is the advantage of using subcontracted parts?

19-2. Explain the relationship of "feeder" production lines to main production lines.

19-3. What types of production are featured when assembly lines are used?

19-4. What type of assembly is involved on custom-produced products?

19-5. Why can special assembly tools be more economically used on production lines than in custom assembly?

19-6. What are jigs and fixtures, and how do they differ?

19-7. What are the advantages of human involvement on a production line?

19-8. What are the problems of human involvement on production lines?

19-9. What are the advantages of automated systems?

19-10. What considerations are made in selecting fasteners and fastening systems for use on assembly operations?

19-11. What are the types of forces that must be offset in assembly operations?

19-12. What is the difference between a *type* of fit and a *class* of fit?

19-13. What is the meaning of the term "fit"?

19-14. What are the advantages of reducing cost by redesign?

19-15. Why are material characteristics included in selecting an assembly process?

CHAPTER 20

HEAT TREATMENT OF MATERIALS

During manufacturing, all metallic materials and some nonmetallic materials are subject to heating processes performed for many different reasons which may not be directly related to heat treatment. Materials may be heated for melting, forming, or fabricating; for vaporizing, baking, or curing; and for heat treatment. Although all heating processes may have heat-treatment effects on the material being heated, controlled heating and cooling to obtain specific properties is a highly specialized process and will be the only one discussed in this chapter. The other heating processes are discussed in other chapters of this book.

Changing or modifying the properties of a material by controlled heating and cooling is referred to as *heat treating*. Heat treating is a broad generic term which includes softening, hardening, and surface-treatment processes. Softening includes such processes as stress relieving, annealing, normalizing and spheroidizing. Hardening includes such processes as martempering, austempering, and hardening. Among the processes included in surface treatment are nitriding, cyniding, carburizing, and induction and flame hardening. These processes are performed for such purposes as increasing strength and hardness, improving ductility, changing the grain size and chemical composition, improving machinability, relieving stresses, hardening tools, and modifying electrical and magnetic properties of materials. Since all these processes involve changes and/or modifications of the structure of the material, complete understanding of them would require extensive knowledge of metallurgy and/or physical chemistry. Therefore, only the simpler aspects of these processes will be discussed in this chapter.

Heat treating includes heating and cooling of the workpiece to affect its structure. To avoid or minimize undesirable effects on the configuration of the workpiece, surface finish and dimensional tolerances, proper preparation, and design of the part before heat treating are necessary steps to be taken. It has been found that the design of a part affects the heat treatment. Therefore, such basic design concepts as balancing of areas of mass; avoiding sharp corners and single internal recesses as keys and keyways; and keeping hubs of gears, pulleys, and cutters of the same thickness should be carefully considered. Basic understanding of the effects of some heat-treating processes on surface finish and dimensional tolerances is another factor to be considered. Selection of the appropriate material for the part to be heat treated by a specific process should also be carefully considered.

BASIC CONCEPTS OF HEAT TREATMENT

Allotropic Changes of Iron

It has been pointed out in Chap. 2 that hardened alloys can be classified as "allotropic" and "nonallotropic." Allotropic alloys are those which change space-lattice structure during the heating-cooling cycle. Most ferrous alloys are of this kind. Heat treatment of these alloys is based primarily on the decomposition of martensite during the heating-cooling cycle. More than 90 percent of all heat treating is performed on ferrous alloys, especially carbon steel. Nonallotropic alloys consist of primarily nonferrous alloys with only a few ferrous alloys classified in this group. Therefore, an understanding of iron's behavior during the heating-cooling cycle would be helpful to understand the behavior of all ferrous alloys.

Figure 20-1 shows the cooling portion of the heating-cooling cycle of iron, the heating portion being the same. This is an approximate cooling curve of pure iron (iron with no carbon content) and the asso-

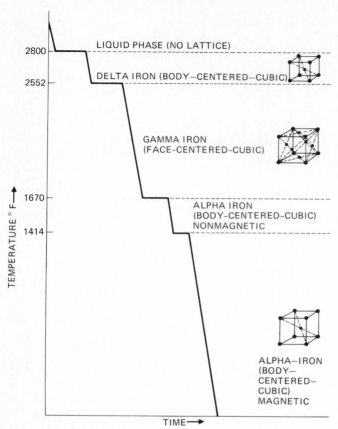

Fig. 20-1 Cooling curve showing the allotropic changes of pure iron.

ciated space-lattice structures. As explained in Chap. 2, pure iron solidified at about 2800°F [1538°C] into a body-centered-cubic (BCC) space-lattice structure known as delta iron. As the temperature drops to about 2552°F [1400°C], the iron atoms rearrange themselves into a face-centered-cubic (FCC) space-lattice structure known as gamma iron. Then, as the temperature drops to about 1670°F (909°C), pure iron changes again into a BCC structure known as alpha iron. In ferrous alloys, however, the space-lattice structure varies with the addition of carbon as an alloying element and with the cooling time (rate of cooling). In addition, the different phases vary greatly in their capacities for holding carbon in solid solution; it is this capacity variation which makes the heat treatment of ferrous alloys possible.

Effect of Cooling Rate on Hardening-Tempering of Steels

When carbon is added to pure iron, an alloy results referred to as *ferrous alloy* or simply as *plain carbon steel*. The carbon content in steels varies from 0.0 to about 1.5 percent. However, carbon steels of less than 0.35 percent carbon content (low-carbon steels) do not respond satisfactorily to heat treatment. It has been previously mentioned that phase transformation of steels is affected by the rate of cooling. Although there are an infinite number of cooling rates possible, slow and fast cooling rates are the two most distinct cooling rates. The resulting products of these two cooling rates are distinctly different. If a piece of plain carbon steel is heated to 50 to 100°F [10 to 38°C] above its critical temperature (transformation temperature) and quenched in water to room temperature, this fast cooling process is called *hardening*. The piece of steel thus cooled will be very hard (about R_c60) and brittle. Its transformed structure will be mostly martensitic. Martensite is a supersaturated solid solution of carbon in iron (Fig. 20-2). It is the principal component of quenched steels and is notable for its hardness and brittleness. If the hardened piece of plain carbon steel is reheated to about the same temperature as before and is not quenched in water but left in the furnace to cool slowly, this slow cooling process is called *furnace annealing*. The piece will be relatively soft (about R_c20) and tough. Its

Fig. 20-2 Martensite formation in an 0.8 percent carbon steel, ×100, picral etch. (*U.S. Steel Corp.*)

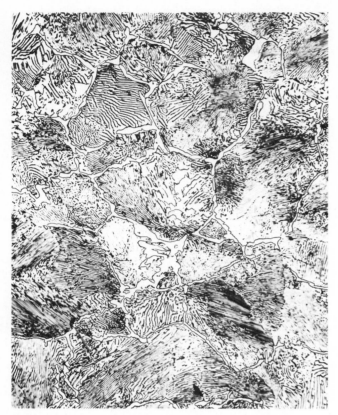

Fig. 20-3 Pearlite and cementite formation in a 1.2 percent carbon steel, ×500, picral etch. (*U.S. Steel Corp.*)

transformed structure will be course pearlite. Pearlite is the eutectoid mixture of ferrite and cementite (Fig. 20-3). It has a lamellar structure with ferrite and cementite arranged in alternate platelike layers. These are the two extremes of the heating-cooling cycle in the hardening-tempering of steels. Other possible cooling rates will be discussed later in this chapter.

Effect of Carbon on Hardening-Tempering of Steels

Although the content of carbon in steels varies up to about 2.0 percent carbon, carbon steels of less than 0.35 percent carbon are not hardenable because the carbon content is not sufficient to produce enough martensite required for hardened steels. On the other hand, steels with high-carbon content which transform to almost 100 percent martensite with quenching become extremely hard and brittle thus losing a great deal of toughness. Therefore, it is desirable that steel has enough carbon content to assure proper martensite structure with quenching, yet retain as much of its toughness as possible. It has been found that steels with about 0.35 to 0.70 percent carbon assure the most desirable combination of hardness and toughness.

Hardenability and hardness are two concepts which are not the same. *Hardness* refers to a property of a solid material and it is affected by the heat-treating process. Thus a material can become harder or softer by heat treating it. Hardness of solid materials is measured by hardness tests as discussed in Chap. 3. On the other hand, *hardenability* refers to the ability of a material to respond to heat treatment, and it is affected by the composition of the material. Not all solid materials have the ability to be hardened by heat treatment, but all have hardness. Hardenability of steels is measured by using the Jominy Test of Hardenability or end-quench test (Fig. 20-4). In this test a bar 1 in [2.54 cm] in diameter and 4 in [10.16 cm] long is heated and hardened by an end-quench method as shown in Fig. 20-5. The gradient hardness of the test bar along its length depends principally on the composition of the bar. Rockwell hardness readings are taken along the length of the bar at $^1/_{16}$-in

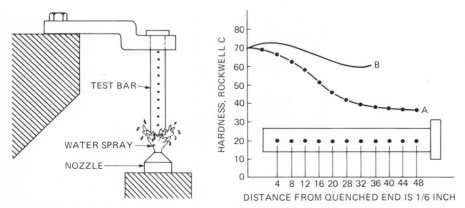

Fig. 20-4 Schematic representation of end-quench (Jominy) test for determining the hardenability of steels. *A* and *B* are both same carbon content; however, *B* has higher alloy content. *B* exhibits greater hardenability.

[0.16 cm] intervals. The readings are plotted in a curve, as shown in Fig. 20-5, which illustrates the hardenability of the material. If the hardenability of the material is zero, its response to hardening will be zero, and the curve would be a straight line on the plot.

HARDENING PROCESSES AND PRACTICES

The primary purpose of hardening is strengthening of the material being hardened. Hardening of materials can be accomplished by a variety of methods. However, hardening of metallic materials by phase transformation and precipitation hardening are the two most widely used methods. In phase hardening, the material is heated above its transformation (critical) temperature and is rapidly cooled to room temperature resulting in a phase transformation producing a structure which substantially increases the strength of the material (Fig. 20-2). This type of hardening is used mostly with ferrous alloys, especially steels. In precipitation hardening, the material is heated to the temperature in which the solute (minor or secondary phase) goes into solution. Then the material is quenched at a rate at which the solute is prevented from being separated, even though the solubility of the primary phase (matrix) is decreased. Thus atoms of the solute are trapped at the atomic level producing a structure which strengthens the material. This type of hardening is used mostly with nonferrous alloys.

Hardening of Steels

The iron-carbon equilibrium diagram, shows the relationship of phase transformation structures and carbon content of steels. Also it shows the solubility ranges of carbon in iron at various temperatures. This diagram may be used as a guide for temperatures for heat-treating operations. It is applied to practical heat-treating operations by incorporating several heat-treating ranges. The tempering (annealing) range is plotted slightly above and parallel to the A_3 points. Thus, as the carbon content increases, the complete carbon-solubility temperatures, represented by line GS, decrease until the eutectoid percentage is reached. From this point on, the line representing the annealing range straightens out, indicating that there is no change in the austenitizing temperature as the carbon content increases above the eutectoid percentage. To harden hypoeutectoid and eutectoid steels, all the carbon must go into solution. On the other hand, to harden hypereutectoid steels, only about 0.80 percent of the carbon need go

into solution. This accounts for the fact that austenitizing temperatures for hypereutectoid steels do not increase with the carbon content.

For a particular operation, the annealing temperature may be slightly above or below that shown by the annealing curve. The same is true of the hardening and normalizing curves. This is why ranges of temperatures are shown on the diagram which are extensively used in heat-treating operations. Thin sections are usually heat treated at the lower side of the proper range, while heavy sections are soaked at the upper side of the range. The steel must be thoroughly soaked at the proper temperature to allow the carbon to completely dissolve into the austenite at that temperature. As the temperature increases, the rate of carbon dissolving into the austenite becomes more pronounced; however, at the same time, the grains coarsen with an increase in either temperature or soaking time. Since fine grains are desirable, the steel should not be heated any higher or soaked any longer than necessary. Some steels contain special alloying elements for control of grain size. In these special alloy steels, there is usually no excessive grain growth until they are heated too high or soaked longer than necessary.

The normalizing range differs from the hardening range in that, to completely normalize steel, all the carbon must go into solution. Therefore, the normalizing range is plotted slightly above the hardening ranges for hypoeutectoid steels, and slightly above and parallel to the PS points for hypereutectoid steels. Thus, the normalizing temperatures for hypoeutectoid steels decrease as the carbon content decreases to 0.8. On the other hand, normalizing temperatures for hypereutectoid steels increase with increases in carbon content.

Therefore, to accomplish satisfactory hardening of steels by the quenching method, three conditions should be met: (1) there should be adequate carbon content in the steel, preferably over 0.35 percent, (2) sufficient time must be allowed at the proper soaking temperature for carbon to dissolve in the austenite, and (3) the cooling rate must be fast enough to prevent the formation of pearlite, but not so rapid as to result in distortion or cracking of the workpiece.

Hardening of steel is based on the transformation of austenite to martensite. Austenite is a solid solution of carbon in FCC gamma iron. It can hold up to about 2.0 percent carbon in solution at 2066°F [1130°C] (Fig. 20-2). If a piece of steel is heated to austenitic state and cooled slowly, the normal transformation from gamma to alpha iron (Fig. 20-3) would take place with cementite precipitating out of solution because of decreased solubility of carbon in alpha iron.

Fig. 20-5 Martensite formation in an 0.8 percent carbon steel, ×1000, picral etch. (*U.S. Steel Corp.*)

However, if the piece of steel is cooled fast, the transformation from gamma to alpha iron will be so rapid that carbon would not have enough time to precipitate out of solution and will be trapped in the solution. This fast transformation and suppression of cementite results in a supersaturated solid solution of cementite in a body-centered-tetragonal structure called *martensite* which is the principal component in steel hardening (Fig. 20-5). Therefore, steel hardening is based on the two-step process: first it is heated to the austenitic state, and second it is rapidly cooled so that the transformation from gamma to alpha iron is suppressed while the transformation from gamma iron to martensite is achieved. This two-step process may be affected by alloying elements present in the steel.

Hardening of Nonferrous Alloys

The two most widely used methods of heat-treating nonferrous alloys are solution and precipitation heat treatment. In *solution hardening* the alloy is usually heated at about 900 to 1000°F [482 to 537°C]. At this temperature, the solute goes into solution in the ma-

trix. Then the alloy is quenched to maintain that phase in the supersaturated solution. Molten baths or air furnaces are used for the heating. Water is the principal quenching medium.

Precipitation hardening usually follows solution heat treatment. The alloy is reheated to about 250 to 400°F [121 to 204°C] for several hours. At this temperature, movement of certain atoms in the solute in the supersaturated solution is accelerated. This process if referred to as *precipitation heat treatment*. If, however, the alloy is left to precipitate naturally at room temperature, which requires considerable time, this process is called *natural aging*. Heating of the alloy for precipitation hardening is usually done in gas or electric furnaces.

Quenching Media and Practices

The primary purpose of quenching is to achieve a rapid cooling rate, thus controlling microstructural changes. To achieve this purpose, the quenching medium (coolant) must have the capacity to remove heat rapidly and uniformly, producing uniform hardening. Heat removal in quenching depends on such factors as the mass and area of the heat-treated part, the type of coolant used, and the type of circulation or agitation used in the quenching bath. In selecting a quenching coolant, such factors as cost, availability, stability, and safety must be considered.

The types of coolants most commonly used are water, oil, air, and brines. Water is probably the most widely used quenching medium. However, because of its low vaporization temperature, water creates a vapor blanket on the surface of the work which affects the cooling rate and uniformity of quench, so agitation or circulation is employed. Quenching mineral oils constitute the most important industrial quenching media, especially suited for quenching at elevated temperatures. Brines have higher cooling rates than water or oils. Air has a relatively low cooling rate and will only produce martensite in air-hardening steels, but it is easy to use and it is cleaner. To increase the cooling rate and provide a more uniform heat removal, liquid quenching media are circulated or agitated by impellers or pumps. If agitation of the media is not provided, the part must be quenched by moving it rapidly in the bath to destroy the vapor blanket.

TEMPERING (SOFTENING) PROCESSES

Hardening constitutes only part of heat treating; the other part is tempering. Almost all hardened alloys undergo some form of tempering to restore such

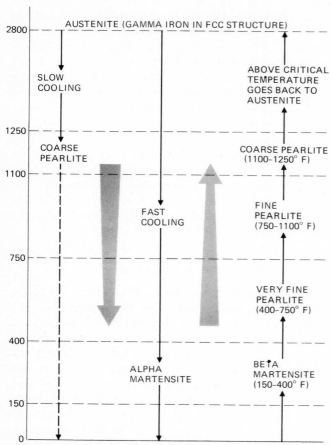

AUSTENITE (GAMMA IRON IN FCC STRUCTURE)

2800

SLOW
COOLING

ABOVE CRITICAL
TEMPERATURE
GOES BACK TO
AUSTENITE

1250

COARSE
PEARLITE

COARSE PEARLITE
(1100–1250° F)

1100

FAST
COOLING

FINE
PEARLITE
(750–1100° F)

750

VERY FINE
PEARLITE
(400–750° F)

400

ALPHA
MARTENSITE

BETA
MARTENSITE
(150–400° F)

150

0

Fig. 20-6 Approximate cycle of hardening and tempering of steels.

properties as ductility and toughness. Tempering is a broad generic term and includes such processes as annealing, normalizing, spheroidizing, austempering, and martempering. Tempering is usually performed to improve ductility, toughness, and machinability; relieve stresses; refine grain size; and improve strength.

Annealing is basically a softening process. The general procedure is to heat the alloy to a specified temperature (Fig. 20-6) depending on the results desired, then cool it slowly to room temperature. Several variations of the annealing process are used, among those are stress relieving, full annealing, normalizing, process annealing, and cycle annealing.

When alloys are welded, cut, formed or heat treated, they develop internal stresses called *residual stresses*. These stresses if high enough to be detrimental to the proper functioning of the part must be removed. Stress relieving is the process used for that purpose. The alloy is heated to about 1100 to 1300°F [593 to 704°C] held at that temperature until uniformly heated, then is slowly cooled to room tempera-

ture. Some dimensional change occurs with stress relieving which can serve as an indication of the degree of stress removal. In full annealing, the alloy is heated above the transformation temperature (Fig. 20-2), held at that temperature until uniformly heated, then cooled slowly in the furnace. Temperature for full annealing depends upon the carbon content of the alloy. Full annealing is performed to achieve such objectives as softening, grain-size refinement, or homogenization.

A spheroidizing anneal differs from stress relieving and full annealing in terms of temperature. When heated to temperatures at about 1325 to 1375°F [718 to 745°C] for long periods of time (four to eight hours), the cementite changes from layers into globular (spherical) form. This process is called *spheroidization*. The globules or spheroids are surrounded by a ferrite matrix which is very soft (Fig. 20-7). Spheroidization results in steel that is in the softest possible condition. This process is widely used to make high-carbon steels more machineable and formable.

With the austempering or isothermal quenching method, the alloy is heated to the austenitic state, then quenched to a temperature of about 500 to

Fig. 20-7 Spheroidized carbon in an AISI 52100 steel, ×1500, picral etch (*U.S. Steel Corp.*)

750°F [260 to 398°C] in a bath. The alloy is held at the bath temperature until complete transformation occurs isothermally, then is air-cooled to room temperature. The advantage of austempering is that the alloy is quenched without going through the entire austenite to martensite transformation. This transformation is circumvented by using the isothermal quenching method. Furthermore, subsequent tempering processes are needed. On the other hand, with the marquenching or interrupted quenching method, the alloy is heated to the austenitic state, then quenched in a bath to about 500°F [260°C] (Fig. 20-9). The alloy is kept in the bath only for a few minutes until the temperature throughout the piece has become equalized, but before any transformation begins. The alloy is then quickly removed and air-cooled or quenched to room temperature. While austempering circumvents the formation of martensite, marquenching does not. In marquenching martensite is allowed to form, but after the alloy has been removed from the bath, at which point martensite forms more uniformly without a high level of thermal gradient stresses. Both austempering and marquenching are used successfully with alloy steels, because the alloy elements in steel slow down the transformation of austenite to martensite. These methods have relatively limited application for plain carbon steels because the transformation time of austenite is relatively rapid. The quenching media used for austemporing and marquenching are oils, molten salts, and low-temperature molten metals.

SURFACE-HARDENING PROCESSES

One of the main purposes of surface-hardening, or case-hardening processes is to increase the carbon content or develop other forms of structure on the surface making it more responsive to heat treatment. Surface hardening is usually applied to workpieces requiring a hard surface but a relatively soft and ductile core. An additional reason for using surface hardening is cost. In situations where only a relatively thin layer of the surface is required to be hardened, surface hardening is less expensive than full hardening. Surface-hardening processes can be classified into two groups. The first group includes the processes used only to harden the surface without changing its composition. These processes are applied to medium- and high-carbon steels and some alloy steels and include induction hardening and flame hardening. The second group includes those processes which change the composition of the surface by increasing the carbon content and include carburizing, nitriding, and cyaniding.

Carburizing involves increasing the carbon content of the surface so that it responds to surface hardening. To increase the carbon content on the surface, both temperature and time are important process variables. The carbon content increases on the surface as the temperature and time increase. Temperatures of about 1675 to 1725°F [912 to 940°C] are used for low-carbon steels. Carbon can be supplied by gas carburizing, pack carburizing, and liquid carburizing. In gas carburizing a controlled atmosphere is produced by using natural gas, manufactured gas, or certain propane gases. The workpiece is heated in the furnace under controlled gas atmosphere and the carbon from the gas penetrates the surface until the desired carbon content and depth of penetration have been reached. The workpiece is then quenched in a suitable quenching medium to be hardened. *Pack carburizing* is by far the simplest of the carburizing processes. The workpiece is packed in a box with a carbon source, usually charcoal and activating salts. Many different carbon compounds are commercially available. The box containing the workpiece packed around with the carbonatious material is sealed and heated to about 1550 to 1700°F [842 to 926°C], depending on depth desired in an electric, oil, or gas furnace for several hours. The time, temperature, carburizing material, and the type of steel determine the depth of surface hardening. After the workpiece is heated for several hours, it is quenched in a suitable coolant. In *liquid carburizing* a molten salt bath of high-carbon content is used as the carburizing material. The workpiece is heated in the salt bath to about 1600 to 1750°F [870 to 953°C], until the desired depth of carburizing is reached, then is quenched to be hardened.

Nitriding is accomplished by heating the work to about 930 to 1200°F in contact with ammonia gas for 20 to 200 hours. At the high temperature ammonia is decomposed into nitrogen and hydrogen, nitrogen is absorbed by the workpiece to form complex nitrides which are hard, producing a hard surface. Nitriding is more expensive and requires closer controls than other surface treatments. It is used with special steels (nitralloy) and in situations where a thin but very hard surface is desired.

Cyaniding is similar in many respects to liquid carburizing. In cyaniding, the workpiece is heated for several hours in a cyanide bath at about 1300 to 1600°F [704 to 870°C], then is quenched in either water or oil to produce a surface hardening of 0.001 to 0.015 in [0.0254 to 0.381 mm] depth. Steel composition, composition of the cyanide bath, temperature, and time are the important process variables. Cyaniding provides an economical means of producing a shallow surface hardening, but it is extremely dangerous due to toxicity of the cyanide used. Therefore, extreme safety precautions are necessary at all times.

Induction hardening is based on heating the workpiece by means of induction heating, then quenching to achieve the desired surface hardening. Unlike other surface-hardening processes, which are used mostly with low-carbon steels, induction hardening is used with medium- and high-carbon steels. There is no change in the composition of the work surface in induction hardening. Induction heating is achieved by subjecting the workpiece to an alternating electric field produced by a high-frequency current which causes eddy currents in the surface and provides the required heating. The heating coil (Fig. 20-8) carries the alternating magnetic field and is the primary of the circuit while the workpiece serves as the secondary. Induction heating is a selective heating process in that it heats the surface of the workpiece only at the area covered by the heating coil. Induction heating is accurate, easy to automate, and produces minimum distortion in the workpiece. Frequency, power, and time are the important process variables. High frequency, high power, and short time produce relatively shallow heating. To harden the surface, the workpiece is heated by induction, then is quenched in water or other suitable coolants. Water is the most widely used quenching medium. Among the most widely used induction-heating equipment are the motor-generator, vacuum-tube, transistor, and spark-

gap types. These units are available in capacities from 0.25 to 2,500 kW (kilowatts), 11 to 800 V, and 1000 to 10,000 Hz (cycles per second) frequencies.

Flame hardening differs from induction heating only in terms of the heating principle. The workpiece is heated by a high-temperature flame, then is quenched in a suitable coolant to achieve the desired surface hardness. The most widely used gases to produce the heating flame are acetylene, city gas, hydrogen, natural gas, or propane. These gases are used either with oxygen or air in proper proportions. Oxyacetylene flames produce the highest heating temperatures, up to 5620°F (3101°C). Two basic heating torches (heads) are used: the oxyacetylene torch and the air-fuel gas torch. With the oxyacetylene torch, oxygen and acetylene are mixed before forcing them through the burning tip of the torch. The air-fuel gas torch is lined with refractory materials to form a preheating surface. The gas is compressed through burning orifices and heats the refractory lining of the torch, which helps to preheat the gas and produce the high temperatures required. Unlike induction heating, which produces a highly localized heating, flame heating heats a larger area of the workpiece and produces stresses and distortion. This is one of the factors which confines flame hardening to relatively thick and large size parts.

HEAT-TREATING EQUIPMENT AND PRACTICES

The complexity and diversity of heat-treating processes discussed above require equipment which will heat and cool the material with precision so that specific procedures and practices can be followed carefully to achieve desirable results. To accomplish this objective, the equipment should provide: (1) size capacity to handle various part sizes, (2) heating capacity to provide the required temperatures, (3) heating atmosphere which is free from scaling, oxidation, and decarburization, (4) cooling capacity to provide the required cooling rates, (5) safety of operation, and (6) reliability of results. The equipment used can be classified as heating, cooling, and heat-controlling devices.

Heating Equipment

Heating equipment is used to generate the required temperatures to heat the workpiece. Heating equipment may be classified according to the method by which heat is transferred to the workpiece. Most heating equipment belongs to the classification which uses hot gases as the medium for transferring heat. Fur-

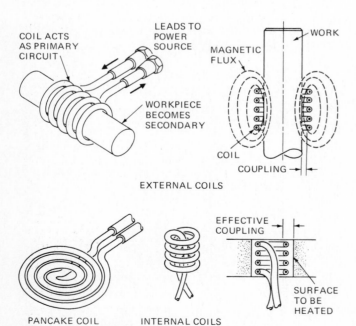

Fig. 20-8 Typical heating coils used for induction heating. (*SME.*)

Fig. 20-9 Batch-type furnaces with controls. (*Ipsen Industries.*)

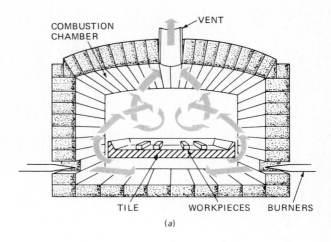

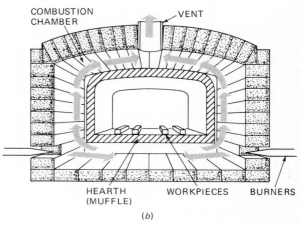

Fig. 20-11 Cross section of typical direct and indirect fired hearth furnaces. (*a*) Direct-fired or semi-muffle-type hearth furnace; (*b*) indirect fired or muffle-type furnace.

naces are used which provide protective gas atmospheres to prevent decarburization or scaling of the workpiece being heat treated. The second broad classification of heating equipment employs a liquid bath as the heat-transferring medium. While salt bath heaters are the most common equipment in this area, other liquids such as oils and molten metals are used. Liquid baths transmit heat fast and uniformly and afford atmospheric protection during the heating cycle.

Furnaces in many shapes, sizes, and designs, using gas, oil, or electricity as heating sources, are commercially available. In terms of production practices furnaces can be classified into batch and continuous types. In each of those two basic classifications, there are many different variations. A *batch furnace* is one in which a workpiece or batches of workpieces are heated at any particular cycle or time (Fig. 20-9). On the other hand, a *continuous furnace* is one in which the workpieces are heated in a continuous manner, one after another, by means of a belt conveyor or rollers that move the workpieces through the furnace.

Batch furnaces are designed to be loaded either horizontally (box furnaces) or vertically (bell furnaces), and they are either of the hearth or box type. A *hearth furnace* is one in which the workpieces are placed in a hearth to be heated (Fig. 20-10) while the

Fig. 20-10 Charge end of roller-hearth annealing furnace.

workpieces are placed in an oil, or molten metal bath, in the bath-type furnace. On the basis of the method of heating the work, batch furnaces may be: (1) the direct-fired type (Fig. 20-11a), where the fuel is burned directly in the hearth containing the work and (2) the indirect-fired type (Fig. 20-11b), where burning of the fuel is accomplished in a separate chamber and does not come in contact with the work. The indirect-fired furnaces may be of the recirculation or retort type. *Recirculation furnaces* are those in which the fuel is burned separately from the work, but the combustion gases are circulated in the hearth of the furnace. On the other hand, in the *retort furnaces,* the fuel is burned separately from the work and does not come in contact with the work. Some hearth-type furnaces have movable hearths in order to facilitate loading and unloading. Two basic designs are used in this type of furnace, the car-bottom (Fig. 20-12), where the hearth is built on wheels, and the rotary, where the hearth rotates.

Fig. 20-12 Car-bottom annealing furnace used to anneal aluminum ingots. (*Surface Combustion.*)

Continuous furnaces are mostly used for high-production rates of relatively uniform parts. In these furnaces the parts move continuously from one end of the furnace to the other while they are being heated (Fig. 20-13). In continuous furnaces, the parts may move by means of a belt conveyor or roller-hearth or power-screw conveyor. These furnaces may be designed with several zones such as heating zone, holding zone, and cooling zone. They also may be designed to operate automatically or semiautomatically.

Molten Bath Furnaces

Molten bath furnaces (heaters) are widely used in heat treating for heating and cooling purposes. The bath is heated until the salt or metal melts and the workpiece is immersed into the bath to be heated. Molten bath heaters have the advantage that the workpiece is immersed into the bath and, therefore, protected in terms of scaling oxidation or decarburization. The bath consists of such salts as sodium and potassium chloride, nitrates, cyanides, or molten

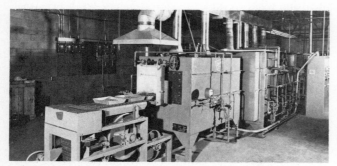

Fig. 20-13 A gas-fired belt conveyor (continuous) furnace. (*Drever Co.*)

Fig. 20-14 Automatic vacuum furnace which provides 2400°F operating temperatures and 0.1 μ vacuum capacity. (*Ipsen Industries.*)

metals, including lead. The working temperature of molten baths varies from 350 to 2400°F (176 to 1314°C), with nitrates and metals used for low temperatures and chlorides and salts for high temperatures. The bath is heated with oil, gas, or electricity. In the case of electricity, resistance-type heating elements are used.

Furnace Atmospheres

To prevent oxidation scaling and decarburization from oxygen, water vapor, and carbon dioxide present in the furnace and to maintain dimensional accuracy and surface finish, the work must be heated in protective furnace atmospheres. The most desirable atmosphere is a vacuum (Fig. 20-14). Vacuum "atmospheres" are becoming more widely used, especially with exotic metals, as vacuum furnaces are perfected. Vacuum atmospheres have certain advantages over conventional atmospheres: vacuum is nonreactive with the workpiece material; vacuum furnaces do not require any protective mixture and thus are cheaper to operate; and vacuum furnaces heat the factory area less since heat loss is minimized in a vacuum.

Many other types of furnace atmospheres are used in heat-treating processes. Two basic types of atmospheres can be produced: exothermic and endothermic. *Exothermic atmospheres* are those in which the combustion of the fuel is complete or partially complete in the furnace. These atmospheres produce reducing rather than oxidizing furnace environments due to high nitrogen content. Relatively low amounts of carbon dioxide and monoxide and hydrogen are also present in these atmospheres. Exothermic atmospheres are widely used for annealing ferrous and nonferrous alloys and are most economical and non-explosive. On the other hand, *endothermic atmospheres* are those in which the combustion of the fuel cannot be accomplished without an external source of heat. These atmospheres produce a furnace environment consisting of nitrogen, hydrogen, and carbon monoxide. Endothermic atmospheres are used for heat treating steels. They are relatively expensive and are explosive. Other atmospheres used are ammonia base, charcoal base, and prepared nitrogen base. The selection of a furnace atmosphere should be based on such factors as composition of the metal being heat treated, temperatures required, type of furnace used, and cost involved.

Cooling or Quenching Equipment

Cooling equipment used in heat treating consists of the heating furnace itself, if the material is to be cooled very slowly, as in the annealing of steels. Salt, oil and lead baths, in addition to being used for heating purposes, are also used for quenching. Water quenching tanks are also very common. Cooling equipment varies widely in design, depending on the volume, agitation, recirculation, and cooling medium. To prevent distortion of the workpiece being heat treated, dies of specific shape to fit the part have been used successfully. Compressed-air cooling systems are also used in specific applications. Heat-treating cooling systems may involve a simple quenching tank or a fully mechanized, automated cooling line. The basic characteristics of a cooling, heat-treating system are cooling rate, production of the workpiece surface, cost, and safety.

Heat-Control Equipment

Temperature-measuring and controlling equipment are very important in heat treating. Temperature-controlling equipment may consist of a manually handled pyrometer to automatically activate thermocouples which operate through amplifying circuits to open and close valves or electric contacts, thus controlling the flow of fuel or current. Commercial furnace pyrometers to measure the temperature are of the indicating, recording, and controlling types. Some automatic pyrometer controllers not only indicate the temperature but also automatically regulate it. These automatic controllers would regulate the cooling as well.

Heat-Treatment of Nonmetallic Materials

Heat-treating processes are used extensively with metallic materials. However, modern technology is using heat-treating processes on certain nonmetallic materials, particularly glass and plastics.

Heat Treating of Glass

Several heat-treating processes are applied to glass and related materials. Among the most commonly used processes are thermal finishing, annealing, tempering, sealing, and flame cutting. However, only annealing and tempering of glass can be considered true heat-treating processes as discussed in this chapter.

During forming processes most glass products develop residual stresses as a result of rapid uneven cooling (since glass has low thermal conductivity). To remove these stresses, the glass productions are annealed in a continuous-type furnace called a *lehr*. The lehr is a controlled heating-cooling furnace employing a conveyor belt to move the parts through the heating-cooling cycle. Temperatures and time vary, depending on the class and type of glass products being heat treated.

Tempering of glass is used primarily to increase its strength. The glass is heated in the furnace to nearly the softening point and then is chilled fast. This rapid chilling (tempering) increases the glass strength two to four times.

Heat Treating of Plastics

The most common heat-treating process applied to plastics is annealing. Annealing is applied to plastics to reduce shrinkage and distortion and to eliminate residual stresses due to finishing and manufacturing processes.

Annealing of plastics consists of prolonged heating at temperatures lower than the mold temperature followed by slow cooling. Plastic manufacturers usually specify the annealing temperature and time. It is advisable to secure and use the manufacturer's specified annealing temperatures to optimize annealing results.

REVIEW QUESTIONS

20-1. Define heat treatment of materials.

20-2. Explain the basic purposes of heat treatment of materials.

20-3. Define and explain the difference between allotropic and nonallotropic alloys.

20-4. Deferentiate between hardness and hardenability of solid materials.

20-5. Explain the procedure used to determine hardenability of steels.

20-6. Explain the process of transformation of austenite into martensite.

20-7. Explain the process of hardening nonferrous alloys.

20-8. Explain the difference between tempering and hardening of ferrous alloys.

20-9. Explain the difference between austempering and martempering of alloy steels.

20-10. Compare induction hardening with flame hardening.

20-11. What are the basic factors to be considered in purchasing heat-treating equipment?

20-12. Distinguish between the direct and indirect fired furnaces. State their advantages and limitations.

20-13. What are the most commonly used quenching media? State their advantages and limitations.

20-14. How are heat-treating processes applied to nonmetallic materials?

SURFACE CLEANING AND FINISHING PROCESSES

Among the requirements that most industrial products must meet are *aesthetic appeal* and *resistance to deterioration*. Aesthetic appeal pertains mostly to surface finish and is an important factor in sales. A pleasing surface finish is usually more attractive to the potential buyer, thus making the product more salable and more competitive in the market. On the other hand, resistance to deterioration pertains to durability of the surface against environmental conditions. Some materials are naturally resistant to environmental conditions while others need to be protected by coating the surfaces to render them durable. For providing pleasing and durable finishes to the many different products manufactured by modern industry, a wide variety of processes and equipment have been developed. Therefore, knowledge of the basic concepts of the cleaning and finishing processes is important in specifying appropriate finishes for particular products. Surface treatments may be classified as *surface cleaning* and *surface-finishing processes*. However, certain surface-cleaning processes may serve both purposes, cleaning and finishing.

SURFACE-CLEANING PROCESSES

In preparing a surface either for protective or decorative purposes, cleaning processes are applied first. The cleaning processes mostly used in modern industry can be classified as *mechanical*, *chemical*, and *miscellaneous* cleaning processes. Among the most important mechanical processes are *abrasive blasting, mass finishing, tumbling, belt sanding, wire brushing* and *polishing*, and *buffing*. Chemical cleaning processes are *alkaline cleaning, solvent cleaning, solvent vapor cleaning, acid cleaning*, and *molten salt cleaning*. Among the miscellaneous cleaning processes are *ultrasonic cleaning* and *steam cleaning*.

MECHANICAL CLEANING PROCESSES

Abrasive Blasting

Abrasive blasting is widely used in modern industry for various purposes. It is used as a cleaning process to remove sand from castings; scale from heat-treated parts; and rust from corroded materials, old paints, carbon deposits, and other soils. It is used as a finishing process to roughen surfaces for application of protective coatings and adhesives, to remove surface irregularities in order to improve surface finish, and to develop or impart various types of mat finishes, especially on relatively soft materials. Abrasive blasting is accomplished by forcing selected abrasive media (Fig. 21-1), dry or suspended in a liquid, against the surface of a part to either clean it or finish it. The dry-abrasive media can be blasted against the surface either by centrifugal force or by air pressure. The centrifugal force of the dry-abrasive media can be produced by a power-driven bladed wheel as shown in Fig. 21-2. This process is becoming more popular because it can be used as a semiautomated or completely automated process for high-production rates. On the other hand, the air-pressure dry-abrasive blasting is manual or semiautomated and used for relatively low production rates. It is considered ideal for small- and medium-size parts with intricate shapes.

The wet-abrasive blasting consists of a slurry formed by mixing a relatively fine abrasive media with chemically treated water. The slurry is forced against the surface by compressed air through appropriate nozzles. This process is more precise than the dry-abrasive blasting, and it is used in situations where dimensional tolerances and surface finish of relatively fragile components are of major consideration.

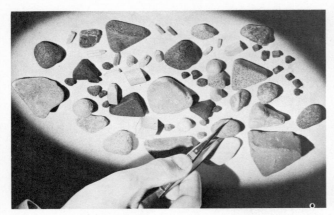

Fig. 21-1 Various types of synthetic and natural abrasive and nonabrasive media used in abrasive cleaning. (*The Wheelabrator Corp.*)

In general, however, abrasive blasting is considered a relatively economical process in terms of work-hour reduction, equipment investment, and recovery of the abrasive media used. A wide variety of natural and manufactured abrasive media are used. Among the most commonly used types are *steel shot, angular grit, slag products, aluminum oxide, silicon carbide, glass beads,* and *sand.* The air pressure for the air blast varies from 40 to 80 psi (2.8 to 5.6 kg/cm²). The angle of blast varies from 30 to 90°, and the distance of the nozzle from the surface of the part varies from 6 to 12 in [15.24 to 30.48 cm] (Fig. 21-3).

In many situations during abrasive blasting, surface damage of the parts may occur. Therefore, optimum

Fig. 21-2 Dry abrasive media forced by a centrifugally operated bladed wheel. (*The Wheelabrator Corp.*)

Fig. 21-3 Dry abrasive hand-blasting process performed in a special blasting room with the operator wearing protective outfit. (*Norton Co.*)

velocity, direction of abrasive flow, equipment, and procedures used should be carefully considered. Because of the damage that may occur on the surface of the part, abrasive blasting should not be considered as an ideal process with parts where uniform surface finish and close dimensional tolerances are critical factors. Abrasive blasting should not be used with gears and threaded parts, with parts containing deep recesses, blind pockets and interiors, with tubular sections, and with parts of relatively thin sections.

Mass finish or tumbling is the most widely used cleaning (or finishing) process for high-volume small-size parts. The main purpose of this process is to improve surface finish, and to remove burrs, sharp edges, tool marks, flash from castings, and scale from heat-treated parts (Fig. 21-4). It has been successfully used to clean parts made of metallic, ceramic, rubber, and plastic materials. The process is used for small and relatively large parts, depending on the size of equipment employed. The process is based on the *scrubbing action* of the abrasive media. Parts are placed in the machine (tumbler) with a suitable natural or synthetic abrasive media, water, and other compounds. The machine provides a revolving, vibratory, or combination of revolving-vibratory motion to the work load in the machine. As the abrasive media pass across the surface of the parts, they impart sliding and impact

Fig. 21-4 Close-ups showing parts before and after mass finishing processes. (*Norton Co.*)

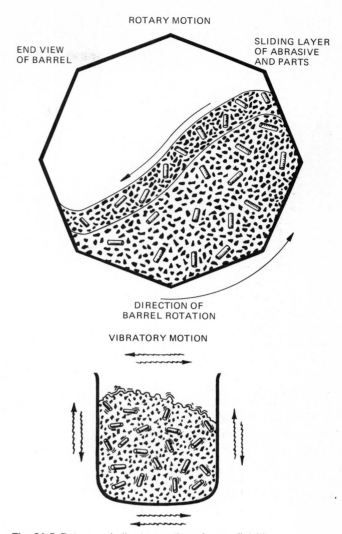

Fig. 21-5 Rotary and vibratory action of mass finishing processes. (*Norton Co.*)

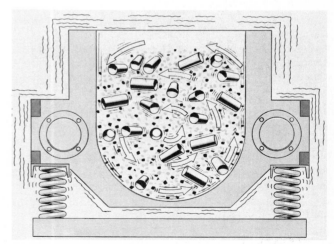

Fig. 21-6 Action of a vibratory-bowl-type cleaning machine (*The Wheelabrator Corp.*)

actions which result in scrubbing the surface of the parts (Fig. 21-5).

The equipment used in this process can be classified as vibratory, rotary, or combination vibratory-rotary. Vibratory machines such as vibratory *tubs, bowls,* and *barrels,* provide an oscillating motion to the work load in the machine (Fig. 21-6). Vibratory machines can be used for large parts. Because of the vibratory motion, they can be used with parts that contain deep recesses and blend pockets. The machines operate at frequencies of up to 3500 vibrations per minute, with adjustable amplitudes of vibration from $^1/_{16}$ to $^1/_4$ in [0.15875 to 0.6385 cm]. On the other hand, rotary machines, commonly called *rotational barrels,* provide a relatively more gentle or less aggressive action than

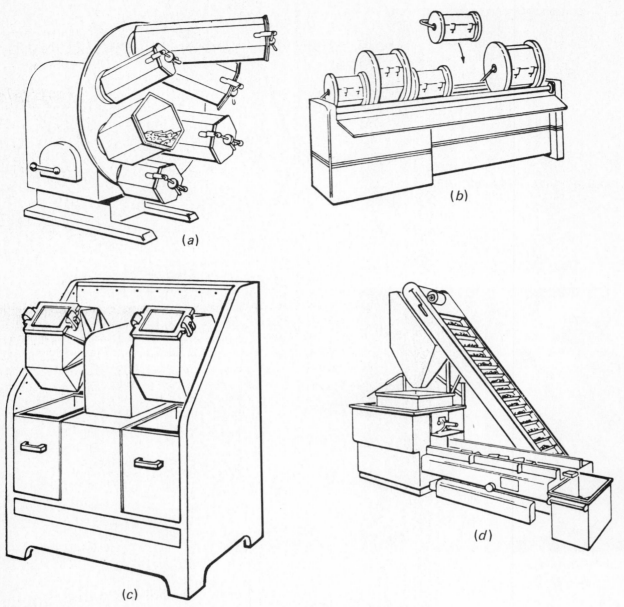

Fig. 21-7 Various types of mass finishing machines used by modern industry: (*a*) multidrum rotational barrel; (*b*) multidrum barrel on rollers; (*c*) twin rotational barrel; and (*d*) automatic vibratory unit.

the other types of machines. These machines are extensively used where time is not a critical factor and work load capacities are relatively large. The scrubbing action in the rotational barrels is accomplished as the upper layer of the work load slides toward the lower end of the barrel as the barrel rotates. The barrels are made in a hexagonal or octagonal shape to maximize the sliding action of the work load. Sizes of the barrels vary from 6 to 46 in [15.24 to 116.8 cm] long and in almost any diameter and rotational speed. The machines are made as single-barrel, multibarrel, or multicompartment machines. The barrels are rotated by a spindle-type drive or on rollers. Barrels are mounted horizontally or in a tilted position to provide an additional movement of the work load, and are either submerged or nonsubmerged (Fig. 21-7*a–d*). The combination (rotary-vibratory) machines provide a combination action to the work load and are made to operate either as rotary machines (rotational barrels) or rotational-vibratory machines at the same time. Therefore, they provide certain advantages which are not provided by either the vibratory or ro-

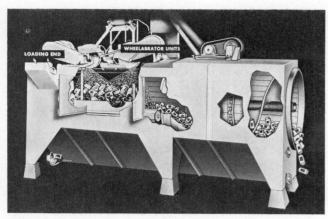

Fig. 21-8 A combination dry-abrasive blasting and tumbling (tumblast) continuous-action cleaning machine. (*The Wheelabrator Corp.*)

tary machines. When run as vibratory rotary units, the speed of rotation is slow and the scrubbing action is accomplished by vibratory and rotational action (Fig. 21-8).

The media used in this process are *manufactured abrasives, metallic products,* and *natural stone* or *sand* or *agricultural materials.* Among the manufactured media are aluminum oxide and silicon carbide. They are manufactured in a variety of shapes, sizes, and compositions to meet the wide variety of needs. The preshaped abrasive media facilitate the process in terms of media selection and work requirements. However, selection of the appropriately shaped media is an important factor inasmuch as the shape of the media has to match the contours of the part's surface (Fig. 21-9). The shapes mostly used are angle-cut triangles, cylinders, stars, cones, diamonds, and spheres. Media consisting of metallic products comes in steel balls, cones,

diagonals, and pin shapes. Among the natural abrasive media used are corundum, granite, limestone, silica sand, and quartzite.

The purpose of the chemical compounds used with the media is to provide accurate control of the cleaning action. Compounds provide lubrication and water softening, keep work and abrasive grains clean, and may promote color and luster on the parts being cleaned. Compounds may be of acid, alkali, or neutral base, depending on the type of material being processed, and are available in powder or liquid form.

Belt sanding and wire brushing are two processes which are used either to finish or clean surfaces. In belt sanding the cleaning or smoothing of the surface is accomplished by pressing the work against a moving belt coated with abrasive material (Fig. 21-10). It is used to clean or finish parts made of such materials as metals, wood, ceramics, plastics, rubber, glass, and others. Belt sanding is used primarily with flat surfaces and is considered a relatively inexpensive process. Belts are available in many sizes, are made of different materials, and are coated with a variety of kinds and sizes of abrasives. On the other hand, wire brushing can be accomplished either by pressing the work against a rotating wire brush or by pressing the rotating wire brush against the work. While belt sanding can remove noticeable amounts of material from the surface of the work, wire brushing removes very little. Wire brushing produces surfaces with very fine scratches which may serve as decorative surfaces, or the surfaces may need further processing to remove such scratches. Among the commonly used types of brushes are wideface, wheel, cup, and end brushes. Brushes are made of steel wire of varying thickness and hardness. Wire brushes can be used with abrasive compounds to produce polished surfaces and im-

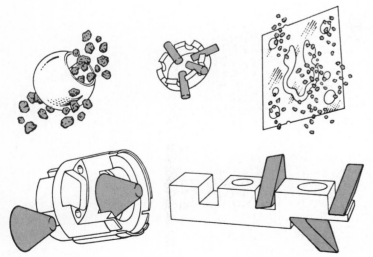

Fig. 21-9 Preformed and random-shaped abrasive media used for dry and wet abrasive cleaning. (*Norton Co.*)

Fig. 21-10 Belt-sanding unit illustrating the basic principle of the process: pressing the work against a moving sand belt. (*Hammond Machinery Builders Inc.*)

prove surface finish. Wire brushes operate at speeds of 4000 to 6000 square feet per minute (sfpm) [360 to 540 square meters per minute (m²/m)].

Polishing and buffing are two rather common processes used with many different kinds of materials. Polishing is strictly an abrading surface-smoothing process used to refine the surface of the work and improve its appearance. Buffing may or may not be an abrading process, depending on whether or not an abrasive compound is used. Buffing removes very little material in comparison to polishing. Both processes, however, are considered cleaning and/or finishing processes, depending on the application. Both processes are based on the principle of abrasive cutting when the work is pressed against a rotating wheel which has been dressed with a fine abrasive (Fig. 21-11).

Polishing wheels are made of *muslin, canvas, felt,* and *leather.* They are made by individual discs of the canvas, or any of the other materials, and are cemented or sewed together to form the wheel, or desired diameter and thickness. Wheels vary in rigidity depending on the material used and the method of construction. Rigid wheels are used for relatively rapid removal of material, while soft wheels are used where rapid removal of material is not a factor. Traditionally, the adhesive used to bond the abrasive grains to the surface of the polishing wheel has been hide glue (Fig. 21-12). In recent years, however, synthetic adhesives and cements have virtually replaced the glue. The synthetic adhesives provide better adhesion, withstand higher working temperatures, are more flexible and easy to work with, and need no special drying since the adhesives are air dry.

The most common abrasives used in polishing are *aluminum oxide, silicon carbide,* and *emery.* Aluminum oxide is sharp, hard, fast cutting, and wears well. Speeds for efficient polishing range from 4000 to 9000 sfpm [360 to 810 m²/pm]. Higher speeds may overheat the wheel and affect the bonding of the adhesive, while lower speeds may affect the abrasive. Lubrication of the cutting face of the wheel is used by applying oil, grease, tallow grease, or special lubricants which are commercially available, such as polishing rouge. One of the critical problems of buffing (to a lesser extent with polishing) is the friction heat created during buffing between the cutting face of the wheel and the work surface. To partly overcome the problem, liquid spray and airless spray buffing have been used. These modified processes have increased the production rates and helped improve the surface quality of buffing. Materials usually polished and buffed are metals, plastics, ceramics, glass, and many others. Buffing and polishing are performed

Fig. 21-11 Multiple head straight-line buffing machine set up for buffing flat-iron covers. (*Hammond Machinery Builders, Inc.*)

Fig. 21-12 Bonding (dressing) the abrasive grains to the polishing wheel with adhesive. (*Norton Co.*)

either by hand (Fig. 21-13) or by semiautomated machines. However, hand operations are expensive due to labor costs.

Electropolishing is another process used to polish metallic materials. Electropolishing is an electrochemical process which works in reverse of electroplating; metal is removed rather than deposited. Since metal is removed rather than deposited, the work is made anodic (positive) rather than cathodic (negative). In this case, cathodes are made by carbon, stainless steel, copper, or lead. Electropolishing is achieved in an immersion tank filled with an alkaline or acid electrolyte. Direct current of densities from 50 to 500 A/ft² is introduced and an anodic film forms over the work. It is the film which is responsible for smoothing and brightening the surface of the work. Electropolishing is considered as a cleaning and finishing process. When used as a finishing process, finishes from satin to mirror-bright are possible. Immersion time and temperature are two important variables in controlling the type of finishing desired. Low temperatures and short immersion times result in satin finishes, while increased time and temperatures produces bright finishes. In addition to electropolishing, the process has been modified to be used for *electrocoloring, electrosmoothing, electroetching, electrodeburring,* and *electromachining* of metallic materials.

Fig. 21-13 Polishing golf irons by hand-polishing operation. (*Norton Co.*)

CHEMICAL CLEANING PROCESSES

Alkaline Cleaning

Alkaline cleaning is a widely used chemical cleaning process. The cleaning action of this process is accomplished by emulsification of the soils by a penetrating solution which consists of alkaline salts as sodium hydroxide, silicates, and carbonates. The type of solution used depends on the kind of material being cleaned, the water used, and the type of equipment employed. To form the solution, in addition to the salts, some sequestering agents, dispersants, and surface-active agents are added. Most alkaline cleaning is done at 140 to 200°F [60 to 93°C]. Cleaning methods used are spray and soak or electrosoak. The process is used to remove soils, smuts, and light scales. After emulsification, the work must be rinsed to remove any residue from the surface. The equipment used is either of the batch or continuous-line type.

Solvent Cleaning

Solvent cleaning is used to remove heavy oils, grease and dirt. Solvents are usually petroleum, chlorinated or emulsifiable type. The purpose of the solvent is to dissolve and remove all or part of the soil from the surface. Petroleum-based solvents are distillates and are used at room temperature. They are less expensive but present a fire hazard. With petroleum solvents, cleaning is achieved by dipping the work in a tank containing the solvent or by brushing the surface of the work with solvent and drying it off with a cloth. Chlorinated solvents are more expensive than petroleum solvents. They are used with vapor-degreasing equipment by soaking, spraying, and/or vapor condensation. Equipment could be of the batch type, where the parts are placed in a wire basket and suspended in vapor formed by the boiling solvent, it could be made of the continuous-line type, which provides soaking, spraying, and vapor condensation (Fig. 21-14). Emulsified solvents are used for loosening oils and greases by dipping the work into the hot, agitated solution. The solution loosens the soil, and as the work is removed from the tank, emulsification takes place. In all solvent cleaning, the work must be rinsed with water at room temperature, or sometimes hot, to remove surface residue. Solvent cleaning is a relatively simple and inexpensive process, but it is limited to removing only those soils which can be easily dissolved. Safety standards and air-contamination levels dictate covering of tanks, exhaust fans, and so on.

Fig. 21-14 Solvent vapor-cleaning system. Parts in basket entering the system at one end of the conveyor travel through the system and exit at the same point of entry. Inset shows typical parts cleaned with this process. (*Detrex.*)

Solvent Vapor Degreasing

Solvent vapor degreasing is a process used to remove solvent-soluble soils from surfaces. This is achieved in a specially designed machine called the *vapor degreaser* (Fig. 21-15). The solvent, *trichlorethylene* or *perchlorethyline,* is heated in the degreaser and forms a vapor zone over the liquid solvent. The vapor of the solvent comes in contact with the cold work which is held in the vapor zone. The vapor condenses on the surface of the work and on the soluble soils dissolved and washed off into the liquid solvent. Excess vapor and the height of the vapor zone are controlled by cooling coils or jackets located at the top of the degreaser's vapor chamber. In modified vapor degreasers, a separate sump has been added to the side of the basic vapor degreaser to collect uncontaminated distillates. The collected clean distillates are sprayed on the work in the vapor zone to loosen up insoluble soils by mechanical action. In overhead conveyorized systems a vapor spray is added to speed up the process. The solvent used in this process is gradually contaminated and must be changed. The degree of contamination of the solvent may be determined by its boiling point; the higher the contamination the higher the boiling point. Solvent-vapor degreasing is relatively rapid, inexpensive, and does not affect the surface of the work. However, it cannot dissolve solid dirt, and it requires sufficient ventilation and safety precautions to operate safely.

Acid Cleaning

Acid cleaning consists of *pickling, deoxidizing,* and *bright dipping.* Acid cleaning is used to remove rust, scales, and other undesirable surface contaminates that cannot be removed by previously discussed chemical-cleaning processes. Before acid cleaning, the work must be cleaned to remove soluble soils from the surface. Acid cleaning is accomplished by immersing the work into an acid solution made of sulfuric, hydrochloric, or nitric, acid. Acid cleaning is a relatively rapid process and lends itself to either batch or continuous cleaning. Continuous cleaning of mill scale from steel products in the steel mill is a widely used application of acid cleaning. *Pickling* is one form of acid cleaning and is used in electroplating and in many other applications. In pickling, the work is dipped into the solution for a specified time and then is thoroughly rinsed. Most pickling is done in hot solutions at temperatures from 150 to 190°F [65.5 to 87.7°C]. After pickling, the work surface must be neutralized to remove any possible acid and prevent rusting. *Deoxidizing* is another variation of the acid cleaning process and used mostly with aluminum. *Bright dipping* is used on nonferrous metals to produce a highly reflective surface.

Molten Salt Descaling

Molten salt descaling is used with metallic materials, particularly alloys, which develop hard, tough, and complex scales during such processes as hot rolling, heat treating, and annealing. These scales cannot be effectively removed with the conventional acid-cleaning processes described above. Neither is it economical to remove such scales by mechanical means, such as grinding or abrasive blasting. For this reason, the molten-salt-bath cleaning process has been used very effectively.

Five variations of the salt-bath process are in use today. Three of the five variations employ an oxidizing salt bath as the cleaning medium. The first is the *high-temperature* oxidizing salt bath which operates at 700 to 850°F [371 to 454°C] and the third is the *low-temperature* oxidizing salt bath which operates at 375 to 425°F [190.4 to 218°C]. The variation in operating temperatures is due primarily to the different chemi-

cal additives used in the bath. This process employs a thermomechanical and chemical action based on the unequal coefficient of expansion of the metallic material and its scale. As the metallic material enters the bath, it is heated instantly to the salt bath's temperature. However, due to the unequal coefficients of expansion, the scale cracks (thermomechanical action) and the bath solution penetrates the cracks, gradually oxidizing the inner layer of the scale which further speeds up the cracking of the scale until complete disintegration and oxidation occurs. These three variations of the molten-salt-bath process are particularly suited for descaling alloys, super alloys, and refractory metals scaling and are used as continuous-strip or batch-type operations.

The remaining two variations are the *reducing-type* salt bath and the *electrolytic* salt bath. The reducing-type bath operates at 700 to 750°F [371 to 399°C] using with caustic soda mixed with 1½ to 2 percent sodium hydride as the reducing agent. It is used primarily for batch-type cleaning and is effective in removing tough, hot-work, and heat-treating scales from chrome alloys. The electrolytic salt bath operates at 800 to 900°F [426 to 482°C] with caustic soda

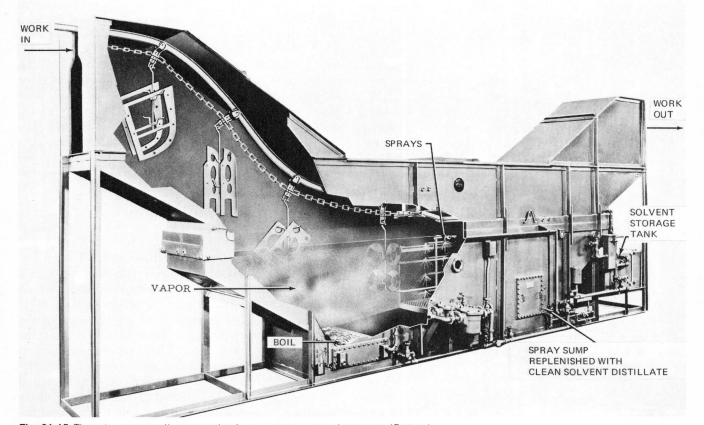

Fig. 21-15 Three-type monorail conveyorized vapor-spray vapor degreaser. (*Detrex.*)

mixed with electrically sensitive catalytic agents. Depending on polarity the salt bath can function as oxidizing or reducing bath. It is suited for descaling continuous hot-rolled and annealed stainless steel, weldments, and iron-steel castings.

MISCELLANEOUS CLEANING PROCESSES

Among the miscellaneous cleaning processes are *ultrasonic cleaning* and *steam cleaning*.

Ultrasonic Cleaning

Ultrasonic cleaning depends on cavitation of the cleaning solution. *Cavitation* is the rapid formation and violent collapse of bubbles or cavities in the cleaning solution at the surface of the product. Cavitation is produced by introducing high-frequency (ultrasonic), high-intensity sound waves into the cleaning solution, by means of a transducer, which convert electrical energy into mechanical energy. The agitation created by cavitation, due to the collapse of the bubbles against the part surface, creates a highly effective "scrubbing" of exposed and hidden surfaces of the work, which is immersed in the cleaning solution. Since ultrasonic energy can penetrate into crevices and cavities, any type of part or assembly can be effectively cleaned.

The three basic components of an ultrasonic cleaning system are the tank for the cleaning solution, the transducer for converting electrical to mechanical energy, and the ultrasonic generator (see Fig. 21-16). However, the heart of the system is the transducer. Two basic types of transducers are employed. The

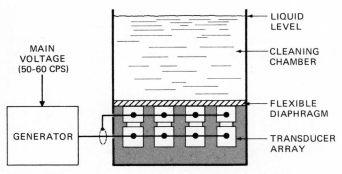

Fig. 21-16 Schematic of an ultrasonic cleaning system. The cleaning tank is filled with an aqueous cleaning solution or solvent. High-frequency current from the generator causes the transducers, which are attached to the flexible bottom of the tank, to vibrate. The vibration is transmitted to the fluid medium, causing cavitation that accelerates parts cleaning. (*American Process Equipment Corp.*)

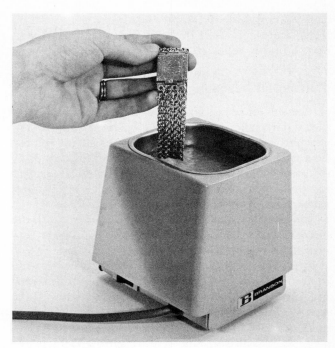

Fig. 21-17 A small bench-type ultrasonic unit used for cleaning jewelry. (*Branson Instruments Co.*)

magnetostrictive type made of materials which change dimensions in a varying magnetic field, and the *electrostrictive transducer* made of materials which change their physical dimensions due to the piezoelectric effect. Transducers are placed at the outside of the tank, or immersed into the cleaning solution, or placed on the bottom of the tank. Cleaning solutions used may be water or solvent. Water is one of the best solutions because it is safe, easy to handle, and relatively inexpensive. Ultrasonic cleaning is best suited for small-size parts (Fig. 21-17). The parts are usually placed in a wire basket or a rack and immersed into the cleaning solution.

Steam Cleaning

Steam cleaning is used to clean relatively large parts which cannot be accommodated with the other cleaning processes. The process is based on forcing a cleaning solution under steam pressure through a gun on the surface of the work. The system consists of the steam-generating units, the cleaning solution, and the delivery hoses and gun. The cleaning medium is usually a detergent solution reinforced by heat of the steam. The combination of heat and impact action provides an effective cleaning process capable of removing heavy oils, greases, and other soils from the work surface.

SURFACE-FINISHING PROCESSES

The major purpose of surface finishing is to protect and/or decorate the work surface. This can be accomplished basically by adding to or removing material from the work surface or by a combination of removing and adding. The material being added to the surface is usually referred to as a *coating*. There are several types of surface-finishing processes which can be classified as *organic, metallic,* and *miscellaneous* finishing processes

Organic Finishes

Organic finishes are used as protective and decorative coatings. As protective coatings, organic finishes protect the work surface from environmental conditions by providing a continuous protective film on the surface of the base material to be protected. Therefore, the degree of protection of the work surface depends upon adhesion of the film to the surface of the base material, durability of the coating film against environmental conditions, and quality of the coating film. Organic coatings are those formed in whole or in part with carbon compounds. They are commercially available in liquid and powdered forms. Among the common liquid form, organic coatings are paints, varnishes, lacquers, shellacs, and enamels. All these coatings utilize resins and pigments in solvents. In addition, two other varieties of organic coatings are the water-reducible coatings and the emulsion-latex variety.

Before applying an organic coating, the surface of the base material must be thoroughly cleaned to improve the adhesion between the film and the surface being coated. Any organic coating lacking a firm bond will fail to perform its expected function. Bonding could be accomplished directly on the surface or through the use of a *primer,* which is an intervening chemical treatment. Therefore, the best practice to produce an organic film with good adhesion, durability, and quality would be to clean the surface thoroughly, apply a coating of a suitable primer (whenever applicable), and finally apply one or more organic coatings as specified.

Paints. *Paints* are basically mixtures or suspensions of finely divided solid pigments in a suitable vehicle with small amounts of drier added. The vehicle consists of drying oil, natural or synthetic resin, and solvent or thinner. Linseed, tung, perilla, soybean, and fish oils are among the most widely used vehicles for paints. The main purpose of the vehicle is to form and harden the film by wetting and drying the pigment. On the other hand, the main purpose of the pigment is to give the paint its desired color and in some cases improve the strength and durability of the organic film. Among the most commonly used pigments are titanium dioxide, white and red lead, and Prussian blue. Driers are added to the paint to improve its drying ability. Driers are usually metallic organic salts such as lead manganese, iron naphthenates, or thallates. Drying of paints is based on evaporation, oxidation, and polymerization. Upon application the solvent (thinner) evaporates and the drying oil which formed the film is gradually hardened by oxidation polymerization with oxygen in the air.

Varnishes. *Varnishes* are produced by cooking a dissolved resin with drying oil. Among the synthetic and natural resins used in varnishes are phenol, alkyd, damar, amber, and kauri. Drying oils used in varnishes are corn, soybean, fish, and castor oil. The consistency of varnishes depends on the amount of oil and resin mixed. The varnish film usually dries by evaporation and oxidation.

Lacquers. *Lacquers* consist of cellulose nitrate dissolved in butylacetate with solvents, thinners, and driers added. Lacquers are air-drying and are used primarily as protective coatings.

Shellacs. *Shellacs* consist of lac mixed in alcohol. Lac is a secretion of small bugs which live in India. Shellacs are used as sealers and coatings of surfaces, primarily for wood.

Enamels. Among the most important types of *enamels* are those which result from the addition of pigments into varnish, the "lacquer enamel," which consists of pigments, cellulose ester, gum, plasticizers and solvents, and the straight enamels which do not contain cellulose derivatives. Enamels dry by a combination of oxidation and chemical conversion of the vehicle, either at room temperature or by baking at temperatures up to 400°F [204°C]. Enamels are water, chemical heat, and abrasion resistant, depending on the type. They are extensively used in modern industry as protective and decorative surface treatments.

Application Methods of Organic Finishes

Among the most commonly used application methods of organic coatings are brushing, spraying, dipping, flow coating, and roller coating, and electrocoating.

Brushing. *Brushing* is probably the most widely used method and requires the minimum of equipment but a great deal of labor. Brushing is a relatively simple method and requires minimum control of the consistency of the coating.

Spraying. *Spraying* is a general industrial method and is done by conventional spray, airless spray, steam, and hot spray. All these methods employ a spray gun. The spray gun is the most important tool of this method, and its appropriate use would determine to a great extent the quality of the coating. The spray gun controls the starting and stopping of the spray, the width of spray, and volume of air and fluid being sprayed. For satisfactory results, the spray gun should be held at a distance of 6 to 12 in [15.24 to 30.48 cm] from the work surface. If the spray gun is held too close, improper atomization will result. If it is held too far, dried particles will hit the work surface and will affect the durability of the film. In conventional spraying the coating material is atomized at the spray gun by compressed air and forced against the work surface with some force. In *airless spray* compressed air is used to atomize the coating material and force it through the nozzle of the spray gun, but it does not force the spray against the work surface. The atomized coating material reaches the work surface by force of gravity. As the small droplets of coating fall on the surface, they flow and join each other, forming a continuous film. In this way airless spray minimizes the "overspray" and waste of coating material associated with conventional spray (Fig. 21-18a and b). Airless spray can be used with most organic coating materials, and it can be very economical when

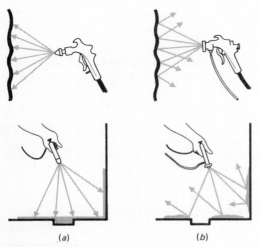

Fig. 21-18 Comparison of airless and conventional spray. (a) Airless spray, no overspray; (b) conventional spray, overspray. (*Graco Inc.*)

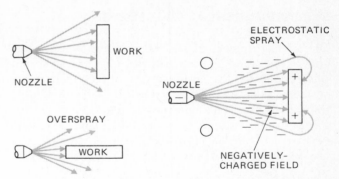

Fig. 21-19 Comparison of conventional spray and electrostatic spray showing overspray associated with conventional spray.

it is combined with electrostatic spraying to form the *airless electrostatic spraying*.

In *hot spraying* the viscosity of the coating material is not achieved by means of a solvent (thinner) as in the conventional spray but rather by means of heat applied to the coating at the time of spraying. In addition to increasing the viscosity of the coating material, the heat helps to speed the drying of the coating by evaporating the solvent. With rapid evaporation of the solvent, the buildup of relatively heavy and fast-drying films can be achieved. Therefore, it is the action of the heat which provides the unique characteristics of hot spraying. The basic limitations of conventional spraying methods, however, are the amount of overspray and pollution (Fig. 21-19).

Unlike the other spraying methods, which are based on air pressure, *electrostatic spraying* is based on the principle of attraction of unlike electrical charges. The coating material is first atomized into a fine mist in an atomizer, and the mist is charged with the same charge as the atomizer and is therefore repelled and forced toward the work surface which has been charged with unlike electrical charges. These charges attract the mist which then covers the work surface and forms a homogeneous film. Electrostatic spraying is used extensively and lends itself easily to fully automated lines (Fig. 21-20) or manual operations (Fig. 21-21). Originally, electrostatic spraying was limited to materials classified as good electrical conductors. However, substances have been found which are used to make nonconductors into conductors and are therefore capable of being coated by electrostatic spraying.

Dipping. *Dipping* is accomplished by immersing the part into a tank containing the coating material. The part is then removed from the tank and the excess of coating material is allowed to drip off the part. Dipping is a relatively rapid method of completely covering the entire part. It is used for parts which do not

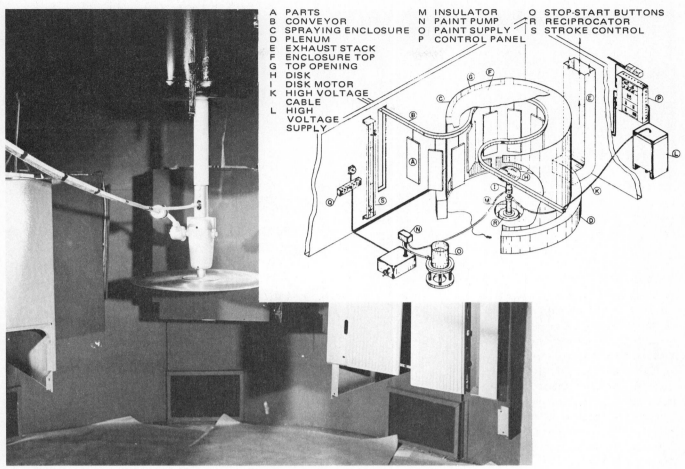

A PARTS
B CONVEYOR
C SPRAYING ENCLOSURE
D PLENUM
E EXHAUST STACK
F ENCLOSURE TOP
G TOP OPENING
H DISK
I DISK MOTOR
K HIGH VOLTAGE CABLE
L HIGH VOLTAGE SUPPLY
M INSULATOR
N PAINT PUMP
O PAINT SUPPLY
P CONTROL PANEL
O STOP-START BUTTONS
R RECIPROCATOR
S STROKE CONTROL

Fig. 21-20 Automated conveyorized electrostatic painting system. Inset shows the basic elements of the system. (*Ransburg Electro-Coating Corp.*)

require high quality finishing because the process usually results in some runs, tears, and sags. Parts which are going to be dip-coated should be designed with that process in mind; otherwise dipping may not be possible. On the other hand, *flow coating* (as the term applies) is basically pouring coating material on the work surface and letting it flow to cover the surface instead of dipping it. The "pouring" of the coating material is accomplished by forcing it through nozzles at relatively low pressure which do not cause atomization. As the coating material reaches the work surface, it is splashed into droplets which flow and form a uniform film. Pressure, viscosity, temperature, and solvent additions are critical factors and must be carefully controlled. Flow coating requires extremely clean surfaces to prevent contaminants from entering the coating and reducing the adhesion of the coating on the surface. It is a highly specialized process and has rather limited application. However, it can be used economically in situations where both sides of

the work are coated, or for work with areas difficult to reach.

Roller Coating. *Roller coating* can be done manually or mechanically. In manual operations a suitable roller is first covered with the coating material and then rolled on the work surface. This is a relatively simple operation and used extensively in the construction industry. Mechanical roller coating is the most automated of all coating methods. This method can be classified as *direct roller coating* (Fig. 21-22) or as *reverse roller coating* (Fig. 21-23). In direct roller coating, the work is passed through two preset rolls. The drive roll drives the work, and the application roll applies the coat on the one side. In reverse roller coating, the applicator roll travels the opposite direction from the work. By adding an additional roll in this method, both sides of the work can be coated simultaneously. Roller coating is used for strip or coil coating of materials.

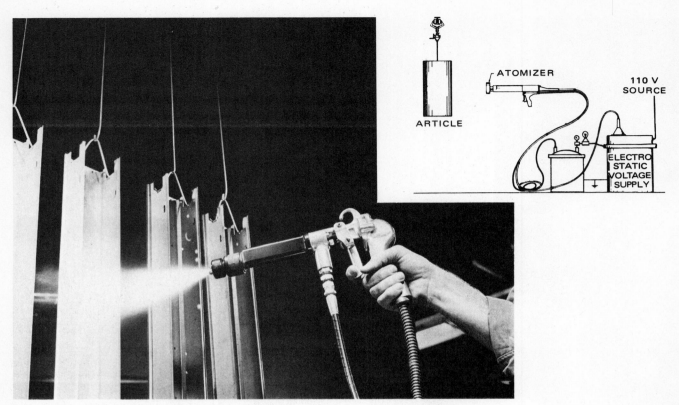

Fig. 21-21 Hand-operated gun of electrostatic painting unit. Inset shows the basic elements of the system. (*Ransburg Electro-Coating Corp.*)

Electrocoating. *Electrocoating* is based on the principle of *electrophoresis*. It involves application of electric current to the coating material. The coating material suspended in an anionic-type vehicle is contained in a tank of the electrocoating unit (Fig. 21-24). The flow of current through the coating material causes migration of the negatively charged pigment and resin particles toward the anode (work) where a destabilization reaction takes place. It is this reaction which results in the deposition of the coating film on the work surface. The electrodeposition on the work (anode) continues until the organic film deposited provides an electrical insulation which prevents further current flow. The cathodes are stainless steel rods immersed into the tank. Electrocoating is a relatively economical process and provides high-quality coatings. Further, it lends itself to automation, which makes it ideal for high-production rates.

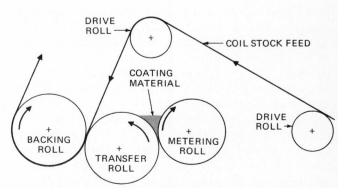

Fig. 21-22 Arrangement of rollers for direct roller coating.

Fig. 21-23 Arrangement of rollers for reverse roller coating.

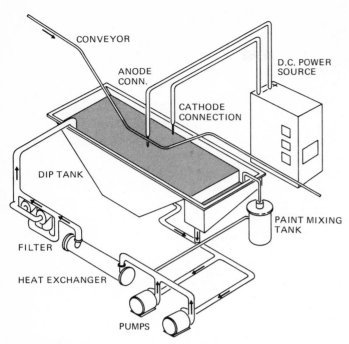

Fig. 21-24 Schematic showing the basic elements of an electrocoating system. (*Binks Mfg. Co.*)

POWDER COATINGS

Coating materials in powder form are relatively new but are becoming popular in many industries. There is a great potential to be realized with the increased use of powder coatings. The powder-coating process is different from the conventional liquid-coating process and many of the application techniques are unique to this process. Basically the powder-coating process involves three elements: the powder, the work surface, and heat. The coating powder is applied on the surface of the work, which may be preheated or not. After the powder has been applied, the work is heated and the powder fused to a uniform film which is durable and chemically resistant. Among the application methods of powder coating are flame spraying, electrostatic powder spraying, pouring or flowing of fluidized powder, rotational coating of pipes, and dipping in nonfluidizing powders. The most commonly used resin-type coatings available are vinyls, epoxies, polyesters, polyethylene, polypropylene, nylon, chlorinated polyethens, and flourocarbons.

Powder coating has certain advantages and limitations compared with the traditional liquid coatings. Among the advantages are that loss of coating material is limited, thick films can be achieved in one operation, powder material is nonvolatile and pollution free, smooth coatings on rough surfaces can be achieved with limited smoothing operations, and no drying is required. Some of the limitations are that thin coatings are difficult to achieve, temperatures required may affect the workpiece, and color matching for maintenance is difficult to achieve.

METALLIC COATINGS

Unlike organic coatings, which are considered relatively soft coatings, metallic coatings are considered hard coatings. Among the most commonly used metallic-coating processes are hot dipping, electroplating, electroless plating, electroforming, anodizing, vacuum metalizing, and metal spraying.

Electroplating

Electroplating is the process by which a metallic or nonmetallic material is coated with a metallic surface. It is the most commonly used metal-coating process. Its unique characteristics are accurate control of coating thickness, which may vary from 0.00002 to 0.01 in [0.0005 to 0.25 mm], and a relatively uniform coating for parts of even surfaces. The process is based on Faraday's law of electrodeposition (Fig. 21-25) which states that one farad would yield one gram-equivalent of plated material. To achieve that in practice, direct current from the anode, which is made of the plating material, ionizes the electrolyte, and positive ions of the plating metal pass through the electrolyte solution to the cathode, which is the part to be plated. As these metallic ions flow current from the anode to the cathode, metal particles in the form of ions migrate

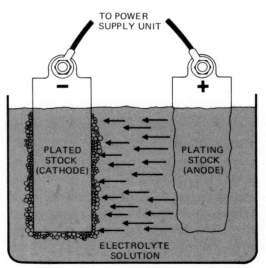

Fig. 21-25 Schematic of an electroplating system showing the deplating of the anode and plating of the cathode. (*Anocut.*)

Fig. 21-26 Partial view of an automated electroplating unit showing the chromium bath and loaded racks removed from the tanks. (*M & T Chemicals, Inc.*)

through the electrolyte from the anode deposit on the cathode, and give up their electrical charge. Therefore, the rate of electrodeposition is determined by the density of the current. The initial electrodeposition on the cathode starts at certain points and gradually covers the entire surface. Surfaces to be plated must first be thoroughly cleaned and scratch-free. Most surfaces are first buffed, cleaned in a solvent bath, rinsed well, and then plated (Fig. 21-26).

Among the most commonly used metals are chromium, nickel, copper, zinc, lead, cadmium, tin, silver, gold, and platinum. Chromium plating is the most widely used metal coating. There are two variations of chromium plating, the *decorative* and the *hard*. Decorative chromium plating is used primarily for improving surface appearance, while hard chromium plating is used for restoring worn-out parts and as hard protective coatings. Parts to be plated must be designed for that purpose and should not have sharp corners and projections inasmuch as corners are difficult to plate and projections usually receive heavier coatings. Nonconducting material such as plastic and rubber can be electroplated by changing the nonconductive surface into a conductive one. This can be achieved by either painting the surface with a conductive paint or by vacuum metalizing.

The main elements of an electroplating system are the cathode; anode; electrolyte, which varies with the metal being plated; direct current; and the necessary tanks for plating, cleaning, and rinsing.

Electroforming

Electroforming is the process by which a thick layer of metal can be deposited on a preshaped mandrel which is later removed from the finished part. Electroforming is based on the same principle as electroplating and employs basically the same equipment. The electroforms produced by electroforming are different from electrodeposited coatings in terms of thickness of the deposit and the removal of the mandrel. There are basically two classes of electroformed products. The first class includes products formed directly on a mandrel, which is removed and leaves the hollow formed part. Fountain-pen caps can be formed this way. The second class includes products formed of composite material. In this case, the thick deposit is deposited directly on the base metal and not on a mandrel. The soft interior of engine bearings are produced this way.

Electroforming can produce parts which are difficult or expensive to produce by other manufacturing processes, parts with unique properties and extremely close dimensional tolerances, or parts with very thin walls which require very fine surface details. The electroforming mandrels are unique elements of the process and are made as permanent or expendable pieces. Materials used to make the permanent mandrels are copper, brass, nickel, stainless steel, and plastics. Among those used to make the expendable mandrels are plastics, metals, waxes, plaster, and wood. The nonconductive materials such as wood, plastics, and waxes can be made electrically conductive by the application of a silver film or conductive paints.

Oxide Coatings

Oxide coatings are applied to metallic materials for decorative and protective purposes. Among the most widely used oxide coatings are *anodizing of aluminum, anodizing of magnesium alloys,* and *black oxide coatings of ferrous metals.* The aluminum-anodizing process has made the most rapid advances in the last few years. The home-appliance and giftware fields and architectural and decorative trims are examples employing the process. Imitation finishes of gold, copper, brass, and many other colors have given the process wide industrial application by adding a dye to the plating solution. The aluminum-anodizing process is based on the electrolytic action of an acid electrolyte which produces a relatively thicker oxide coating than the natu-

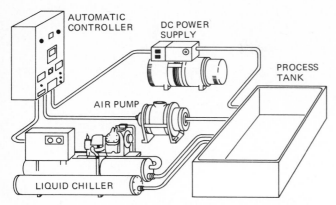

Fig. 21-27 Schematic showing the basic elements of an anodizing system. (*Sanford Process Corp.*)

ral inert aluminum oxide (Fig. 21-27). The anodizing system consists of a tank with heating and/or cooling elements, compressed air for agitation of the electrolyte solution, and a direct current source. The parts to be anodized are made the anode, and the tank the cathode. A relatively low voltage is applied through the solution. The thickness of the oxide film is almost directly proportional to the ampere-hours passed. The film thickness varies between 0.0001 and 0.0005 in [0.0025 and 0.0127 mm]. The anodizing solution temperature is a critical factor since it affects the softening and porosity of the film. Higher temperatures produce relatively softer and more porous films which are capable of absorbing dye more readily, thus producing darker colors. Among the aluminum anodizing methods most commonly used are the chromic acid, sulfuric acid, and hard anodizing. All these methods require different solutions and produce anodizing finishes. The anodizing finish without dye is usually gray and dull in appearance, but it can be electropolished or bright dipped to change it to a finish that cannot be produced by other processes.

Vacuum Metalizing

Vacuum metalizing is the process used to deposit a metallic surface, under high vacuum conditions, on metallic and nonmetallic materials. Metalized parts made from nonmetallic material are gradually replacing metal stampings and die castings, realizing cost, weight, and scope of styling advantages. The process of metalizing is relatively simple. The parts to be metalized are first coated with a conductive paint, a lacquer, or varnish mixed with such conductive pigments, such as graphite, copper, or silver. After the conductive treatment, the parts are placed in a vacuum chamber (Fig. 21-28). A relatively small amount of the metalizing material is also placed in the vacuum

chamber on the tungsten filaments of the chamber. The chamber is evacuated, and the tungsten filaments are heated until the metalizing material is evaporated and radiates throughout the chamber. The metallic vapor condenses on the parts in the chamber forming a coating of brilliant finish. A further protective coating is then applied. Among the nonconductive materials which have been successfully metalized are aluminum oxide, berylium oxide, glass, wood, rubber, plastics, ceramics, paper, and textiles.

Metal Spraying

Metal spraying is the process used to spray melted metal on metallic or nonmetallic material. Metal spraying employs a spray gun (Fig. 21-29) which melts the metal by means of oxyacetylene flame and forces it against the work surface in the form of small droplets. The small droplets stick to the work surface by means of mechanical adhesion and form a metal coating. Metal spraying is used primarily as a protective coating. Although many different metals can be sprayed, the most common are zinc and aluminum.

PHOSPHATE COATING

Phosphate coatings are conversion or transformation of the work surface by means of a complex chemical

Fig. 21-28 High-vacuum-metalizing chamber ready to be unloaded of finished toy parts. (*Norton Co.*)

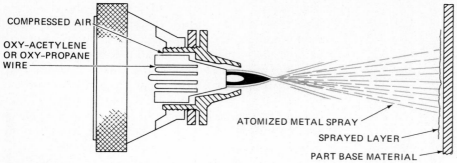

COMPRESSED AIR

OXY-ACETYLENE OR OXY-PROPANE WIRE

ATOMIZED METAL SPRAY

SPRAYED LAYER

PART BASE MATERIAL

Fig. 21-29 Schematic showing the cross section of a metal spraying nozzle (gun).

reaction caused by the phosphatizing solution and the metal being coated. The resulting surface has non-metallic and nonconductive properties. Phosphate coatings are used to prepare surfaces for organic coatings and as finishes. By far the most important use of phosphate coatings is the preparation of metallic surfaces for painting. Phosphatizing solutions consist of metal phosphates dissolved in phosphoric acid. For phosphatizing to take place, the free acid concentration in the solution must be high enough to attack the metal and form metallic ions. To achieve this, the acid concentration is reduced as soon as the solution comes in contact with the work surface. As a result, the acid concentration is reduced to a liquid-metal inerface and iron is dissolved forming the metallic ions, hydrogen is evolved, and phosphate coating is precipitated on the work surface. This conversion of the work surface forms the phosphate coating which is integrally bonded to the base metal.

The four most commonly used phosphate coatings are iron, zinc, heavy zinc, and manganese. Phosphate coatings can be applied by spray or immersion, the immersion method being the most common.In addition to functioning as base for paints and plastics, phosphate coatings are used for metal forming and lubricating, rust prevention, improvement of uniformity of surface texture, and prevention of surface scratching or scarring.

Porcelain Enameling and Ceramic Coatings

Porcelain enameling (vitreous) *processes* produce hard coatings which are thoroughly fused to the base metal. To form the enamel coating or mixture of fused refractories with fluxes and metal oxides called the *frit* is ground and mixed with water and clay to form a thin slurry called the *slip*. The slip is applied on the clean work surface by dipping or spraying and allowed to dry. The part is then placed in an enameling furnace and fired at about 1500°F [815°C] until completely fused into a hard scratch-free, heat- and chemical-resistant coating. Enamels are available in

many colors and in three basic finishes: gloss, matte, and semimatte. Porcelain enamels are used extensively in bathroom accessories and kitchen appliances. Parts to be porcelain enameled should be designed for that purpose in order to realize better results.

REVIEW QUESTIONS

21-1. List the main purposes of surface cleaning and finishing processes.

21-2. State the advantages and limitations of mechanical and chemical cleaning processes.

21-3. Compare abrasive blasting and tumbling in terms of cost, quality of work, and extent of application in modern industry.

21-4. Why are manufactured abrasive media considered better than natural media used in cleaning processes?

21-5. Is electropolishing a more effective process in cleaning metallic materials than buffing? Explain.

21-6. Why is solvent cleaning not used to remove rust from metallic materials?

21-7. Describe the process of bright dipping.

21-8. Can ultrasonic cleaning be used to clean parts made of plastic? Explain.

21-9. Are water paints considered organic finishes? Explain.

21-10. Why does electrostatic spraying not require compressed air to spray the coating on the surface?

21-11. Are powdered coatings considered the future coatings of modern industry? Explain.

21-12. How can metallic coatings be used on nonmetallic surfaces? Explain.

21-13. Compare the phosphate coatings with oxide coatings in terms of cost, quality of work, and extent of application in modern industry.

CHAPTER 22

INTRODUCTION TO MASS PRODUCTION AND AUTOMATION

Mass production is a method of producing a large number of products rapidly and efficiently. In order to manufacture consumer products at the tremendous volume and profitability required in industry, mass production is necessary.

Automation on the other hand refers to devices which are highly efficient, self-controlling, and independent of human input. Automated devices are used to produce, transfer, inspect, and assemble the consumer products of industry. True automation is based on the concept of feedback and self-regulating controls. While true automation systems are used in industry, a completely automated plant does not exist.

MASS PRODUCTION

Mass production techniques are based on the concept of interchangeability of standardized parts. Eli Whitney developed the mass production concept while producing muskets for the government. Whitney spent considerable time in planning his system and building specialized tools and devices required to implement his concept. Congressional leaders became disturbed because the order for muskets had not been fulfilled. Therefore, Whitney was called to Washington, D. C., to defend his position and actions regarding the lack of production. During this visit he demonstrated the mass production concept while assembling the muskets. As with Whitney, mass production techniques in modern manufacturing requires considerable planning, design, and construction of specialized production and assembly devices.

Two types of manufacturing employed in industry are the continuous and noncontinuous production methods. *Noncontinuous* manufacturing is used primarily for low-production items which are not feasible for continuous-production techniques. For exam-

ple, a company wins a contract with the government to manufacture 100 supersonic planes. The planes are unique in design and construction, requiring specially built components. Some parts needed for the plane may be standard items readily available from other sources. However, a number of the components are unique to this particular plane. It would be economically unfeasible to develop specialized machines to manufacture and assemble the components.

Noncontinuous production lends itself to small-batch production or special production orders. Standard machine tools are used to produce the required components of the product. Products manufactured by the noncontinuous method have a high cost per unit because skilled machine operators and technicians are required. An automated assembly line is impractical for this type of production.

Continuous or mass production is used for manufacturing a high volume of consumer products. The majority of consumer items, such as home appliances, automobiles, and so forth, are produced on a continuous-production line. Since a large number of these products will be marketed, the manufacturer is warranted in spending considerable time in planning the most efficient production system. Specialized machines to produce, transfer, assemble, and inspect the product needed to be developed at considerable cost. The time and expense is hopefully recovered by the product's sales.

Continuous-production methods utilize specialized machines to perform a number of important aspects in the manufacturing of a product. The machines utilize special loading, unloading, holding, indexing, and controlling devices to rapidly and accurately perform the required operations on the part. Transfer devices such as conveyors, belts, chains, slides, and tubes, move the workpiece from one work station to the next continuously. Inspection of the component parts and the final product is required to assure that the standards have been met and that the product performs as designed. When a large volume of products is to be produced, specialized methods and equipment are designed to inspect the product on a continuous basis. The final assembly of the product must be as efficient as possible. Therefore, an assembly line utilizes nonskilled or semiskilled workers and requires semiautomated or automatic equipment.

A prime example of a continuous-production line is the assembly of an automobile. Component parts and subassemblies (engine, radio) are received at the assembly plant from other specialized plants and thousands of suppliers. The assembly plant consists of main and secondary assembly lines. *Secondary lines* are continuous-production lines which are primarily responsible for assembling various sections (body, drive train) of the automobile. Each secondary line feeds the continuously moving main assembly line at a predetermined time and sequence with the appropriate assemblies. The *main line* starts with the car's frame to which all other units are attached either directly or indirectly. As the frame progresses through the line, the automobile takes shape.

The automobile assembly plant utilizes a number of automatically controlled operations. Computers are used extensively in the automobile's assembly. The various body styles, engines, colors, and options must meet the main assembly line at precisely the right time in order to manufacture the car ordered. Without the computer's assistance, this task would be an enormous undertaking, if not impossible, in the modern automotive assembly plant.

AUTOMATION

Automation, a term commonly misunderstood by many individuals, is not a new concept. The word "automation" is derived from the term "automatic," which refers to performing some task or operation automatically. The term "automation" became the commonly accepted substitute for the term "automatization" because of pronunciation difficulties encountered by individuals.

Since the late forties the term "automation" has gained acceptance while generating concern for most individuals. The idea that machines could replace human effort created fear among many industrial workers, while in effect automation has created new jobs and fields of endeavor. The concept of automation is not new, but has continuously grown since early cave man days.

The early cave man had basic needs such as food and shelter. In order to provide for these needs, he developed a means of using sticks and stones to perform basic life-essential task. Early tools were used to hunt for food, cut wood, and grind grain. Through evolution and advancing knowledge, simple hand tools were constructed from metal to perform basic operations such as grinding, cutting, milling, drilling, and turning. Early tools and hand tools of today require the individual to exert some form of muscular effort. The first industrial revolution relieved much of the physical effort needed to perform the basic machining actions by using some form of power to drive the machine. Results of the industrial revolution did not degrade or dehumanize workers. Rather workers became more productive and able to produce better products for sustaining life with less physical effort.

Power-driven machine tools were further developed and improved by adding automatic feed devices and multiple tool holders in order to reduce the physical effort required in the machine's operation. This type of machine requires the operator to make the necessary settings and change the workpiece and tools as needed.

Technological advances provided for further reduction of physical effort by using automatic devices such as loading and unloading devices, tool changers, and mechanical controls—prime developments in the power-driven machine tool. Cams, stops, and slides provide a means of mechanically commanding a machine to perform specific operations. Devices were used to move or transfer workpieces to the next work station.

In the next stage of machine develoment, the machine was referred to as truly automatic. A truly automatic machine does not require an operator to make decisions on such questions as, Should a tool be replaced? or Is the proper dimension achieved? In an automatic machine, an internal feedback system is the prime component. *Feedback systems* have units or components which continuously monitor the machine's function. One type of feedback system constantly monitors the dimensional size of the workpiece and compares it to a standard value. The unit calculates the differences and feeds the information back to the machine through electronic devices. The machine then makes self-correcting adjustments and proceeds as directed. This phase of automation probably has created the greatest fear and concern for the worker. However, this system of automation is used everyday in the home. The most common example of a truly automated system is the home thermostat which regulates heating and cooling.

The thermostat contains a monitoring device which senses the differences between the predetermined setting and the room temperature. The monitoring device then determines what action is to be taken —either to increase or decrease the system's output. Signals are then sent to the heating or cooling unit starting the sequence required to achieve the desired room temperature. Constant monitoring and feedback provides for a relatively constant room temperature.

A number of consumer products utilize the feedback concept to provide for a completely automated system. Typical home appliances such as the oven and hot-water heater are preset to the desired temperatures and through a feedback monitoring system provide hot water or properly done food. Electronic sensing devices are used to monitor light in order to regulate camera exposure or the brightness of a TV screen. In the automobile, electric eyes adjust headlight beams and alternators and regulators provide needed current and properly charge the battery. Cruise control in the car allows for a relaxing drive at a constant speed while saving fuel. Automatic devices in the home and industry are endless and provide for less physical exertion or annoyance while improving the system's output.

A device that added another degree of automaticity to manufacturing is the computer. In addition to performing one function, such as monitoring and controlling the room temperature, computers are capable of performing a number of activities simultaneously. Computers are capable of monitoring and controlling a number of factors on machine tools such as dimensional accuracy, tool wear, and cutting-tool path. In addition, computers are capable of creativity by designing a new or improved system or mechanism.

Nonproduction Automation

Automatic or *electronic data processing* is used extensively in the nonproductive aspects of industry. Computers in the office are used to perform routine activities such as bookkeeping, developing schedules, sales forecasts, processing orders, and issuing payrolls. Virtually any routine information-handling problem can be handled by the computer and a data processing system. Other automated support equipment used in data processing includes: printing machines, tabulators, microfilming, check writers, and addressing machines.

Computers are used extensively in the research laboratories to analyze data in a fraction of a second. Calculations formerly done by analyst and scientist are performed in a fraction of the time previously required. The computer is used to analyze input data and select the best possible alternatives for designing

systems or a new product. The potentials of the computer have not yet been fully exploited, but it has already proven to be a most remarkable device.

Manufacturing Automation

Advances in technology leading to the machines and systems previously discussed in the evolution of automation relate primarily to manufacturing applications. Automation in manufacturing is concerned with four prime areas of production, including production equipment, transfer devices, inspection, and assembly devices.

Production Equipment. Improved production in manufacturing is the direct result of applying various automated or semiautomated devices in production lines. Advances in production equipment includes work-holding devices, machine tools, and controlling devices.

Work-holding devices include vises, clamps, jigs, and fixtures. Vises and related holding devices have been designed for adequately holding the workpiece during various production operations. Some vises are capable of rotary and axial movements allowing multiple positioning of the work. Clamps of various sizes and shapes are used in the various operations to hold and support the workpiece. Jigs and fixtures are devices used to hold the workpiece in position for the production operation. While vises, clamps, and other conventional holding devices are used in a multitude of situations, jigs and fixtures are designed for specific applications and workpieces. Primarily their purpose is to save set-up time and aid in producing consecutive parts which are exactly alike. Human error is reduced when a holding device will hold the workpiece in the exact position for the operation. The major difference between a jig and a fixture is that a *jig* guides a tool relative to the workpiece, while a *fixture* holds the workpiece in position for the tool. Their prime function is to provide for quick loading, unloading, and accurate repeated positioning of the workpiece.

Machinery used in production operations includes chip-producing machines, molding, forcing, electro, chemical, and fabrication equipment used in modern manufacturing. Chip-producing equipment is used for operations such as turning, sawing, milling, planing, grinding, drilling, and their related counterparts. In this type of production operation, a cutting tool removes material in the form of a chip yielding the desired shape or surface. Molding-production equipment such as die casting, vacuum forming, blow forming, and casting provide for production of prod-

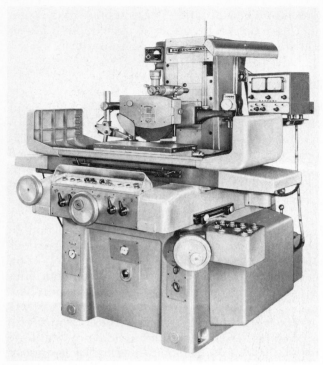

Fig. 22-1 Surface grinder with automatic grinding gauge. (*Source: Sheldon Machine Company, Inc.*)

ucts from liquid or semiliquid materials. Forcing operations utilize machinery capable of shaping material into useful products by applying force. Pressing, stamping, bending, spinning, punching, drawing, and rolling are common production operations requiring equipment to force the material into the desired shape. Electro-type operations and chemical operations such as electrodischarge, chemical milling, electrolytic machining, plating, plasma erosion, welding, and ultrasonics are extensively used as production operations in manufacturing.

Production operations regardless of their main characteristics employ some means of physically changing or altering a material's appearance shape and size to yield a finished product.

In addition to the work-holding devices and the type of machinery required in production operations, the controlling devices employed influence the degree of automation. Highly automated systems reduce the need for human control. Essentially human control of the production process has been replaced by either mechanical or electrical/electronic devices. Mechanical control of machine functions is not a revolutionary development. Stops located on the machine stop the operation when designated positions are achieved. Cams, gears, and mechanical linkages provide a means of mechanically controlling the ma-

chine's operations. Electrical/electronic control of machines is accomplished by using standard switching devices, limit switches, micro switches, and various other sensing devices. In addition to standard controls, punch tape, fluidics, and computer-control devices further automate the system.

Numerical control units utilize a preprogrammed set of instructions on a tape or continuously computer-generated instructions to regulate the various operations. Although numerical control units were primarily designed for machine tools, their usefulness and desirability has expanded to all areas of automated manufacturing. The preplanned program sequence controls the speed, feed, indexing of the part, and tool selection of numerically controlled machines. Computer-generated programs are used for continuous-path operations such as turning an irregular contour. In addition to contouring capabilities of the machine, sensing devices can feedback information regarding tool wear and surface quality of the workpiece, at which time the computer can make immediate machine adjustments. Figure 22-1 illustrates a surface grinder which automatically grinds the workpiece to a predetermined size. The electronic probe contacts the part being ground and senses its exact dimension. When the correct size is obtained, the grinding wheel retracts and the machine is automatically stopped. Figure 22-2 illustrates a typical numerically controlled machine tool.

Fig. 22-2 Numerically controlled machine tool. (*Source: Brown and Sharpe.*)

Materials Handling. The movement of the workpiece from one operation to the next work station or operation is of prime importance in automated manufacturing. Transfer devices including moving tables, conveyors, hoist, hoppers, and pneumatic tubes are used to move workpieces from the initial production phase through the various operations and assemblies to the final stage or destination. Materials handling in addition to using transfer devices uses indexing attachments and loading and unloading devices to increase the production rate.

Automatic machines have provisions for moving the workpiece through various machine cycles. Indexing is the process of moving the workpiece or changing the tool on a machine. Indexing of the workpiece is used to intermittently move the workpiece from one position to the next position for processing. Indexing may be the simple rotation of the workpiece on its axis to turn the part over for subsequent operations. Another form of indexing is to move the workpiece from one work station to the next work station. Two types of indexing are the rotary and straight line. Rotary indexing in turn may be either horizontal or vertical. The type of indexing (horizontal or vertical) is designated by the plane in which the workpieces are moved. During the indexing and subsequent operations, the workpieces are securely held in fixtures specifically designed for the item. Workpieces are loaded and unloaded, either manually or automatically, and indexed through the production cycle. A number of machines are located around and over a horizontal rotating table which provides the workpiece movement. A disadvantage of rotary indexing is that all operations should take ap-

Fig. 22-4 Automatic numerically controlled tool changer. (*Source: Cincinnati Machine Tool Company.*)

proximately the same time to perform. The workpiece cannot be indexed to the next station until all the operations have proceeded through a cycle. Figure 22-3 shows a typical machine set up for horizontal indexing.

Cutting tools can be indexed to reduce the production time. A simple example of tool indexing is the turret located on many types of machine tools. The tools, rather than the workpiece, are indexed into position for performing the required operations. Numerically controlled tool selectors are another means of selecting the needed cutting tool and locating the tool in the drive unit of the machine (Fig. 22-4).

Straight-line-indexing production lines are referred to as transfer lines. The transfering device moves the workpiece in a straight line motion intermittently past a series of machine tools. Rather than a standard machine tool, a special power-tool head assembly is used for flexibility and improved space utilization. The power-tool heads are designed specifically for high-production lines.

The power-head production unit consists of a base on which is mounted a powered spindle used to drive the rotating cutting tool. A feed unit usually hydraulically operated feeds the entire power-head assembly into position for the operation. Feed of the cutting tool is also accomplished by movement of the power-head assembly. Individual components of the power-head production machine are available in many sizes and shapes which are interchangeable. The interchangeability of these component parts makes this type of machine flexible and easily adaptable for special production-line operations.

Cutting tools such as drills, reamers, milling cutters, countersinks, and taps are the usual type of tool

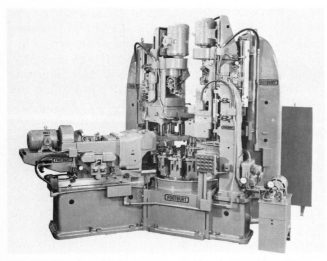

Fig. 22-3 Horizontal indexing on a multistation machine setup. (*Source: Footburt Division, Reynolds Metals Co.*)

used in the production-head assembly. Each head assembly contains a number of tools which are held in special multiple-spindle assemblies. Tools are changed periodically to ensure proper cutting action and lessen the down time of the line when tools wear or break sporadically.

The workpiece is moved intermittently by a number of transfer devices. A fixture is used to hold the workpiece securely in the proper position during the transfer and subsequent operations. Sometimes fixtures are not used, but the workpiece is automatically positioned at each work station. Transfer of the workpiece and holding device can be accomplished by a continuous conveyor system, supported on rails and pulled along the line, pushed by hydraulic-piston assemblies or moved by hoist type conveyor systems. Figure 22-5 illustrates an in-line production unit. This particular line is capable of producing 92 cylinder blocks per hour. Located throughout the line are 29 stations, with 85 spindles containing multiple-tool heads. The transfer device utilizes both hydraulic and mechanical screw-feed units. Operations performed on the line include: drilling, reaming, milling, tapping, counterboring, and inspection.

Another type of automated materials-handling system is the continuous-travel system. Either rotary or in-line continuous workpiece motion is utilized in this type of system. The workpiece moves past a cutting tool during the operation rather than being stopped

Fig. 22-6 Continuous rotary-grinding machine. (*Source: Blanchard.*)

as in the intermittent-type production line. An advantage to this type of system is the increase in production in a shorter period of time. It should also become evident that a few restrictions and disadvantages relative to the type of work being performed exist in this type of production system. Operations are restricted to broaching, milling, and grinding. Drilling, reaming, and so forth, must be performed on an intermittent-type system. Another restriction is that it is difficult to load and unload parts because of the continuously moving work-holding devices. Figure 22-6 illustrates a continuously moving rotary-grinding unit. Figure 22-7 shows the fixture, workpiece, and chain-feed unit utilized on a straight-line broaching machine.

Various types of work-holding and loading devices are used to speed the production process. Conveyors, vibratory feeds, magazines, gravity chutes, transfer arms, barrel hoppers, turrets, and chain-type units are only some of the many loading devices utilized in an automated system. Figures 22-8 to 22-12 illustrate some of the common methods utilized in loading and unloading the workpiece onto a machine. Figure 22-8 uses a turret-type system to load and unload gears into a gear-rolling machine. Figure 22-9 illustrates a gravity-feed magazine system. An automatic tapping machine is illustrated in Fig. 22-10. Figure 22-11 illustrates an automatic high-speed tapping machine with a five-row loading magazine. The rear view of the machine shows the 10-spindle tapping head and workpiece discharge chute. Finally Fig. 22-12 illustrates a hydraulically operated transfer through lift and carry mechanism used to move the workpiece on a straight-line lapping machine.

Fig. 22-5 In-line transfer machine. (*Source: Footburt Division, Reynolds Metals Co.*)

Fig. 22-7 Continuous in-line broaching machine. (*Source: The Lapointe Machine Tool Co.*)

Assembly of Products. The previous two sections dealt with the machinery and materials-handling methods used to either produce a finished product or prepare component parts for a larger assembly. In most cases manufacturing consists of producing com-

Fig. 22-8 Automatic-gear rolling machine employing a turret-type fixture for loading and unloading the gear blands. This machine is capable of producing 760 gears per hour. (*Source: Landis Machine Tool Company.*)

Fig. 22-9 Gravity-type feed mechanism. (*Source: The Lapointe Machine Tool Co.*)

ponent parts which must be assembled to form the finished product.

Two methods are employed in assembly of products, a *continuous line* or a *single station* assembly. Low-volume-production or very large assemblies are usually assembled at a single station. The component parts are delivered to one location and assembled. Assemblies at a single station may be done by one or a

Fig. 22-10 An automatic tapping machine used to thread couplings; note the gravity-feed unit and hydraulic positioner. (*Source: Landis Machine Tool Company.*)

(a)

(b)

Fig. 22-11 High-speed tapping machine, five-unit loading magazine, and discharge chute. (*Source: Ettco Tool and Machine Co. Inc.*)

number of individuals working in a team. A continuous-line assembly utilizes transfer machines to move the assemblies through a series of assembly work stations. Assembly at the stations may be done by individuals or automatically. A number of individuals perform the assembly operation by hand or with the aid of power-assisted devices. A completely automated line utilizes special work-holding devices or automatic workpiece positioners to assemble the products.

Assembly operations require that the products be completed by attaching component parts in a prescribed order. A number of methods are used in attaching the components, such as chemical, mechanical, and thermal methods. Chemical means such as gluing and solvent cementing are commonly used to assemble wood and plastic components. Thermal methods used in assembly include: spot welding, arc welding, gas welding, hot-gas plastic welding, pressure welding, soldering, brazing, and others. In mechanical assembly, items such as rivets, screws, keys, pins, sewing, bolts, crimping, nails, and so on, are employed. Special methods may also be employed in assembly operations such as slip welding in attaching ceramic components. Figure 22-13 on page 369 illustrates an automatically controlled switching-gear-assembly machine.

Inspection. A prime area of automation which assures a quality product is the inspection. Inspection procedures are conducted throughout the production and assembly stages of manufacturing. Assembly

Fig. 22-12 Lapping machine utilizing hydraulic lifting mechanism. (*Footburt Division, Reynolds Metals Co.*)

Fig. 22-13 Sheffield with automatic-gear assembly machine. (*Automation and Measurement Division, Bendix Company.*)

of the final product would be hampered considerably if the component parts were not within acceptable tolerances. Inspection and quality control devices were discussed in Chap. 2. Some of the common devices or methods used for inspection include: linear-measurement instruments, pneumatic electrical and mechanical gauges, ultrasonic units, penetrants, and X-ray units. Figure 22-14 shows a multiple pneumatic gaug-

Fig. 22-14 Pneumatic gauging station. (*Source: Automation and Measurement Division, Bendix Company.*)

Fig. 22-15 Automatic Zyglo testing unit employed to inspect bearing races. (*Source: Magnaflux Corp.*)

ing unit used to check an automobile's transmission part. Figure 22-15 shows an automatic transfer device used in the Zyglo method of inspection.

Automation is a vast and expanding area of manufacturing. Considerable planning is required and extensive use of automatic equipment is employed in this form of manufacturing. A completely automated plant does not exist; however, we are getting closer to realizing total automation. Probably the chemical and petroleum industries are the closest to being fully automated.

REVIEW QUESTIONS

22-1. Distinguish between automation and mass production.

22-2. Compare noncontinuous and continuous types of manufacturing systems.

22-3. Develop a flowchart which traces the history of automation.

22-4. List and describe various automated systems which employ a feedback control system commonly found in the home.

22-5. Develop a list of equipment used for: *a.* Production equipment, *b.* Transfer machines, *c.* Inspection equipment, *d.* Assembly devices.

22-6. Describe the purpose and operation of transfer devices.

22-7. Explain the differences between a jig and a fixture.

The International System of Units, SI, is a modernized version of the metric system established by international agreement. It provides a logical and interconnected framework for all measurements in science, industry, and commerce. Officially abbreviated SI, the system is built upon a foundation of seven basic units, plus two supplementary units. Multiples and submultiples are expressed in a decimal system. Use of metric weights and measures was legalized in the United States in 1866, and since 1893 the yard and pound have been defined in terms of the meter and the kilogram. The base units for time, electric current, amount of substance, and luminous intensity are the same in both the customary and metric system. Chart 1 in this Appendix illustrates the base and supplementary units. Units for all other quantities are derived from these nine units.

Appendix Table 1 lists 17 derived units with special names which were derived from the base and supplementary units in a coherent manner, which means, in brief, that they are expressed as products and ratios of the nine base and supplementary units without numerical factors. All other SI derived units are similarly derived in a coherent manner from the 26 base, supplementary, and special-name SI units (see Appendix Tables 2 and 3). For use with the SI units, there is a set of 16 prefixes to form multiples and submultiples of these units (see Appendix Table 4).

In order to convert from the traditional system to the metric system and vice versa, common conversions are listed in Appendix Table 5. Additional information on the SI system of measurement can be found in the following documents:

1. National Bureau of Standards Special Publication 304A, for sale by the Superintendent of Documents, U. S. Government Printing Office, Washington, D. C., 20402. SD Catalog No. C13.10:304 A, price 25 cents.

2. NBS Special Publication 330, 1972 edition, International System of Units (SI), available by purchase from the Superintendent of Documents, Government Printing Office, Washington, D. C. 20402, order as C13.10:330/2, price 30 cents.

3. ASTM Metric Practice Guide E380-72, available by purchase from the American Society of Testing and Materials, 1916 Race Street, Philadelphia, PA 19103. Price $1.50 a copy, minimum order $3.00.

4. Rules for the Use of Units of the International System of Units, order as ISO Recommendation R1000. $1.25 a copy from the American National Standards Institute, 1430 Broadway, New York, N. Y. 10018.

SOURCE: Irwin H. Fullmer, *Dimensional Metrology*. U. S. Department of Commerce, National Bureau of Standards. Publication 265, 1966. (This source is for the above, all tables, and the chart in this Appendix).

SI SYSTEM – BASE AND SUPPLEMENTARY UNITS

meter -m LENGTH

The meter (common international spelling, metre) is defined as 1 650 763.73 wavelengths in vacuum of the orange-red line of the spectrum of krypton-86.

1 METER — 1 650 763.73 WAVELENGTHS — ONE WAVELENGTH — An interferometer is used to measure length by means of light waves. — 86 Kr ATOM

The SI unit of area is the square meter (m²).

The SI unit of volume is the cubic meter (m³). The liter (0.001 cubic meter), although not an SI unit, is commonly used to measure fluid volume.

kilogram -kg MASS

The standard for the unit of mass, the kilogram, is a cylinder of platinum-iridium alloy kept by the International Bureau of Weights and Measures at Paris. A duplicate in the custody of the National Bureau of Standards serves as the mass standard for the United States. This is the only base unit still defined by an artifact.

U.S. PROTOTYPE KILOGRAM NO. 20

The SI unit of force is the newton (N). One newton is the force which, when applied to a 1 kilogram mass, will give the kilogram mass an acceleration of 1 (meter per second) per second.
$$1N = 1kg \cdot m/s^2$$

1N — ACCELERATION of 1m/s²

The SI unit for pressure is the pascal (Pa).
$$1Pa = 1N/m^2$$
The SI unit for work and energy of any kind is the joule (J).
$$1J = 1N \cdot m$$
The SI unit for power of any kind is the watt (W).
$$1W = 1J/s$$

second -s TIME

The second is defined as the duration of 9 192 631 770 cycles of the radiation associated with a specified transition of the cesium-133 atom. It is realized by tuning an oscillator to the resonance frequency of cesium-133 atoms as they pass through a system of magnets and a resonant cavity into a detector.

Schematic diagram of an atomic beam spectrometer or "clock." Only those atoms whose magnetic moments are "flipped" in the transition region reach the detector. When 9 192 631 770 oscillations have occurred, the clock indicates one second has passed.

TRANSITION REGION (CAVITY) OSCILLATING FIELD — DETECTOR — DEFLECTION MAGNET — CESIUM SOURCE — DEFLECTION MAGNET — OSCILLATOR — NBS ATOMIC TIME SCALE SYSTEM

The number of periods or cycles per second is called frequency. The SI unit for frequency is the hertz (Hz). One hertz equals one cycle per second.

The SI unit for speed is the meter per second (m/s).

The SI unit for acceleration is the (meter per second) per second (m/s²).

Standard frequencies and correct time are broadcast from WWV, WWVB, and WWVH, and stations of the U.S. Navy. Many short-wave receivers pick up WWV and WWVH, on frequencies of 2.5, 5, 10, 15, and 20 megahertz.

ampere -A ELECTRIC CURRENT

The ampere is defined as that current which, if maintained in each of two long parallel wires separated by one meter in free space, would produce a force between the two wires (due to their magnetic fields) of 2 × 10⁻⁷ newton for each meter of length.

1A — 1A — FORCE = 2×10^{-7} N — 1m — 1m

The SI unit of voltage is the volt (V).
$$1V = 1W/A$$
The SI unit of electric resistance is the ohm (Ω).
$$1\Omega = 1V/A$$

kelvin -K TEMPERATURE

The kelvin is defined as the fraction 1/273.16 of the thermodynamic temperature of the triple point of water. The temperature 0 K is called "absolute zero."

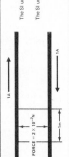

TEMPERATURE MEASUREMENT SYSTEMS
°F: 212, 98.6, 32, -40 — Water Boils, Body Temperature, Water Freezes
°C: 100, 0, -40
K: 2045 Platinum Freezes, 273.15, 0 Absolute Zero
FAHRENHEIT °F — °C CELSIUS — KELVIN

On the commonly used Celsius temperature scale, water freezes at about 0 °C and boils at about 100 °C. The °C is defined as equal to 1 K, and water at 0 °C is defined as 273.15 K.

1.8 Fahrenheit degrees are equal to 1.0 °C or 1.0 K; the Fahrenheit scale uses 32°F as a temperature corresponding to 0 °C.

The standard temperature at the triple point of water is provided by a special cell, an evacuated glass cylinder containing pure water. When the cell is cooled until a mantle of ice forms around the reentrant well, the temperature at the interface of solid, liquid, and vapor is 273.16 K. Thermometers to be calibrated are placed in the reentrant well.

THERMOMETER (ELECTRICAL RESISTANCE TYPE) — WATER VAPOR — ICE — WATER — REENTRANT WELL — REFRIGERATING BATH — TRIPLE POINT CELL

mole -mol AMOUNT OF SUBSTANCE

The mole is the amount of substance of a system that contains as many elementary entities as there are atoms in 0.012 kilogram of carbon 12.

When the mole is used, the elementary entities must be specified and may be atoms, molecules, ions, electrons, other particles, or specified groups of such particles.

The SI unit of concentration (of amount of substance) is the mole per cubic meter (mol/m³).

candela -cd LUMINOUS INTENSITY

The candela is defined as the luminous intensity of 1/600 000 of a square meter of a blackbody at the temperature of freezing platinum (2045 K).

CAVITY — FREEZING PLATINUM — INSULATING MATERIAL

The SI unit of light flux is the lumen (lm). A source having an intensity of 1 candela in all directions radiates a light flux of 4 π lumens.

A 100-watt light bulb emits about 1700 lumens.

TWO SUPPLEMENTARY UNITS

radian -rad PLANE ANGLE

The radian is the plane angle with its vertex at the center of a circle that is subtended by an arc equal in length to the radius.

ONE RADIAN

steradian -sr SOLID ANGLE

The steradian is the solid angle with its vertex at the center of a sphere that is subtended by an area of the spherical surface equal to that of a square with sides equal in length to the radius.

Area r^2 — ONE STERADIAN

Appendix Chart 1

Appendix Table 1 **SI derived units with special names**

Quantity	Name	Symbol	Expression in Terms of Other Units	Expression in Terms of SI Base Units
		SI unit		
Frequency	hertz	Hz		s
Force	newton	N		$m \cdot kg \cdot s^{-2}$
Pressure, stress	pascal	Pa	N/m^2	$m^{-1} \cdot kg \cdot s^{-2}$
Energy, work, quantity of heat	joule	J	$N \cdot m$	$m^2 \cdot kg \cdot s^{-2}$
Power, radiant flux	watt	W	J/s	$m^3 \cdot kg \cdot s^{-3}$
Quantity of electricity, electric charge	coulomb	C	$A \cdot s$	$s \cdot A$
Electric potential, potential difference, electromotive force	volt	V	W/A	$m^2 \cdot kg \cdot s^{-3} \cdot A^{-1}$
Capacitance	farad	F	C/V	$m^{-2} \cdot kg^{-1} \cdot s^4 \cdot A^2$
Electric resistance	ohm	Ω	V/A	$m^2 \cdot kg \cdot s^{-3} \cdot A^{-2}$
Conductance	siemens	S	A/V	$m^{-2} \cdot kg^{-1} \cdot s^3 \cdot A^2$
Magnetic flux	weber	Wb	$V \cdot s$	$m^2 \cdot kg \cdot s^{-3} \cdot A^{-1}$
Magnetic flux density	tesla	T	Wb/m^3	$kg \cdot s^{-2} \cdot A^{-1}$
Inductance	henry	H	Wb/A	$m^2 \cdot kg \cdot s^{-2} \cdot A^{-2}$
Luminous flux	lumen	lm		$cd \cdot sr$ [a]
Illuminance	lux	lx	lm/m^2	$m^{-2} \cdot cd \cdot sr$ [a]
Activity (radioactive)	becquerel	Bq		s^{-1}
Absorbed dose	gray	Gy	J/kg	$m^2 \cdot s^{-2}$

[a] In this expresion the steradian (sr) is treated as a base unit.

Appendix Table 2 **Examples of SI Derived Units Expressed in Terms of Base Units**

Quantity	SI Unit	Unit Symbol
Area	square meter	m^2
Volume	cubic meter	m^3
Speed, velocity	meter per second	m/s
Acceleration	meter per second squared	m/s^2
Wave number	1 per meter	m^{-1}
Density, mass density	kilogram per cubic meter	kg/m^3
Current density	ampere per square meter	A/m^2
Magnetic field strength	ampere per meter	A/m
Concentration (of amount of substance	mole per cubic meter	mol/m^3
Specific volume	cubic meter per kilogram	m^3/kg
Luminance	candela per square meter	cd/m^2

Appendix Table 3 **Examples of SI Derived Units Expressed by Means of Special Names**

Quantity	Name	Symbol	Expression in Terms of SI Base Units
		SI Units	
Dynamic viscosity	pascal second	Pa·s	$m^{-1} \cdot kg \cdot s^{-1}$
Moment of force	newton meter	N·m	$m^2 \cdot kg \cdot s^{-2}$
Surface tension	newton per meter	N/m	$kg \cdot s^{-2}$
Heat flux density, irradiance	watt per square meter	W/m^2	$kg \cdot s^{-3}$
Heat capacity, entropy	joule per kelvin	J/K	$m^2 \cdot kg \cdot s^{-2} \cdot K^{-1}$
Specific heat capacity, specific entropy	joule per kilogram kelvin	J/(kg·K)	$m^2 \cdot s^{-2} \cdot K^{-1}$
Specific energy	joule per kilogram	J/kg	$m^2 \cdot s^{-1}$
Thermal conductivity	watt per meter kelvin	W/(m·K)	$m \cdot kg \cdot s^{-3} \cdot K^{-1}$
Energy density	joule per cubic meter	J/m^3	$m^{-1} \cdot kg \cdot s^{-2}$
Electric field strength	Volt per meter	V/m	$m \cdot kg \cdot s^{-3} \cdot A^{-1}$
Electric charge density	coulomb per cubic meter	C/m^3	$m^{-3} \cdot s \cdot A$
Electric flux density	coulomb per square meter	C/m^2	$m^{-2} \cdot s \cdot A$
Permittivity	farad per meter	F/m	$m^{-3} \cdot kg^{-1} \cdot s^4 \cdot A^2$
Permeability	henry per meter	H/m	$m \cdot kg \cdot^{-2} \cdot A^{-2}$
Molar energy	joule per mole	J/mol	$m^2 \cdot kg \cdot s^{-2} \cdot mol^{-1}$
Molar entropy, molar heat capacity	joule per mole kelvin	J/(mol·K)	$m^2 \cdot kg \cdot s^{-2} \cdot K^{-1} \cdot mol^{-1}$

Appendix Table 4 **SI Prefixes**

Factor	Prefix	Symbol	Factor	Prefix	Symbol
10^{18}	exa	E	10^{-1}	deci	d
10^{13}	peta	P	10^{-2}	centi	c
10^{12}	tera	T	10^{-3}	milli	m
10^9	giga	G	10^{-6}	micro	μ
10^6	mega	M	10^{-9}	nano	n
10^3	kilo	k	10^{-12}	pico	p
10^2	hecto	h	10^{-13}	femto	f
10^1	deka	da	10^{-18}	atto	a

Appendix Table 5 **Common Conversions**

Symbol	When You Know	Multiply By	To Find	Symbol
in	inches	*25.4	millimeters	mm
ft	feet	*0.3048	meters	m
yd	yards	*0.9144	meters	m
mi	miles	1.609 34	kilometers	km
yd²	square yards	0.836 127	square meters	m²
	acres	0.404 686	†hectares	ha
yd³	cubic yards	0.764 555	cubic meters	m³
qt	quarts	0.946 353	‡liters	l
oz	ounces (avdp)	28.349 5	grams	g
lb	pounds (avdp)	0.453 592	kilograms	kg
°F	fahrenheit	*5.9 after subtracting 32	Celsius	°C
mm	millimeters	0.039 370 1	inches	in
m	meters	3.280 84	feet	ft
m	meters	1.093 61	yards	yd
km	kilometers	0.621 371	miles	mi
m²	square meters	1.195 99	square yards	
ha	†hectares	2.471 05	acres	yd²
m³	cubic meters	1.307 95	cubic yards	yd³
l	‡liters	1.056 69	quarts (lq)	qt
g	grams	0.035 274 0	ounces (avdp)	oz
kg	kilograms	2.204 62	pounds (avdp)	lb
°C	Celsius	9/5 then add 32	Fahrenheit	°F

* Exact.

† Hectare is a common name for 10,000 square meters.

‡ Liter is a common name for fluid volume of 0.001 cubic meter.

NOTE: Most symbols are written with lower case letters; exceptions are units named after persons for which the symbols are capitalized. Periods are not used with any symbol.

Allen, Dell K.: *Metallurgy Theory and Practice,* American Technical Society, Chicago, 1969.

Amber, George H., and Paul S. Amber: *Anatomy of Automation,* Prentice-Hall, Englewood Cliffs, N. J., 1962.

American Foundrymen's Society: *Introduction to the Cast Metals Industry,* Des Plaines, Ill., 1971.

Anderson, James, and Earl E. Tatro: *Shop Theory,* 5th ed., McGraw-Hill, New York, 1968.

Andrews, R. S.: *Shell Process Foundry Practice,* American Foundrymen's Society, Des Plaines, Ill., 1963, and rev.

Ansley, Arthur C.: *Manufacturing Methods and Processes,* rev. ed., Chilton Book Company, Philadelphia, 1968.

Ayers, Chesley: *Specifications: An Introduction for Architecture and Construction,* McGraw-Hill, New York, 1975.

Baird, Ronald J.: *Industrial Plastics,* Goodheart-Willcox Company, South Holland, 1971.

Baker, Glenn E. and L. Dayle Yeager: *Wood Technology,* Bobbs-Merrill, Indianapolis, Ind., 1974.

Barrett, Craig R., William D. Nix, and Alan S. Tetelman: *The Principles of Engineering Materials,* Prentice-Hall, Englewood Cliffs, N. J., 1973.

Black, P. H.: *Theory of Metal Cutting,* McGraw-Hill, New York, 1961.

Boston, O. W.: *Metal Processing,* Wiley, New York, 1951.

Burghardt, Henry D., Aaron Axelrod, and James Anderson: *Machine Tool Operation: Part I,* 5th ed., McGraw-Hill, New York, 1959.

Burghardt, Henry D., Aaron Axelrod, and James Anderson: *Machine Tool Operation: Part II,* 4th ed., McGraw-Hill, New York, 1960.

Chaplin, Jack W.: *Metal Manufacturing Technology,* McKnight, Bloomington, Ill., 1976.

Crandall, Keith C., and Robert W. Seabloom: *Engineering Fundamentals in Measurements, Probability, Statistics, and Dimensions,* McGraw-Hill, 1970.

Datsko, Joseph: *Material Properties and Manufacturing Processes,* Wiley, New York, 1966.

Degarmo, E. Paul: *Materials and Processes in Manufacturing,* 4th ed., Macmillan, New York, 1974.

Dellow, E. L.: *Measuring and Testing in Science and Technology,* Davis & Charles, London, 1970.

Dieter, G.: *Mechanical Metallurgy,* McGraw-Hill, New York, 1976.

Doyle, L. E.: *Manufacturing Processes and Materials for Engineers:* Prentice-Hall, Inc., Englewood Cliffs, N. J., 1969.

Doyle, L. E.: *Metal Machining,* Prentice-Hall, Englewood Cliffs, N. J., 1953.

Duncan, Acheson J.: *Quality Control and Industrial Statistics,* Irwin, Homewood, Ill., 1965.

Dunlop, John T. (ed.): *Automation and Technological Change,* Prentice-Hall, Englewood Cliffs, N. J., 1962.

Edgar, Carroll: *Fundamentals of Manufacturing Processes and Materials,* Addison-Wesley, Reading, Mass., 1965.

Feirer, John L.: *SI Metric Handbook,* The Metric Company, Kalamazoo, Mich., 1977.

Fullmer, Irwin H.: *Dimensional Metrology,* U. S. Department of Commerce National Bureau of Standards. Publication 265, 1966.

Gerrish, Howard H.: *Technical Dictionary,* Goodheart-Willcox Company, South Holland, Ill., 1968.

Grant, E. L., and R. S. Leavenworth: *Statistical Quality Control,* 4th ed., McGraw-Hill, New York, 1972.

Gregor, T. G.: *Manufacturing Processes: Ceramics,* Prentice-Hall, Englewood Cliffs, N. J., 1975.

Habicht, F. H.: *Modern Machine Tools,* Van Nostrand, Princeton, N. J., 1963.

Heine, R. W., et al.: *Principles of Metal Casting:* McGraw-Hill Book Company, New York, 1967.

Hurschhorn, Joel S.: *Introduction to Powder Metallurgy,* American Powder Metallurgy Institute, Princeton, N. J., 1969.

Johnson, Harold V.: *Manufacturing Processes: Metals and Plastics,* Bennett, Peoria, Ill., 1973.

Kazanas, H. C.: *Properties and Uses of Ferrous and Nonferrous Metals:* Prakken Publications, Inc., Ann Arbor, Mich., 1979.

Kazanas, H. C., and Lyman Hannah: *Manufacturing Processes: Metals,* Prentice-Hall, Inc., Englewood Cliffs, N. J., 1976.

Kazanas, H. C., and D. F. Wallace: *Materials Testing, Laboratory Manual,* Bennett, Peoria, Ill., 1974.

Kazanas, H. C., Roy S. Klein, and John R. Lindbeck: *Technology of Industrial Materials,* Bennett, Peoria, Ill., 1974.

Keyser, C. A.: *Materials of Engineering,* Prentice-Hall, Englewood Cliffs, N. J., 1965.

Klemm, Friedrick: *A History of Western Technology,* M. I. T., Cambridge, Mass., 1964.

Krar, S. F., and J. W. Oswald: *Grinding Technology,* Delmar Publishers, New York, 1974.

Lindberg, Roy A.: *Processes and Materials of Manufacture,* Allyn and Bacon, Boston, 1964.

Ludwig, O. A., et al.: *Metalwork Technology and Practice,* McKnight, Bloomington, Ill., 1975.

McCarthy, Willard J., and Robert E. Smith: *Machine Tool Technology,* McKnight, Bloomington, Ill., 1968.

Moore, Harry D., and Donald R. Kibbey: *Manufacturing Materials and Processes,* Irwin, Homewood, Ill., 1965.

Norton, F. H.: *Elements of Ceramics,* Addison-Wesley, Reading, Mass. 1952.

Patton, W. J.: *Materials in Industry,* Prentice-Hall, Englewood Cliffs, N. J., 1968.

Patton, W. J.: *Modern Manufacturing Processes and Engineering:* Prentice-Hall, Inc., Englewood Cliffs, N. J., 1970.

Perry, John: *The Story of Standards,* Funk & Wagnalls, New York, 1955.

Pollack, Herman W.: *Materials Science and Metallurgy,* Reston Publishing, Reston, Va., 1973.

Porter, Harold W., Orville D. Lascoe, and Clyde A. Nelson: *Machine Shop Operations and Setups,* American Technical Society, Chicago, 1967.

Roberts, A. D., and R. C. Prentice: *Programming for Numerical Control Machines,* McGraw-Hill, New York, 1968.

—— and Samuel C. Lapidgo: *Manufacturing Processes,* McGraw-Hill, New York, 1977.

Rusinoff, S. E.: *Manufacturing Processes: Materials and Production,* American Technical Society, Chicago, 1963.

Simonds, Herbert R., and James M. Church: *A Concise Guide to Plastics,* Reinhold Booh Corp., New York, 1963.

Smith, C.: *The Science of Engineering Materials,* Prentice-Hall, Englewood Cliffs, N. J., 1969.

Society of Manufacturing Engineers: *Tool Engineers Handbook,* 3rd ed., McGraw-Hill, New York, 1976.

Swinehart, H. J. (ed.): *Cutting Tool Material Selection,* Society of Manufacturing Engineers, Detroit, Mich., 1968.

Thomas, Geoffrey G.: *Engineering Metrology,* Butterworths, London, 1974.

Wright, R. Thomas, and Thomas R. Jensen: *Manufacturing: Material Processing, Management, Careers,* Goodheart-Willcox, South Holland, Ill., 1976.

How to Run a Lathe, Sound Bend Lathe Company, South Bend, Ind.

International Standards, International Organization, American National Standards Institute, New York, 1973.

Panorama of Lubrication, Shell Oil Company, New York, 1953.

Practical Treatise on Milling and Milling Machines, Brown and Sharpe Manufacturing Company, 1947.

A Treatise on Milling and Milling Machines: Section Two, 3rd ed., Cincinnati Milling Machine Company, Cincinnati, Ohio, 1946.